Life's Solution
Inevitable Humans in a Lonely Universe

The assassin's bullet misses, the Archduke's carriage moves forward, and a catastrophic war is avoided. So too with the history of life. Rerun the tape of life, as Stephen J. Gould claimed, and the outcome must be entirely different: an alien world, without humans and maybe not even intelligence. The history of life is littered with accidents: any twist or turn may lead to a completely different world. Now this view is being challenged. Simon Conway Morris explores the evidence demonstrating life's almost eerie ability to navigate to the correct solution, repeatedly. Eyes, brains, tools, even culture: all are very much on the cards. So if these are all evolutionary inevitabilities, where are our counterparts across the Galaxy? The tape of life can run only on a suitable planet, and it seems that such Earth-like planets may be much rarer than is hoped. Inevitable humans, yes, but in a lonely Universe.

SIMON CONWAY MORRIS is Professor of Evolutionary Palaeobiology at the University of Cambridge. He was elected a fellow of the Royal Society in 1990, and presented the Royal Institution Christmas lectures in 1996. His work on Cambrian soft-bodied faunas has taken him to China, Mongolia, Greenland, and Australia, and inspired his previous book *The Crucible of Creation* (1998).

Pre-publication praise for *Life's Solution*:

'Having spent four centuries taking the world to bits and trying to find out what makes it tick, in the twenty-first century scientists are now trying to fit the pieces together and understand why the whole is greater than the sum of its parts. Simon Conway Morris provides the best overview, from a biological viewpoint, of how complexity on the large scale arises from simple laws on the small scale, and why creatures like us may not be the accidents that many suppose. This is the most important book about evolution since *The Selfish Gene*; essential reading for everyone who has wondered about why we are here in a universe that seems tailor-made for life.'

John Gribbin, author of *Science: A History*

'Are human beings the insignificant products of countless quirky biological accidents, or the expected result of evolutionary patterns deeply embedded in the structure of natural selection? Drawing upon diverse biological evidence, Conway Morris convincingly argues that the general features of our bodies and minds are indeed written into the laws of the Universe. This is a truly inspiring book, and a welcome antidote to the bleak nihilism of the ultra-Darwinists.'

Paul Davies, author of *How to Build a Time Machine*

'Is intelligent life in the Universe common or incredibly rare? Are even planets like the Earth rare? We won't really know until our searches are further advanced, but until then these debates pivot on the tension between contingency and convergence. Advocates of the first point to the unlikelihood of particular historical paths, while those favoring the second emphasize multiple paths to similar functional outcomes. In *Life's Solution* Conway Morris argues that the evidence from life on Earth supports a variety of paths leading toward intelligence. Our searches for life elsewhere are informed by such insights into life here.'

Christopher Chyba, Stanford University and the SETI Institute

Life's Solution

Inevitable Humans in a
Lonely Universe

SIMON CONWAY MORRIS

University of Cambridge

CAMBRIDGE
UNIVERSITY PRESS

CAMBRIDGE UNIVERSITY PRESS
Cambridge, New York, Melbourne, Madrid, Cape Town, Singapore, São Paulo

Cambridge University Press
40 West 20th Street, New York, NY 10011-4211, USA

www.cambridge.org
Information on this title: www.cambridge.org/9780521827041

First published 2003
Reprinted 2003, 2004 (with corrections), 2005, 2006

Printed in the United States of America

A catalogue record for this publication is available from the British Library

Library of Congress Cataloguing in Publication data

Conway Morris, S. (Simon)
Life's solution: inevitable humans in a lonely universe / Simon Conway Morris.
   p.   cm.
ISBN 0 521 82704 3
1. Evolution (Biology) – Philosophy.   2. Convergence (Biology)   I. Title.
QH360.5 .C66   2003
576.8'01–dc21   2002035069

ISBN-13   978-0-521-82704-1 hardback
ISBN-10   0-521-82704-3 hardback

ISBN-13   978-0-521-60325-6 paperback
ISBN-10   0-521-60325-0 paperback

For Zoë, with love

Contents

Preface. The Cambridge sandwich

Writing in the *New York Review of Books*,[1] John Maynard Smith, one of Britain's greatest biologists, remarked 'If one was able to re-play the whole evolution of animals, starting at the bottom of the Cambrian (and, to satisfy Laplace, moving one of the individual animals two feet to its left), there is no guarantee – indeed no likelihood – that the result would be the same. There might be no conquest of the land, no emergence of mammals, and certainly no human beings'.[2] This review, written with characteristic flair and economy, was addressing three books on evolution, two by S. J. Gould and the third by E. Mayr. Maynard Smith was raising this issue because both the authors under review have been forthright in claiming that the emergence of human intelligence during the course of evolution has a vanishingly small probability. The logic of the argument, that because we are unique on this planet then nothing like us can occur elsewhere, is gently checked by Maynard Smith: 'This argument seems to me so manifestly false that I fear I must have misunderstood it'.[3] However, he, Mayr and Gould, and I imagine almost anyone else, would agree that the likelihood of 'exactly the same cognitive creatures – with five fingers on each hand, a vermiform appendix, thirty-two teeth, and so on'[4] evolving again if, somehow, the Cambrian explosion could be rerun is remote in the extreme.

What, however, of the emergence of more general biological properties? In considering some earlier views of R. C. Lewontin, who was uncertain as to whether 'general principles of biological organization'[5] existed, Maynard Smith was more upbeat: 'In seeking a theory of biological form, I would probably place greater emphasis than Lewontin on the principles of engineering design. I suspect that there are only a limited number of ways in which eyes can possibly work, and, maybe only a limited number of ways in which brains can work. But I agree that it would be good to know whether such principles exist, and, if so, what they are'.[6] Even though neither Lewontin nor Maynard Smith thought 'A description of all the organisms that have ever been'[7] could settle this issue,[8] *Life's Solution* sets out to demonstrate that what we already know gives some strong indicators of what must be: even in this book pigs don't fly.

The central theme of this book depends on the realities of evolutionary convergence: the recurrent tendency of biological organization to arrive at the same 'solution' to a particular 'need'. Perhaps the best-known example is the similarity between the camera-like eye of the octopus and the human eye (or that of any other vertebrate). As we shall see in this particular instance, where the camera-like eye has evolved independently at least six times, Maynard Smith's premise that 'only a limited number of ways in which eyes can possibly work' is amply confirmed. If this book happens to serve no other purpose than act as a compilation of evolutionary convergences, be it head-banging in mole rats and termites or matriarchal social structure in sperm whales and elephants, then that will be sufficient. But, of course, the net is in pursuit of a much bigger prey. Its main, but not ultimate, aim is to argue that, contrary to received wisdom, the emergence of human intelligence is a near-inevitability. My purpose is not to demonstrate the inevitability of a five-fingered organism, although in this context it is amusing to note that the famous panda's 'thumb' is, in one sense, convergent.[9] Nor is it my aim to find repeated examples of species with 32 teeth, even though we might note that there are a number of fascinating examples of dental convergence. And it is this that matters, not five of this or 32 of that, but the recurrent emergence of various biological properties.

This book has its anecdotes, from baboons operating railway signals to a harbour seal that spoke like an inebriated Bostonian, but there is a serious argument that takes us from the apparently arcane, such as the natural (and convergent) gyroscopes of insects, through to the convergences of the sensory modalities (vision, of course, but also olfaction, hearing and echolocation, electroreception, and so on) to agriculture, brain size, and culture. And there are four conclusions. First, what we regard as complex is usually inherent in simpler systems: the real and in part unanswered question in evolution is not novelty *per se*, but how it is that things are put together. Second, the number of evolutionary end-points is limited: by no means everything is possible. Third, what is possible has usually been arrived at multiple times, meaning that the emergence of the various biological properties is effectively inevitable. Finally, all this takes time. What was impossible billions of years ago becomes increasingly inevitable: evolution has trajectories (trends, if you prefer) and progress is not some noxious by-product of the terminally optimistic, but simply part of our reality.

There is, however, a paradox. If we, in a sense, are evolutionarily inevitable, as too are animals with compound eyes or tiny organelles that make hydrogen, then where are our equivalents, out there, across the galaxy? After all, the Milky Way has been available for colonization for at least a billion years, so in Enrico Fermi's famous words concerning putative extraterrestrials: 'Where are they?' To paraphrase much of this book, life may be a universal principle, but we can still be alone. In other words, once you are on the path it is pretty straightforward, but finding a suitable planet and maybe getting the right recipe for life's origination could be exceedingly difficult: inevitable humans in a lonely Universe. Now, if this happens to be the case, that in turn might be telling us something very interesting indeed. Either we are a cosmic accident, without either meaning or purpose, or alternatively . . .

Enough of backgrounds; what specifically is this book about? Here is a brief outline. Overall it is a sandwich. The central meat on convergences is in Chapters 6 to 10, flanked by thinner expositions in the form of the first five chapters and two end chapters, the last very short indeed. So, the first two chapters are introductory. They look at two extraordinarily effective biological systems. The first concerns the genetic code, how the building blocks of protein, the amino acids, are read off the DNA. This code is eerily effective, indeed it has been argued to be 'one in a million'. This raises the question of how life navigates to such precise end-points, an analogy being how the Polynesians in the great diaspora across the Pacific ever managed to find that remote speck of land that we call Easter Island. This is followed in the second chapter by a consideration of DNA, a molecule of iconic if not totemic significance. But for all its familiarity, DNA also turns out to be one of the strangest molecules in the Universe. A rather useful invention.

The next two chapters (3 and 4) consider how easy it is to make the molecules necessary for life, but paradoxically how difficult it is to make life itself. To some the universality of organic material, from immense interstellar gas clouds rich in carbon compounds to questing bipeds plodding around out-of-the-way planets, almost suggests the cosmos 'breathes' life; a Universe seeded with vital possibilities. Maybe so, but the trillion upon trillion tonnes of interstellar organics may still be a universal 'goo'. To be sure, they could be the essential ingredient for getting life started in terms of basic supplies, but the question of just how inanimate became animate has proved stubbornly recalcitrant. It should all be rather simple, especially if you worship at the crowded shrine of self-organization. Yet, somewhere, somehow the right question has not yet been asked, and not for want of trying.

So confident, however, is the majority that the emergence of life is a pre-ordained inevitability that the question of whether beyond the Earth there are any planetary homes available has only recently emerged. Thus Chapter 5 looks at what we know of the many peculiarities of our Solar System. Planets there will be aplenty, but suitable abodes for organic evolution might require very special sets of circumstances. This is an area that has been reviewed by such workers as Peter Ward and Donald Brownlee[10] and Stuart Ross Taylor,[11] but here I take the argument further as the ferment of discussion continues.

Chapters 6 to 10 are, as already mentioned, the heart of the book. They effectively track the story of evolutionary convergence, starting with the classic cases familiar to biologists as well as some very intriguing experiments, using bacteria, which allow evolutionary history to be rerun. That provides a framework of a sort, but the goal is to argue for the inevitable emergence of sentience. This is achieved by first (Chapter 7) looking, in some detail, at the sensory modalities. Eyes provide a superb story, but so too in their different ways do such features as balance, hearing, olfaction, echolocation, and electrogeneration: all are rampantly convergent. These complex systems can arise from very different starting positions, but again and again converge on the same evolutionary solution. Chapters 8 and 9 develop the story by seeing how certain features that we believe are peculiarly human, such as agriculture, human brains, and even advanced culture are each convergent.

This is not, emphatically, to say that humans are the only evolutionary outcome worth considering: clearly they are not. And this leads to the last two chapters (10 and 11), and a brief coda (Chapter 12). Too often evolutionary convergence is regarded as simply anecdotal, good for a bedtime story. Its importance is surely underestimated, and for two reasons. The first is scientific. Ideas on evolution about such features as adaptation and trends have been under fierce attack, especially by those who believe that if contingent happenstance dogs every step of evolution then assuredly the emergence of humans is a cosmic accident, leaving us free to make the world as we will, with such happy results as are plain to see. Yet convergence tells us two things: that evolutionary trends are real, and that adaptation is not some occasional cog in the organic machine, but is central to the explanation of how we came to be here. In principle, such ideas are in themselves so unremarkable as to require no comment, were it not for the ferocious attacks by such writers as S. J. Gould. What, one wonders, did he get so excited about, and how, one may ask, has our understanding of evolution really changed despite more than forty years of polemic?

Yet, convergence also opens another door. If the emergence of our sentience was effectively inevitable, then perhaps we should take rather more seriously the sentiences of other species? So too perhaps we should stand back and consider what a very odd set-up it is we inhabit, from the eerily efficient genetic code, to the deeply peculiar molecule DNA, to a set of biological organizations that repeatedly throw up complex structures, not least the brain. The late Fred Hoyle, no friend of most biologists, carried some strange ideas about the origins of biological complexity to his grave, yet his remark that the Universe was a set-up job rings strangely true. Having said that, if you happen to be a 'creation scientist' (or something of that kind) and have read this far, may I politely suggest that you put this book back on the shelf. It will do you no good. Evolution is true, it happens, it is the way the world is, and we too are one of its products. This does not mean that evolution does not have metaphysical implications; I remain convinced that this is the case. To deny, however, the reality of evolution and more seriously to distort deliberately the scientific evidence in support of fundamentalist tenets is inadmissible. Contrary to popular belief, the science of evolution does not belittle us. As I argue, something like ourselves is an evolutionary inevitability, and our existence also reaffirms our one-ness with the rest of Creation. Nevertheless, the free will we are given allows us to make a choice. Of course, it might all be a glorious accident; but alternatively perhaps now is the time to take some of the implications of evolution and the world in which we find ourselves a little more seriously. If you haven't put *Life's Solution* back on the shelf, please read on.

Acknowledgements

'To copy one paper is plagiarism, to copy many is scholarship': few academics are unfamiliar with one or other version of this gentle jibe. Moreover, given that my one area of vague scientific knowledge concerns fossil worms from the Cambrian it will be self-evident that to have been able to write this book I am heavily dependant on the expertise, knowledge, and enthusiasm of hundreds of workers. For this reason I have drawn upon a number of their quotations, which are of course fully acknowledged. This is not to say that the researchers I have cited would necessarily agree with the overall theme of *Life's Solution*, but I trust that in each case the context is clear and fair. Thus I hope that a book that flits from extraterrestrial amino acids to dolphin brains, from the eyes of spiders to the discovery of a Roman terracotta head in pre-Columbia Mexico, or Francis Galton calculating by smells, is understood as an exploration along a common theme rather than simply a jumble of half-digested facts. So first I must acknowledge the many authors whose work I have drawn on liberally. So too I thank the following friends for reading one or other section, and in a few cases the entire draft at one stage or another. Thus I record my gratitude to the following friends: Ken Catania, Stephen Clark, Rob Foley, Stephen Freeland, Jack Lissauer, Ken McKinney, Lori Marino, Ulrich Mueller, and Nick Strausfeld for their detailed critiques. In addition, many other colleagues provided illustrative material (also acknowledged in specific figures), particular insights, and information. Again, I am most grateful, and specifically I thank Rachelle Adams, Tim Bayliss-Smith, Curtis Bell, Yfke van Bergen, Quentin Bone, Graham Budd, Hynek Burda, John Chambers, Jenny Clack, Rod Conway Morris, James Crampton, Cameron Currie, Nick Davies, Eric Denton, Laurence Doyle, Doug Erwin, Albert Eschenmoser, Richard Felger, Russ Fernald, Larry Field, Siegfried Franck, Adrian Friday, Linda Gamlin, Liz Harper, Carl Hopkins, Ken Joysey, Harvey Karten, Jeyaraney Kathirithamby, Richard Keynes, Kuno Kirschfeld, David Kistner, Mike Land, Charley Lineweaver, Ken McNamara, Charles

Melville, Eviator Nevo, Dan Nilsson, Euan Nisbet, David Norman, Ray Norris, Beth Okamura, Art Popper, Christopher Pynes, Simon Reader, Neill Reid, Michael Ruse, John Taylor, Nigel Veitch, Tom Waller, Michael Wilson, and Rachel Wood.

It is oxymoronic to say that the mistakes that remain are mine: of course they are. Nor can I promise that everything is up to date; it can't be, nor are my references intended to be exhaustive; they aren't. I hope, however, they are sufficient for the interested reader to begin to explore the literature.

The source of this book was the invitation by Trinity College, Cambridge, to deliver the Tarner Lectures for 1999, and I thank the Master and Fellows, especially Boyd Hilton, for their encouragement and support. I owe an enormous debt to several other people. First, I wish to thank wholeheartedly Sandra Last for her patience and stamina as smoothly and flawlessly draft after draft emerged. Next, I owe a debt of gratitude to the University of Cambridge and especially the Department of Earth Sciences, for allowing me time for such an enterprise. I also specifically wish to thank Sharon Capon and Dudley Simons for assistance with drafting and photography, and also to acknowledge the superb libraries in many departments, the University Library, and the unfailingly helpful librarians. So, too, I give thanks to Cambridge University Press, especially Sally Thomas, Alison Litherland, and Robert Whitelock, and to Bruce Wilcock and his skills in disentanglement.

In one way or another, support has been provided by my college in Cambridge, St John's, the Leverhulme Foundation, the SETI Institute, NASA–Ames, and the Royal Society. Finally, I want to thank my wife Zoë for her interest as I droned on about star-nosed moles, dolphin brains, or electric fish. To maintain such an attention span tells me something rather important, and to her I dedicate this book.

Abbreviations

GENERAL

ATP   adenosine 5'-triphosphate, the triphosphate of the
       nucleotide adenosine, which plays a key role in the
       energetics of the cell. See also p. 25.
BP   before the present; by convention taken as before 1950.
CHZ   the Circumstellar Habitable Zone, the zone surrounding a
       star in which the evolution of life is both possible, and can
       be maintained for protracted intervals of time. See also pp.
       83, 99–100.
DNA   deoxyribonucleic acid, the nucleic acid that forms the basis
       of genetic inheritance in nearly all organisms. See also
       pp. 4, 23–24.
EOD   electric organ discharge; the discharge of electricity from
       specialized tissues in fish. See p. 184.
IDO   the enzyme indoleamine 2,3-dioxygenase.
JAR   jamming avoidance response, exhibited by fish that use
       electrogeneration. See p. 186.
K/T   the boundary between the end of the Cretaceous (K) period
       and the beginning of the Tertiary (T) era at about 65 Ma
       ago. The K/T event that occurred at this time resulted in a
       mass extinction. See pp. 94–95.
LPTM   the late Paleocene thermal maximum, a warm interval that
       occurred during the Paleocene period at c. 55 Ma.
OZMA (Project)   the first radio-telescope project to search for
       extraterrestrial signals, so named by Frank Drake in
       reference to organisms as strange as the Wizard of Oz. See
       p. 231.
PAHs   polycyclic aromatic hydrocarbons; organic compounds with
       a carbon "network" that are abundant in the Universe. See
       p. 43.

RNA   ribonucleic acid, a polynucleotide that conveys genetic information to the proteins in the cell. There are three forms: messenger RNA (mRNA), ribosomal RNA (rRNA), and transfer RNA (tRNA). See also pp. 4, 13, 44.

SETI   the Search for Extraterrestrial Intelligence. See pp. 231–232.

TNA   threo-nucleic acid. See pp. 52–53.

UV   ultraviolet; electromagnetic radiation in the range between visible (violet) light and X-rays, i.e. with wavelengths from about 400 nm to 4 nm. Ultraviolet radiation is invisible to the human eye, but not to many animals.

SOME ABBREVIATIONS FOR UNITS

Length

cm   centimetre, one-hundredth of a metre (0.3937 inch)

m   metre (39.37 inches)

km   kilometre, 1000 metres (approx. 0.621 mile)

nm   nanometre, 10^{-9} metre, i.e. one millionth of a millimetre (0.03937 millionths of an inch).

AU   astronomical unit, equal to the mean distance of the Earth from the Sun; 1.496×10^8 km or approx. 93 million miles.

Mass

g   gram (0.03527 ounce)

kg   kilogram, 1000 grams (2.2046 pounds)

Time

s   second

Ga   billion years (10^9 years)

Ma   million years (10^6 years)

Frequency

Hz   herz, frequency per second

kHz   kiloherz, 1000 Hz

MHz   megaherz, 10^6 herz.

Temperature

°C   degree Celsius (0 °C is the freezing point of water, 100 °C is the boiling point of water).

K   temperature on the Kelvin (thermodynamic) absolute scale (with 0 K as absolute zero); 1 degree K = 1 degree C; 0 °C is about 273 K and 100 °C is about 373 K.

Pressure

Pa   Pascal, SI unit of pressure, equivalent to the pressure produced by a force of one newton applied (uniformly) over an area of one square metre; 10^5 Pa (100 kPa) is equivalent to 1 bar, or roughly 1 atmosphere.

I   Looking for Easter Island

I am a bipedal hominid, of average cranial capacity, write my manuscripts with a fountain pen, and loathe jogging. Thanks to years of work by innumerable biologists I, or anyone else, can tell you to a fair degree of accuracy when the ability to walk upright began, the rate at which our brain increased to its present and seemingly astonishing size, and the origin of the five-fingered forelimb whose present versatility allows me to hold a pen, not to mention the fishy origin of those lungs that make such a noise as the joggers pass me early in the morning on Cambridge's Midsummer Common.

It is obvious that the entire fabric of evolution is imprinted on and through our bodies, from the architecture of our bony skeleton, to the proteins carrying the oxygen surging through our arteries, and our eyes that even unaided can see at least two million years into the past – the amount of time it has taken for the light to travel from the Andromeda Galaxy. In every case – whether for hand or brain – we can trace an ancestry that extends backwards for hundreds of millions, if not billions, of years. Yet, for all that, both the processes and the implications of organic evolution remain controversial. Now at first sight this is rather odd, because it is not immediately clear what is being called into question. Certainly not the fact of evolution, at least as a historical narrative: very crudely, first bacteria, then dinosaurs, now humans. More specifically in terms of process, Darwin's formulation of the mechanisms of evolution is not only straightforward, but seemingly irrefutable. Organisms live in a real world, and evolve to fit their environment by a process of continuous adaptation. This is achieved by a constant winnowing through the operation of natural selection that scrutinizes the available variation to confer reproductive success on those that, by one yardstick or another, are fitter in the struggle for survival.

So is that all there is to say? The recipe for evolution just given is a decidedly bald summary. One intuitively senses that it is an inherently feeble response to an extraordinarily rich history that has brought forth an immense coruscation of form and diversity. Among living forms this ranges across many scales of complexity, from bacteria that build colonies like miniature trees[1] to immense societies of ants whose populations run into the millions and, independent of us, have stumbled across the advantages of agriculture (Chapter 8). And it is a history that is by no means confined to the complexity of colonies or the limpidity of a geometric shell. It is as much in the range, scope, and acuity of living organisms. They may be mere machines, but consider those owls whose hearing can pinpoint within a two-degree arc the rustling made by a mouse,[2] the navigational abilities of albatrosses across the seemingly trackless Southern Ocean[3] (Fig. 1.1), or even Nellie the cat that smelled Madagascar across more than two hundred miles of ocean.[4] But despite our admiration, wonder, and – if we are candid – even awe, surely we can still offer the following paraphrase: evolution happens, this bone evolved from that one, this molecule from that one. To be sure, not every transformation and transition will be elucidated, but we are confident this is because of a lack of information rather than a failure of the method.

Yet despite the reality that, as it happens, we humans evolved from apes rather than, say, lizards, let alone tulips, the interpretations surrounding the brute fact of evolution remain contentious, controversial, fractious, and acrimonious. Why should this be so? The heart of the problem, I believe, is to explain how it might be that we, a product of evolution, possess an overwhelming sense of purpose and moral identity yet arose by processes that were seemingly without meaning. If, however, we can begin to demonstrate that organic evolution contains deeper structures and potentialities, if not inevitabilities, then perhaps we can begin to move away from the dreary materialism of much current thinking with its agenda of a world now open to limitless manipulation. Nor need this counter-attack be anti-scientific: far from it. First, evolution may simply be a fact, yet it is in need of continuous interpretation. The study of evolution surely retains its fascination, not because it offers a universal explanation, even though this may appeal to fundamentalists (of all persuasions), but because evolution is both riven with ambiguities and, paradoxically, is also rich in implications. In my opinion the sure sign of the right road is a limitless prospect of deeper knowledge: what was once baffling is now clear, what seemed absurdly important is now simply childish, yet still the journey is unfinished.

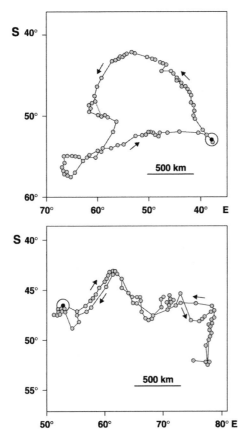

FIGURE 1.1 Two trackways, obtained by satellite monitoring, of the Wandering Albatross across the Southern Ocean. Dots indicate data intercepts, and arrows direction of travel. The upper panel is a departure from South Georgia, on its 13-day trip it passed the Falkland Islands and subsequently Tierra del Fuego. Apart from the distance covered, note the near-straight-line intercept for home. Lower panel is an excursion from Crozet Islands; note how close are the outward and return pathways. (Redrawn from fig. 4b of P. A. Prince *et al.* (1992), Satellite tracking of wandering albatrosses (*Diomedea exulans*) in the South Atlantic, *Antarctic Science*, vol. 4, pp. 31–6 (upper panel) and fig. 8A of H. Weimerskirch *et al.* (1993), Foraging strategy of Wandering Albatrosses through the breeding season: A study using satellite telemetry, *The Auk*, vol. 110, pp. 325–42 (lower panel), with permission of the authors, Cambridge University Press, and *The Auk*.)

One such ambiguity is how life itself may have originated. As we shall see (in Chapter 4) there is no reason to doubt that it occurred by natural means, but despite the necessary simplicity of the process, the details remain strangely elusive. Life itself is underpinned by a rather simple array of building blocks. Most notable are the four (or more accurately five) nucleotides (that is molecules, such as adenine, consisting of a ring of carbon atoms with an attached nitrogen, a phosphate, and a sugar) that comprise the DNA (and RNA). The other key building blocks are twenty-odd amino acids that when arranged in chains form the polypeptides and ultimately the proteins. Yet, from this, by various elaborations, has arisen the immense diversity of life. At first sight this would seem to encapsulate the entire process of evolution, yet it soon becomes clear that we hardly understand in any detail the links between the molecular substrate and the nature of the organism. To be sure, there is some crude correlation between the total number of genes and the complexity of the organism, but when we learn that the 'worm' of molecular biologists (the nematode *Caenorhabditis elegans*), which has a relatively simple body plan with a fixed number of cells, has more genes than the 'fly' (the fruit-fly *Drosophila*) with its complex form and behaviour, then there should be pause for thought (see Chapter 9).

One response is to reconsider what we mean by 'the gene'. In particular, it is time to move away from a crippling atomistic portrayal and rethink our views. As has been pointed out by numerous workers, the concept of the gene is without meaning unless it is put into the context of what it is coding *for*, not least an extremely sophisticated biochemistry. Nor are these the only complications. It is well known that significant quantities of DNA, at least in the eukaryotic cell (that is a cell with a defined nucleus and organelles such as mitochondria), are never employed in the process of coding. Pejoratively labelled as 'junk DNA' or 'parasitic DNA', it may be just that, silent and surplus DNA churned out by repeated rounds of duplication of genetic material, like an assembly line commandeered by lunatic robots.[5] Such a view fits well with the notion that evolution is a process of blind stupidity, a meaningless trek from primordial pond to glassy oceans dying beneath a swollen Sun.

So, beyond the brute fact that evolution happens, the mechanisms and the consequences remain the subject of the liveliest debate and not infrequently acrimony. But, contrary to the desires and beliefs of creation 'science', the reality of evolution as a historical process is not in dispute. And whatever the divergences of opinion, which as often as not have a tacit ideological agenda concerning the origins of human uniqueness, there is a uniform consensus that vitalism was safely buried many years ago, and the slight shaking of the earth above the grave marking the resting place of teleology is certainly an optical illusion. But is it an illusion? Perhaps as the roots and the branches of the Tree of Life are more fully explored our perspectives will begin to shift. Evolution is manifestly true, but that does not necessarily mean we should take it for granted: the end results, be it the immense complexity of a biochemical system or the fluid grace of a living organism, are genuinely awe-inspiring. Could it be that attempts to reinstall or reinject notions of awe and wonder are not simply delusions of some deracinated super-ape, but rather reopen the portals to our finding a metaphysic for evolution? And this in turn might at last allow a conversation with religious sensibilities rather than the more characteristic response of either howling abuse or lofty condescension.

INHERENCY: WHERE IS THE GROUND PLAN IN EVOLUTION?

Although much of this book will be concerned with retelling the minutiae of biological detail in support of the general thesis of the ubiquity of evolutionary convergence and, what is more important, its implications, here is a brief overview of what strike me as the basic tensions in evolution. The first is what, for want of a better name, I might term 'inherency'. A hard-boiled reductionist will dismiss this as a non-problem, but I am not so sure. Perhaps the first obvious clue was the result, surprising at the time, of the minimal genetic difference between ourselves and the chimps. In terms of structural genes the much-quoted difference amounts, it is said, to about 0.4%. If there were any residual doubt of the closeness between *Homo* and *Pan*, then other indicators of similarity, such as the fact that the string of amino acids that make up the protein haemoglobin is identical in number and sequence, are surely a sufficient indicator of our evolutionary proximity. This, of course, confirms the obvious: we and the chimps share an ancestor, probably between about 6 and 12 million years ago, and indeed there is much we have in common. But in other respects we are poles apart. I'm told that chimps driving cars (or at least go-karts) have the time of their lives, but we are neither likely to see a chimp designing a car, nor for that matter mixing the driest of Martinis, let alone being haunted by existentialist doubts.

This problem of inherency, however, is far more prevalent and pervasive than the local quirk that chimps and humans are genomically almost identical, but otherwise separated by an immense gulf of differences. Let us look, for example, at a much deeper stage in our evolution, effectively at the time of the ancestors of the fish. Enter the moderately undistinguished animal known as the lancelet worm or amphioxus (*Branchiostoma* and its relatives, Fig. 1.2). By general agreement this beast is the nearest living approximation to the stage in evolution that preceded the fish, which in turn clambered on to land, moved to using the egg, grew fur, and in one lineage developed into socially alert arborealists. All these changes and shifts must have been accompanied by genetic changes, but if we look back to amphioxus we see a genetic architecture in place that seemingly has no obvious counterpart in its anatomy. To give just one example: the central nervous system of amphioxus is really rather simple. It consists of an elongate nerve cord stretching back along the body, above the precursor of the vertebral column (our backbone, consisting of a row of vertebrae) and a so-called brain. The brain can only be described as a disappointment. It is little more than an anterior swelling (it is called the cerebral vesicle) and has no obvious sign in terms of its morphology of even the beginnings of the characteristic threefold division seen in the vertebrate brain of hind-, mid- and fore-sections. Yet the molecular evidence,[6] which is also backed up by some exquisitely fine studies of microanatomy,[7] suggests that, cryptically, the brain of amphioxus has regions equivalent to the tripartite division seen in the vertebrates.

The clear implication of this is that folded within the seemingly simple brain of amphioxus is what can almost be described as a template for the equivalent organ of the vertebrates: in some sense amphioxus carries the inherent potential for intelligence. Quite how the more complex brain emerges is yet to be established. The evidence that a key development in the molecular architecture of the vertebrates was episodes of gene duplication,[8] that is, doubling up of a gene, could well give one clue. This is because the 'surplus' gene is then potentially available for some new function. It could alternatively be claimed that amphioxus is secondarily simplified (the condition sometimes referred to as *regressive*), but it retained genes for

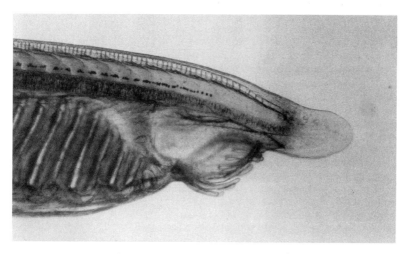

FIGURE 1.2 The amphioxus animal. Upper, entire animal. The anterior end is to the right, with the 'brain' located towards upper side. Prominent white units are gonads. Lower, detail of anterior with prominent feeding (buccal) tentacles and more posteriorly gill bars. The notochord is the longitudinal structure slightly above the mid-line, with closely spaced vertical lines. The nerve cord lies above the notochord, with minimal enlargement at the anterior. (Courtesy of Dale Stokes, Scripps Institution of Oceanography (upper) and Thurston Lacalli, University of Victoria, British Columbia (lower).)

vital functions, although ones no longer specifically connected to the coding for a complex brain. Unfortunately the rather limited information on the earliest amphioxus-like animals, from the Cambrian period (c. 545–500 Ma (million years) ago)[9] does not extend to seeing their brains. In general, however, the genomic evidence suggests that the living amphioxus is not in some sense degenerate but is genuinely primitive.

Revealing the foundations of the molecular architecture that underpins our brains and sentience gives us not only a feeling of emergence, but underlines how little we really know about why and how organic complexity arises. Nor is this example of the amphioxus brain and its molecular inherency in any way unusual. Equally instructive examples can be culled from the most primitive animals, such as the sponges and *Hydra* (the latter is a relative of the sea-anemones and corals), in which genes (or proteins) that are essential for complex activities in more advanced animals are already present. Doubtless they have their functions, but what these are and how they have been redeployed, co-opted, or realigned in more advanced animals is for the most part still unknown. The unravelling of these evolutionary stories is going to be one of the most fascinating episodes in recent biological history, but what will almost certainly be more extraordinary is how much of organic complexity will be seen to be latent in more primitive organisms. Or perhaps not that extraordinary: it is sometimes forgotten that the main principle of evolution, beyond selection and adaptation, is not the drawing of new plans but relying on the tried and trusted building blocks of organic architecture.

THE NAVIGATION OF PROTEIN HYPERSPACE

Life, then, is full of inherencies. We might legitimately enquire whether there is anything in the human condition that could prefigure some future level of complexity that with the virtue of hindsight will, no doubt, seem to have been inevitable, but to us remains unimaginable. Yet whatever privileges exist for untrammelled speculation, there is a story to be told which will occupy the rest of this book. My critics will, I imagine, complain at its eclectic, if not unorthodox, nature; and given that the topics covered will include such matters as extraterrestrial amino acids and ants pursuing warfare it is advisable to try to explain the underlying thread of the argument. Here we can do no better than to look at a stimulating and thoughtful essay written by Temple Smith and Harold Morowitz,[10] which is an exploration of the tension between the predictabilities of physical systems and the seemingly contingent muddle that we call history. In brief, and their paper contains many other insights, they remind us of the simplicity of the basic building blocks of life, by which is meant such molecules as amino acids (which go to build the proteins, perhaps familiar as collagen or haemoglobin), or sugars (which when joined together can form carbohydrates). In the case of amino acids, however, even with the rather modest total of the 20 available variants and taking a relatively simple protein – consisting, say, of 100 such building blocks – it is immediately apparent that the potential number of combinations in which this protein could be assembled is absolutely gigantic. Specifically it is 20^{100}, which is equivalent to 10^{130}.

This is an uncomfortably large number,[11] and, as Smith and Morowitz see it, this immensity of possibilities confers an inherent unpredictability on the process of evolution. Taking the figures given above, that is the 20 available amino acids and their random inclusion in a protein composed of a total of 100 amino acids, Smith and Morowitz then apply some apparently stringent criteria to the selection of those proteins that will actually work. The specific function they have in mind for proteins is as natural catalysts, that is, to function as the organic molecules known as enzymes,[12] which serve to accelerate metabolic processes. The alternative, of course, is that a hypothetical protein will be non-functional, failing in one way or another. We know that in principle this is perfectly feasible, because there are many examples known where only a handful of changes, and sometimes even the substitution of a single amino acid for another one, will render the entire protein inoperative and thereby biologically useless. Let us then suppose that only one in a million proteins will be soluble, a necessary prerequisite for the watery milieu of a cell. Let us further suppose, and again the figure seems reasonable, that of these again only one in a million has a configuration suitable for it to be chemically active. How many potentially enzymatically active soluble proteins with an amino acid length totalling 100 could we expect to be available to life? A few thousand, perhaps even a few million? In fact, the total far exceeds the number of stars in the universe.

As Smith and Morowitz dryly note, 'It is quite clear from such numerology that the domain of possible organisms is enormously large if not infinite',[13] especially when we recall that many proteins are substantially longer than 100 amino acids. The only way we can begin to envisage such a protein domain is in the abstract terms of a kind of hyperspace. Mathematically this will encompass all the measurements that together serve to define the totality of this 'protein space'. As Smith and Morowitz point out, with such an immense number of potential possibilities the number of proteins known to exist on Earth can only be an infinitesmally small fraction of this vast total. As they say, notwithstanding 'the immensity of the dimensionality of the descriptive hyperspace', the world we know and the evolutionary processes that define it have 'produced a very sparsely sampled hyperspace in the actual living world'.[14]

One inference that might be immediately drawn from this is that in principle the likelihood of any other world employing an area of 'protein space' that is even remotely close to that found on Earth should be vanishingly small. At this early stage of the argument we can leave aside, for the moment, the distinct likelihood that Earth-like planets are going to be in exceedingly short supply (Chapter 5), and simply remind ourselves that even as our net of exploration spreads first across the Milky Way and then from galaxy to galaxy, so each time a protein chemist steps on to the surface of a new planet only another tiny fraction of this immense 'hyperspace' will be documented. The combinatorial possibilities are so much more immense than all the planets with all their biospheres that most proteins will for ever be only hypothetical constructs. That, at least, is the expectation and it would seem difficult to refute. All other worlds might be expected to be truly alien, at least in so far as the occupation of protein 'space' is concerned. That is, at least, the assumption.

THE GAME OF LIFE

Nevertheless, despite Smith and Morowitz's calculus of immensity, matters are probably much less alien than might at first be imagined. This is because at one level the strings of amino acids and their exact sequence are irrelevant, so long as the protein works effectively. To be sure, specific regions of a protein may be exceedingly sensitive to which amino acid is present, but we also know that various proteins have evolved independently of each other to perform a similar, if not identical, function. Such examples of molecular convergences are examined in more detail later (Chapter 10). What matters here is that these convergences emphatically do not depend on arriving at a closely similar sequence of amino acids, which given the size of protein 'hyperspace' would be almost a miracle. Navigation through this 'hyperspace' depends rather on two principal properties that, as it happens, underpin all life. The first concerns the remarkable specificities of particular sites within the protein that confer the necessary function, for example in those microbial pathogens whose existence depends on precise molecular mimicry to outwit a host's defences. The secondary property is that the complex functions that characterize proteins depend not only, in many cases, on highly specific sites, but also on particular architectural forms that are highly recurrent.[15] As we shall see in at least some protein designs, such as those that render tissue transparent (as in an eye lens), transport or store oxygen (for respiration), or are sensitive to light (rhodopsin and cryptochromes), the same solution to the biological need has been arrived at independently several times.

The implications are far-reaching, because the 'nodes' of biological possibility may, because of physical constraints, be much more limited than is usually supposed. As Michael Denton and Craig Marshall remark,

> If forms as complex as the protein folds are intrinsic features of nature, might some of the higher architecture of life also be determined by physical law? The robustness of certain cytoplasmic forms ... suggests that [they] may also represent uniquely stable and energetically favoured structures ... If it does turn out that a substantial amount of higher biological form is natural, then the implications will be radical and far-reaching. It will mean that physical laws must have had a far greater role in the evolution of biological form than is generally assumed. And it will mean ... that underlying all the diversity of life is a finite set of natural forms that will recur over and over again anywhere in the cosmos where there is carbon-based life.[16]

I agree. Not all is possible, options are limited, and different starting points converge repeatedly on the same destinations. Any such evolutionary journey, including navigation through protein 'hyperspace' must presuppose intermediary stages. And here there may be further constraints because seemingly 'sensible' paths may turn out to be non-functional.[17] The 'landscape' of biological form, be it at the level of proteins, organisms, or social systems, may in principle be almost infinitely rich, but in reality the number of 'roads' through it may be much, much more restricted.

This is not to say that there are no alternatives: patently there are, and the world is a diverse place. Smith and Morowitz remind us that despite these potential immensities the actual 'Game of Life', as they call it, is still going to be played the same way everywhere. Here are the four basic rules, which incidentally presuppose variation (which is offered by the different alleles (determining characters) of a gene) and subsequent process of selection.

(1) Hindsight and foresight are strictly forbidden. Of course we are fully entitled to hug ourselves with delight as we trace, for example, the multiple evolutionary origins of the electric organs

in certain fish (a topic I return to in Chapter 7), but we can only retrodict and not predict.

(2)   Minor changes are easier than major changes. That's something all biologists recognize,[18] and why, for example, there is a deep-seated distrust of macroevolutionary 'jumps' that allow a fully fledged body plan to emerge from some strikingly dissimilar ancestor.

(3)   Resources are not unlimited: the world is finite, and ultimately energy and space are in restricted supply.

(4)   Life has no option but to carry on; it must always play the best hand it can no matter how poor and disastrous the hand might be, and no matter who or what offers the challenge.

In this way Smith and Morowitz neatly encapsulate what evolution is all about. They suggest that given these four basic rules for the Game of Life we should not be surprised to see the emergence both of evolutionary trends and of emerging complexity; Smith and Morowitz also remind us that symbioses[19] and sex are two good ways of speeding up the game. And that is all there is to it? Not quite. Trying to keep the surprise out of their joint voices, Smith and Morowitz then continue:

> There is at least one major evolutionary trend not immediately explained by our strategy rules [i.e. their Game of Life]. That being the numerous examples of morphological convergence. Why, in the sparsely sampled genetic space, have there been so many cases of apparent convergence or parallelism? It is surprising in the light of the high probability for novelty to find, even in similar niches, high morphological similarity in distinctly different genetic lines ... there may be additional rules operating at coarser levels of the genetic space which are less statistical than those discussed.[20]

This prescient statement prefigures the main purpose of this book, that evolution is indeed constrained, if not bound. Despite the immensity of biological hyperspace I shall argue that nearly all of it must remain for ever empty, not because our chance drunken walk failed to wander into one domain rather than another but because the door could never open, the road was never there, the possibilities were from the beginning for ever unavailable. This implies that we may not only be on the verge of glimpsing a deeper structure to life, but that it matters little what our starting point may have been: the different routes will not prevent a convergence to similar ends.[21]

EERIE PERFECTION

The understanding of the genetic code was, after the elucidation of the structure of DNA with its four bases and famous double helix, the next triumph in the field of molecular biology. As already noted, proteins are built from the twenty available amino acids,[22] although it has long been known that particular examples, such as the protein collagen that goes to form such structures as tendons (Achilles' heel) or the silk proteins that form the spider's web, are enriched in particular amino acids which reflect, in ways that even now are not completely understood, the functional and structural properties of these and other proteins. Thus collagens are enriched in such amino acids as proline, while spider-silks possess notable quantities of alanine and glycine.

Each of the amino acids is coded for by a set of three nucleotide base pairs, accordingly known as a triplet. The original code is, of course, stored in the DNA of the chromosomes, but the actual synthesis of the amino acids occurs through the agency of the RNA in minute structures within the cell known as the ribosomes. Thus, in RNA the four bases are adenine (A), cytosine (C), guanine (G), and uracil (U), the last of which substitutes for thymine (T), which is found in DNA only. With a triplet code and four base pairs there are of course 64 possible combinations. This implies that with only 20 amino acids there is a considerable degree of redundancy, even with the assignment of certain codons to signal 'Start' and 'Stop'. In fact we see that only two amino acids (methionine (abbreviated M) and tryptophan (W)) rely on a single codon each (respectively coded for by AUG and UGG), whereas the remaining 18 amino acids are able to call upon from two to six codons. (For example, histidine (H) uses either CAC or CAU; arginine (R) employs CGU, CGC, CGA, CGG, AGA, and AGG.) It has long been known that this redundancy means that mistakes in coding may not be detrimental; if a substitution within the codon fails to result in the identical amino acid, it stands a good chance of producing another amino acid with similar properties. Amino acids with similar properties, of which their affinity to or repulsion from water (the property of polarity) is particularly important, also tend to have similar pathways of biosynthesis. Here, too, if errors occur then the mistake need not be lethal. For these and other reasons, therefore, it is clear that the genetic code is excellently adapted to the needs of reliably providing the amino acids that underpin protein construction.

But how good is good? The rule of thumb in evolution is 'good enough to do the job in most circumstances', but not to waste time building a Rolls-Royce of an organism, or, to put it more flippantly, no supersonic albatrosses. Even so, measuring this 'goodness' for purpose is not so easy: organisms themselves are rubbery, slippery, and pliable and non-invasive techniques of investigation are time-consuming and often difficult. One way to address this problem is to look at the design tolerance of an organism, that is, to see the margins of safety built into such a structure as a bone. A powerful analogy, as Jared Diamond reminds us,[23] is to think of a lift in a prestigious building dedicated to the serious accumulation and worship of money. 'Room for one more', says the lift attendant, before the cage shuts, shoots skywards towards the 59th floor, which it never reaches because at the 48th floor the cable snaps ... Such instances are, in the absence of malice, mercifully rare because the safety factor of such a lift cable, measured as the ratio between its ultimate capacity and maximum load imposed in normal use, is almost 12 times. The equivalent ratio for a cable in a dumb waiter ascending with its cargo of brown Windsor soup and claret is about five; for a bridge, engineers are content to allow a safety factor of only two. In this last case, however, Henry Petroski reminds us that the safety factors for some modern bridges may in reality be perilously small.[24]

It is perhaps not surprising that by and large the safety factors adopted by organisms[25] are closer to those of the dumb waiter and the bridge. Thus the silk dragline of a spider has a modest safety factor of only 1.5, whereas the factor for the leg bone of a kangaroo hopping through the Australian outback is 3. There is an additional and quite important point that many safety factors may in themselves be sub-optimal – spider silk does snap and kangaroos can break their legs – but the margins of safety are necessarily a compromise between strength and many other vital functions in the organism. Even so, over-design does provide an important safety margin, especially when an organism encounters an unpredictable and rare circumstance. In assessing this and other reasons for such safety margins Carl Gans also makes the point that such tolerances may facilitate the occupation of a hitherto untested adaptive zone.[26] One of the examples he gives is the New Zealand parrot known as the kea. This is a fascinating bird with highly adaptable feeding habits. The kea also has a penchant for trashing cars, and its behavioural characteristics include delinquent gangs of young birds.[27] In passing I should also mention that notwithstanding the overwhelming evidence for adaptation and functional demands faced by organisms there remain some examples of structures whose significance still baffles biologists. John Currey gives a nice example in the form of the rostral bone in the snout of Blainville's beaked whale (*Mesoplodon densirostris*).[28] As the species name suggests, this bone is incredibly dense, but why? One can speculate that it might be employed in fighting, but this rostral bone is very brittle, a consequence of its very low organic content. Alternatively, it might act as ballast,[29] but Currey is candid when he writes, 'At the moment, its function, in this rarely found whale, is a mystery'.[30]

By this stage you will be wondering what possible connection could exist between the safety factors of a kangaroo, let alone the rostral bone of a rare whale, and the efficiency of the genetic code. The point, simply, is that given the realities of the physical world and adaptation, organisms and their components should be designed to do the job adequately, but no more. Humans shudder at the prospect of hurtling to their doom down a lift shaft, and so incorporate a safety margin that seems to be found very seldom in organisms. And at first sight this is what we should see in the genetic code: it certainly isn't random; in fact it is really rather good. But in recent years a group of molecular biologists, notably Steve Freeland and Laurence Hurst, have been trying to arrive at a more precise answer.[31]

Their approach is computer-based, and the basic aim is to randomize the genetic code and then compare the efficiency of a certain fraction of the vast number of alternative codes the computer can generate with the real one, here on Earth. There is, of course, the implicit assumption that a genetic alphabet composed of two base pairs (that is AT/CG),[32] as well as the system of triplet codons and the 20 amino acids[33] available for protein construction found in all terrestrial life represents some sort of norm. Alternatives to codon usage and the number and type of amino acids can, of course, be envisaged, but Arthur Weber and Stanley Miller have gone so far as to suggest that 'If life were to arise on another planet, we would expect that ... about 75% of the amino acids would be the same as on the earth.'[34] Naturally we need to be cautious in assuming that even if proteins are universal they necessarily depend on the terrestrial mechanism of codons[35] and the same battery of amino acids. Yet there still may be constraints. Codons built as doublets, i.e. only two base pairs (e.g. AA or AU) to code for an amino acid, would probably be rather vulnerable, while quartet or quintet (e.g. AAAA or AUAUA) codons might be getting cumbersome. There are, of course, many more amino acids known than are actually employed in the proteins and, as we shall see (Chapter 3), some of these are best known from meteorites and have no biological equivalents. Even so, given that the simplest amino acids (such as glycine, serine, and alanine) are probably the most readily synthesized anywhere in the Universe, it is possible that they predispose the biosynthetic pathways that lead to the more complex amino acids.[36] So, perhaps both the genetic code and protein construction 'out there' are not so very different.

There is, however, a second difficulty in deciding just how effective the terrestrial code might be. This is because randomizing the existing genetic code leads to an astronomical number of alternative possibilities: Freeland and his co-workers suggest a figure of about 10^{18}, which, as they helpfully remind us, is ten times as many seconds as have elapsed since the formation of the Earth. It is another big number (see note 11), and echoes the point I raised in discussing the essay by Smith and Morowitz (see note 10), that with the immensity of a protein, or in this case, genetic 'hyperspace', it would not only be *a priori* exceedingly unlikely that any two biospheres – separated also by a gulf of many light years – would arrive at the same evolutionary solution, but it would be even more fantastically improbable that the solution achieved was not only good (the process of natural selection should see to that) but in fact the very best. Yet, this appears to be the implication in the work by Freeland and his colleagues.

Their work, as is customary, has proceeded in several stages. Well aware of the preceding work already indicating the general efficiency of the genetic code, they examined a million alternative codes (Figure 1.3). To the first approximation the distribution can be compared to the familiar bell-shaped curve that, it is said, describes the distribution of human intelligence (IQ): a few stupid people and equally few geniuses, with most of us somewhere in the middle. So, too, with the distributions of alternative genetic codes: there is a wide range of efficiencies; some alternatives are extremely inefficient ('disastrous') and, perhaps not surprisingly, the majority are quite efficient but not

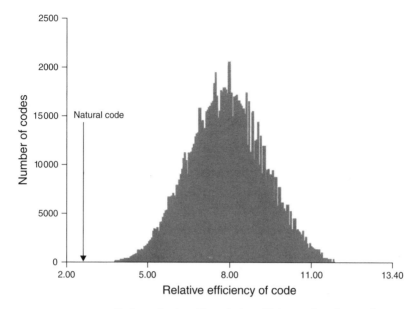

FIGURE 1.3 Eerie perfection. The relative efficiency of randomized genetic codes, ranging from disastrous on the right to increasingly competent to the left. Note the approximately bell-shaped curve: most codes are pretty good, a few terrible, and a few very good. Also note where this planet's genetic code falls: far, far to the left. (Reproduced with permission from *Journal of Molecular Evolution*, from the article The genetic code is one in a million, by S.J. Freeland and L.D. Hurst, vol. 47, pp. 238–248, fig.7; 1998, copyright Springer-Verlag, and also with the permission of the authors.)

remarkably effective. Very few of the alternatives are really impressive, but note where in Figure 1.3 the real or natural code falls. Freeland and Hurst have difficulty in keeping the surprise out of their report, even given the proviso that their approach necessitates a number of assumptions. They write: 'the natural genetic code shows *startling* [my emphasis] evidence of optimization, two orders of magnitude higher than has been suggested previously. Though the precise quantification used here may be questioned, the overall result seems fairly clear: under our model, of 1 million random variant codes produced, only 1 was better ... than the natural code – our genetic code is quite literally "1 in a million".' [37]

This result, however, needs to be put into a wider context, because the million (10^6) alternatives that Freeland and Hurst looked at is only a small fraction of the total number of possibilities, which, as already noted, they estimate to be about 10^{18}. On this basis there could still be an astronomically large number of alternative genetic codes, each of which in its 'local' context could also prove to be very good indeed when compared to a randomly chosen set of a million other codes. In their analysis of the million alternatives Freeland and Hurst specifically noted that the one code that in principle might be better than the natural one had, as one might expect, little similarity to the one used by life on Earth. It seems, however, that the potential figure of 10^{18} alternatives is, in reality, inflated. This is because not all the biosynthetic pathways used to construct the 20 different amino acids are in themselves viable. In a subsequent analysis Freeland and his co-workers suggest that the number of alternative codes that over-all are realistically functional is relatively small. They estimate that this number might be about 270 million; and taking into account the similarities between certain amino acids they conclude, again in my opinion startlingly, 'that nature's choice [on Earth] might indeed be the best possible code'.[38]

In one way we should hardly be surprised at the efficiency of the genetic code.[39] It is difficult to believe that the genetic code is not a product of selection, but to arrive at the best of all possible codes selection has to be more than powerful, it has to be overwhelmingly effective. The reason for saying this is that with some minor, and ev-idently secondary, exceptions,[40] the genetic code is universal to life: you, the primrose on the table, and the bacteria in your gut all employ the same code. The earliest evidence for life is about 3.8 billion years ago and these forms are presumably directly ancestral to all groups still alive today. If so, this indicates that whatever changes occurred as the genetic code evolved towards its stable state must have been achieved still earlier; the genetic code would not otherwise be uni-versal. Yet, as we shall see (Chapter 4), life itself may not be older than about 4 billion years. Two hundred million years (and possibly much less) to navigate to the best of all possible codes, or at least from the 270 million alternatives? Part of the explanation, as is so often the case in evolution, may be to look for a step-like arrangement: once one stage is achieved, other things then become so much more likely.[41] Yet, there is also a sense that given a world of DNA and amino acids, then perhaps the genetic code we know is more or less an inevitable outcome. And if this is true, then what else might be inevitable, both here on Earth and elsewhere?

This is not the only way to look at inevitabilities in evolution. The argument from the genetic code looks to a potentially gigantic 'hyperspace' of alternative possibilities, yet the evidence suggests that rapidly and with extraordinary effectiveness a very good, perhaps even the best, code is arrived at. It is as if the Blind Watchmaker takes off her sunglasses and decides to visit her brother Chronos. Off she sets, crossing streets roaring with traffic driven by psychotics, through the entrails of the subway system of a megalopolis, and, after catching a series of intercontinental express trains with connection times of two minutes each, she arrives at Chronos' front door at 4 p.m. prompt, just in time for a relaxing cup of tea.

FINDING EASTER ISLAND

Hence to an explanation for the title of this first chapter. Easter Island is the remotest speck of land on Earth, surrounded by the vastness of the Pacific Ocean. At first sight it seems quite extraordinary that it could have been encountered by the seafaring Polynesians, however audacious. Surely, one would suppose, it was a chance discovery, perhaps by mariners who had been blown far off course, which led to the prows of the first canoes accidentally grating onto a beach of Easter Island perhaps some 1500 years ago. Another quirk of history? Very probably not. Easter Island may have marked one of the furthest points in this great human diaspora, but its discovery was inevitable given the sophisticated search strategy of the Polynesians. As Geoffrey Irwin has shown,[42] not only were these people superb navigators, but they developed a method of quartering the ocean that aimed to find new lands. Century by century their net of exploration widened. When a particular season failed in the objective they had a sure way of finding their way home to safety. Their vessels were designed for protracted journeys, but the key to their success was to head against the prevailing winds on the outward journey. At the limit of their range on any one journey, the sternward winds rapidly returned them towards their home and safety. And how was home, another speck in the ocean, arrived at? In the sky above the boats the net of stars provided the clues to celestial navigation, and as the constellations fitted into place so an increasingly familiar starlit sky provided the beacons for a successful homecoming.

So, too, in evolution. Isolated 'islands' provide havens of biological possibility in an ocean of maladaptedness (Fig. 1.4). No wonder

FIGURE 1.4 A metaphorical view of protein 'hyperspace', in which functional proteins project above an immense 'ocean' that submerges non-functional alternatives. (Reprinted from *Journal of Molecular Biology*, vol. 301, D.D. Axe, Extreme functional sensitivity to conservative amino acid changes on enzyme exteriors, pp. 585–596, fig. 5. Copyright 2000, with permission from Elsevier Science, and also with the permission of the author.)

the arguments for design and intelligent planning have such a perennial appeal. Whether it be by navigation across the hyperdimensional vastness of protein space, the journey to a genetic code of almost eerie efficiency, or the more familiar examples of superb adaptation, life has an extraordinary propensity for its metaphorical hand to fit the glove. Life depends both on a suitable chemistry, whose origins are literally cosmic, and on the realities of evolutionary adaptation. The chemistry is acknowledged but largely ignored; the adaptation is often derided as a wishful fantasy. As with the audacious and intelligent Polynesians, so life shows a kind of homing instinct. Its central paradox revolves around the fact that despite its fecundity and baroque richness life is also strongly constrained. The net result is a genuine creation, almost unimaginably rich and beautiful, but one also with an underlying structure in which, given enough time, the inevitable must happen.

To conclude this chapter – and to anticipate the main theme of this book – let us accept that the genetic code must be spectacularly efficient, driven to a one-in-a-hundred-million alternatives by the remorseless action of selection. All life shares this one code, but this commonality has not stifled the creative potentials of life, as both the fossil record and the exuberance of the living world so clearly demonstrate. Yet for all this exuberance and flair there are constraints: convergence is inevitable, yet paradoxically the net result is not one of sterile returns to worn-out themes; rather there is also a patent trend of increased complexity. Some cosmologists like to speculate that the Universe is designed to be the home of life, to which some biologists might add 'Yes, and not only that but we have a pretty shrewd idea of what was on the cards.' But to see how the hand is played, we need first to see how life itself might have originated – and a very odd story is now emerging.

2   Can we break the great code?

One of the paradoxes of science is that its very greatness as an intellectual adventure is perversely mirrored by a crippling diminution of what it is to be human. Emerging from the slime, our animal instincts barely controlled, we are informed in gloating terms of our complete and utter insignificance.[1] Trapped in an out-of-the-way solar system, the galaxy around us may well pulsate with sentient activity set in a dazzling array of civilizations, but if we are ever visited it will be either to view one more zoo (or lunatic asylum) or to stock up on a depleted larder. This is hardly an encouraging view, but, as many others have pointed out, our genocidal and destructive tendencies may make a plea for a lenient sentence sound hollow.[2] And, in principle at least, this possibility should be taken seriously. After all, the scientific view is that the emergence of life is surely inevitable, and there is no shortage of planets on which the long climb from pond-scum to shopping must surely be taking place – or, happy thought, perhaps twinkling just out of sight is a planet now covered with shopping malls, parking lots, and internet cafes.

The question of whether such a climb need necessarily lead to a sentient species is the topic of a later chapter (9), although conceivably your reading of this section may be interrupted by the roar of descending spacecraft or, perhaps more plausibly, the dramatic news of the success of the SETI enterprise as the long-awaited extraterrestrial signals finally confirm that indeed we are not alone. Alternatively, it may be that sentience is rare indeed, and pond-scum is the rule.[3] No matter: orthodoxy states that life is ubiquitous. Indeed, it may be elevated to a cosmic principle.[4] Christian de Duve, for example, writes, 'the Universe was pregnant with life ... we belong to a Universe of which life is a necessary component, not a freak manifestation.'[5]

In the absence of direct evidence for extraterrestrial life, the question of the ubiquity of life in the cosmos really hinges on discovering how easy it might be to synthesize life and on reassuring ourselves that there are mansions a-plenty to allow a near-infinitude of evolutionary experiments. Yet, as I shall try to demonstrate in Chapter 4, the scientific annexe dedicated to the problem of the origin of life has not been marked by a series of sweeping and spectacular advances – the norm for science – but has lurched indecisively across a landscape dotted with stumbling blocks and crevasses. Nor does the discovery that the processes of organic chemistry permeate the galaxy offer any more than a crumb of hope. The next chapter (3) is entitled 'Universal goo', and with good reason. And is there really a plurality of worlds, a trillion planets endowed with an unimaginable range of biological diversity, while on a favoured handful – let's be conservative and propose a mere ten thousand – millions of alien 'children' wait for the 'bedtime' story as the spectacular sunset of Threga IX[6] floods the landscape? Distant solar systems are now being discovered at a remarkable rate (see Chapter 5), but, as we shall see, the results so far are distinctly discouraging: worlds without number, but strange, hostile, and most probably uninhabitable.

THE GROUND FLOOR

What then might be ground rules for life, here or anywhere? At first sight this second chapter may seem something of a diversion. This is especially so because it looks to what preceded even the genetic code, discussed in the previous chapter: that is, to consider the structure of DNA. But there is a purpose to this approach because it can be argued that DNA shows some very peculiar features. This, in turn, is the first hint that in terms of its fundamental building blocks there may be few, if any, alternatives available to life, here on Earth or at the far end of the Universe. Yet the basic building blocks of DNA (composed in essence of the nucleotides and a sugar), and for that matter those of proteins (the amino acids), are really rather simple. There is a paradox here, which Maitland Edey and Don Johanson rightly regard as 'breathtaking'.[7] Their words have a peculiar resonance in this book. Speaking of DNA and life itself they write:

> on the one hand [there is] the endless complexity of the process, on the other the simplicity of the principle. To make everything that can be called alive, to monitor the development of every fern and feather on earth, to direct their growth, to enable them to function, to replace worn-out parts, to turn things on, to turn them off – for all those activities to be orchestrated by just four kinds of

small molecules [that is, the nucleotide bases] is awe-inspiring. It is that magisterial power of DNA, the power to direct, that commands our attention. Its molecules are absurdly simple. They are not alive. In many ways they resemble crystals. But they can do things no crystal ever dreamed of ... They wave their instructional wands ... proteins grow, enzymes snip and patch. Lo, there emerges a bacterium, a flower, a fish, a Frenchman.[8]

From remarkable simplicity arises immense complexity, yet a basic theme still emerges which confers on evolution a broad predictability. If indeed we can delineate the architecture of life, then two tantalizing prospects arise. Perhaps we can really begin to explore the reality of alternatives, of evolutionary counterfactuals. And possibly, to anticipate the last chapter (12), we shall discover in the end that there are none. And, despite the almost crass simplicity of life's building blocks, perhaps we can discern inherent within this framework the inevitable and pre-ordained trajectories of evolution? This second topic, of life's directions, will occupy the second half of this book, but now let us look briefly at some of the underpinnings of life.

At a fundamental level matters seem to be rather simple. The principal elements required for life, at least as we know it, are carbon, hydrogen, oxygen, nitrogen, and phosphorus. These are all readily available, and carbon certainly has an almost uncanny knack of arranging itself in configurations that are both flexible and robust. Not that we should fail to consider alternatives. Iris Fry, for example, has an interesting discussion of both the definitions of life and the likelihood of radical alternatives that regard the 'carbaquist' (i.e. carbon plus water) foundation as hopelessly parochial.[9] An ever-popular topic in this area is whether the element silicon, whose most familiar natural manifestation is probably as quartz crystals, could ever act as a substitute for carbon in providing the backbone of molecular architecture, be it in some sort of equivalent to the carbohydrates, lipids, or proteins of life. Silicon certainly possesses some advantages over carbon, such as its much greater abundance. Yet these seem to be more than offset by a number of drawbacks.[10] For example the silicon–silicon bond is unstable in the presence of water and oxygen, which is something of a disadvantage for a water-based aerobic organism. In addition, silicon has a tendency to unite with oxygen to form very stable silicate polymers. Fry reminds us that speculations on alternatives need not be restricted to hypothetical silicon-based life forms. Robert Shapiro and Gerald Feinberg go even further and conjure up life in liquid ammonia, molten silicates, the interiors of stars, or within interstellar clouds.[11] Fry reminds us that however far-fetched these ideas might be 'the formal characteristics of hypothetical extraterrestrial life [still derive] from the principles of organization of life as we know it,'[12] and she continues,

> Every living system, as revealed particularly at the molecular level, is organized in a much more complex way than any ordered physical system known to us. The unique character of this complexity lies in the ability of an organism to maintain and reproduce its organization according to specific internal instructions, or information, manifested in specific macromolecules. This character is connected with the purposeful, functional nature of biological organization, in which each part serves the survival of the whole. The Shapiro and Feinberg definition [effectively free flow of energy and emergence of order] is silent on those prominent biological features.[13]

These comments are important in the context of what is, and is not, likely in alien 'biospheres'. It is important to stress that Fry does not regard non-carbaquist life forms as impossible, but in some contexts, e.g. stellar interiors, their detection is problematic, and even planetary-based systems that might involve liquid ammonia or silicon 'backbones' (note 10) run into difficulties. Thus, liquid ammonia requires very low temperatures (less than -33 °C) and its ice sinks, unlike water ice, which is a remarkable solid in that it floats on its own liquid. Fry concludes, together with most workers, that life elsewhere, if there is any, is most probably carbon-based.

Somewhat similar arguments may also apply to a number of elements vital for the operation of life, perhaps most notably phosphorus.[14] In the chemical form of phosphate it plays key, and probably irreplaceable, roles. Of central importance is its employment in the construction of the nucleic acids, the building blocks of DNA and RNA. Phosphorus is also integral to the molecule ATP (adenosine triphosphate), which is central to the storage and transfer of energy within the cell. As with the case of the merits of silicon versus carbon, it needs to be emphasized that it is not that alternatives simply cannot be envisaged. Indeed, if humans had been detailed to get the whole process going, the eager occupants of the First Laboratory might well have passed by phosphates to consider such biochemical alternatives as arsenic and silicic acid (note 14). As is repeatedly the case in looking at the basis of life, however, these alternatives are substantially less fit for the purpose. As we shall also see when discussing some current problems in understanding the origin of life (Chapter 4), it is probably just as well that humans or their equivalents were not recruited to the First Laboratory: my guess is that they would still be hard at work.

There is another twist to the way life uses the elements of the periodic table. Consider, for example, the rather unfamiliar element known as molybdenum. Some years ago Francis Crick and Leslie Orgel came up with a particularly intriguing idea that nicely reflected their ingenuity and lateral thinking.[15] In essence their argument was as follows: molybdenum is an essential element for life. It plays, for example, a key role in various enzymes. On Earth, however, the availability of this element seems to be drastically limited, occurring as it does in only minute concentrations. Crick and Orgel's imaginative leap was to argue that this dependence on molybdenum was not an unfortunate, if brute, fact, but actually it indicated that our ancestry was far from Earth, derived from life forms where the precious molybdenum was more freely available. The idea that we might represent marooned colonists – perhaps from a long-dead planet engulfed in some stellar catastrophe – has a romantic appeal that taps a recurrent root in humans of displacement and longing. Not, of course, that these hypothetical colonists would be anything more than bacteria or some such equivalent. In any event, the history of life provides no evidence (although perhaps it should) of any subsequent visitation, let alone intervention, by extraterrestrials. Of course, getting even bacteria across the interstellar wastes, those cubic parsecs of hard vacuum drenched in radiation, is in itself so problematic that it may be reasonable to suppose that if panspermia (that is, transport from one star system to another[16]) occurs at all it can only be by a directed, that is, an intelligent, activity. And this is what Crick and Orgel suggested: from a simple observation, too little molybdenum, to a bold hypothesis of a directed panspermia. The latter notion may still be credible, but, alas, the original estimates of the amounts of molybdenum were too pessimistic. It looks as though even on Earth there is actually enough.[17]

It seems, therefore, that not only is life definitely terrestrial (unless it is from Mars; see Chapter 5), but is firmly grounded in a number of key elements of which carbon is pre-eminent. There are, however, few elements in the periodic table that are not employed somewhere or other in the panoply of life.[18] With the exception of carbon it would be stretching matters, at least from a biochemical perspective, to describe the elements themselves as the real building blocks of life. The standard view is that this epithet is better applied to the nucleotide base pairs of DNA (and RNA), as well as the twenty-odd amino acids that are employed to build the proteins, not to mention the sugars, the basis for the carbohydrates and more complex polymers such as cellulose and chitin, and the fatty acids (based on a hydrocarbon chain) that are the key component of the lipids employed, for example, in the cell wall.

The DNA is, of course, the basis of the genetic information; and through the further agency of RNA it provides the basic instructions that give rise to the scaffolding of life, most notably in the proteins. There are two points to be made at this juncture. First it seems reasonable to suppose that in unravelling at least one component of the mystery of the origin of life (Chapter 4) we must ask not only how the nucleotides and amino acids were synthesized in some credible prebiotic milieu – be it as different as an interstellar cloud or a tidal pool – but also how they came to be in a position of mutual association. As we shall see, there remains a yawning gulf between this unremarkable expectation and reality, at least in so far as experimental data are concerned. The second point actually pertains to the central theme of this book: how does evolutionary complexity emerge and what viable alternatives are there to the world we know?

DNA: THE STRANGEST OF ALL MOLECULES?
Consider again the nucleotides. As individual building blocks these molecules are rather unremarkable, yet when they combine to form DNA, they suddenly show some very strange properties indeed. Consider, for example, these comments by Christopher Switzer and colleagues:[19] 'The interaction between the two complementary oligonucleotide strands [that entwined form the famous double helix] remains one of the most remarkable examples of molecular recognition known to chemistry, especially because several features of nucleic acids appear *a priori* to suit them poorly for this process,'[20] that

is their quite superb performance in the process of replication, the *sine qua non* of DNA. It is difficult to believe that any terrestrial chemist, if handed those molecules and asked to make something useful, would feel encouraged. As Switzer *et al.* point out the whole problem is fraught with ambiguities. First, the oligonucleotides are a type of polymer that is effectively unidimensional and also flexible. Stringing them into the double helix involves unexpected distortions to the polymer. In the technical language of the original article they experience a 'substantial loss of conformational entropy'.[21] Put simply, polymers like these shouldn't really behave like this. Moreover, the double helix itself springs various surprises. The specificity of the helix, with the pairing between the oligonucleotides (in DNA adenine matches to thymine, and guanine to cytosine) is mediated by hydrogen bonds; yet, as Switzer *et al.* again point out, such an arrangement is decidedly 'problematic in an aqueous environment'.[22] Here, too, a hypothetical designer might be expected to construct a replicatory system that would not face, potentially at least, disablement in the watery interior of the cell. Yet despite these and other ambiguities[23] the action of DNA is almost uncannily effective.

Nor does the versatility of the DNA molecule end here: it has a whole series of other strange properties. Not only can the DNA strand readily coil and twist, but it can also be assembled into so-called Borromean rings. These are named after the motif of a noble Italian family, and consist of several interlocked rings so arranged that if any one ring is broken then the whole structure disintegrates.[24] DNA might be most familiar as a double helix, but it can also be arranged in two-dimensional sheets,[25] and even configured into a sort of octahedron.[26] Not only that, but DNA can act as a template upon which metals can be seeded, forming, for example, tiny gold crystals[27] and silver wires.[28]

The very strangeness of DNA, of course, immediately begs the question: all right, granted DNA is certainly a decidedly odd molecule, but surely we can envisage alternatives? The answer to this question is that indeed we can inasmuch as considerable progress has been made in constructing various sorts of 'artificial' DNA (and RNA). The crucial question is whether, even if such a DNA shows superior properties, however defined, such an alternative configuration could have arisen by natural processes. The picture is still emerging, but what is not in doubt is that the five bases (adenine (A),

cytosine (C), guanine (G), thymine (T, in DNA)/uracil (U, in RNA))
found in nature can be replaced in the laboratory by other base pairs,
which for obvious reasons are referred to as 'non-standard'. Quite a few
are available to the experimenter; typical examples are the molecules
isocytidine (iso-C) and isoguanosine (iso-G). When incorporated into
a short strand of DNA, these novel base pairs do not usually alter
the overall geometry of the double helix but the arrangement of the
hydrogen bonds does differ. Potentially we could envisage a new type
of DNA with not the four and possibly optimal (note 32, Chapter 1)
standard 'letters' (ACGT/U), but six (or perhaps even more?). Switzer
and his colleagues even wonder whether 'In another, more whimsi-
cal form, we might ask whether the (iso-C).(iso-G) base pair might be
found as part of a coding system in extraterrestrial life.'[29]

This is not the first time we shall encounter apparent whimsy
as covering a potentially interesting evolutionary question. I would
argue that these remarks by Switzer et al. can be taken seriously:
it is surely important to explore at least some of the alternatives.
This is for at least two reasons. First, it may illuminate how DNA
(or more probably RNA if one follows the popular view of a prior
'RNA world') emerged in the prebiotic milieu. Second, the effort in
imagining alternatives may assist in the exercise of trying to conceive what may approximate to an alien biochemistry. The fact that
these so-called 'non-standard' base pairs can be woven, by suitable
laboratory manipulation,[30] into the double helix certainly excites
our curiosity. As Joseph Piccirilli and his co-workers[31] have pointedly remarked, however, despite their special interest to chemists
such novelties 'apparently never have been pursued by nature.'[32] At
present it seems that the reason why we do not find these alternative DNAs, at least on Earth, is not due to chance factors but for two
specific reasons. First, these 'unnatural' base pairs[33] seem to be significantly more difficult to prepare in plausible prebiotic conditions.
Thus Piccirilli et al. also note that such non-standard base pairs represent 'Structures [that] are undoubtedly more difficult to prepare under prebiotic conditions ... perhaps explaining their absence from the
repertoire of natural nucleotides. But such molecules can be obtained
by synthetic chemistry which is not subject to prebiotic constraints.'[34]
Second, and perhaps more significantly, serious difficulties then
arise in replicating faithfully at least some of the non-standard
substitutions.[35]

Introducing peculiar base pairs is, of course, not the only way of meddling with DNA (and RNA) to see what alternatives might exist, either in the laboratory or extraterrestrially. The question can, in any event, be tackled from several directions. Here, too, the results are neither comprehensive nor complete, but for the most part they seem to suggest that finding viable alternatives to DNA in any sort of natural context may be difficult, and perhaps even impossible. For example, if a series of base pairs are linked to make a short strand, in a process known as oligomerization, then if this is carried out in the test tube, the results are interestingly different from those obtained within the living cell. Although the molecules that are employed to build DNA are relatively simple, the association of the various components is precisely determined. For example, the phosphate unit and the sugar (known as a ribose), both of which are essential components of DNA, are each attached to the other principal unit, i.e. the nucleotide (e.g. adenine) at two specific locations of sites. In 'real' DNA, which is made in organisms, the linkage is at sites 3' and 5' (the superscripts refer simply to positions on the molecule). In contrast, in the artificial situation in the laboratory, the linkage positions are now at 2' and 5'.[36] Perhaps the former arrangement (3',5') evolved from the latter (2',5'), abiotic leading to biotic. If it did, this may not be surprising, because DNA showing the 2',5' linkage seems to be perceptibly weaker and, at least under laboratory conditions viable only when the salt content of the surrounding solution is elevated.[37] In the case of RNA matters are somewhat different. This is because here an 'abiotic' 2',5' linkage does not seem to be prejudicial to its operation until, that is, DNA is involved in the system. Then everything collapses.[38]

What else can we tinker with? Another key element is the sugar, ribose in RNA and deoxyribose in DNA. As we shall see subsequently (Chapter 4), one of the particular hurdles in understanding how life itself might have originated is to find a believable pathway by which ribose might be synthesized, at least in sufficient quantities to be of any use. At this juncture all we need to ask, in the context of alternative worlds, is whether ribose (or deoxyribose) is the only molecule in town? In fact, this topic has been explored in considerable detail by Albert Eschenmoser and his team, in what is appropriately termed an aetiology of nucleic acids,[39] and what has been very appropriately described as 'a chemical *tour de force.*'[40]

The overall aim of Eschenmoser's team is to find viable alternatives to the genetic code by the synthesis of what they call homo-DNAs. Such work is important for at least two reasons. First, it might conceivably give us some glimpse of an alien biochemistry or alternatively confirm that, in a way analogous to the employment of opsins in visual processes (see p. 170), DNA is the molecule of choice – on Earth, across the galaxy, and at all points beyond. Second, given the complexity of DNA and the improbability that it evolved *de novo* from the prebiotic 'soup', alternatives that incorporate simpler building blocks could give us some clue as to how DNA might have emerged. In brief, many of the constructs devised by the Eschenmoser team seem to be distinctly sub-optimal, if not catastrophically inept.[41] For example, if the sugar hexose is substituted for the normal ribose the nucleic acid strands are effectively inoperable.[42] On the other hand, employment of a chemical derivative of a simpler sugar (a threose), with four carbon atoms, as against the five of ribose[43] (and six of hexose), does produce a viable analogue of DNA.[44] The synthesis of a key compound, α-threofuranosyl, however, involves a complex laboratory procedure that is far removed from any credible prebiotic milieu, a problem that is returned to in Chapter 4.

Other types of homo-DNA show quite robust features that seem to match and sometimes exceed, at least in certain respects, the demanding design specifications set by the natural product. It is still too early to say whether genuine alternatives to DNA, either in the nucleotide–sugar field or on some wilder shore of biochemistry, really exist. Certainly some of the molecules used in the natural world are strangely suitable. Also we need to remember, as Eschenmoser notes, that the aim is optimization not maximization. Thus he writes how the 'moderate base-pairing strength ... in RNA and resulting from the high conformational flexibility of the ribofuranose backbone, was essential for the evolution of a rich diversity of nucleic-acid-related biological functions.'[45] Even if any of these modified DNAs and RNAs are found to show the full versatility of the original, it has yet to be established whether such an alien replication system could emerge from the primeval broth on some distant planet. But before we taste any soup it is to deep space we first need to turn in our pursuit of life's origins, to wallow in goo.

3   Universal goo: life as a cosmic principle?

For centuries, if not millennia, we have looked towards the night sky and with the glory of the Milky Way, now enhanced beyond earlier imagination by the extraordinary pictures captured by such great telescopes as the Hubble and Keck, felt a deep resonance, a wider and wilder reality, a glimpse of a cosmic architecture. Naturally enough such an awestruck view has also enfolded the prospect of other habitable worlds. Yet it is only quite recently that a deeper, more pervasive view has emerged that looks beyond the question of extraterrestrials to invoke organic activity on a truly universal scale, life as a cosmic imperative. In the previous chapter I mentioned Christian de Duve,[1] and perhaps he best of all encapsulates this wider and more optimistic view that the Universe is not a howling wilderness, but part of our home. Thus he writes, the 'Universe is a hotbed of organic syntheses leading among others, to amino acids and other typical building blocks of life. This 'vital dust' permeates the entire Universe and most likely represents the chemical seeds from which life arose.'[2] Nor is de Duve the only exponent of life as a cosmic principle. Cyril Ponnamperuma, for example, has stated that 'You look at the interstellar molecules and you see cyanide and formaldehyde. These two can provide the pathway for everything else. There is a simplicity in the whole scheme – so much so that you practically feel that the whole universe is trying to make life.'[3] Such a cosmic dimension has a strong resonance with our sense of belonging somehow to a wider order, but it still could be a serious misreading of the evidence. Here I shall take a contrary view and shall argue on the basis of several different approaches that 'Life may be a universal principle, but we can still be alone.'[4] In this regard, major stumbling blocks might be the apparent difficulty in understanding how life actually originated (Chapter 4) and the possible dearth of habitable planets (Chapter 5), but first in pursuit of de Duve's vision let us look into outer space.

A MARTINI THE SIZE OF THE PACIFIC

Among the many surprises of radio astronomy has been the detection in the huge interstellar molecular clouds of various spectra that are evidently compatible with the presence of a significant number of organic molecules.[5] Outer space being what it is, the quantities available are truly gigantic. The chemist Robert Shapiro[6] remarks that there is enough alcohol to make any number of dry martinis, each the size of the Pacific Ocean. But what is the real significance of this interstellar material? The first problem, of course, is that many organic compounds may escape detection either because their spectra are masked by the stronger vibrations of other molecules or because they occur in very low concentrations and are difficult to recognize. It would not be at all surprising, for example, if amino acids were to be detected (and the presence of amino acids in carbonaceous meteorites is clear enough evidence for their extraterrestrial formation), but so far not even the simplest amino acid, known as glycine, has been definitely recorded.[7]

However, not all the organic molecules in deep space are simple. Another surprise in this adventure has been the detection of rather more complex forms. This includes the so-called PAHs, that is, the polycyclic aromatic hydrocarbons.[8] Being organic compounds they have a carbon 'backbone', which typically shows a texture like chicken wire. PAHs are probably the most abundant organic molecules in the Universe, entirely dwarfing the total biomass of the Earth, which weighs in at a mere 6×10^{27} grams. This molecular group has quite a wide diversity of configurations, a number of which have been detected in deep space. One of these is a relatively familiar compound, napthalene. Indeed, one might derive wry amusement from the fact that the abundance of napthalene means that the entire Universe smells faintly of mothballs. As with the other interstellar organic compounds there is, of course, no question of these PAHs being made by some sort of organism, although on Earth they may be a chemical end-product of once-living organisms. For example, the napthalene in those old-fashioned mothballs is a derivative of the chemical product coal tar, which in turn ultimately comes from the plants and algae of the distant Carboniferous swamps.

There is also a famous case where the source of the PAHs is excitingly ambiguous. This is connected to the controversy surrounding the famous Allan Hills meteorite from Antarctica (ALH 84001). This has been claimed to yield evidence consistent with Martian life, and one of the lines of evidence is the recovery of undoubted PAHs.[9] The other lines of evidence include (a) tiny crystals of magnetite,[10] said to be similar to those made by certain terrestrial bacteria that act as tiny intracellular magnets; (b) putative fossils; and (c) the geochemistry,[11] especially of the carbonate,[12] and the related question of the temperature at which it was formed. Unfortunately the enthusiasm of the original team investigating this meteorite has not been matched by most other scientists. Whatever the significance of these Martian PAHs, which are also well known in carbonaceous meteorites,[13] the general consensus is that, as for the other lines of evidence, non-biological processes are much more likely to be responsible. So, too, on a much grander scale, the interstellar equivalents may tell us little of a Universe pregnant with life, but may rather point to the existence of a universal 'goo'. These PAHs are incredibly stable, going nowhere, an end-product with no future. Is this too pessimistic a view? Not according to received opinion, which not only embraces life as a cosmic principle but also seizes upon the only strictly tangible evidence we have for extraterrestrial organic matter, which is in the form of meteorites.

GOO FROM THE SKY

So far as human activities are concerned, meteorites tend to be regarded principally as potential agents of destruction: nobody fancies even a few kilograms of iron falling out of the sky on to an aeroplane or garden party. Even so, it is sometimes claimed that there is no definite evidence of death by impact. There is, of course, the ill-fated dog that is said to have been killed in 1911 by the Nakhla meteorite,[14] which like the Allan Hills meteorite was subsequently shown have come from Mars. Human casualties are usually thought, with some reservation, to be limited to a handful of possible examples, including a Franciscan monk in seventeenth-century Milan.

Yet, as Kevin Yau and his colleagues[15] have shown, a study of Chinese records, representing an exemplary bureaucracy, paints a far more sombre picture of sudden destruction as tons of material rain from the sky. One Ming document, for example, refers to events in 1490 when 'Stones fell like rain in the Ch'ing-yang district of Shansi province ... They struck dead several tens of thousands of people.'[16] Yau *et al.* question the size of the casualty list, and cannot rule out entirely the alternative agency of hailstones, but there is certainly no reason to doubt that multiple deaths and injuries have been caused by meteorites, although in many instances the casualties probably resulted from the collapse of the building rather than from a direct hit. Moreover, a more recent survey[17] provides a list of buildings hit (one through the observatory of an amateur astronomer), near misses, and several deaths. There is little doubt that further mining of historical records (see also note 15) will reveal more examples of meteoritic casualties. So, too, one may expect, will archaeological evidence. Of course, it is tempting to ascribe the collapse of great civilizations to such cosmic interruptions, but to date there is little, if any, evidence to support such ideas. There is, however, evidence for a major impact (the Kaali event) on the Baltic island of Saaremaa (Estonia) during the local Bronze age[18] (c. 400–800 BC; a possible report by the famous traveller Pytheas, who originated from the Greek colony of Marseille, might narrow this event to c. 350 BC). The energy released was about equivalent to the atomic bomb that destroyed Hiroshima, and as is evident both from the crater and the pollen record the impact must have led to widespread devastation.

Yet, from the prospect of the origins of life, certain meteorites, and especially a rather peculiar class known as the carbonaceous chondrites, are now being taken as a key creative agency. As such, these meteorites provide a terrestrial glimpse of the vital principle that, it is suggested, permeates the entire Universe. Carbonaceous meteorites are quite rare, but the most famous examples (such as the Allende, Murchison, and Orgueil) have proved treasure troves of information on extraterrestrial organic chemistry. There is, however, a potentially major problem because not only do they quickly disintegrate, but there is the ever-present danger of contamination from terrestrial sources. Not surprisingly, new discoveries, for example, the Tagish Lake fall,[19] excite considerable attention. The Tagish Lake find was especially propitious because the meteorite came down at a low angle and its fall was observed; and its rapid recovery from an ice-covered lake in northern British Columbia ensured a remarkably pristine sample. It has yielded many surprises. The fact that its chemistry is very different from that of other carbonaceous meteorites implies that our overall understanding is far from complete.[20] Most striking perhaps is the low concentrations of amino acids, some of which are clearly terrestrial contaminants.[21] Even so, many of the organic molecules discovered in this and other meteorites are indigenous and represent

the material necessary for life to emerge – in other words those building blocks that are basic to life on Earth and presumably anywhere
else. So, in contrast to the blocks of iron and stone wreaking destruction, here we have a potentially rich brew of chemicals that, at the
dawn of Earth history, might have helped to initiate the whole process
of organic evolution.

As we shall see, the importance of the carbonaceous meteorites,
as well as comets, in enabling the Earth to become habitable should
not be underestimated. A closer look, however, suggests that to regard
these meteorites as some sort of primitive crucible out of which life
could take its first steps may not merely be simplistic: it could be seriously misleading. Consider, for example, the amino acids[22] recovered
from various carbonaceous meteorites, such as the famous Murchison
fall that landed near the town of the same name, in Australia in 1969.
Certainly these compounds occur in some abundance, but it is very
significant that only the simplest types of amino acids are found and
that they represent two groups (the monoamino monocarboxylic and
monoamino dicarboxyl classes). There is, moreover, another equally
important observation that, of the various molecular configurations
that can exist in these two groups, effectively *all* are found in the
meteorites.[23] The message is simple. These amino acids are by definition organic, but the processes of formation are patently abiotic
and unrelated to the ways in which amino acids are made in a biological cell. Exactly the same remarks apply to the more recent
discovery in carbonaceous meteorites of compounds related to the
sugars.[24]

The restriction to the simpler types of molecule and the exploration of all possible variants is non-selective and in marked contrast
to the activities of life. These extraterrestrial amino acids are the product of an unconstrained chemistry, with no hint of life's strange specificities and subtle avenues of synthesis. Within the restricted range
of the simple amino acids, all available types have been produced in
these carbonaceous meteorites by the random combination of yet simpler precursors, but then the chemical process comes to a full stop.
In such an environment it is scarcely surprising that we see an expo-
nential decline in the abundance of amino acids with a larger number
of carbon atoms in their framework. Again the message is abiologi-
cal: even slightly more complex molecules are much more difficult to
assemble.

These extraterrestrial amino acids show other clear signatures of an origin free from life's influence or interference. For example, a number of the amino acids, such as α-amino isobutyric acid and isovaline, recovered from carbonaceous meteorites have effectively no terrestrial counterparts.[25] In addition, the isotopic signatures of such elements as nitrogen, carbon, and hydrogen are also greatly at variance with those found in living systems.[26]

The problems with latching on to the carbonaceous meteorites as some sort of cosmic guide as to how the foundations, if not the first floor, of life are put in place do not stop with the amino acids. Together with these compounds there is an enormous range of other organic compounds, but most of them have no terrestrial counterparts.[27] These include very complex mixtures of hydrocarbons, including both the aliphatic and aromatic varieties (effectively defined as to whether the carbons are in a chain or ring respectively). Given this welter of forms it is hardly surprising that special enthusiasm is reserved for molecules, however rare, that do play a key role in life. Take, for example, the group known as the nitrogen heterocycles. They are certainly important because they include the compounds known as the purines and pyrimidines, which are the main building blocks of DNA (and also RNA). But once again any optimism that floating in space is a sort of proto-genetic code needs, so to speak, to be brought firmly down to Earth.[28]

It is, of course, understandable that the discovery of the key components of proteins and DNA in carbonaceous meteorites reinforces the notion of life permeating the Universe. Yet this is at best a selective reading. As already noted, most of the compounds have no parallels in living systems, at least on Earth. More seriously, most of the carbon occurs as a sort of featureless 'gunk': highly stable, insoluble, this complex entanglement of organic compounds is going nowhere – it is another variant of the universal 'goo'. The occurrence of this extraterrestrial muck closely parallels, as it happens, many of the origin of life experiments in terrestrial laboratories (Chapter 4) where a plethora of unwanted by-products and tar-like goo give at least the sceptic pause for thought.

But carbonaceous meteorites continue to exert their fascination. Surely they are telling us something? Many years ago there was a flurry of excitement when seeming fossils were found, but their biogenicity was never confirmed, except that is from some structures that were clearly terrestrial contaminants. But are there other ways in which more complex structures might emerge: perhaps some of these extraterrestrial organic compounds exhibit self-organizing properties that could give us at least a glimpse as to how life emerged from a distinctly discouraging mess? Nearly all are agreed that at some point a key step in making life had to be the invention of some sort of en- velope, a proto-cell, in which the nascent biochemical machinery of replication and molecular synthesis could be shielded from the chem- ical chaos of the surrounding milieu. And there are some intriguing hints as to how this might have happened. Among the hydrocarbons found in the carbonaceous meteorites are some compounds quite like the lipids. These are the fatty compounds that typically make up part of the wall of a living cell. Experimental manipulation of such ex- traterrestrial compounds typically leads to the formation of simple droplets, as you might see if you splash olive oil into water. Under cer- tain conditions, however, genuine vesicles, apparently with a double- walled membrane (as in living cells), can be formed.[29] These artificial vesicles are very small, typically in the micron range and so compa- rable in size to bacteria. The main snag, however, is that although hydrocarbons of a suitable size (that is with chains ranging in length from 12 to 20+ carbon atoms) occur in the carbonaceous meteorites, as yet no plausible prebiotic milieu has been devised in the labora- tory to synthesize hydrocarbons of a comparable length. The obvious implication is that life might not have emerged unless these extrater- restrial hydrocarbons had been delivered to the surface of the early Earth.

BACK TO DEEP SPACE

Carbonaceous meteorites therefore represent the product of a com- plex chemical factory that generated a cocktail of organic molecules, of which only a handful might be relevant to the origin of life on Earth. But there is a more profound problem: where exactly did these compounds form? The general idea is to look at events early in the history of the Solar System when a multitude of small planets formed before being destroyed in larger collisions. Many of these fragments were incorporated into the planets as we know them but others re- mained free, and of these some have since been captured as meteorites by the Earth. Early in the history of the Solar System, however, it is likely that at least some of these micro-planets were warm and wet, allowing a ferment of chemical activity to take place. Not only were organic compounds formed in abundance, but perhaps life itself flickered into existence before another planet-shattering event abruptly terminated one more experiment. So we should not necessarily dismiss the possibility of one day finding extraterrestrial fossils in some of these meteorites.

But before rekindling too much optimism we need to have another look at these extraterrestrial organic compounds. Rather surprisingly, it is now becoming increasingly clear that at least some of what we regard as the building blocks of life were probably synthesized long before the Solar System itself formed. They are products of an interstellar environment far removed from anything we could envisage as remotely habitable. Here the processes of organic synthesis take place in conditions of extreme cold, a 'hard' vacuum that is bathed in radiation. The evidence of this dramatic shift in location – from a warm cradle in the early Solar System to interstellar wilderness – comes from a surprising discovery concerning some of the extraterrestrial amino acids. Like many organic molecules, most amino acids are asymmetrical. This means that they can adopt one of two forms, each the mirror image of the other. Our two hands, left and right, are a more familiar example. The technical name for these mirror variants (also assigned to a left and a right) is *enantiomorphs*, based on the Greek words for opposite and shape. Despite their unfamiliar name, enantiomorphs are far from being some arcane area of chemistry, with dome-headed, white-coated scientists frowning at banks of equipment, and a silent, breathless huddle of assistants waiting for the meter to sink below 0.82. In fact enantiomorphs are quite familiar, if not domestic. The example that is probably most often quoted is the smell of oranges and lemons, which is said to be the product of the same molecule, but as respective mirror images. As it so happens, this is not quite correct. The molecule in question belongs to an interesting group known as the terpenes, and is specifically known as limonene. In oranges it is present as (+)-limonene but, as Nigel Veitch of Kew Gardens informs me, its mirror counterpart (−)-limonene, is actually found in pine needles and not lemons.

So what does all this have to do with extraterrestrial amino acids, and where they were originally synthesized? In general the abiotic production of amino acids produces equal proportions of either enantiomorph of a given molecule. This result applies as much to the amino acid content of carbonaceous meteorites as it does to laboratory experiments on Earth. This fifty–fifty arrangement, of equal numbers of left- and right-handed molecules, is known as a *racemic mixture*. One of the puzzles of life on Earth is that with very few exceptions the amino acids are consistently of one type, specifically the left-handed variety.[30] If by some sleight of hand we were to persuade a cell to employ the identical amino acid, but in its right-handed enantiomorph, the effects would be very deleterious: Alice passes through the looking-glass at her peril. Once life embarked on a left-handed history, in so far as the amino acids were concerned, matters were presumably fixed. It is much less clear, however, whether the original 'decision' was an accident of history or whether there is some tendency to favour, even slightly, the left-handed enantiomorphs. At first sight the latter possibility seems rather remote because, as already mentioned, racemic mixtures are the rule in abiotic syntheses. The general implication is that when life 'chose' the left-hand option it was some time after amino acids had begun to be synthesized. It now seems possible that the 'choice' was not only made much earlier, but before life itself evolved.

When the discovery of non-racemic ratios of extraterrestrial amino acids in carbonaceous meteorites was first announced there was understandably considerable scepticism.[31] All prior expectations of abiotic organic chemistry pointed towards racemic mixtures. In the case, however, of at least two amino acids there is an excess of one enantiomorph. Even more interestingly, it is the left-handed variety that is in the majority.[32] The crucial point to appreciate is that it seems to be generally agreed that such a racemic imbalance cannot have arisen by any known chemical process that could occur on some proto-planet swirling around its newly formed sun. Rather, the shift away from a racemic mixture must have been the product of interstellar processes that occurred long before the Solar System formed. It has to be admitted that exactly how these amino acids reached a preponderance of the left-handed enantiomorph (values are *c.* 7–10%) remains an unsolved question. One intriguing idea is that polarized ultraviolet radiation from nearby stars is responsible.[33] There is, therefore, a hint that interstellar processes might stamp a handedness on all life, wherever it occurs. Many suspect that the DNA/protein basis of life is universal, and this too may apply to left-handed amino acids.

Whether or not this is correct will depend in part on the success in finding if there are other plausible mechanisms to break the racemic ratio, a problem which continues to exercise a perennial fascination. There has been some progress inasmuch as enantiomorphic excesses have been produced by various mechanisms, which in the laboratory include the operation of a magnetic field.[34] The problem is whether these experiments have much relevance to what may have happened originally in deep space. This is not only because the compound investigated (a chromium complex) is probably not relevant to biological processes, but more importantly because it is not clear whether natural magnetic fields could produce sufficient quantities of one enantiomorph, given that the laboratory experiments produced only a very small excess of either enantiomorph (depending on the direction of the magnetic field). Other approaches to this problem include specific chemical reactions,[35] as well as sorting on crystal surfaces.[36] The latter approach, of course, excites the interest of those who believe the selection of amino acid handedness might have occurred in the surface of the early Earth. The use of a mineral surface, in this case calcite, is important because, as we shall see in the next chapter, there is considerable interest in the possibility that such surfaces play an important, if not a key, role in the origin of life. Even so, experiments using calcite yield only a very small (of the order of 10^{-3}) enantiomeric excess, in this case L- or D-aspartic acid.

The point of this discussion is not to pour cold water on the search for an explanation of biological handedness; there is a strong sense that the choice of left-handed amino acids is unlikely to be a simple accident and a satisfactory explanation may well constrain other problems in the origin of life. There are, however, two general points relevant to both life in the Universe and the origin of life on Earth (see Chapter 4). The first difficulty is the wildly unconstrained theatre of operation: opinion varies between those who regard much of the basis of life emerging as the Solar System formed, if not earlier, while others regard the heavens as almost irrelevant as they peer into one sort of muddy pool or another. The second difficulty is that despite the attraction of using analogies, many of the experiments designed to explain one or other step in the origin of life are either of tenuous relevance to any believable prebiotic setting or involve an experimental rig in which the hand of the researcher becomes for all intents and purposes the hand of God.

A LIFE-SAVING RAIN?

It seems that however and wherever extraterrestrial organic material was, and still is being, formed it may be of limited relevance to understanding how the building blocks of life came to be assembled and coordinated on the early Earth. Even so, there are still strong arguments that without chunks of extraterrestrial material, rich not only in organic compounds but also in water (as ice) and other volatiles, in the form of both carbonaceous meteorites and comets, then life on Earth might never have arisen. This in turn would have precluded some billions of years of evolution during which at least one sentient species could evolve to a position where it could begin to think about the whole business.

So why should carbonaceous meteorites and comets play such a key role? By general agreement the Solar System originated from a huge cloud of interstellar dust that, perhaps nudged by the blast from a nearby supernova, subsequently condensed into a rapidly spinning disc. Out of this emerged the central star and a series of planets. In our system only one body, our Sun, achieved sufficient mass to initiate the thermonuclear reactions that provide the sunlight. The rapid processes of accretion, and especially the gravitational maw of the newly forming Sun, means that even as the planets continued to collide and grow, the central star itself achieved a large enough size to begin to shine. The prospect for those planets nearer to the Sun was not good, at least for the future of life. This is because the effect of the Sun's radiation was to drive volatile compounds (necessary to form an atmosphere and ocean) out to the cooler regions of the spinning disc, to or beyond the so-called 'snowline' which approximates to the present-day position of Jupiter. Stripped of their volatiles, planets like the Earth would be reduced to balls of bone-dry rock, with the elements that are essential to life either driven far beyond the orbit of Mars or locked up in crystal lattices within the interior of the planet. There would certainly be no oceans, and presumably no chance for life to emerge.

So what are we doing here? One possibility is that such a planet as Earth did not accrete in its present position, but formed in an orbit much further out where the volatiles were ready to provide it with the mantle of atmosphere and ocean. Subsequently, under the influence of the chaotic system of planetary movements, the Earth was flung inwards, to occupy an orbit much closer to the Sun and – by strange, good fortune – comfortably within the so-called habitable zone (see Chapter 5): not too close to the Sun to be reduced to a searing inferno of a runaway greenhouse, as befell Venus, nor too remote to end up as a sort of super-tundra, as was the fate of Mars.

An alternative, and perhaps more plausible, suggestion to the Earth being catapulted across the Solar System is that it remained close to the Sun, but was 'rescued' by a massive influx of volatiles as carbonaceous meteorites, and more particularly comets,[37] bombarded the early Earth. Without this cosmic infall which provided the seas and an atmosphere it is difficult to imagine how life could have arisen. Some workers take an almost 'manna from heaven' view of this process, in which the Earth was drenched in a rain of fully formed organic molecules: a mizzle of amino acids, hydrocarbons, and other complex organic molecules such as the PAHs and derivatives like the quinones[38] wafting as a rain to the surface, accumulating in warm seas and coating tidal flats. There is certainly some evidence to suggest that some quite complex molecules might survive even catastrophic impacts.[39] The more general consensus, however, is that comets brought in, not the building blocks of life, but the basic raw materials, such as carbon, water, and nitrogen. Four billion years ago the natural laboratory was here on Earth, and now owing to human ingenuity we can replicate these processes. Never mind the universal goo: in laboratories all over the world we are on the verge of seeing how the spark of creation transmuted the inanimate to the animate. Nothing could be further from the truth.

4   The origin of life: straining the soup or our credulity?

This chapter is a story of abject scientific failure. It is also a story of omission and simplification, but how else could one encapsulate this enormous and unresolved area? Perhaps by way of introduction I should mention two items. First, there are some excellent books devoted to the topic of life's origins.[1] There are also, of course, many review papers in the literature,[2] although almost invariably they reflect a particular school of thought so that while opposing schools receive raking fire the inadequacies of their own stance tend to receive more muted criticism. Second, here I make effectively no mention of the so-called 'RNA world'.[3] This 'world' is widely, but not universally, accepted as predating our familiar DNA-based world, and is taken to represent a key staging post that attempts to solve a sort of 'chicken-and-egg' question as to what came first: a system of replication to code and transmit instructions or a chemically controlled metabolism to channel energy using catalysts (a.k.a. enzymes). Proponents of the 'RNA world' solve this conundrum by answering this question with the reply: 'simultaneously'. This is on the basis of earlier speculation, combined with subsequent experimental proof of two facts: first, that RNA acts as an agent of replication (in nearly all organisms this is on the basis of instructions coded in the DNA and then transferred to the RNA); second, that in some circumstances RNA acts in an enzymatic capacity (ribozymes), a role that in practically all living cells depends on particular proteins (enzymes). So why is there no discussion here of this putative 'RNA world'? Simply, because by the time you have reached this stage in the early evolution of life you are probably home and dry. Getting there, however, remains much more problematic. The remarks concerning the 'RNA world', made by David Bartel and Peter Unrau[4] are particularly apposite when they write, 'its status as a hypothesis is easily forgotten. Problems remain, particularly the implausibility of prebiotic RNA synthesis and stability'.[5]

    And that gently questioning tone is the gist of this chapter, drawing as it does on a sceptical tradition articulated most cogently by

Robert Shapiro (see note 16). Another individual for whom I have a particular admiration in this context is Leslie Orgel, not only because he is a key individual in the study of the origin of life, but as importantly because, in contrast with the sunny optimism displayed in so much of the literature, his papers reveal a traveller who knows he might be on the high road to success, but will not be totally surprised if the path leads to a fetid bog. This stance is neatly encapsulated in one his review articles,[6] where he writes:

> The problem of the origin of life on the earth has much in common with a well-constructed detective story. There is no shortage of clues pointing to the way in which the crime, the contamination of the pristine environment of the early earth, was committed. On the contrary, there are far too many clues and far too many suspects. It would be hard to find two investigators who agree on even the broad outline of events.[7]

Yet, the study of life's origins is set in a paradox because the underlying principles could not be more straightforward. This is because the first chemical steps that led to the processes of self-replication (ultimately held in the thrall of DNA) and controlled cycles of chemistry (what we call metabolism) *must* have been simple. Knowing these foundations it surely cannot be that difficult to add storey to storey, ultimately to remove the scaffolding, and so reveal the functioning cell, a near-miracle of encapsulated design but arising by unremarkable processes?

To be sure we must acknowledge preconditions, including the fact that if the precursors of life had not rained from the heavens on to the surface of the early Earth, supplying it again and again with volatiles, then quite possibly there would have been no oceans or atmosphere. Without this cosmic drizzle it is quite possible that life would never have had either the building blocks or the milieu necessary for its origination. On the other hand, such an influx of water and organic compounds is hardly relevant to the actual question, how did life begin? This is because the simplest of the molecules required for the beginning of organic chemistry, compounds such as formaldehyde and hydrogen cyanide, are themselves readily synthesized in at least some circumstances that might have occurred on the early Earth.[8]

In other words, it doesn't really matter where such compounds form: what does matter is what happens next. So, too, with the next stage, that is, the assembly of such basic building blocks of life as the amino acids, sugars, and hydrocarbons, as well as the nucleotides that are fundamental to DNA. Such syntheses are widely regarded as inevitable, and images of warm ponds, seething volcanic springs, and massive thunderstorms rumbling across a deserted yet pregnant landscape are used to feed the imagination. At this stage, life has yet to exercise its peculiarly specific grip, its spinning of the genetic code, its weaving of biochemical complexities; but no matter: despite the vast pot-pourri of resultant chemicals, the nascent processes of Darwinian selection are already winnowing and reaping, the inappropriate is steadily devoured in chemical competition with the winners. Metaphorically the molecules slug it out by tooth and claw. Cycles develop, life emerges, and four billion years later one species invokes the marvels of autocatalysis and emergent properties to cap the argument. These ideas are the bread and butter, so to speak, of a substantial part of the origin-of-life industry. They are well encapsulated in the proposals of Manfred Eigen.[9] An important part of his argument is the emergence of primitive cycles that combine, as so-called hypercycles, to permit replication. A critique of Eigen's ideas is offered by John Maynard-Smith,[10] who suggests that the mechanism of hypercycles becomes plausible only if it is housed within a cell, and the origin of the cell ... ? Even so, the drums and trumpets that regularly sound in support of self-organization and autocatalysis as appropriate mechanisms that could lead to stable and self-replicating systems form a regular musical accompaniment to those who believe that these precepts are a highly desirable, if not necessary, goal,[11] even though there are painfully few experiments.[12]

And yet, something is clearly missing: life cannot be created in the laboratory, nor is there any clear prospect of it happening. When we return to the simpler stages of organic synthesis, the apparently crisp outlines begin to blur. Even the arrangement of the experimental apparatus, the literal disposition of the test tubes, can spell the difference between success – of a sort – and yet another failure. And yes, certain compounds can be generated, but as often as not in minute quantities, mixed up with all sorts of other goo and under conditions which, to be frank, bear precious little resemblance to anything conceivably like the early Earth.

FINDING ITS PATH

It is certainly not my intention to suggest that the origin of life is a scientifically intractable problem, but at this stage of the proceedings simply to register mild surprise at the relative lack of experimental success. There has, of course, been a succession of ingenious experiments that aim to throw light on one or other facet of prebiotic chemistry. The real problem is getting past this first stage, from the early organic 'soup' (however envisaged) to the metabolic and biochemical highway, with a functioning cell as its destination. Self-evidently this end product – the cell – is an immensely complex chemical factory. The path leading to the living cell must have required simpler predecessors, and most probably can be envisaged as analogous to building with something like a modular construction kit. Parts that were formerly independent could then be 'bolted on' at various stages. The difficulty remains: exactly which path, which kit? These questions seem to be virtually intractable, at least at the moment.

Now to me this difficulty appears to be rather strange. Clearly the conditions on the early Earth are difficult to judge with any accuracy. Just how much methane was in the atmosphere? How salty was the ocean? Were there ice caps? And so on. In any of these cases, and many others, the precise values or conditions are elusive. Yet the early Earth must have had some sort of atmosphere, a sea bed with certain clays, volcanic hot springs, and so forth. Strictly speaking the number of variations is very large indeed, but many plausible combinations can be assembled. Perhaps we need to get just the right balance: so much carbon dioxide and no more; a particular mix of clay minerals. ('Steady Trevor with that beaker of kaolinite!'); just a smidgeon of a phosphatic mineral. Maybe, but this is almost getting us into the position of arguing that life was effectively a fluke-like happening, a one-in-a-trillion chance. Perhaps it was. Perhaps the emergence of life was 'almost a miracle' (to anticipate a phrase; see note 89) and given enough time, patience (and money), so too may we stumble on a similar solution. Yet, the general consensus is that far from being a fluke the emergence of life on any Earth-like planet is more or less a foregone conclusion, an inevitability inherent in the self-organizational properties of organic chemistry. So what is the evidence?

Certainly the area dealing with the origin of life has no shortage of enthusiasts: zeal of purpose is combined with an unwavering naturalism. Yet, as Klaus Dose[13] reminds us, the research programme, which was initiated by such giants of the field as the Soviet scientist Alexander Oparin, the maverick genius J. B. S. Haldane, and subsequently the chief rejuvenator of this field Stanley Miller, is still fraught with problems. It should all be rather simple. Here are the building blocks, like bricks: today we build something rather simple, say a garden wall, and tomorrow we aim higher, perhaps a cathedral. Despite the many disagreements about the most appropriate experiments and about the environmental milieu, there is a deeply shared belief in the power of the self-organization of prebiotic molecules. In fact this belief in self-organization not only implies a sort of 'biochemical predestination' but, as Dose wryly notes, the research programme has led to an 'era of unconfessed vitalism'. Dose continues with some apt words of warning: '[It has become] abundantly clear that the power of self-organization inherent in macromolecules synthesised in cells is based on extremely subtle physical and chemical, and particularly stereochemical, properties [which]', he continues in near-understatement, 'have never been observed in this highly organized form in prebiotic molecules.' He concludes bleakly, 'It appears that the field has now reached a stage of stalemate.'[14]

Others have reached similarly pessimistic conclusions. John Horgan[15] and more notably Robert Shapiro[16] have discussed how difficulties crop up at every stage and each step. Not only that but, as Horgan makes clear, the proponents of the various rival schemes of investigation are happy to express scepticism about alternative approaches to the problem, yet for the most part retain quiet optimism that somehow their chosen methodology – with a little bit of luck and another injection of research money – will crack the problem. For many years the cries have been repeatedly 'Breakthrough! Nearly there! A major step! Most significant!!' Let us hope they are right, even if the shouts of acclamation are beginning to sound somewhat hollow. At least Stanley Miller is candid enough to tell Horgan that 'The problem of the origin of life has turned out to be much more difficult than I, and most other people, envisioned.'[17]

Yet, turning to the mainstream literature we see hardly any inkling of these largely submerged doubts. The practitioners fully realize that the problem cannot be solved by a single master-demonstration, but will yield only to a patient, chip-by-chip assault on the edifice. The research workers seem always to be frustrated in their quest for what they refer to as the 'one-pot reaction', in which chemicals go in at one end and a replicating system emerges from the other. A survey of this area, which occupies the remainder of this chapter, reveals a picture that can only be described as distinctly discouraging. Indeed, there is a quite general pattern of problems.

PROBLEMS WITH EXPERIMENTS

One difficulty is that the great majority of experiments connected to one or other aspect of the origin of life do not actually work. Not surprisingly, such results are seldom reported. This is not any sort of dishonesty; it is simply that in exploring prebiotic chemistry most avenues are going to prove to be dead ends, tracks leading nowhere. Well, that is quite typical in science, yet in the field of the origin of life this problem indicates some more pervasive difficulties. An illuminating example is provided by Anthony Keefe and Stanley Miller,[18] who consider the crucial role of the element phosphorus. As is well known to biologists, phosphates play a key role in the biochemistry of all organisms, perhaps most notably by the agency of the compound ATP (adenosine triphosphate), which is usually referred to as the 'currency' for energy transfer. This is on the basis of its ability to store substantial amounts of energy, specifically in the branch of the molecule that contains the phosphoanhydride bonds. It seems almost certain that before the 'invention' of ATP there must have been some sort of prebiotic phosphate compound capable of a similar, if less efficient, function. It had been long appreciated that the production of such compounds was fraught with difficulties, at least if the experiments were to be conducted in a believable prebiotic milieu. Keefe and Miller therefore embarked on an extensive series of experiments that took as their starting points a credible range of chemicals that one might reasonably expect to exist on the early Earth. The net result of this programme was effectively a catalogue of disasters, and although there were local successes Keefe and Miller were forced to conclude that none of the new processes they had investigated would be 'sufficiently robust to have been of importance in the prebiotic ocean'.[19]

One of the problems Keefe and Miller encountered was that in most cases the yields of the favoured compounds were very low. This is one of the recurrent features of such experiments into the origin of life: again and again a compound that is believed to hold a key role in some significant path of biochemistry is indeed synthesized, but typically the reactions are highly inefficient and the quantities produced are quite negligible. To be sure, one or more agencies of distillation or concentration may be invoked, but unless life itself is asked to give a helping hand – defeating, of course, the purpose of the exercise – then the mechanisms employed are highly artificial, if not contrived. There is also a flip side to the production of potentially important pre-biotic molecules. Not only are the yields often disappointingly low, even minuscule, but typically (and unsurprisingly) the experiments produce a wide range of other chemicals that seemingly have no relevance to the origin of life. In some instances a substantial quantity of the organic material synthesized forms a tar-like 'gunk', reminiscent of the heterogeneous 'goo' found in the carbonaceous meteorites (Chapter 3).

Many of the difficulties just referred to are exemplified by what has come to be known as the 'ribose problem'. Ribose is a sugar, one of a large group of molecules related to the molecular family known as aldopentoses. In its right-handed, or dextral, form it is one of the key ingredients of life, most notably as a key component of the backbone of the genetic code, namely deoxy*ribo*nucleic acid (DNA) and *ribo*nucleic acid (RNA). Given this crucial role, it is hardly surprising that the successful laboratory synthesis of ribose is a prime objective for understanding the origin of life. The starting point is formaldehyde, a simple molecule (H_2CO), which is readily synthesized abiotically.[20] Starting with formaldehyde, the next step is to engineer a process that will lead to a polymerization, that is, a stringing together of simple molecules, whereby more complex sugars can be formed. In certain circumstances one of these should be ribose.

So what is the 'ribose problem'? In a paper reviewing this area Robert Shapiro casts a withering eye on a number of earlier experiments in which success was claimed.[21] In discussing one proposed pathway Shapiro bluntly remarks that the investigators are to be congratulated for their 'vivid use of imagination'.[22] Among the various procedures that can be adopted, particular attention has been paid to what is called the formose reaction.[23] This was first identified by a chemist, A. M. Butlerov, and is effectively concerned with the way in which complex compounds can be produced from the polymerization of formaldehyde under certain conditions. The limits within which this reaction operates are at least broadly consistent with what we

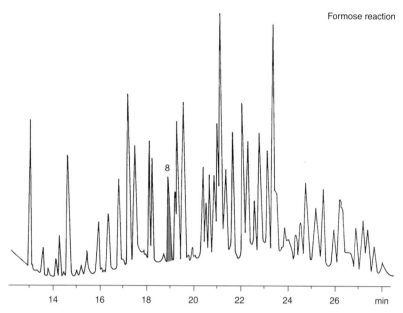

Formose reaction

FIGURE 4.1 The 'ribose problem' exemplified. The graph (a gas
chromatograph) shows the many products of the formose reaction, of
which ribose – an essential component of DNA – is marked by peak 8.
Such reactions, with a plethora of compounds, are the norm in prebiotic
reactions, raising the question as to how separation and sufficient
abundance of key compounds, such as ribose, was achieved. (Reprinted
from *Journal of Chromatography*, vol. 244, P. Decker, H. Schweer, and
R. Pohlmann, X. Identification of formose sugars, presumably prebiotic
metabolites, using capillary gas chromatography/gas chromatography–
mass spectrometry of n-butoxime trifluroacetates on O/V–225, pp. 281–
291, fig. 11.5. Copyright 1982, with permission from Elsevier Science.)

might expect on the early Earth, and indeed ribose is produced. But
there is a snag: the yields are very low and the formose reactions
produce a disturbingly large array of other compounds (Fig. 4.1). To
compound the difficulties the ribose is rather unstable[24] and degradation is particularly rapid if the solution becomes more alkaline.
It is, of course, possible to tinker with the reaction to engender a
greater degree of selectivity, but unfortunately this does not in itself
boost ribose production. So we seem to have been led into an impasse.
Nothing daunted, chemists have devised reaction pathways that can
produce reasonable quantities of ribose,[25] but the sheer complexity of
the process and the careful manipulation of the many steps during the
reaction make one wonder about its applicability to the origin of life.

A key element in the hunt for chemical pathways that might be relevant to prebiotic evolution is the search for natural catalysts that could not only accelerate the reactions but ideally could also impose a degree of specificity that might circumvent the problem of producing a cocktail of molecules, if not more goo. Perhaps this is the way to crack the ribose problem? G. Zubay[26] tackled this question by looking at a wide array of potential catalysts that might direct the formose reaction in favour of particular sugars, including ribose. On his own admission this work was in part spurred on by the disappointment and frustration experienced in investigating whether widely available compounds such as calcium or magnesium hydroxides could act as catalysts. He found that they could not, but a search among other potentially suitable catalytic agents was equally futile, with the single exception of lead. This element certainly occurs in natural compounds such as the sulphide mineral galena, but Zubay himself was cautious as he added, 'we have pondered the legitimacy of lead as a prebiotic catalyst.'[27]

The most obvious way forward from such an apparent impasse is to find a viable precursor that is chemically simpler and, ideally, easier to synthesize. This is an important and more general point. If such a stage could be identified, it might even begin to show how nascent life, or its molecular antecedents, began to exert its strange yet characteristic specificity, its ability to select the one-in-a-million chemicals, by metaphorically dipping into the organic 'soup' and selecting any molecule that improves, be it ever so slightly, one or other biological process. Perhaps we should stop worrying about ribose, and look towards a more believable predecessor. Such might be the case with a compound known as α-threofuranosyl. This can be derived from a sugar (threose) with only four atoms (i.e. tetrose) and thus is simpler than ribose and perhaps easier to synthesize. Of equal significance is that α-threofuranosyl can be used in the building of a short molecular strand that is directly analogous to DNA (and RNA). This molecule, referred to as TNA (T for 'threo'), can form a double strand (a duplex), and, rather remarkably, can also pair with RNA and DNA. TNA could thus, at least potentially, be capable of replication, and thereby act as a genetic code.[28] If so, this could obviously a very important step forward, but it raises two general problems. It is all very well to by-pass the difficulty of making ribose, but we need to ask whether the pathways of threose sugar synthesis (and those of the other component molecules of TNA) can themselves be reconstructed in a credible pre-biotic milieu. The sequence of steps undertaken in this laboratory investigation is certainly very far removed from any 'warm, little pond'. The second problem is that while life must have originated in a series of step-like processes, as one moves to ever simpler compounds, so the real difficulty, that of bridging the gap between complex organic chemistry and any sort of functioning system we choose to call living, actually grows.

ON THE FLAT

Mineral surfaces have played a key role in many speculations concerning the origin of life, and for rather simple reasons. One is the potential for the binding and selective entrapment of key molecules, possibly with the added bonus of orientating them in some sort of regular array. Of particular importance would be a prebiotic process of stringing together simpler building blocks, a process known as oligomerization. What might be a demanding and problematic process in a free aqueous environment, with its continuous molecular jostling and recurrent danger of chemical interference by such processes as hydrolysis, might be neatly short-circuited on the surface of some mineral such as a clay or the phosphorus-rich apatite. Mineral surfaces potentially have other advantages. These include an enormous surface area, especially if the mineral grains are small. Thus a cubic centimetre of clay will contain particles whose total surface area, so Christopher Jeans tells me, could, in principle, reach an astonishing 1800 m^2, roughly equivalent to nine tennis courts. Moreover, if the mineral grains are not too tightly packed then an interconnecting system of pores exists so that fresh solutions can be swept in to supply a growing molecule. And there is an added bonus if the mineral happens to have some sort of catalytic property, capable of accelerating chemical reactions. It all sounds very promising.

The experimental work is certainly interesting. Although most efforts have been expended in persuading amino acids to link together,[29] the necessary step to form first a polypeptide and ultimately a protein, there have also been investigations into the oligomerization of nucleotides,[30] which would be analogous to building a strand of DNA. In these experiments typically only a very limited variety of the potential building blocks is used, and quite often it is the oligomerization of only one type of molecule that is attempted.

Moreover, as is so often the case in research on the origin of life, it is difficult to avoid a strong artificiality in the experimental set-up. Thus, on occasion, analogues of 'real' nucleotides are employed, while the chosen substrate is of a chemical purity unlikely to be found in a natural prebiotic environment.[31] Given all this, it is perhaps not surprising that the results of oligomerization experiments are variable. Some substrates bind efficiently but others do not. The experimental process typically entails successively feeding the solution across the surface. In this way chains of up to 50 or so monomers can be linked into an oligomer. There is, however, a snag inasmuch as the binding to the mineral surface is typically an irreversible process: the oligomer remains locked in position. It is tempting, of course, to see this as a sort of staging post to life itself, with the mineral-bound strand then reading off multiple copies that then jostle for selective advantage as some sort of protolife. Perhaps so, but the investigators Rihe Liu and Leslie Orgel note that concerning such oligomerizations 'Many more experiments will be needed.'[32]

It is also worth remarking that while the role of clays in the origin of life has been seen largely in a positive light – be it as catalytic surfaces or in terms of Graham Cairns-Smith's ingenious ideas of a clay-based information system preceding that of DNA[33] – these minerals may have unexpected drawbacks. Earlier in the history of the Earth there were many radioactive isotopes that have now disappeared because of their short half-lives. Of particular significance, perhaps, is that the radioisotope potassium-40, which would have been much more abundant at the time life was emerging. Given that potassium is an integral component of many clays, then, as has been pointed out,[34] if these minerals 'acted as important nucleation sites, the protobiological synthetic reactions had to proceed much more rapidly than deleterious decomposition reactions to avoid the long-term effect of radiolysis and hydrolysis within days.'[35]

But not all minerals are so radioactive, and the potential role of their surfaces as the key to understanding the origins of life has an almost mesmeric attraction that can be traced back to that uneasy genius J. D. Bernal.[36] There has of late been considerable interest in the mineral pyrite, popularly known as 'fool's gold' a compound of iron and sulphur (FeS_2). This interest concerns both its chemistry and its surface properties, as well as a possible setting in active hydrothermal vents. For many people this is by far the most encouraging avenue in the pursuit of life's origins, although it comes as little surprise that those who have devoted their research energies in peering into other crucibles adopt a duly critical stance.[37] The chief exponent of the possible role of pyrite is Gunter Wächtershäuser, of whom the generally (and correctly) sceptical Robert Shapiro[38] wrote, 'When I examined his longest paper, I felt as if someone had handed me a biochemistry text from the late twenty-first century. Most of the chemicals were familiar, but they were organized into an evolutionary pattern I had never seen before.'[39]

The gist of Wächtershäuser's approach is to argue that when pyrite is formed from the pre-existing ferrous sulphide (FeS) in combination with hydrogen sulphide energy is made available (by the release of hydrogen). This energy could act to reduce carbon dioxide, so freeing the carbon for involvement in organic reactions. An important corollary to this hypothesis is that the surface of the pyrite crystal also plays a key role in controlling the ensuing chemical reactions that are the first steps in a primitive metabolism.[40] It is certainly true that there have been some encouraging experimental results,[41] although the experiments were performed under closely controlled conditions, and other research workers have arrived at more discouraging conclusions.[42] In addition, attempts to portray a longer series of reactions in this hydrothermal setting look rather more precarious: a number of key steps remain hypothetical.[43] Still, these are early days and there is a widespread sense that for all its uncertainties the Wächtershäuser school is on to something important. One obvious attraction is the abundance of the main substrate, pyrite. As will be discussed below, the likelihood that the overall setting was in a volcanic hydrothermal system is also attractive, especially as the heat budget of the early Earth was substantially in excess of that of today. Another reason for (cautious) enthusiasm is the possibility of identifying very ancient biochemical traits and pathways: to watch a metabolism grow out of a mineral surface. Even so, it is worth bringing to mind some more general remarks of Andrew Ellington[44] when he comments on

a disturbing trend in modern thinking about abiogenesis and the evolution of metabolism ... the results of these experiments [on the origin of life] are often interpreted as being proof of a particular scenario, rather than as general support for a more vague view of our origins. While it would be extremely exciting to establish that the RNA world evolved to a particular scenario, it is unfortunately unlikely that sequence minutiae have remained unchanged and unscathed following a several billion year journey through multiple different types of organisms and biochemistries.[45]

The notion that pyrite (and other metal sulphides, especially nickel) could play the key role in the initiation of life has attracted attention for another potentially important reason. This is because such sulphides are abundant in hydrothermal systems, most famously in the form of the 'black smokers' found on oceanic spreading ridges. As potential sites for the origin of life, these systems are attractive for several reasons. These include their dynamic nature with strong temperature gradients, the possibility of active mineral growth, and a strong flux of both heated water and various chemicals. Such settings are now the focus of investigations into various prebiotic syntheses that might be alternative sources for some of the major building blocks of life.[46] Hydrothermal systems have another advantage in that, of all the regions of the early Earth, they would have been the most immune to the searingly powerful destructive forces released by a series of violent impacts early in the history of our planet (see Chapter 5).

The possible association between very hot environments and the earliest life has, in turn, led to a potentially fascinating possibility that the most primitive types of bacterial life are those still found today living in hot springs,[47] such as those in both the mid-ocean ridges adjacent to the 'black smokers' and other volcanic regions such as the famous Yellowstone geyser and fumarole pools. This does not necessarily mean that the very first life emerged in such torrid circumstances, although proponents of pyrite-based metabolism and hydrothermal systems will offer their own warm support. The fact that nucleotides require temperatures nearer to freezing for long-term stability (outside, that is, the functioning cell) has already been mentioned (see note 37), and this is not the only difficulty with a hot start to life.[48] More interesting is the possibility that it was only those bacteria capable of resisting extreme temperatures, imposed by the series of colossal meteorite impacts (see Chapter 5), that could survive these 'thermal bottlenecks' to repopulate the planet.[49] Attractive and popular as this idea might be, there is also a vociferous group[50] who reject the notion that bacteria adapted to extreme heat, the hyperthermophiles, represent the common ancestor of all extant life. This group argues that these adaptations to high temperatures must be a secondary feature. Even so, despite this scepticism and more general reservations,[51] at the moment the idea that life's origins are to be found in ancient environments nearer to the traditional depictions of Hell – sulphur and boiling temperatures – is probably the most influential idea in town.

Yet even a cursory knowledge of the investigations into the origin of life invites some caution. We have been here before: a new locale, apparently realistic conditions, some intriguing results that indeed confirm the possibility of replicating certain early stages – say stringing amino acids into peptides – of the evolution of living systems. It is all very encouraging, but twin questions remain. First, can we move from the carefully controlled environment of the laboratory bench ('all solutions were prepared from doubly distilled water'[52]), to a more realistic setting? Second, research on the history of the origin of life may repeat itself. Those first stages certainly require chemical skill and imagination, but the question remains as to whether the further and crucial stages to reach a functioning metabolism are so easily achievable. To put it bluntly, the very success of one system in one context, say pyrite, makes it questionable whether a very different set of reactions will be catalysed with anything like the same efficiency. Thus, while finding much to admire in Wächtershäuser's work, Leslie Orgel[53] points out that there are difficulties in extrapolation. He writes,

> One must expect the results of mineral catalysis to be highly idiosyncratic; most minerals will probably catalyze some reactions and many reactions will no doubt be catalyzed by some minerals ... Nor would the situation be changed if the proposed participants in a complex cyclic reaction scheme were synthesized *in situ* on a mineral surface. If the products are mobile on the surface, the situation is identical to that for adsorbed molecules. If they are not, one must postulate a series of remarkable coincidences to conclude that all of the reactions are catalyzed on the same mineral and that each intermediate product is formed in the correct position and orientation to become the substrate of the next regiospecific reaction of the cycle. The self-organization of a complex cycle, such as the reductive citric acid cycle, this time on the surface of FeS/FeS_2, although logically possible, is very unlikely.'[54]

In conclusion, it would be most surprising if some, perhaps many, minerals did not play important roles in catalysing in one way (e.g. oligomerization) or another (e.g. a simple cycle) some of the first steps to life, but getting to the next stage seems to be as elusive as ever.

BACK TO THE TEST TUBE
So such problems as trying to synthesize the sugar ribose (or even a precursor) or building chains of molecules in a credible prebiotic environment are fraught with problems. What then of the many other experiments that have been carried out in the attempt to decipher one or other stage in the otherwise mysterious journey by which life may have emerged? It is certainly the case that some do seem, at least at first sight, to have a ring of reality. Such is often claimed for the famous experiments by Stanley Miller and Harold Urey, in which electrical discharges in a primitive atmosphere led to the formation of a striking panoply of organic molecules. Yet, as we shall see, the elegance of this set of experiments may conceal less well publicized difficulties. There is, moreover, a rather general problem, already alluded to in the context of other experiments: are the conditions, such as those chosen by Miller and Urey, even remotely relevant to the early Earth? At first sight, this seems to be a harsh criticism. It is, of course, understandable that the investigators will take short cuts. In particular, they will use reagents that are conveniently free of impurities, ones that are made available by processes that in turn rely on complex industrial pathways, and are often used in concentrations that in any prebiotic milieu seem improbable. The methods employed are carefully explained, not least to guide others who might wish to duplicate or extend the procedure, but it is very seldom that more than a passing remark is offered as to whether any environment on the early Earth (or Mars; see Chapter 5) even approximates to what is going on in the laboratory. There are rather few moments of candour, such as those connected to a series of experiments investigating oligomerization of nucleic acid-like compounds, where the investigators remark, 'However, it is unlikely that chemically pure substrates of this kind, for example 2-MeImpG ... [the chemical 2-methylimidazolide guanosine used as the template for oligomerization] could have accumulated on the primitive earth.'[55]

This question of how far chemical analogies can plausibly be extrapolated in studying the origin of life certainly makes it difficult to judge when experimental convenience is given too much precedence. Certainly, one is entitled to wonder whether, for all its elegance, a study of the self-assembly of nucleotides in the form of monolayers via the catalytic agency of molybdenum sulphide on highly orientated pyrolytic graphite[56] has a direct relevance to the question of how DNA was polymerized. Indeed the authors are candid when they write that the 'substrate surfaces were used because of their well-defined crystallographic orientations and their convenience in electrochemical and scanning probe microscopy studies.'[57] They go on to remark that 'The prebiotic relevance of self-assembled monolayers of purine and pyrimidine bases is one of conjecture,'[58] although nothing daunted they extend their hypothesis by commenting that 'It is attractive to suggest an interaction between the purine and pyrimidine monolayers and amino acids as the physicochemical hypothesis for origin of the genetic code.'[59]

It is indeed difficult to distinguish what will develop into an attractive hypothesis rather than mere wish-fulfilment. There are many other examples where the question of using a convenient substrate or procedure seems difficult to reconcile with any prebiotic plausibility. To the example of molybdenum sulphide one could add, for example, the use of minute droplet-like structures, known as micelles, composed of such chemicals as sodium caprylate,[60] cetyltrimoethyl ammonium bromide (CTAB),[61] or lithium chloride[62] as possible models for the process of self-replication or oligomerization. Concerning the last of these substrates, Leslie Orgel shows characteristic caution when he writes, 'Clearly this particular replication system is unlikely to have operated on the primitive Earth. But the model is very suggestive, and may encourage work along the same lines using more plausible prebiotic substrates.'[63] Orgel goes on to remark that the process of self-copying may be a rather general phenomenon, but he acknowledges that designing a primitive system of replication is difficult. Thus he continues, 'The subunits of the self-replicating polymer must be potentially prebiotic molecules, that is, molecules that could have accumulated on the Earth without the intervention of organic chemists. Different authors will draw the line between plausibly and implausibly prebiotic molecules in different ways ... However generously "prebiotic" is interpreted, it is obvious that it will be harder to realize a replicating system if we are restricted to prebiotic molecules and cannot use molecules specially designed for replication.'[64] With this distant whisper of teleology, the perennial problem of the chicken and egg returns. In evolution we know the egg came first, but in the origin of life?

Those familiar with investigations into the origin of life may be surprised that no detailed mention has been made yet of the now-classic experiments by Stanley Miller and Harold Urey,[65] which are still routinely cited (especially in that zoo of good intentions, the undergraduate textbook) and seem to be dramatic evidence of the ease with which simple organic compounds can be synthesized under conditions closely similar to those found on the early Earth. All the appropriate ingredients were there: from the spotless glassware handled by the dedicated and intelligent student of a Nobel Laureate, to the seething organic broth emerging beneath the flicker of artificial lightning. Yet it is now clear that the elegance of these and subsequent experiments[66] may have little, if anything, to do with what may have happened on the early Earth. The difficulties fall into several categories (notes 16 and 66), and I want to re-emphasize how pervasive the difficulties really are.

First is the general assumption that these experiments form some sort of continuum with extraterrestrial organic chemistry, with the implicit assumption (at least in some quarters) that the emergence of life is a cosmic process (Chapter 3). Thus, it is said that the profile of the amino acids produced in these experiments closely mirrors those found in carbonaceous meteorites. But this is a rather selective reading of the evidence. First, even though some amino acids, including not surprisingly simple ones such as glycine, formed in the experiments also occur in meteorites, the similarities are not that precise. More important, perhaps, is the evidence (reviewed in Chapter 3) that at least some of the extraterrestrial amino acids formed under conditions of interstellar chemistry are very different from those of Miller's experiments. Another well-known feature of these experiments is that the enantiomorphs (see pp. 39–40, Chapter 3) of the amino acids occur in equal proportions. Indeed, this racemic equivalence was used by Miller to refute the possibility of biological contamination of the system. This is not to contest the fact that how, when, and where amino-acid chirality was broken in favour of the left-handed enantiomorphs used by all life are, as we saw in the last chapter, all very conjectural questions. Even so the most optimistic interpretation of Miller's results is that if his experiments really do have a bearing on the early synthesis of organic molecules, including amino acids, then somehow the left-handed choice was arrived at a later stage of prebiotic history.

Having reviewed a few of the experiments on the origin of life earlier in this chapter, it comes as little surprise that notwithstanding their classic status the success of individual experiments of the Miller–Urey type was highly variable. Even the best results, or at least those giving the highest yields of amino acids, were remarkably inefficient.[67] These revealed a whole series of complexities, such as variably declining carbon yields after no more than a few days, meaning that deciding what is the most propitious 'atmosphere' is fraught with problems. Perhaps more surprisingly the experiments also depended critically on the physical arrangement of the laboratory apparatus. Thus, the simple interchange of a condenser (a piece of equipment for condensing gases to the liquid state) and the energy source (a sparker) resulted in hardly any amino acids being produced, although quite substantial quantities of hydrocarbons were formed. Moreover, the choice of a spark discharge was determined not so much to represent the effects of lightning flickering across a lifeless landscape but because, as Miller and his colleagues note, 'The ease of handling and high efficiency ... are factors favoring its use.'[68] Thus sparking is very efficient at synthesizing hydrogen cyanide and cyanoacetylene, two simple but key chemicals in the pathway to the subsequent synthesis of the amino acids. Yet the use of what were effectively spark plugs may be an imperfect analogy to lightning because experiments in which the electric charge was allowed to build up and suddenly flash curiously produced very few organic compounds. No doubt, then, thunderstorms rumbled across the lifeless seas of the ancient Earth, but as even a potential energy source to drive the prebiotic reactions the importance of lightning would have been far outweighed by the ultraviolet (UV) radiation from the Sun. Unfortunately, yields of amino acids seem to be much lower when UV radiation is used, and to complicate matters it has also been pointed out that if this were the source of energy then many of the chemical reactions would have taken place high in the atmosphere, with the risk of decomposition as these compounds floated down to the surface of the Earth.

The key criticism, however, that has been levelled at the Miller–Urey experiments concerns whether the gases put into the flasks were in any way equivalent to the atmosphere of the early Earth. At the time of these investigations it was widely assumed that this early atmosphere was devoid of oxygen. This was a reasonable enough assumption given that life, specifically photosynthesis by algae and plants, is the main source of this toxic and corrosive gas.[69] It was also assumed that the early atmosphere was quite rich in two key gases: ammonia (NH_3) and methane (CH_4). Only a small amount of ammonia is required (at 25 °C about a millionth of an atmosphere), but as Miller himself remarked, 'The details of the ammonia balance on the primitive earth remain to be worked out.'[70] In the experiments the levels of methane were much higher, typically up to 0.2 of an atmosphere. Again Miller noted, 'This pressure is used for convenience, and it is likely, but never demonstrated, that organic synthesis would work at much lower partial pressure of methane.'[71] Nobody disputes that establishing the nature of the early atmosphere is problematic, but the general consensus now is that so far as the Earth was concerned methane, and even more so ammonia, were present only in small, perhaps negligible, quantities. Rather than being strongly reducing, the air was probably only slightly so, being composed of a mixture of carbon dioxide, carbon monoxide, water vapour and nitrogen. In these conditions attempts to produce not only amino acids but also compounds such as the key precursor hydrogen cyanide (see note 8) are effectively failures. It is hardly surprising, therefore, that other ways of generating amino acids abiotically have been sought. As we saw in the previous chapter, one hypothetical route is simply by cometary impact ('extraterrestrial infall') and suchlike. A popular alternative, already mentioned, is to consider hydrothermal systems, such as mid-oceanic ridges, where amino acids can be generated at surprisingly high temperatures, at least in experimental analogues,[72] although with some important provisos.[73]

Miller himself shows considerable candour in discussing the research programme he has inspired. Despite the iconographic status of his early experiments in sparking a methane-rich gas mixture, he and his colleagues have no illusions that they are in any stronger position to arrive at that long-sought-after 'one-pot reaction': chemicals in at one end, cells out the other. The origin of some of the hydrocarbons is a case in point. Here the usual invocation is to draw attention to a process known as the Fischer–Tropsch reaction, where at quite high temperatures (c. 200–400 °C) high yields of hydrocarbons are obtained when a mixture of carbon monoxide and hydrogen flows over a catalyst such as iron and nickel. This is a well-known industrial process (the Mond process), but in the natural world it is likely to occur only in a volcanic environment. The reaction can, however, make straight hydrocarbon chains up to only a length of 16 carbon atoms. The longer chains, which are critical for building the hydrocarbons used to construct a cell wall must be made by some additional process, the details of which are yet to be established. One can see, therefore, the attraction of deriving at least some of the organic material needed for life from an extraterrestrial source (Chapter 3). Although most of the interest has been in the dumping of raw materials, the synthesis of so-called amphiphilic compounds, with carbon chains and a propensity to organize themselves into tiny vesicles in which prebiotic chemical reactions could continue,[74] has an obvious relevance to this problem.

A SCEPTIC'S CHARTER

The description of this bringing together of the basic ingredients for life has been reduced to near parody by Robert Shapiro,[75] but, however amusing, he is making a serious scientific point. The nub of his argument, which I paraphrase below, is that even ignoring the complex chemical reactions with their plethora of irrelevant by-products, the varying stabilities of the resultant chemical compounds, the frequently pathetically low yields, not to mention the various laboratory protocols that necessarily insist on convenience of chemical purity and efficiency of human-directed reactions, the range of molecules necessary for life to emerge can be synthesized only in widely, if not wildly, different environments. To be sure, in certain cases important molecules, such as the simpler amino acids, can be generated in a series of quite different settings. In other examples, however, synthesis depends on rather specific pathways, while the stability of a particular molecule is often precarious and sometimes dependent on a particular environment. The sugar ribose, as we have seen, is decomposed in only a few weeks, but most notorious in this respect are the nucleotides, such as adenine, which for long-term survival ideally require conditions well below freezing.[76] Inconveniently, the reverse holds for other molecules, at least in some high-temperature hydrothermal settings. So let me give the gist of Shapiro's scenario for what might be needed to bring the prebiotic ingredients together to ignite the vital spark. He asks us to imagine a volcanic island, large enough to generate a series of distinct climatic zones. In this volcanic environment not only are there various minerals and muds, but also hot thermal springs. The Big Island of Hawaii might be a modern analogue. Thus we can envisage a rainy side, with abundant thunderstorms, and near the summit sufficient snowfall to generate glaciers. The atmosphere lacks oxygen and in the thunderstorms hydrogen cyanide is generated and so accumulates in the glacier ice. Suppose, Shapiro continues, that one year the glaciers advance and in doing so displace the water from partly frozen alkaline lakes, located at the glacier snouts, into a more temperate zone. Here the hydrogen cyanide (and its reaction products) happens to run into an acidic hot spring. Behold, adenine is formed. As Shapiro points out, the quantity is small and the adenine would not survive for very long, but by good fortune the water flows over alkaline ground, so neutralizing the reaction. Elsewhere formaldehyde is formed, and from this a small proportion of the sugar ribose is synthesized as the mixture is swept into a series of hot pools. This time, however, the water is neither acidic nor alkaline (i.e. it has neutral pH). Happily the necessary reaction time to make the ribose is just right before it is carried further downstream in ice-cold water. This is helpful: the ribose would otherwise have been destroyed.

Already even the simple compounds of adenine and ribose are available by several happy chances. But good luck is our constant companion. The separate streams containing these two chemicals now merge, and, as Shapiro reminds us, the next product, adenosine (the precursor of one of the nucleotides of DNA), needs another set of special conditions to form, including heat and salt. No problem: the water containing the adenine and ribose is carried in a waterfall to sea level, where hot salty pools, lined with appropriate minerals, receive the chemicals. One stroke of good fortune follows another. As Shapiro continues, 'It was a very hot day. The sun evaporated the remaining water in the pool and heated the adenine and ribose in the presence of salt, converting them in part to the nucleoside adenosine ... the tides returned to the tidal pool in a rush, sweeping out its contents and transporting them farther inland to ... (the) Darwin Pond ... No sooner had the adenosine reached Darwin Pond when successive waves, each flowing from a different direction, brought in supplies of the other nucleosides to make RNA.' So Shapiro concludes, 'Had these chemicals only been human, they would have embraced at the joy of their first meeting, and in anticipation of the glorious future that lay ahead of them. They would then have taken turns, each describing the marvellous and different series of events that led to its own creation.'[77]

Nor does the story end there. Before life itself can emerge in Shapiro's tale more fortunate accidents involving phosphate-rich minerals, convenient catalysts, and floods are required; but his point has already been made well enough. It might have been made, but has it been heard? To show that Shapiro's thinking is by no means off limits, consider these remarks by two of the most respected experts, Gerald Joyce and Leslie Orgel,[78] in a section headed ' The prebiotic chemist's nightmare:

> Scientists interested in the origins of life seem to divide neatly into two classes. The first, usually but not always molecular biologists, believe that RNA must have been the first replicating molecule and that chemists are exaggerating the difficulties of nucleotide synthesis. They believe that a few more striking chemical 'surprises' will establish that a reasonable approximation to a racemic version of the molecular biologist's pool could have formed on the primitive earth, and that further experiments with different activating groups and minerals will solve the enantiomeric cross-inhibition problem. [A reference to an observation that template oligomerization of nucleotides proceeds efficiently if one enantiomorph, e.g. the left-handed form, is present, but is severely inhibited if the other handed molecule is present[79]]. The second group of scientists is much more pessimistic. They believe that the *de novo* appearance of oligonucleotides on the primitive earth would have been a near miracle. (The authors subscribe to this latter view). Time will tell which is correct.[80]

I should emphasize that Joyce and Orgel are equally careful to point to encouraging lines of research, but near the end of the chapter they remark, 'There remains a lingering doubt that we are on the right track at all.'[81]

How then can we escape these dilemmas? The sheer unlikelihood of a 'one-pot reaction' emerging in any natural and primitive environment helps to explain the paradox of both frustration at repeated failure and hope springing eternal that some environment – pyrite?

clouds?[82] hydrothermal springs? – will bring the magic ingredients together. Present ignorance need not be a reason for pessimism. The general principles of evolution remind us that advantages need only be slight; systems can act as templates and molecules may be co-opted for new uses. And that, perhaps, is the central dilemma in biology. This book is really about evolutionary navigation, be it through the immensities of protein hyperspace (Chapter 1) or the arrival at intelligence (Chapter 9). Everything we know about biology argues that it is seeded with inevitabilities; in principle, there is an incomprehensibly enormous universe of possibilities, but in reality the number of destinations – metaphorically the Easter Island of Chapter 1 – is a vestigially minuscule fraction of the theoretical possibilities that can never be visited, not because the journey would never be made, but because the journey would never be possible in the Universe we happen to inhabit.

And this should apply with equal force to the steps involved in the origin of life. As Manfred Eigen wrote many years ago, 'The evolution of life, if it is based on a derivable physical principle [in this case selective value] must be considered an *inevitable* process ... not only inevitable "in principle" but also sufficiently probable in a realistic span of time. It requires appropriate environmental conditions (which are not fulfilled everywhere) and their maintenance. These conditions have existed on earth and must still exist in many planets in the universe.'[83] The importance of law-like processes in the study of the origin of life is also emphasized by Iris Fry.[84] In particular she offers an analysis of the extent to which such hypotheses as those advanced by Manfred Eigen to explain the origins of replication can be shifted from a supposedly random basis to one that presupposes 'a deterministic element in the evolution of molecular self-organization',[85] and thereby an imposed directionality. Echoing this, Kenneth James and Andrew Ellington remark, 'While Stephen Gould is fond of saying that re-running the tape of life will always give different outcomes, it is not clear whether this thought experiment is as true for molecules as it is for multicellular organisms [so far as the latter are concerned, the exact opposite is the principal argument of this book]. The strictures of prebiotic chemistry may have narrowly defined the first self-replicating organic molecules [so] ... it becomes possible to at least hypothetically chart some of the intermediate forms that life may have taken.'[86]

Quite so, and one should be optimistic. Despite their difficulties, as already seen the imaginative hypothesis of Wächtershäuser may be the most encouraging way forward. Somehow, we must assume that the problems of combining those molecules that prefer hot springs as against glaciers, not to mention the myriad of often unwanted chemical by-products all ready to gum up a reaction, will be surmounted. Perhaps it is impractical to do more than stumble round the edges of a problem that may have taken millions of years to solve. Yet, the suspicion, as articulated by James and Ellington, is that there may be only a narrow way. Not only here, but everywhere. At present, however, the problem is that the simplest prebiotic systems are pretty elementary, and there is no obvious or straightforward way that seems to lead easily to the specificity of chemical reactions that is so characteristic of life. Experiments are contrived and frankly for the most part unrealistic, and as often as not the end-result is failure or goo, or both.

There seems to be a feeling that there is some general and quite simple principle that keeps on being overlooked. Leslie Orgel reminds us[87] that practically all the gem-quality rubies in the world come from one locality near Mogok in Myanmar (formerly Burma), where presumably the chemical conditions in the host marbles (metamorphosed limestones) were 'just right'. In this context, it is surely interesting that Francis Crick[88] can write 'An honest man, armed with all the knowledge available to us now, could only state that in some sense, the origin of life appears at the moment to be almost a miracle, so many are the conditions which would have had to have been satisfied to get it going.'[89] Crick is careful to continue by pointing out that the time available (but see note 16, Chapter 5), the diversity of habitats, and combinatorial possibilities of chemistry do not exclude life originating 'by a perfectly reasonable sequence of fairly ordinary chemical reactions'.[90] More than two decades on from Crick's ruminations, however, it still remains the case that the notion of an infinitesimally unlikely series of chemical reactions – that from our perspective can be described only as a 'near miracle' (or 'happy accident', a curious term, given that it echoes traditional Christian teaching on the nature of the Fall) – remains the unbidden and silent observer at much of the discussion of how life originated. Yet, as Iris Fry (note 85) reminds us, such terminology is effectively that of creationism. Put this way, nearly everyone will ask that the now unwelcome guest should vanish through the adjacent wall, and agree that for all their differences (soup, clay, clouds ...) they share the common hypothesis that the steps from inert to vital must be those of an unbroken continuity.

It would be an uncomfortable corollary if the series of meetings, interactions, and reactions of those few chemicals that led to the origin of life were little more than a series of fortuitous and happy flukes. If so, a scientific campaign for understanding the origin of life is not, to put it mildly, going to be straightforward. It is hardly surprising that George Wald[91] wrote, 'One has only to contemplate the magnitude of this task [of making life] to concede that the spontaneous generation of a living organism is impossible. Yet here we are – as a result, I believe, of spontaneous generation.'[92] This quotation has become justifiably famous, but it is sometimes forgotten that in its original context there was a very strong underlying assumption that such a process of spontaneous generation could be possible only if there were enough aeons of time. As we shall now see, there probably were not, and to make matters worse what obviously did happen in our Solar System may itself be either a very rare occurrence, or, dangerous thought, possibly unique.

5   Uniquely lucky? The strangeness of Earth

Wandering though ruins, along deserted thoroughfares and past top-
pled columns, can be a melancholy, if not disquieting, experience. For
some it is the sense of vast futility of vanished and vanquished kingdoms, of which one might read in Shelley's *Ozymandias*. But ruins
open other prospects, not least that sense of majesty and awe such
as the Company felt as they swept along the River Anduin past the
Argonath, the giant sentinels of Númenor. For some today such emo-
tions are either a dangerous or a worthless currency, a price to be
paid for our disenchantment. Yet for innumerable generations men
and women have stood before temples and shrines, now derelict if
not entirely vanished, to scry the Moon, perhaps as a heart-catching
sliver at dusk or an immense harvest orb pulling itself above the horizon. The potency of the Moon needs no emphasis. Yet it has only
recently come to be realized that it, too, is a gigantic and lifeless
ruin, a shattered world, but a satellite whose existence, we now realized, has had a profound effect on the Earth and its four-billion-
year-old cargo of life. Indeed, the peculiarities of the formation of the
Moon help to epitomize the principal theme of this book, the odd
fortuitousness of the world in which we find ourselves, where again
and again matters seem to be remarkably well arranged. To some,
this is just a happy string of coincidences, but the strangeness of the
Moon, which perhaps we take too much for granted, serves to open
a series of vistas into the equally peculiar construction of our Solar
System.[1]

THE SHATTERED ORB
To return to the beauty of the Moon: in what way is it a ruin? Even a
glance reminds us that it is far from homogeneous in appearance.[2] To
a first approximation, in the reflected sunlight much of the Moon's
surface is bright, but there are also conspicuous and more-or-less circular dark areas, once thought to be seas across which sailed the lunar
inhabitants or selenites. These areas are still referred to as the maria
(singular *mare*). The Moon has no atmosphere and there is thus no

weathering of its surface, nor is there any crustal recycling akin to terrestrial plate tectonics. Even the Moon's volcanism is long extinct. Today the ancient surface of the Moon is modified only by the impact of meteorites and comets. The large craters, such as Copernicus, Tycho, and, on the far side of the Moon, Giordano Bruno, exhibit dramatic rays that in radiating from the point of impact provide graphic evidence for collisions. The two craters visible from Earth are, in comparison with the age of the Moon, relatively young. Copernicus was formed about one billion years ago, but Tycho is much younger. It was formed perhaps 50 or so million years before a similar object hit the Earth, spelling out the doom of the dinosaurs. Just like the Earth, the Moon continues to receive high-velocity visitors. One of these, it has been claimed, was recorded as a dramatic eye-witness account in a medieval manuscript, written by the chronicler Gervase of Canterbury in 1178. Possibly it was this impact that formed the Giordano Bruno crater.[3]

Such impacts are relatively infrequent, and it is agreed that the surface of the Moon was largely shaped by intense episodes of bombardment early in the history of the Solar System. The principal bombardment (Fig. 5.1) literally pulverized the surface with repeated impacts, churning the surface to rubble and bestrewing the Moon with craters. As the meteoritic material was swept up by the Moon (and, as we shall see, the other planets) so the intensity of the impacts diminished. There was, however, a last, violent interval when the Moon experienced a series of very energetic impacts.[4] The standard view is that these bodies were presumably also products of continued accretion, and so would have had diameters of the order of a hundred kilometres or so.[5] It was the impact of these massive bodies[6] which led to the formation of the darker maria, such as the Mare Tranquillitatis, where the astronauts of *Apollo 11* first stepped on to the Moon. The relative smoothness of the maria, which have rather few craters, shows that they must have been formed after the main bombardment episode. These maria are effectively huge pools of basalt, now solid rock, but molten lava that flowed into the circular depressions formed by the impacts. It appears that the basaltic magma was not formed as a direct result of the collision, but was formed by melting deeper within the Moon and then ascended through marginal cracks to fill the giant craters. In some instances, as in Mare Imbrium, the tops of mountains formed earlier still protrude above the lava plains.

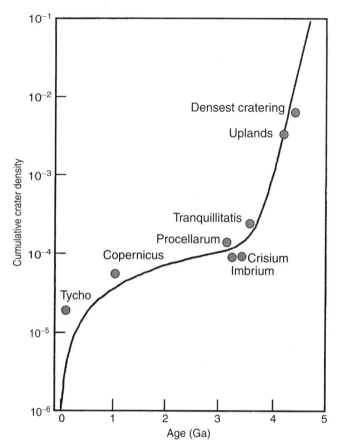

FIGURE 5.1 The inferred bombardment history of the Moon in terms of cumulative density of cratering since the formation of the Moon about 4500 million years ago. Note the intense rate of early cratering, before about 4000 million years, which then declines with the formation of the huge maria, of which four are listed here. Thereafter, the rate of impacts gradually falls off, with a steep decline from about 1000 million years ago. In this interval infrequent giant collisions include those that led to the formation of Copernicus and Tycho. (Reprinted from *Icarus*, vol. 92, C. F. Chyba, Terrestrial mantle siderophiles and the lunar impact record, pp. 217–233, fig. 1, copyright (1991), with permission of Elsevier Science, and also the author.)

BATTERING THE EARTH

So it is that when we look at the Moon riding in the sky we see inscribed upon it events of almost unimaginable violence, inflicted during the early history of the Solar System. Even so, both the Moon and those ancient times may seem remote and scarcely of relevance to life on Earth. In fact, the opposite is the case. Consider that if the Moon was subject to so heavy a bombardment, then the much greater gravitational attraction of the Earth, with a mass almost one hundred times greater than that of the Moon, would almost certainly have made it subject to an even more severe bombardment. Unlike the Moon, the Earth now shows no direct trace of this bombardment: the ceaseless processes of erosion and crustal recycling have obliterated all the evidence. Many of the impacts must have been large, and a few were cataclysmic. What happens when something very big hits the Earth?[7] The amount of energy released is enormous, so great in fact that the heat generated was probably sufficient to evaporate the entire world ocean. The net result is that the surface of the Earth would then have become an inferno. Initially there would be an atmosphere of rock vapour, but this would cool quite quickly. This rock would rain down in a couple of months to form a layer carpeting the Earth to a depth of several hundred metres. The Earth would still be a searing desert, and the atmosphere would remain a boiling mass of water vapour and other gases: a scene in some ways resembling the present-day Venus, where the surface temperatures are an uncomfortable 480 °C. Venus has become trapped in a permanent furnace, too close to the Sun and with an atmosphere that acts as a powerful and irreversible greenhouse. On the early Earth, however, after each giant impact the excess heat was radiated into space, the atmosphere cooled, and it began to rain, for perhaps 3000 years as the oceans refilled.[8]

It is estimated that during the later bombardments the Earth may have been pounded by five or so of these monster impacts, each capable of evaporating the ocean and presumably sterilizing the surface. In the case of the lunar maria, which as noted above, were also formed in the late bombardment episode, it appears that although they formed during a fairly substantial interval, the largest impacts are quite closely clustered in time and, more significantly, terminated rather abruptly about 3800 Ma ago. Now this happens to be a rather significant date in Earth history, because it is the age of the oldest reasonably well-preserved sediments.[9] These are the famous sediments from the Isua Supergroup of central – west Greenland. Despite metamorphism, a wide variety of lithologies, including waterlain conglomerates, are known. Although fossils are lacking[10] the presence in these sediments of carbon, altered by metamorphic heat to the mineral graphite, is possibly, if controversially, indicative of life.[11] The main reason for thinking this is that there is evidence of fractionation in the graphite in favour of the lighter isotope carbon-12, a process widely accepted as evidence of ancient photosynthesis.[12]

Earlier rocks are known,[13] but these have been so distorted and altered by the processes of metamorphism that any evidence for life will have been destroyed. We may never obtain direct evidence from the geological record of when life either emerged on Earth or, as some propose, arrived as unintentional Martian colonists (see pp. 75). In one sense, however, this ignorance of whether life emerged here or came from Mars may not matter overmuch. If we assume that the end of the late stage bombardment episode is correctly dated at about 3800 Ma, then the fact that the first identifiable sediments also date from this interval is more or less what we would expect: before this the surface of the Earth was repeatedly harrowed by giant impacts. Before this time, the impacts may have seriously impeded, if not entirely frustrated, the emergence or survival of life.[14]

As already noted, the actual history of the Earth's bombardment does not survive, but the scarred surface of our daughter satellite provides its own mute testimony. One estimate, for example, suggests that the Earth received in total between 17 000 and 22 000 major impacts,[15] of which a few were truly colossal. The fact that the bombardment drew to a close at about the same time as the deposition of the oldest known sedimentary rocks, containing carbon that is perhaps of organic origin, leads to three interesting possibilities. In brief, they are as follows: (a) no sooner were the conditions benign enough on Earth than life evolved, very quickly; (b) the bombardment episode was severe in the extreme, but life had evolved appreciably earlier and hung on; or (c) life evolved, but on Mars, and a regular traffic ensured that as soon as it was safe the Martian bacteria settled down to enjoy their new home.

Thus, the first possibility (a) is that the degree of devastation, with at least several immense collisions, would have released sufficient energy to evaporate the oceans and sterilize the surface of the Earth. This implies that the time available for life to emerge may have been remarkably short,[16] given the evidence already noted for life being present more or less at the time this catastrophic interval ended. One such estimate suggests that the time life needed to emerge was as little as 20 Ma,[17] and some origin-of-life enthusiasts seem quite content with even 10 Ma.[18] This, incidentally, is not to say that life did not emerge earlier. Quite possibly it did; in principle it could have evolved several times,[19] but was repeatedly snuffed out when the oceans boiled and billions of tons of rubble descended to a fiery surface.

What of the other two possibilities mentioned above? It is scarcely surprising that an estimate of a mere 20 Ma (or even 10 Ma) for life to originate is sometimes greeted with considerable unease, if not disbelief. As we saw at the end of the previous chapter (see p. 68), it is a common assumption that life will certainly emerge at least in an Earth-like milieu, but it needs not only time, but aeon stacked upon aeon as the chemicals blindly combine and re-associate, with the thin red thread of vitality worming its way past uncounted blind alleys and entropic pits. One response is that life is indeed an abrupt process of self-assembly, by implication a cosmic inevitability, as natural as water being wet. Quite possibly so, but the experimental morass reviewed in the preceding chapter might give us pause for thought. So can we find a loophole, to allow more time for the emergence of life?

There are certainly two possibilities. The first is that despite the violence of the giant impacts, life not only evolved much earlier, rocked by wave after wave of planetary disaster, but still managed to cling on. Nobody doubts the tenacity of life, but not until the discovery of deep-sea hydrothermal ecosystems with bacteria that incredibly are able to function at temperatures close to 100 °C was there any hint as to how early life might have survived repeated thermal catastrophe. It is certainly interesting that many of the bacteria that appear to fall closest to the root of life are adapted to function at very high temperatures, and hence are known as the extreme thermophiles.[20] This could indicate either the ancestry of life on a hot Earth, or more probably an ability to survive a series of thermal 'bottlenecks', each imposed by the vast quantities of energy released by a giant impact. Despite the widespread popularity of these ideas (a sort of bacterial equivalent of Benvenuto Cellini's story of the fire-salamander basking in the flaming coals), some scientists continue to argue that the most primitive forms of life were adapted to much more moderate temperatures.[21]

So perhaps life is even more ancient than we suppose, with its origins in the great heat of hydrothermal vents and its early survival set against a backdrop of mega-impacts and searing temperatures far beyond any human tolerance. But there is a third way: in the beginning our world was still far too dangerous, and by the time conditions were stable enough to allow life to begin to evolve, it was far too late. Hence, the third of the possible explanations for life appearing no sooner than the great bombardment had ceased: life had indeed evolved, but elsewhere, on Mars. Terrestrial proto-life didn't stand a chance.

THE MARS EXPRESS

At first sight the alternative to removing life's origins from the fiery heat of the early Earth to the arguably more benign habitat of Mars seems incredible. After all, Mars is today so inhospitable that few expect to find living organisms there. But early in the history of the Solar System, it may have possessed a major advantage over Earth. First, its smaller size (it has about ten times less mass) means a weaker gravitational field, which in turn would have ameliorated the episode of giant impacts.[22] Moreover, it is generally thought that early in its history the surface of Mars was considerably more benign than it is now: warmer, wetter, perhaps even with an ocean.[23] So if Mars is the cradle of life, then how did we, or more specifically, our bacterial ancestors, get here? The recognition of more than twenty Martian meteorites,[24] including the now-famous Allan Hills (ALH 84001)[25] find and the dog-destroying Nakhla fall (Chapter 3, p. 34) is clear enough evidence that pieces of Mars can be blasted from the surface by impacts. Indeed, calculations suggest that the vast majority of Martian meteorites arriving on Earth escape detection because on average one such fall per month would be predicted.[26] Geologically the transit time from planet to planet is relatively short: a Martian meteorite typically spends about a million years in space. But it can be much less, and given a particular trajectory of launching about one in every ten million would arrive in only a few years. Evidence from the exposure of these meteorites to cosmic rays as they drifted through outer space shows that their times of arrival on Earth are relatively recent, all of them having landed in the past 15 Ma. It is equally conceivable, however, that about 4 billion years ago some such chunk of rock, but with a cargo of microbes, landed safely on Earth.[27] One obvious problem, whether life forms such as bacteria could survive both the large accelerations and sudden changes in velocity as a meteorite was blasted off the surface of a planet like Mars, has been addressed experimentally, with the conclusion that an abrupt departure from Mars presents few problems for survival.[28] So we may really be Martians.

Meanwhile, on Mars itself the environmental conditions steadily deteriorated: the ocean dried out, the surface temperatures plummeted, so that today in the Martian winter the ice caps grow as carbon dioxide is precipitated. So, too, Mars lost its atmosphere to outer space, with the result that now the air there is far thinner than it is at the top of Everest. On the Earth, the story was different. The likelihood is that before too long we shall have information as to whether life did evolve on Mars. If it did emerge, or just conceivably still survives, then we should learn whether it had (or has) any biochemical similarity to terrestrial cells. It needs to be remembered, of course, that if the Earth were fertilized from Mars, the reverse is in principle possible. It seems, however, that the return journey, while not impossible, is still considerably more difficult.[29] It also remains possible that evidence for the most ancient crust of the Earth will be pushed back into the time of major bombardment, with an outside chance of finding remains of organic activity. If, by some remote chance, very ancient terrestrial life were found, this might make a Martian colonization less likely.

At present, however, we face many other intangibles. Perhaps, especially if it eventually transpires that Mars was never an abode to life, we shall simply have to accept that life on Earth did evolve very rapidly. This begs many questions, but it also offers some intriguing possibilities. As already noted in Chapter 4, the attempts to delineate a plausible set of prebiotic pathways have met with limited success. Laboratory investigations may target, or just as likely, stumble across, a neglected possibility. Nevertheless, the present situation is ironic. The fact is that as and when we do make a living organism from a prebiotic brew, and in principle this cannot be impossible, it will be constructed by intelligent and directed intervention. In other words, it will have to be designed or, if you wish to duck teleological terminology, at least engineered. From our present perspective, it might be no more unreasonable to invoke participation by a visiting squad of alien molecular biologists. Alternatively, and more scientifically, it may be found that when life evolves it can only do so by one route. There is a famous remark by George Wald[30] that is particularly apposite, along the lines that so long as you can teach terrestrial biochemistry, then your students will have no fears as they wander into the Great

Examination Hall of Arcturus to sit Biochem. IV. So, too, writing on 'The universal nature of biochemistry'[31] Norman Pace remarks,

> it seems likely that the basic building blocks of life anywhere will be similar to our own, in the generality if not in the detail. Thus, the 20 common amino acids are the simplest carbon structures imaginable that can deliver the functional groups used in life ... Similarly, the five-carbon sugars used in nucleic acids are likely to be repeated themes ... Further, because of the unique abilities of purines and pyrimidines to interact with one another with particular specificity, these subunits too, or something very similar to them, are likely to be common to life wherever it occurs.[32]

MAKING THE SOLAR SYSTEM

Let us now imagine that at least the potential for life is more or less universal. In doing so we shall conveniently neglect, but not entirely forget, the difficulties in understanding how it might have been assembled. I shall assume, perhaps incorrectly, that life is confined to planets with liquid water. So far as the search for planets beyond our Solar System is concerned, until a few years ago, matters were not very encouraging. Now, with the discovery of tens of what are referred to as extra-solar planets, everything is changing.[33]

Although remote planetary systems had long been postulated, the first clear hint of what would be found came with the discovery of discs of gas and dust surrounding stars. It is widely believed that planets nucleate from such discs. Of these structures perhaps the best known is the one associated with the Southern Hemisphere star β Pictoris, where an immense disc extends into space for a distance many times greater than the diameter of our Solar System. It is conjectural whether planets are forming in this particular disc,[34] but in any event with the existing telescopes no direct image[35] of any extra-solar planet is obtainable. How then is it possible to detect their presence? Nearly all the recent series of extraordinary discoveries depends on a method known as high resolution stellar spectroscopy. The principle of the method is based on the fact that although the star is much more massive than the planet, the latter also has a mass and so exerts a gravitational pull that slightly affects the star. Not surprisingly, given the enormously disparate masses of star and planet, the effect on the star is very small. Nevertheless, as the planet moves round in its orbit so the star is pulled in different directions. As a result, the wavelengths of the light from the star as seen by observers on Earth are very slightly shifted because of the relative motion between the source and the observer (the Doppler effect). The shift can be measured by interferometry, and periodic shifts in the spectral lines indicate that a distant planet has been detected.

The measurements are very sensitive, and only massive planets capable of imposing measurable perturbations on their stars can be detected. In addition, the measurements need to be made over a protracted period, sometimes several years, before the evidence is sufficiently compelling. Even so, the sensitivity of the technique and the precision of the fits are both quite remarkable (Fig. 5.2). About 90 extra-solar planetary systems have so far been detected (Fig. 5.3). In most instances the available data indicate that there is one planet orbiting each star. This is not to say there are no other planets in each system. Quite possibly there are, but they are too small to exert a measurable influence on the star. There are, however, a number of observations where the data indicate a two-planet system (e.g. stars HD 12661, HD 37124, HD 38529), and in a few cases even a three-planet system.[36] In Upsilon Andromedae, for example, the spectroscopy of the star indicated a planet orbiting the star every 4.6 days, but there was a residuum of data; when analysed further this revealed the presence of two more planets (Fig. 5.4), respectively taking about 240 days and 3.5 years to orbit. So, too, 55 Cancri is now known to possess three planets: two close to their sun, and the third, substantially larger, in an orbit somewhat beyond the equivalent position of Jupiter.

The results to date are spectacular, but they are also deeply sobering. These remote planetary systems are patently very different from our Solar System. So far as the search for extraterrestrial life is concerned, this seems discouraging, but some qualification is needed. First, the limitations of the detection method mean that only very large planets can be discerned. These are typically somewhat larger than Jupiter, which itself is about 300 times more massive than the Earth. A number of the extra-solar planets are truly gigantic (and it is important to remember that the estimate is a minimum because it is the mass times the sine of the inclination of the orbital plane relative to the observer). The planets circling HD 162020 and HD 202206, for example, are each estimated to have a mass equivalent to about

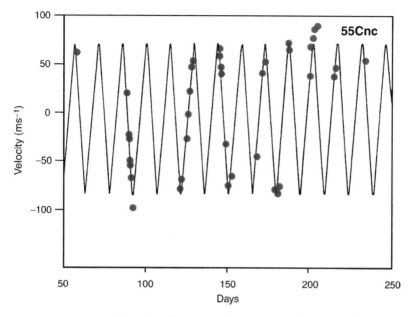

FIGURE 5.2 Detection of an extra-solar system, in this case a planet orbiting the 55 Cancri. The dots represent measurements of the Doppler velocities that arise as a result of minute shifts in the spectral absorption of the starlight as the gravitational tug of the orbiting planet distorts very slightly the shape of its star. The sinusoidal line is the predicted pattern; notice the goodness of fit. These data suggest the planet has a mass slightly less than Jupiter (0.84), orbits the star in about a fortnight (14.64 days), and is located ten times closer to the star than is the Earth (at 0.11 Astronomical Units). Note that subsequently evidence has emerged of 55 Cancri possessing two more planets; see Fig. 5.3. (Redrawn with permission from fig. 1 of R. P. Butler *et al.* (1997) Three new '51 Pegasi-type' planets. *Astrophysical Journal*, vol. 475, pp. L115–L118, copyright of the authors and the American Astronomical Society.)

14 Jupiters, while one of the planets in HD 168443 approaches 17 Jupiter masses. On such planets gravity would, by terrestrial standards, be immensely powerful,[37] and outside any oceans (whose existence, as we shall see, may in any event be problematic) life would probably be only microscopic and present as thin films: a difficult place for a human to visit. A shower of rain would be like standing under flying gravel, and walking down steps would probably result in multiple fractures. In reality many of these enormous planets may be more similar to gas giants like Jupiter,[38] effectively without a solid surface and where any sort of terrestrial ecology is probably

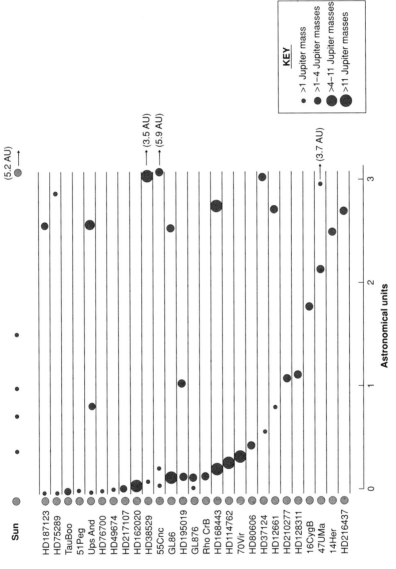

FIGURE 5.3 A sample of extra-solar planets, scaled to our Solar System and the size relative to Jupiter. (An astronomical unit is the distance from the Earth to the Sun, and is equivalent to 1.49×10^{11} m.) Data taken from websites given in note 33.

unimaginable. As we shall see later (Chapter 6), this has not prevented speculation about possible forms of life, although such hypotheses are unlikely to be compatible with the extreme turbulence of such giant worlds.

The real surprise, however, is not so much the enormous size of these newly discovered planets but how closely some of them orbit their stars (hence their nickname of 'hot Jupiters'). Some are substantially nearer to their stars than Mercury is to the Sun, whirling around their stars in a few days. The searing surface temperatures that are inevitable so close to the star almost certainly preclude the possibility of life. To complicate matters, many of these planets have eccentric orbits. Even on the Earth, with a rather mild eccentricity (0.0167) the variation in distance from the Sun is clearly discernible in climatic changes. For extra-solar planets with eccentricities up to 0.93 (as for HD 80606), the thermal regime at the surface will be much more variable.

Although with present-day techniques we can detect only giant planets, the continuing torrent of discoveries strongly suggests a trend towards the majority of planets being relatively small. For example, the planet orbiting the star HD 76700 has a mass approximately one-fifth that of Jupiter, while that of HD 49674 is even smaller. As more observations are made, planets more distant from their stars, which take longer to complete an orbit, will also be detected. As noted above, the third planet of 55 Cancri is remote from its star, taking more than 14 years to complete its orbit. It is important to stress that much smaller planets, similar in size to the Earth, are still beyond the limits of detection. There is so far no reason to think that such planets cannot exist, somewhere. Telescopes suitable for their detection are already being designed. And there is some cause for optimism. Some earlier computer models of solar system formation were able to produce terrestrial planets, but there were difficulties. For example, both the eccentricity and the inclination of the orbits were substantially greater than those of the Earth, and there were also problems in 'making' smaller planets, comparable to Mars and Mercury.[39] Still, refinements continue to be made, both as a result of new knowledge of extra-solar systems and faster methods of computation. The results obtained by John Chambers are impressive,[40] inasmuch as it is now possible to produce a range of hypothetical solar systems (Fig. 5.5) equipped with a variety of terrestrial planets which 'are reasonable

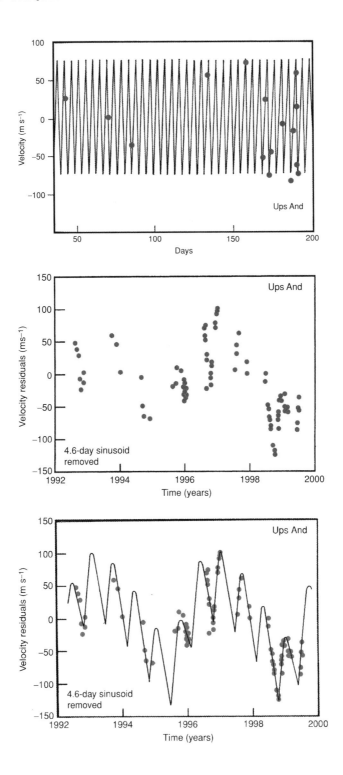

analogues for Earth and Venus',[41] although conjuring up Mars and Mercury is still proving more difficult. In addition, in these models the gas giants Jupiter and Saturn are already in place, and their presence is integral to the evolving story of planetary formation closer to the Sun. Not surprisingly, there is a diversity of systems, and it seems possible that up to six planets could be formed, although they were not produced in these simulations. That would be a fine sight on a starry evening. Whether there would be anybody to appreciate them is another question.

These simulations demonstrate that while broadly analogous systems can develop, an exact counterpart to our Solar System may be much less likely. How important this might be in constraining the emergence of life, let alone of intelligence, is rather conjectural, but the answers depend in part on the position of the so-called Circumstellar Habitable Zone (CHZ) (see below). It is worth noting, however, Chambers's closing remarks on these simulations, to the effect that 'the high concentration of mass in the region occupied by Venus and Earth is not reproduced in any of the simulations, and this discrepancy shows no sign of disappearing as a result of using different initial conditions. This represents perhaps the most important outstanding problem for theories of the final stage of terrestrial planet formation.'[42] It is probably the models that will require further tweaking; after all the reality entails starting with billions of orbiting and colliding bodies, rather than about 150 planetary embryos, and time intervals of millions of years, not a month or so of computer time. But the alternative is also possible: solar systems with terrestrial planets occur, perhaps with some frequency, but our Solar System is still very atypical. Time will tell.

FIGURE 5.4 (*opposite*) Detection of an extra-solar system in Upsilon Andromeda. Upper panel shows detection of the principal planet, which orbits the star in less than a week (4.61 days). Once the 4.6-day sinusoidal pattern was removed, there was a residuum of Doppler velocities, and these are shown in the middle panel. The lower panel shows the best fit, consistent with the presence of two more planets. (Redrawn with permission from fig. 4 of R. P. Butler *et al.* (1997) Three new '51 Pegasi-type' planets, *Astrophysical Journal*, vol. 475, pp. L115–L118 (upper panel), fig. 2 of R. P. Butler *et al.* (1999), Evidence for multiple companions to *v* Andromedae. *Astrophysical Journal*, vol. 526, pp. 916–927 (middle and lower panels), copyright of authors and the American Astronomical Society.)

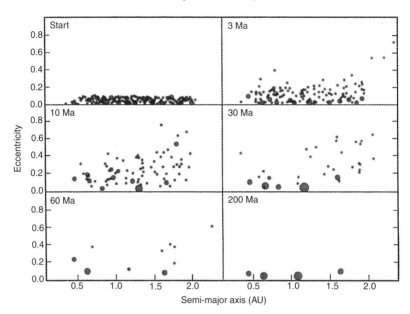

FIGURE 5.5 Making terrestrial planets. Panel above shows a simulation with 154 planetary embryos with a bimodal distribution of masses: half the mass is housed in 14 larger embryos, and the other half in the remaining 140 miniature planets. In this simulation the basic structure of the solar system with four main planets has emerged by about 30 Ma, although about 25 smaller planetismals remain, of which many plunge into the star. The panel opposite shows the end-product of sixteen of the simulations, in comparison with our Solar System (top). Simulation 23 is the end result of the panel above. Relative sizes of planets are shown; also shown are their perihelion and aphelion distances (the horizontal line), i.e. the nearest and furthest distances to the star that thus define the eccentricity of the orbit. Spin axes of each planet are depicted with an arrow. Reprinted from *Icarus*, vol. 152, J. E. Chambers, Making more terrestrial planets, pp. 205–224, figs 3 and 10, copyright (2001), with permission from Elsevier Science, and also the author.

Such models give tantalizing glimpses of what might be the reality of other worlds, but at present they remain computed constructs, virtual possibilities that for one reason or another might never be manifest. On the other hand, there is, as we have already seen, a known reality of extra-solar systems (within the limits of how best to interpret minute fluctuations of distant starlight). What needs particular emphasis is that the configurations of these newly discovered systems

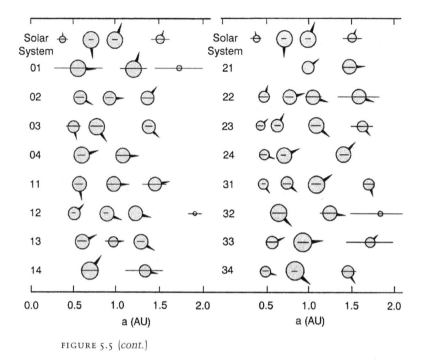

FIGURE 5.5 (cont.)

were to a large extent unpredicted. These new results are forcing radical reconsiderations of how the nebular discs collapse into planets. What to us is a familiar arrangement in our Solar System of an inner series of small rocky planets and an outer group of gas and ice giants may be very much less common than was once thought, and may be very rare indeed.

Perhaps the most dramatic implication is the strong likelihood, indeed near certainty, that the giant planets detected so far in the ninety-odd extra-solar systems actually formed much further away from their stars, and only subsequently were moved into their present position of proximity.[43] The reason for thinking this is that it is effectively impossible for a planet to accrete so close to the star, and mechanisms exist to explain why the planet should spiral inwards. It will be obvious that even if life evolves on such planetary giants it will be doomed as its home migrates closer and closer to its sun. Indeed, in many instances it is likely that the planet's ultimate destination is the star itself, to be swallowed in the thermonuclear maelstrom. To observe such a catastrophe is beyond our present technology, but evidence for such an engulfment already exists.[44] If the planet continues to circle the star, the earlier process of migration explains the otherwise puzzling eccentricity of the orbit. There is also some evidence that over time this orbit may become more circular. The inward migration of a giant planet has, however, another ominous consequence: should the system contain two or more of these giant planets, then the likelihood is that at least one will be ejected into the wastes of interstellar space.[45] As already noted, the existence of smaller planets, perhaps Earth-like, in such extra-solar systems is still conjectural. If they turn out to be much more scarce than is hoped, this may be because they were slung-shot into the interstellar wilderness early in the history of the solar system when the giant planet spiralled inwards, rather than because smaller planets were not formed in the first place. On a lost planet the prospects for life surviving seem grim as its parent sun slowly recedes to become ultimately just another twinkling and hopelessly remote star. As Stuart Weidenschilling and Francesco Marzari remark, 'Many potential systems of terrestrial-type planets may be disrupted by a gas giant in their midst; we would owe our existence to the possibly fortuitous circumstance that only two cores in our own system, Jupiter and Saturn, grew large enough to accrete gas'.[46]

This is not to say that we should take too Earth-bound a view of where life may be allowed to evolve. In some quarters hope seems to spring eternal. Thus, no sooner had the extra-solar super-Jupiter planets been discovered than the possibility of life inhabiting smaller moons circling these monsters was raised.[47] The wish-list for such a world is interesting. Apart from the problem of whether giant planets can retain any moons during their inward migration, the desiderata include a body large enough to retain an atmosphere, an effective magnetic field if embedded in a powerful magnetosphere (as possessed by Jupiter), and association with a giant planet that does not have an eccentric orbit around the star. There is another interesting thought connected to the possibility that before inward migrations the moon might have been very well endowed with volatiles if it accreted near the 'snow-line'. These volatiles might serve to form a global ocean when moved closer to the star. So life is a possibility in such a regime, although this idea is not free of difficulties.[48] In part, the inspiration for this proposal was news from Europa, which rivals volcanic Io as the most remarkable of Jupiter's satellites. Images from the spacecraft *Galileo* revealed an icy crust, with clear evidence of dynamic motion.

Geophysical measurements indicate that beneath the ice, which may be more than a hundred kilometres in thickness, there is probably an ocean,[49] and it is speculated that life may have evolved within its depths. So, if the extra-solar planets are generally uninhabitable, intelligent life might conceivably emerge on such moons, with their counterparts to Kepler and Galileo struggling to make sense of planetary dynamics on a moon slung beneath the huge orb of their parent planet.

RARE MOON

At present the search for extraterrestrial life would receive immense encouragement if a solar system at least similar to ours were to be detected. Building a broad facsimile, however, may only be a first step, because on closer examination our Solar System possesses a series of apparent oddities (note 1) that may play crucial roles in the emergence of life. Here our Moon, distant Jupiter, and the rare spectacle of a comet all indicate that once again the Solar System may represent a very special arrangement, and Earth an equally special abode.

Let us return to our most familiar planetary companion, the Moon. As we have seen, it is scarred with a record of the early and cataclysmic history of the Solar System, mutely reminding us of the cataclysmic beginnings to the Earth's history and the precarious, and perhaps repeatedly frustrated, hold that life then had. Although there have been a number of theories as to how the Moon came to be associated with the Earth, it is now widely agreed that our satellite was itself a product of a violent impact between the newly formed Earth and another body, approximately the size of Mars.[50] The reasons for thinking that the Moon is a genuine daughter rather than a chance capture are various; they include certain geochemical coincidences and the nature of the dynamical relationship between the two bodies. The Moon itself, however, is very old; it records the early bombardment episodes, and rocks collected by the *Apollo* missions date back close to the dawn of the Solar System.[51] Given that the Earth had formed by about 4500 Ma ago, perhaps 100 Ma after the initiation of the Solar System,[52] then the mega-impact that led to the Moon could have been only a little later.[53] Some evidence suggests that at this early stage the Earth may have been somewhat smaller than at present, and that the impact added substantially to its mass.[54] Alternatively, the collision could have occurred somewhat later as the Earth approached its existing size.[55]

The collision must have been truly cataclysmic, and has been analysed by the application of some remarkable computer models. These give a glimpse of just how catastrophic the event must have been, but more importantly they reveal that making the Moon may be far from a straightforward process. Not surprisingly, modelling this impact is computationally intensive, and delineating the exact nature of the collision depends on a number of variables that in turn are difficult to constrain. These include the mass of the collider, its composition (for example, how iron-rich it was), the amount of angular momentum available, and whether the collision was glancing or head-on. In one such simulation[56] the impactor plunges through the mantle of the proto-Earth, settling on the core. As it does so the massive release of energy ejects the existing atmosphere and replaces it with one made of vaporized rock. As these investigators dryly note the hemisphere which received the impact was 'devastated', while the opposite hemisphere fared little better when a wall of molten rock flowed across it. Other simulations show the early Earth deforming as the result of the impact, like a high-speed photograph of a bouncing tennis ball.

A key result of this giant impact, however, was the detachment of sufficient material to form the Moon. It is here that we may begin to run into difficulties. The simulations tend to suggest that building a single large Moon from the material blasted off by the giant impact is difficult.[57] The alternatives might be either a series of small moonlets or even a ring of material, the latter analogous to those encircling Saturn (and also, as recently discovered, Jupiter, Neptune, and Uranus), being more likely to end up orbiting the devastated Earth. At first sight a ring of debris around the Earth would seem a promising start towards the aggregation of a larger moon. Some of the models are not, however, very encouraging: either the single body is too small or the disc inconveniently aggregates into several moons.[58] This simulation of the possible methods of coalescence of impact debris into larger bodies is necessarily simplistic and looks at the behaviour of only 37 'particles', each with a radius of approximately 500 km. Even this was computationally extremely demanding, but, as the investigators remark, they 'identify only a few scenarios in which a single Moon results'.[59] Further analyses[60] are somewhat more optimistic, although a recurrent difficulty is that a single large body tends to form near to the Earth, close to the Roche limit, the point at which the gravitational field of the Earth would tear the new Moon to pieces. Other simulations tended to form several smaller moons, and the conditions necessary to form our Moon seem to remain very sensitive to starting conditions. The problem of propelling a large mass of material clear of the Roche limit is also stressed by A. G. W. Cameron (see note 57). One way to achieve this is a collision with high angular momentum. The question then arises as to how this angular momentum was subsequently lost. It is important to stress that increasingly sophisticated simulations are serving to constrain some of the boundary conditions from which the Moon could form. Various sensitivities still remain, and the work done so far suggests that a wide variety of possible end-points different from the actual configuration are equally feasible. A comprehensive simulation may explain much of the history of the actual impact, but apparently minor variations could have led to a profoundly different end-point.

As Jack Lissauer has remarked, 'It's not easy to make the Moon'.[61] This is brought into sharp focus by some earlier remarks by A. G. W. Cameron and W. Benz.[62] Thus, concerning the origin of the Moon they wrote:

> We deal with a unique event. In the absence of the Giant Impact, it seems plausible that the Earth would have developed with a massive atmosphere comparable to that possessed by Venus. Just as in the case of Venus, it seems unlikely that life would have developed in such an environment. Thus, there may be a very large selection effect which means that reasoning creatures will speculate about the probability of a Giant Impact only on Earth-like planets which have suffered a Giant Impact and lost most of their primordial atmosphere.

They continue, 'This issue lies well beyond the present reach of science.'[63]

Although not everyone would agree that the nature of the Earth's atmosphere was critically controlled by the Giant Impact, it still remains the case that even given such an immense collision, forming the Moon was difficult, possibly extraordinarily difficult. Yet, without our daughter satellite this world would certainly be a very different place, and arguably much less benign.[64] One advantage of having a Moon is that it confers considerable stability to the axis of rotation of the Earth, making it difficult to shift it appreciably from the existing inclination of about 23°.[65] To be sure there are controversial hypotheses proposing that in the geological past there have been major shifts in obliquity, with the Earth falling in effect on its side, to explain climatic anomalies such as glaciations in equatorial regions.[66] The notion of profound changes in obliquity has not been well received, but there is little doubt that without the Moon the Earth would have been just as subject to chaotic shifts in obliquity as have occurred in the other planets.[67] At the least such shifts in obliquity would have had major, possibly catastrophic, climatic consequences (Fig. 5.6).

But having the Moon has other advantages. Neil Comins,[68] for example, reminds us that the coupling between the Earth and its large Moon leads to the transfer of angular momentum from the larger to the smaller body, because of such factors as tidal friction. This has the result during geological time of both increasing the separation between the Earth and Moon and, more importantly, decreasing the Earth's rate of rotation. The most obvious result of this slowing in rotation is an increasing day length and a corresponding smaller number of days in the year, for which the fossil record shows some confirmatory evidence.[69] If, however, the rate of rotation had not decreased, then Comins suggests, by analogy with other rapidly rotating planets, that the wind speeds on the Earth would be persistently higher, with many more violent and protracted storms. A planet with life, perhaps, but with winds tearing across the world at an average speed of 200 km per hour it would be an uncomfortable place to exist. Marine life would face continuously stormy seas, while on the continents vegetation would be tough and ground-hugging and most animals would perhaps be burrowers.

Having the Moon therefore seems to confer significant advantages. Without it life would presumably still be possible, but it might be more stressed and probably very different; and the emergence of intelligence (Chapter 9) might well be compromised. Early in the history of the Solar System other giant impacts on Mars and Venus might have spalled off material to make moons. The disproportionately large size of the metallic core of Mercury has been used as an argument for another giant impact that smashed away most of that planet's outer layer.[70] Some scientists[71] have consequently tended to argue that the Earth's Moon is not so exceptional: perhaps, they say, Venus lost any satellites it had by the action of solar tides, while Mars was too small and gravitationally weak to retain any such moons.[72] The evidence

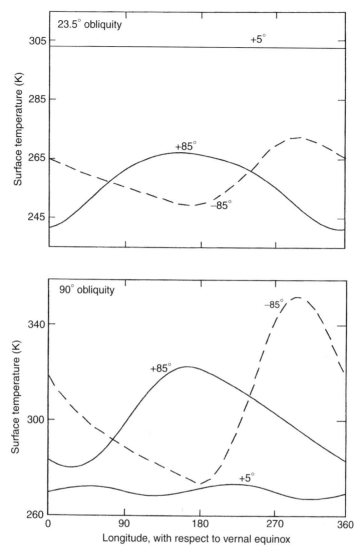

FIGURE 5.6 A model of the climatic effects of an Earth without a Moon, so allowing massive changes in obliquity, perhaps from the existing and stable value of 23.5° to 90°. Shown here are the respective temperature cycles for near-equatorial (5°) and near-polar (85°) latitudes. Solid lines are for northern latitudes; broken lines represent southern latitudes. Note the contrast between the relatively benign seasonal swings of our seasons, as against the wild oscillations in an Earth with an obliquity of 90°. In particular, Antarctica would vary in a year from freezing to a temperature of 350 K (equivalent to 80 °C), beyond the lethal limit of most organisms. [Figure from 'Earth–Moon interactions: Implications for terrestrial climate and life' (fig. 4) by D. M. Williams and D. Pollard in *Origin of the Earth and Moon*, edited by R. M. Canup and K. Richter, © 2000 The Arizona Board of Regents. Reprinted by permission of the University of Arizona Press and with the permission of the authors.]

from the modelling of the Giant Impact that is thought to have produced the Moon, however, suggests that the impactor may have been as large as Mars, and possibly substantially bigger. Even in the early history of the Solar System, such a collision would have been very rare, and as we saw above even size alone may not guarantee a Moon. Also, it may well be that had this Giant Impact not occurred then the Earth would have been a substantially smaller body. Is that so important? It turns out that it is.

JUST THE RIGHT SIZE

In a fascinating discussion of alternative scenarios Jack Lissauer[73] invites us to consider Earth-like planets, but somewhat smaller and larger than our familiar home (Fig. 5.7). In neither case is life necessarily excluded, but if it were present the biosphere would probably be impoverished as compared with what we enjoy, or at least should enjoy. On a smaller planet gravity is weaker, the consequences of which are much more precipitous mountains, a thinner atmosphere, and substantially lower surface temperatures. The internal heat budget of this smaller planet would be reduced, and its more rigid crust might rule out plate tectonics. There would probably not be large oceans, and possibly only shallow seas. On this rather undynamic planet, life might be constrained, not least because levels of atmospheric oxygen would be substantially lower than on Earth.

What of a planet substantially larger than the Earth? It would have stronger gravity to be sure, leading to a much more subdued topography and probably a globe mostly, if not entirely, covered in oceans. As we saw earlier, such a possibility was also raised for an excessively water-rich moon that might orbit an extra-solar giant planet. While aquatic life might develop in either situation it would be a problematic habitat because without continents there would be a greatly reduced influx of nutrients and minerals derived from weathering, although volcanic and hydrothermal activity might provide some compensation. Even if intelligent life, perhaps dolphin-like, emerged, the transition to a technology based on metallurgy and controlled combustion would probably be impossible. So, if we wish to meet conceptualizing beings we might concentrate on a planet pretty much the size of the Earth with a large Moon. It would be a further curious coincidence if the Moon were of a size and at a distance to eclipse exactly the nearby sun during the time a sentient species was becoming

FIGURE 5.7 A comparison of Earth-like planets, with our Earth in the middle flanked by smaller (left) and larger (right) versions. Smaller means drier and precipitous landscape, while bigger means an ocean-world with little, if any, emergent land. (Reproduced from fig. 13.9 of *Planetary sciences*, by I. de Pater and J. J. Lissauer (Cambridge University Press, Cambridge, 2001) with the permission of the authors and Cambridge University Press.)

interested in astronomy.[74] Perhaps that is not asking too much – until, that is, we look at the Solar System's goalkeeper, mighty Jupiter.

JUPITER AND THE COMETS

Notwithstanding the dramatic images of its turbulent cloud belts, transmitted back to Earth by the *Galileo* spacecraft, to most of us Jupiter remains remote and very far removed from terrestrial circumstances. Probably at about the time Hitler was climbing to power in the 1930s Jupiter happened to capture a comet. This was Shoemaker–Levy 9, now famous because on one of its approaches to Jupiter, it disintegrated into a string of fragments that ultimately fell on the planet in July 1994.[75] Dramatic pictures revealed the violence of the

impact, but still in a world far away. Yet this capture and destruction actually appear to have a direct relevance to our own existence.
As George Wetherill[76] reminds us, the powerful gravitational 'well'
provided by Jupiter serves to deflect and capture innumerable comets
that would otherwise continue to hurtle towards the inner Solar System. Not only that, but other estimates[77] suggest that about half the
comets entering the planetary Solar System from the more remote
Oort Clouds are then flung into interstellar space by the slingshot
effect of Jupiter's gravity. The importance of this observation will be
returned to below. This still leaves plenty of comets roaming around
the Solar System in a variety of orbits, but Wetherill estimates that
without the shielding influence of Jupiter the comet flux to the Earth
would be about a thousand times greater. That would be bad news for
life on Earth and, as Wetherill observes, 'this could frustrate the evolution of organisms that observe and seek to understand their planetary
system.'[78]

The danger of impacts has been brought into sharp focus, of
course, by the famous end-Cretaceous extinctions, the so-called K/T
event.[79] Given the protracted time that ecosystems require to recover
from such disasters, typically a couple of million years,[80] it is difficult
to believe that a repeatedly traumatized biosphere, suffering recurrent
and frequent impacts, would be similar to that of our own Earth. More
than a hint of this grim conclusion comes not only from the mayhem
associated with the extinctions, but also from some rather startling
evidence concerning the aftermath of perhaps the two greatest catastrophes, specifically the end-Permian (c. 250 Ma) and end-Cretaceous
(c. 65 Ma) events. Here the ecological disruption seems to have been
so severe that it almost seems as if the evolutionary clock had been
wound back to a time when there were no animals; a 'return' to Precambrian ecosystems dominated by microbial activities, with bacteria
and various mostly single-celled eukaryotes. Such a world, before animals literally imprinted themselves on the biosphere, was evidently
very different from the one familiar to us. Thus, following the Permian
debacle there is some intriguing evidence for the wide-scale return
both of stromatolites[81] and other microbially bound sediments.[82]

Analogous evidence for profound ecological disruption after the
end-Cretaceous catastrophe comes from a study of the sedimentary
carbon. The aftermath of this catastrophe has been something of
a paradox because although the marine algae, and thereby primary

productivity in the oceans, recover fairly quickly as a whole, the marine ecosystems remain in a state of protracted crisis for much longer. The resolution of this problem appears to be the continued near-absence of those animals that can graze or otherwise harvest this plankton.[83] The net result of an absence of such grazers is readily detectable because of a distinctive change in the geochemical imprint of the organic carbon. The recycling of the carbon is then effectively dependent on bacterial metabolisms. This indicates a state of affairs in the post-catastrophe ocean that is reminiscent of the Precambrian oceans,[84] a time, of course, before grazers had evolved. These various lines of evidence, indicating how sluggish were the responses of post-catastrophe biotas, thus suggest that on a comparable planet with greatly elevated rates of extraterrestrial bombardment the ecosystems would remain in a state of near-permanent turmoil. Even if a sentient and cultural species emerged, the fragility of its civilization and the delicate nature of the associated infrastructures would be such that frequent major collisions would be, to put it mildly, challenging.[85]

Paradoxically, the occasional giant impact may be beneficial, at least in the longer term. First, it needs to be remembered that even if Jupiter acts as the 'goal-keeper' so far as comets are concerned, it is indirectly responsible for steering meteorites towards the Earth. This is because the great majority of these rocks are derived from the so-called asteroid belt, which in effect represents a 'failed' planet, the coalescence of which was prevented by the massive presence of Jupiter. Moreover, it is the gravitational resonances set up by the planets, especially, as it happens, the one connected with Saturn, that are responsible for perturbing the asteroid orbits so that there is a continuous supply of bodies directed towards the inner Solar System. Their time there is relatively short-lived, and most will ultimately plunge into the Sun or be captured by Jupiter, but before that there is the ever-present possibility that one of them may intersect the path of the Earth.[86] How then might such impacts be beneficial? In principle, by wiping the biological slate largely clean, so that new groups are given fresh ecological opportunities. The example most usually quoted is the demise of the dinosaurs following the end-Cretaceous (K/T) event, which made possible the rapid radiation of the mammals. No bolide, the argument runs, no adaptive burst among the mammals and so ultimately no humans. This is an apparently cogent supposition, especially if your world-view is that history is driven as much by accident and blunder as by any inherent tendencies that in a biological realm might be driven by adaptation and competition.[87] But does the demise of the dinosaurs as a portal to mammalian success really pass muster? Possibly not, and for two reasons, one particular and one general, which are returned to in more detail below. First, unbeknown to the survivors of the K/T catastrophe, in due course the world was to head into a time of global refrigeration, culminating in the present-day ice ages. With the dramatic cooling of the temperate and high-altitude zones, it is likely that the warm-blooded groups, notably the mammals as well as the birds, would have been at a selective advantage. Thus it seems likely that the mammalian radiations would have been delayed, rather than never happening; in other words, postponed rather than cancelled. The second and more general reason is the main theme of this book, that of evolutionary convergence. As all the principal biological properties that characterize humans are convergent, then sooner or later, and we still have a billion years of terrestrial viability in prospect,[88] 'we' as a biological property will emerge.

Barring a daring technology in the future, human civilization will be hard-pressed to destroy or deflect incoming bolides and so avoid impact-driven catastrophes. Perhaps a solar system unencumbered with roaming comets and asteroids would be a safer place? Possibly so, but paradoxically it is arguable that without these bolides there might not have been any terrestrial observers to admire the beauty of a passing comet, let alone know the fear that one day an asteroid might repeat the K/T catastrophe. The reason for this has already been touched on earlier (Chapter 3, pp. 42–43). It is only necessary to remind ourselves that very early in the history of a solar system, as the star begins to shine and the volatiles, such as water, are driven outwards to the 'snowline' (approximately where Jupiter presently orbits), the inner rocky planets are in serious danger of remaining dry and, in the absence of water, presumably sterile. It is generally thought that the fact that Earth has an ocean is due to the delivery of large quantities of volatiles early in the history of the planet. (Mars and Venus probably originally also had oceans, but for different reasons things there went badly wrong.) It is in this context that once again the presence of a Jupiter may become very important for the future habitability of the Earth. This is because early in the history of the Solar System the volatiles the Earth would ultimately 'need' for such things as dolphins, ants, and mystics, were far removed, out at the 'snow-line'. This, as mentioned, is adjacent to the orbital position of Jupiter and it was Jupiter that was in a position to displace the water-rich asteroids[89] back towards the inner Solar System. If, however, Jupiter had formed somewhat inward of its present position, then the Earth might have remained sterile and dry. This is because Jupiter would have still efficiently scattered the volatile-rich asteroids, but in the wrong direction, outwards and away from the inner rocky planets.[90]

But what of the comets? Given that they appear to be largely composed of ice, would they not be an ideal source of the Earth's volatiles?[91] So far as water is concerned, however, it seems that the comets may have made only a minor contribution. The reason for thinking this is that the ratio in water between the isotope of 'heavy' hydrogen, known as deuterium, and the 'light' hydrogen is, in the few comets yet examined, very different from that of the Earth's oceans. Hence the preferred source from the asteroids (note 89). This does not mean, however, that comets played no part in ensuring the habitability of the Earth; far from it. The principal source of the organic compounds, the raw material for the origin of life, was probably delivered to the Earth at a late stage of its accretion, by a cometary bombardment.

So, if Jupiter had been in the wrong position, perhaps there would have been no oceans, and perhaps if the comets had not arrived, no life? In this context it is therefore intriguing to learn that cometary systems around other stars may be less common than might be expected, and perhaps even rare. There are several reasons for thinking this.[92] In the case of our Solar System the estimated population of comets runs into the trillions, perhaps about 10^{13}. Broadly these fall into two categories: those located somewhat beyond the orbit of Neptune (the Kuiper Belt); and much further out a vast number, occupying an immense volume of space, far beyond the planets, in what is known as the Oort Cloud (see note 77). It is thought that the Kuiper Belt and Oort Cloud supply respectively the so-called short- and long-term comets. So what does this have to do with comets from other solar systems? The explanation is that on the edges of our Solar System the gravitational pull of the Sun is so weak that any nearby disturbance, perhaps a passing star (not in itself an unlikely occurrence given that the outer Oort Cloud extends about half-way to the nearest star), will exert its own gravitational field and so disrupt the position of many comets. Some will be displaced to fall inwards towards the planets and the Sun, but others will be hurled outwards into deep space. Now the curious thing to note is that it appears that all the comets that have been observed by astronomers derive from either the Kuiper Belt or the Oort Cloud. Not one seems to be a visitor displaced from an extra-solar system. This is slightly puzzling, because calculations suggest that if equivalents to the Oort Cloud occur elsewhere then there should have been a small number of such cometary visitors to our Solar System. These interlopers would be recognized, because they would swing in on hyperbolic orbits rather than the customary tighter parabolic trajectories.[93]

To be sure, there is some evidence of what may be analogous to cometary infall occurring in the giant disc around β Pictoris.[94] These bodies are inferred from variable and sporadic observation of spectral absorption lines, especially those of magnesium, calcium, and aluminium. They also appear to be larger than the Solar System comets, and therefore might be better classified as infalling planetismals that evaporate as they approach the star. In a comparable way extra-solar equivalents of the Oort Cloud might be detected by evidence of vaporization of engulfed comets as an ageing star swells towards its red giant stage.[95] Preliminary searches have not been encouraging,[96] but recently the presence of Kuiper-like comets has been inferred around an ancient star whose increasing luminosity is evidently vaporizing the icy comets, which is revealed by the detection of the resulting water vapour.[97] The comets of the equivalent to the Oort Cloud, if it exists, are too remote from this star to suffer the same fate, and at the moment the number of extra-solar systems with a large reservoir of volatile-rich comets (or similar bodies) is necessarily conjectural.

It would certainly be surprising if no equivalents of a cometary Oort Cloud existed elsewhere.[98] The possibility, however, is that even if extra-solar systems are abundant the differences from our Solar System may include not only the planetary configurations, but also an absence of volatile-rich bodies. Even if such bodies exist, with their latent reservoirs of water and other light elements, there still needs to be a method of delivering them to any suitable planet where life can arise. But in one sense that is only the beginning of our problems. Life may conceivably arise in many Solar Systems but then, because of one exigency or another, it is snuffed out. The evolution of intelligence may be an inevitability, but on Earth at least this process has taken billions of years. For this to happen demands the persistence of a zone of safety, that is a Circumstellar Habitable Zone (CHZ). To us humans, of course, the concept of a habitable zone is exceedingly narrow: without technology we can neither fly nor burrow, and swimming beyond very modest depths requires elaborate paraphernalia. Even the biosphere, from the deepest bacteria to the migrating storks drifting past turbo-props at 20 000 feet (6000 metres), occupies an exceedingly thin skin, wedged between the intense heat and pressure of the Earth's interior and the sterile vacuum of outer space. Yet despite these flanking hostilities beneath our feet and above our heads, the Earth itself occupies a zone of comfort, a planetary Habitable Zone, and it is to this topic we now turn.

JUST THE RIGHT PLACE

What we now know of the infernos of Venus and the super-arctic wastes of Mars is a clear indication that being either too near or too far from the Sun may have perilous consequences. The geological record of the Earth, on the other hand, has shown that whatever the vicissitudes of the surface environment, perhaps most spectacularly manifested in terms of the episodes of major glaciations ('snowball' earths[99]) and super-torrid equatorial regions,[100] the surface temperature has been remarkably constant. Whether or not this is a reflection of self-regulating homeostasis mediated by biotically driven feedback (as most influentially promoted by James Lovelock[101]) is less material to this chapter than the wider question that needs to be posed: how likely are planets to find themselves in a zone which ensures long-term habitability?

The question of how to define the Circumstellar Habitable Zone (CHZ), how wide (or narrow) it might be, and how it might evolve, was addressed some years ago by Michael Hart.[102] The fate of our nearest planetary neighbours is a clear enough indication that this habitable zone must be relatively narrow (somewhere between the orbits of Mars and Venus), and most calculations suggest that it is substantially narrower. The principal problem in arriving at precise parameters is that the Sun's luminosity is increasing steadily, but various feedback processes on the planet, such as those linked to the amount of atmospheric carbon dioxide and water vapour (both greenhouse gases), mean that the calculations are not simple. Assuming that a planet's orbit is fixed, which may not always be a reliable supposition,[103] the general conclusion has been that the inner boundary of the CHZ will change little, but the outer boundary will extend quite markedly.[104] There are further complications in these estimates that need to be taken into account: the type of star, and hence its brightness; whether it produces a drenching of ultra-violet radiation; its lifespan; and its size. For example, if the star is two-and-a-half times as large as our Sun, then it will burn very quickly, probably too fast for life to get much beyond the first stage.[105]

Nor do the complexities end there, because planets like the Earth are not only dynamic systems but they also change through geological time. The net result of all these factors is a complex set of possibilities, in terms of stellar and terrestrial histories, that mean that both the definition and the evolution of the habitable zone are not easy to constrain and must depend on various assumptions. An accurate estimate, however, will have implications not only for the long-term habitability of our Earth, but also for the likelihood of finding near-equivalents elsewhere in the Galaxy. One recent analysis is of particular interest. Siegfried Franck and his colleagues suggest that not only has the habitable zone contracted significantly in the past billion years, but that in another billion years it will have moved beyond the Earth (and also as it happens will have narrowed significantly) (Fig. 5.8), so that higher life at least will have disappeared under the increasingly intolerable temperatures (Fig. 5.9).[106] Although the long-term prognosis for the Earth is not good, indeed Franck and his colleagues simply remark, 'In about 500 Myr [500 Ma] – the biosphere ceases to exist',[107] their calculations of both the evolution of the habitable zone and the number of Earth-like planets are used to suggest that in our Galaxy the total of such planets might be 500 000 (give or take). How many are suitable for the emergence of life, let alone civilizations, is much more of an imponderable. The strong hint, on the basis of what I have already said, is that of the half-million, 499 999 are not suitable.

A further indication that this assessment, which might be regarded as vaguely discouraging, might be realistic comes from a rather different line of evidence. This indicates that even if earth-like planets are relatively common, most of them will be significantly older than the Earth, on average by a rather remarkable 1.8 Ga.[108] Plenty of time for civilizations to evolve and develop a taste for interstellar

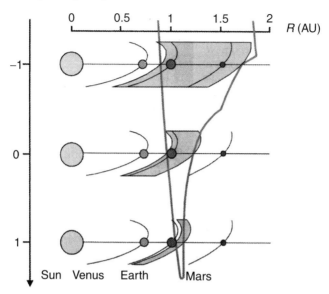

FIGURE 5.8 Evolution of the circumstellar habitable zone (CHZ; shaded) from 1 Ga (1 billion years) ago, through today to 1 Ga in the future, relative to the orbits (distance in Astronomical Units (AU)) of Earth, Venus, and Mars. Note that 1 Ga ago Mars was well within the HZ, but 1 Ga in the future the CHZ has shrunk dramatically and the Earth's oceans are boiling. (Reproduced with permission from *Naturwissenschaften*, from the article Planetary habitability: is Earth commonplace in the Milky Way?, by S. Franck, A. Block, W. van Bloh, C. Bounama, I. Garrido, and H.-J. Schellenhuber, vol. 88, pp. 416–426, fig. 5; 2001, copyright Springer-Verlag, and also with the permission of the authors.)

travel? The argument for this conclusion is important because it has an immediate bearing on the detection of the extra-solar planets and the so-called 'hot Jupiters'. The key observation is that the stars that are known to possess planetary systems are what astronomers call metal-rich.[109] By this it is meant that they contain elements heavier than helium, which is the first product of the thermonuclear burning of hydrogen within the stellar interior. Metallicity, in an astronomical context, is, therefore, a wider concept than the more familiar notion of metals such as nickel and iron. The fact that metal-rich stars also possess planets is probably no coincidence, and such elements as silicon, oxygen, and aluminium are essential, not only for the rocky planets like the Earth, but also for the cores of the gas- and ice

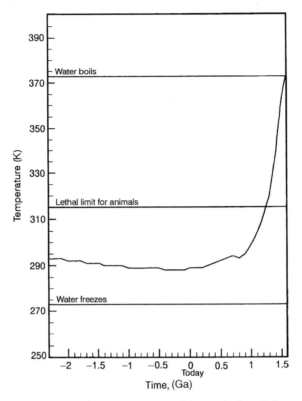

FIGURE 5.9 Surface temperatures of the Earth, from 2 Ga ago and looking forward 1.5 Ga; from a billion years on things are not looking so good. (Reprinted from *Chemical Geology*, vol. 159, S. Franck, K. Kossacki, and C. Bounama, 'Modelling the global carbon cycle for the past and future evolution of the earth system', pp. 305–317, fig. 9, copyright 1999, with permission of Elsevier Science.)

giants. The converse is also true: stars of low metallicity will generally lack planets.[110] It also seems likely, however, that stars with a proportionally higher metallicity are more likely to be accompanied by 'hot Jupiters'. But, as we have already seen, as such a giant migrates inwards from its original orbit to position very close to its star so it will either destroy or eject any earth-like planets. There is therefore a fine balance in a solar system between having a star with sufficient metallicity to ensure the existence of earth-like planets and having a star with too high a metallicity that will lead to the spawning of destructive 'hot Jupiters'. These observations can then be combined with calculations of the rate of star formation (and thus, in suitable circumstances, the formation of planets) and the assumption of a steady build-up of metallicity. This last assumption is based on the fact that the metals (in the astronomers' sense) are synthesized within the star, and may subsequently be blasted into interstellar space during a supernova (where more metals are made). Remnants of such supernovae eventually reaggregate as new stars, which will tend to be more metal-rich. Hence the overall conclusion is that, given these trends, the majority of planets similar to the Earth will have been formed at a substantially earlier stage. Plenty of time not only for life, but also for intelligence to evolve? So where are these planets?

The irony that the destructive cataclysm of a supernova is a necessary prerequisite for life, because only thus can the elements essential to biology be broadcast into space and ultimately gathered together in a new solar system, has often been remarked upon. Yet any biosphere unlucky enough to be relatively close to a neighbouring star that does explode will face a traumatic interval. Nor is this the only risk. Sudden and extremely energetic bursts of gamma-ray radiation are regularly detected by Earth-based instruments, although their origin is still rather mysterious. Perhaps they result from the collision of two neutron stars. Such events release some 100 000 times more energy than a supernova, a colossal 10^{53} ergs,[111] so placing biospheres thousands of light years away at risk. Now it is curious that calculations by Ray Norris[112] suggest that such events should afflict the Earth roughly every 200 million years, yet here we are. Even if this risk factor (as well as that of more proximal supernovae) is decreased it does not easily explain the apparent immunity of our biosphere. One explanation (if that is the word) is that if 'we have already survived for some 20 times the mean interval between catastrophes ... [then we are] very lucky indeed.'[113] If that is correct then one conclusion might be that 'we are alone in the galaxy.' James Annis,[114] however, offers an interesting twist to this prospect, suggesting that while these gamma-ray bursts would reset the evolutionary clocks across much of any galaxy that experienced such an intense pulse of radiation, over time the frequency of these catastrophic events would decline, and ultimately the emergence of intelligence would become possible. On Earth the fossil record indicates that this process took about 10^8 years. What applies here should, on the Copernican principle of mediocrity (that is, there is nothing special about the Earth) apply elsewhere. Thus Annis suggests that as the Galaxy becomes relatively safe, so

'A previously forbidden configuration is now allowed. It is likely that intelligent life has recently sprouted up at many places in the Galaxy and that at least a few are busily engaged in spreading. In another 10^8 years, a new equilibrium state will emerge, where the galaxy is completely filled with intelligent life.'[115]

There is another turn to the story of habitable zones that might also narrow the likelihood of finding a real counterpart to Earth. This is not in the context of distance from a star, but within the galaxy as a whole. Guillermo Gonzalez and colleagues suggest that galaxies like our Milky Way also have relatively restricted zones of habitability.[116] There are several reasons for this that range from the potentially violent nature of the galactic centre with its great density of stars, some of which will inevitably explode, not to mention the resident black hole, to possibly more frequent perturbations affecting the equivalents of the cometary Oort Cloud, so greatly enhancing the bombardment rates of the inner planets. In addition, as we have already seen, the relative metallicity of a star is correlated to the likelihood of its possessing both planets and destructive 'hot Jupiters'. Metallicity of stars, however, declines away from the galactic centre, and this also helps to define, at least approximately, the habitable zone of a galaxy.

That certainly still leaves millions of candidate stars, but, as Gonzalez and his colleagues argue, this leaves the Sun (and thereby the Earth) occupying 'an especially comfortable region of the Milky Way'. Recalling the paradox of the Fermi question ('Where are they?'), they continue, 'Any civilization seeking a new world would, no doubt, place our solar system on their home-shopping list.'[117] Not only are we probably alone in our Galaxy, but looking further afield may suggest that any neighbours are very, very remote. As Gonzalez and his colleagues also remark, 'The broader universe looks even less inviting than our galaxy. About 80 per cent of stars in the local universe reside in galaxies that are less luminous than the Milky Way. Because the average metallicity of a galaxy correlates with its luminosity, entire galaxies could be deficient in Earth-size planets. Another effect concerns the dynamics of stars in a galaxy. Like bees flying around a hive, stars in elliptical galaxies have randomized orbits and are therefore more likely to frequent their more dangerous central regions. In many ways, the Milky Way is unusually hospitable: a disk galaxy with orderly orbits, comparatively little dangerous activity, and plenty of metals.'[118]

A COSMIC FLUKE?

Earlier I remarked on the almost gleeful abasement of humans, not least to inform us that we are insignificant worms in the cosmic drama. One powerful ingredient of this dreary world picture is the Copernican triumph, of Earth the Insignificant. Perhaps so, but it could be that our planet and its Solar System are both very much odder than is realized. Life may be a universal principle, but we can still be alone. Suppose, at least for the sake of the argument, if not humility, we really are. Suppose also that the Earth is genuinely a cosmic accident, a chance fluke arising from spinning clouds of dust and gas. Such a view is now widely accepted, but so, too, are such principles applied with equal conviction to the history of life. From its starting point, however and wherever that strange event might have been, orthodoxy states that evolution can potentially explore a million different trajectories. Even if somewhere there is a planet like the Earth, so the argument continues, there may well be life but assuredly no biped writing lines similar to these. In the next few chapters I shall try to persuade you to take another view.

6   Converging on the extreme

One of the boasts of the physicist is that if she is standing on an Earth-like planet on another galaxy, then the apple she tosses to her male colleague will follow a trajectory whose course may be followed by precise equations. The biologist, however, could not even begin to address the probability of there being sexes, let alone hands and apples (not to mention serpents) on other worlds.[1] That is certainly the widely accepted view: rewind the tape of life, as S. J. Gould repeatedly claimed, and let it replay: assuredly next time round the world will be a very different place, with a vanishingly small prospect of anything like a human emerging. I have already argued forcibly against such a position,[2] and the purpose of much of the rest of this book is to develop in more detail why the trajectories of evolution are much more severely constrained than is sometimes supposed. Nevertheless, this is not the popular view. The present consensus is that, first, this world is only one of many similar ones, although as we have already seen (Chapter 5) such optimism is open to some doubt. Second, and more significantly, it is widely agreed that notwithstanding the unremitting processes of biological diversification even the most trivial differences in the starting conditions would have led to entirely different evolutionary histories, each with a radically different outcome. Nothing like a butterfly, a daisy, or a dolphin and certainly, as we are repeatedly informed, nothing like a human. Not only do these alternative histories debar the emergence elsewhere of anything that is familiar from this world, but even, it is argued, of more general biological properties. Thus, it has been repeatedly proposed that even intelligence itself may be just one more chance end-product of the innumerable possibilities thrown up by evolution. In due course I shall attempt to show the exact opposite.

UNIVERSAL CHLOROPHYLL?
At first sight the problem of addressing the question of evolutionary alternatives seems almost intractable. How can we begin to compare something of which we know nothing with that with which we are

familiar? Self-evidently, but also trivially, the history of life is unique,[3] and perhaps the absence of extraterrestrial communication[4] is simply because intelligence is just another evolutionary quirk. Yet constraints on the alternative possibilities and trajectories through time, at least in evolution, can be demonstrated by at least three avenues. These are respectively: (a) to make a genuine attempt to consider the alternatives, (b) to devise experiments that at least in a restricted way rerun evolution, and (c) to look again at the history of life and enquire whether the ubiquity of convergence, be it in the anatomy, behaviour, or molecules, is anything more than a biological curiosity.

In the first chapter I mentioned the very surprising degree of efficiency of the genetic code. In principle the combinatorial immensity of the 'code hyperspace' means that there may be many other equally effective genetic codes, but as also explained one can impose a number of additional restrictions that suggest that the code adopted on Earth really is not just adequate, but quite remarkably good. There are, moreover, a number of other ways in which we can consider biological alternatives. An appropriate place to start is at the base of the planet's life-support system: that is, photosynthesis, and specifically the molecule that serves to trap the energy of the sunlight: chlorophyll. Chlorophyll must be a very ancient invention,[5] because it occurs not only in the familiar plants but also in various bacteria. Of these, one group, usually referred to as the cyanobacteria (or sometimes blue-green algae), is of particular note. This is because, unlike the other photosynthetic bacteria, the oxygen it produces, as a by-product of the photosynthetic process, was (and, of course, still is) released into the surrounding water and thence to the atmosphere. Green land plants and eukaryotic algae, the latter most familiar as the various sorts of seaweed, also release free oxygen. This is, however, a consequence of the photosynthetic organelles within the cell, known as the chloroplasts. They were once free-living gram-positive bacteria, similar to the living cyanobacteria, but which merged symbiotically with eukaryotic cells, ultimately to lose all independence.[6] In this way, billions of years ago, the atmosphere of the Earth, originally free of oxygen, slowly became oxygenated, and as a result increasingly corrosive. Thus, the planet started to rust.[7] The principal episode of oxygenation is usually identified as occurring at about 1.9 Ga ago. This is inferred principally on sedimentary evidence, notably the disappearance of mineral grains susceptible to oxidation, notably those of pyrite and uranium oxide (uraninite), and the development of red beds, whose colour signifies the onset of atmospheric oxygenation.

As oxygen levels climbed, so organisms had to adapt to this toxic and reactive molecule. But the process of oxygenation started much earlier. Just how geologically ancient photosynthesis is likely to be is hinted at in the oldest known well-preserved sedimentary rocks. These are from west Greenland (Chapter 5, p. 72) and as already noted contain carbon. The carbon has, however, been heated by temperature and pressure within the Earth and is now in the form of graphite,[8] but it may still retain a signature of very ancient photosynthesis. To understand how this is possible, one needs first to recall that photosynthesis is a process whereby carbon in the atmosphere, in the form of carbon dioxide, is combined with water and ultimately transmuted to sugars, using sunlight as the source of energy. The carbon itself occurs in the form of two isotopes, one slightly heavier than the other: carbon-13 and carbon-12.[9] The processes of photosynthesis 'prefer' to use the slightly lighter carbon-12, and hence the carbon stored in the plant tissue is slightly enriched in the lighter isotope. Photosynthesis thus imprints an isotopic signature, and provided that the carbon survives in the sedimentary record one can infer the ancient activity of chlorophyll, even if no other evidence for photosynthesis survives.[10] And although altered to graphite, the carbon in the most ancient sediments, dated at about 3.8 Ga, looks as if it, too, might once have passed through the photosynthetic process.

The chemical processes of photosynthesis are very complex. Exactly how the Sun's photons are captured to yield the necessary energy within the photosynthetic 'factory', located within the chloroplasts of the green leaf, is still not completely understood. Despite the apparent miracle of sunlight pouring down on orchards and lagoons to produce apples and sea grass, it is clear that the process of photosynthesis is far from perfect. There are many chemical steps, and if an engineer had been in charge of the design process, the accountants and administrators would by now have been asking awkward questions. In particular, the activity of one key enzyme (known as D-ribulose-1,5-bisphosphate carboxylase, or RuBisCO for short) is severely hampered in the presence of oxygen, which ironically it also produces by catalysing another chemical reaction that accordingly competes with the process of photosynthesis. The net result is that valuable energy is consumed in an unavoidable metabolic process known as photorespiration.[11] It is as if the crew on a sinking ship are ordered into the hold to stem the flood, and are all issued with sieves.

Nor is this the only problem the photosynthesizers face. An essential prerequisite of the process, of course, is carbon dioxide. Suppose, however, that the amount of this gas in the atmosphere were to plummet drastically. This might be due to a number of natural processes, such as episodes of massive mountain-building that expose vast areas of rock and scree where the carbon dioxide can be 'soaked' up by the enhanced rates of rock weathering. Plants would be faced with an extremely serious problem, and it was one that actually started about 15 million years ago. At this time the levels of atmospheric carbon dioxide began to decline precipitously, in part probably because of the uplift of a series of huge mountain belts, most notably the Himalayas. The plants' solution was to modify the steps in the photosynthetic process, leading to a transition from so-called C_3 photosynthesis to a C_4 mode (the numbers refer to the number of carbon atoms in the first compound to be formed). The details of this C_4 photosynthesis are considered in Chapter 10, but what is worth noting here is that its evolution is rampantly convergent.

The processes of photosynthesis are, therefore, hedged in with many constraints, but even so chlorophyll is a remarkable molecule that effectively underpins the entire biosphere. It is thus rather surprising to learn that not only does chlorophyll fall short in such matters as the effectiveness of its RuBisCO, but it also seems to suffer from rather more general difficulties. George Wald, in particular, has pointed out that if one compares the absorption spectra of the various types of chlorophyll to the available visible light spectrum of our Sun the match is, to put it mildly, disappointing (Fig. 6.1). Clearly the chlorophyll has to absorb some sunlight or it simply would not work. In addition, different types of chlorophyll vary somewhat in their absorption spectra: chlorophyll d, for example, shows a rather remarkable shift towards absorption of red light.[12] Accordingly chlorophyll has some latitude in the wavelengths of light it is best adapted to absorb, but it is all the more remarkable that the lion's share of the Sun's radiant energy remains largely untapped. And from this apparent anomaly Wald draws a very interesting inference. He argues that, however desirable a 'perfect' chlorophyll might be, it is simply not attainable: as the initial supplies of prebiotic 'soup' run out, so life must move to the situation where it can synthesize its own food

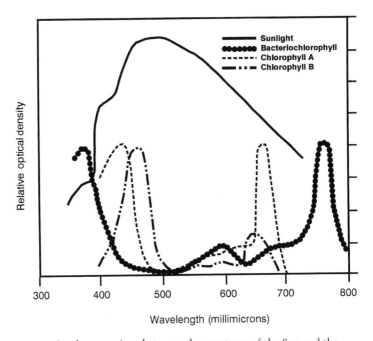

FIGURE 6.1 A comparison between the spectrum of the Sun and the absorption spectra of three types of chlorophyll. Note how the spectra of the chlorophylls differ, but none is able to intercept the maximum output of the Sun. An 'ideal' chlorophyll would mirror the outline of the Sun's spectral profile, leading George Wald to suggest that not only is terrestrial chlorophyll the best available, but anywhere else in the Universe on a planet the same chlorophyll will occur. (Redrawn from figure by Emai Kasai on p. 96 of G. Wald's article 'Life and light', *Scientific American*, vol. 201 (4), pp. 92–108. Copyright © 1959 by Scientific American, Inc. All rights reserved.)

from the light of its sun. Accordingly the necessary machinery, including chlorophyll, must evolve. On this basis Wald draws a bold conclusion, remarking that 'When that time comes, it seems to me likely that the same factors that governed the exclusive choice of the chlorophylls for photosynthesis on the Earth might prove equally compelling elsewhere.'[13]

We have already seen that the chances of encountering habitable planets may be much lower than is generally supposed, but what if we do locate an alien biosphere? Wald's arguments suggest that however many light years we may be from the Earth, if we discover a planet and wander along its remote shores with their entanglements of seaweeds or explore its immense forests, sure enough there will be the same and all-too-familiar chlorophyll soaking up the light of an alien star.[14]

THE WHEELS OF LIFE?

Nor is chlorophyll the only way to look at alternatives in terms of evolutionary possibilities. Another approach is to assess the diversity of life and ask whether there is anything one might reasonably expect to see, but seemingly has failed to evolve. The most popular of these thought-experiments is to ask why no organism with wheels has ever evolved. While this is certainly a biomechanically demanding concept, it does not seem to be intrinsically impossible. The nearest approach in nature to our familiar technology of bicycles and prams seems to be the rotary mechanism associated with the base of the whip-like process or flagellum that serves to propel certain bacteria through a watery medium.[15] This organic motor is built of a series of proteins, and the arrangement of rings and sockets that holds the base of the rotating flagellum is extraordinarily machine-like. The speeds of rotation are astonishingly high: some bacterial flagella turn at more than 1000 revolutions per second.[16] Not surprisingly the complexity of assembly[17] has attracted the attention of so-called 'intelligent de-sign' theorists, who argue that in such cases of irreducible complexity (i.e. remove any component and the mechanism falls to bits) one must conclude that the theory of evolution ultimately fails, so necessitating a teleology controlled by either a designer or Designer.[18] While we should not underestimate the difficulty in explaining how such a flagellar motor might have evolved, everything else we know about evolution indicates that the pathway to construction will involve the twin processes of cobbling together and co-option, with at least some of the proteins being recruited in quite surprising ways from some other function elsewhere in the cell. This is not to deny that the question as to how the flagellar motor was assembled is still unsolved, but in principle its origin should ultimately be no more inscrutable than explaining any other complex organic structure. Interestingly, this gyrating bacterial flagellum is not the only type of molecular motor; almost unbelievably even a part of a single molecule is able to rotate freely.[19] Even so, and fascinating as these molecular motors may be, there seems to be no way in which they could simply be scaled up to provide the equivalent of a wheeled giraffe, or whatever.

The only other way in which at least some animals can loco-mote in this fashion seems to be to turn the entire body into a wheel. Perhaps the most remarkable of such examples are certain shrimps (stomatopods) that can execute up to 40 consecutive rolls,[20] and a pangolin[21] on the island of Siberut, west of Sumatra, which has been observed to roll itself up and then go crashing through the under-growth of a steep slope.[22] Other examples, such as cartwheeling spiders and sea anemones,[23] not to mention the so-called circumro-tatory corals,[24] are interesting in their own right, but scarcely qualify as wheeled organisms. Organic wheels, therefore, seem to be an ex-cluded possibility, and it is not difficult to see why. In a thoughtful study Mike LaBarbera points out that, among other factors, any ani-mal wheels of reasonable size would presuppose effectively flat and continuous surfaces.[25] Hence our need for roads as well as, more lo-cally, ramps for wheelchairs. In the natural world as often as not, and especially on sea floors, this means acres of mud and other soft, sticky, substrates, ideal for getting bogged down. On this planet, and perhaps everywhere, any wheeled organisms are going to be bipeds cycling to the pub.[26]

FORTEAN BLADDERS

Is there any other neglected possibility? R. A. Fortey, for example, has pondered whether any major ecological niche remains untenanted, and suggests that one such niche might exist.[27] Thus he writes that 'High in the atmosphere there is a stream of air which transports in-sects and spiders, like some plankton of the ether. Could an aerial 'whale' have evolved to harvest this stratospheric protein: a light, flying animal with a wide feeding gape, an animal that could cast a shadow across the sky?'[28] Fortey is careful to note that there might be good explanations why such an animal could not exist, but for whatever reason he does not choose to address them.

In principle, such a Fortean bladder,[29] named in honour of that master of intangible aerial phenomena, Charles Fort (see also note 49), should be biologically unproblematic.[30] Perhaps the best analogy is the swim-bladder found in many fish, a gas-filled organ that confers buoyancy.[31] The ever-present problem of leakage of gas is controlled by deposition in the swim-bladder wall of guanine crystals[32] that evidently confer impermeability. In looking at the swim-bladder of the conger eel Eric Denton remarked that removal of the silvery layer, containing the guanine, reduced the permeability by about one hundred times.[33] This arrangement is augmented by an ingenious net-like arrangement of blood vessels (known as the rete) that provides a counter-current system so that the gas dissolving into the blood as it exits from the gas gland of the swim-bladder is transferred across the rete and so returns to the bladder.[34] Nor is a gas-filled bladder unique to the fish; the females of the octopus *Ocythoe* possess an equivalent structure.[35] A somewhat similar arrangement also occurs in the complex 'jellyfish' known as the siphonophores,[36] most familiar perhaps as the Portuguese man-of-war. Here, too, buoyancy is conferred by a gas-filled bladder, with an associated gas gland that releases carbon monoxide.[37] The versatility of these siphonophore bladders is quite remarkable, given that one type, *Thermopalia*,[38] secretes gas despite living in a hydrothermal ridge habitat that lies at an ocean depth of about 2.5 km, where the hydrostatic pressure is equivalent to a crushing 260 kg per cm^2. Mention should also be made of the gas-filled vesicles found in numerous species of bacteria, especially aquatic cyanobacteria, where they evidently confer buoyancy.[39]

The gas employed in the fish swim-bladders is mostly a mixture of oxygen and nitrogen, and is perfectly suitable for conferring buoyancy. In principle, however, the biological generation of the lighter-than-air gas hydrogen should be possible.[40] Hydrogen, after all, is extremely abundant, albeit combined chemically with oxygen in the form of water. It is known, for example, that some microbes[41] (and microbial mats[42]) can generate hydrogen as a side effect of the fixation of nitrogen. More specifically, in trying to assess how realistic a Fortean bladder might be it is worth considering some specialized structures found in the cells of some microbes. These are a type of organelle[43] known as the hydrogenosome which, as the name suggests, generates hydrogen.[44] The process is connected with the generation of the molecule ATP (which is involved with energy metabolism), and entails the reduction of protons (i.e. hydrogen ion, H^+) to form molecular hydrogen. In most other circumstances involving ATP the electron acceptor is oxygen, but the hydrogenosomes are found in organisms living in anoxic conditions, for example, stinking muds or the rumens of cows, where oxygen is excluded. The key enzyme involved with hydrogen production is, unsurprisingly, referred to as hydrogenase.[45] Not only is it central to the operation of hydrogenosomes, but it has also been found in organisms that are normally aerobic. In this instance the enzyme evidently kicks in during times of oxygen crisis.[46]

There seems, therefore, to be no biological reason why a Fortean bladder could not employ hydrogen-producing microbes, recruited by symbiosis, and store them in an area guarded against any influx of destructive oxygen. Such mass packaging of bacterial symbionts is by no means unusual. One can think, for instance, of the deep-sea hydrothermal rift-worms (known as the vestimentiferans), which possess a spectacular organ, the trophosome, that is packed with sulphide-oxidizing bacteria. These bacteria are integral to the metabolism of this worm, not least because it lacks a gut. This example is also of interest in the context of Fortean bladders because the worms (like practically all animals) require oxygen, that is, they are aerobic, yet they have an ingenious metabolism that is able to transport the normally toxic reduced sulphur to the symbionts without poisoning themselves.[47] Another example, in a less extreme environment, is the so-called bacteriome, such as occurs in the aphids and other insects.[48] In the aphids and their relatives the bacteriome is a specialized organ, which as the name suggests is packed with bacteria. Like the trophosome it is a symbiotic association, but here the insect augments its diet with bacterially derived metabolites, because the food otherwise consists of nutrient-poor plant sap. After considering these various contrivances and the steps necessary to envisage an animal with a gas-filled bladder, rendered impermeable by a guanine lining and containing a protected hydrogenosome-packed organ pumping out hydrogen, we might conclude that such a structure does not seem beyond the realms of possibility.

There are, however, cogent biological reasons why we are no more likely to see a Fortean bladder gliding over the Alps than a flotilla of flying pigs.[49] There seem to be two problems. First, as Steve Vogel[50] has pointed out, there is a simple area-to-volume problem. The incipient evolutionary stage of the Fortean bladder would almost certainly be rather small, and here the restricted volume in comparison to a relatively large surface area would make lift-off problematic, especially since the bladder wall would be very thin and prone to leak the small hydrogen molecule. There is, moreover, another problem, which is that any such organism that ever got afloat would soon starve to death. While small animals may be swept high into the air, the idea of some sort of 'plankton of the ether' is a fantasy because above the ground the concentration of organisms drops off very rapidly with increasing height.[51] This is, of course, why on land and in the air there is no equivalent to the aquatic suspension feeders that strain the much more viscous water for tiny particles; apart from birds such as swallows, the nearest approach to terrestrial suspension feeders are the web-spinning spiders.

A SILKEN CONVERGENCE

Spiders, of course, are celebrated for their ability to extrude liquid protein through the narrow nozzles of their spinnerets, and thereby draw out a slender thread of silk.[52] Although it is generally agreed that these spinnerets are modified appendages the origins of the silk glands are still rather mysterious.[53] Even so, spider's silk itself is probably most familiar from the orb-webs, even though it is also employed in a number of other ways. The ogre-faced spiders, for example, capture their prey with a gladiator-like net of sticky silk (see also Chapter 7, p. 156, for a discussion of the huge eyes of these spiders). The mechanical properties of spider silk, not least its strength, are quite remarkable. It is also well known that the orb-weavers (the araneoids) can secrete several types of silk,[54] each with its own combination of physical properties, including mechanical properties (e.g. tensile strength, toughness), and degree of stickiness, that make it suitable for different parts of the web. These physical properties arise from the behaviour of the silk proteins,[55] known as fibroins, and the key role of a few amino acids, notably alanine and glycine. The proteins are, of course, coded for by specific genes,[56] and some of the structural units in the fibroins are remarkably conservative, apparently maintaining the same organization for more than 100 Ma. Clearly, therefore, functionality cannot be jeopardized and this is a good example of what is called stabilizing selection.[57] It is a curious thought that an insect blundering into a spider's web is encountering a molecular predicament that is unchanged since the time of the dinosaurs. Not only that, but, as Catherine Craig and her colleagues remark, the evolution of these proteins has been by 'selection for functional properties specific to the ecological purposes for which they are used.'[58] These comments were made in the context of discussing such properties as ultraviolet reflectance. This determines how visible the silk is to a passing insect and thereby appears to explain the dramatic evolutionary success of the spiders, especially the familiar orb-web spinners. Craig *et al.* suggest that these innovations have led to a 37-fold increase in their species numbers, as well as invasion of new habitats.

Silk production is integral to a spider's existence, and it might be thought to be an invention unique to these remarkable animals. Not a bit of it. Many other examples of silk production in a wide variety of other arthropods[59] are also known. These include, of course, the silk moth whose cocoon is the source of raw silk for fabrics.[60] This and the other arthropod silks are all convergent on that of spiders.[61] In some ways the differences among these silks, including their extrusion from a wide variety of secretory glands and their various functions (e.g. as lifelines[62]) are as interesting as the similarities. Spiders alone produce large and well-organized entrapment webs, but many insects use silk to trap their prey. In New Zealand and Australia, for example, the larva of a fungus-gnat (*Arachnocampa*) is luminescent (i.e. it is a type of glow-worm) and so attracts its prey towards vertical threads of sticky silk,[63] a feature also found in some other flies.[64] Silken nets are also employed by larval insects in aquatic circumstances, in such groups as the midges (chironomids)[65] and caddis-flies (trichopterans).[66] Typically these provide an extremely fine silk mesh, suitable for trapping tiny food particles. Certain types of caddis-fly (e.g. *Plectrocnemia*), however, construct nets that are not only, for the size of the larva, surprisingly large,[67] but also aim to trap prey in a spider-like 'sit-and-wait' strategy.[68] Interestingly, this mode of capture occurs where the density of suspended food is too low for suspension feeding, thus echoing the improbability of the Fortean bladders straining after gnats and other such aerial 'plankton'.

Other insects are also adept weavers, and some spin a silk whose molecular organization (a β-parallel sheet) is closely parallel to that of the spiders. Of all the examples of arthropod weaving, perhaps the most remarkable is the ant *Oecophylla*.[69] This, the weaver ant, deserves a mention, not only because of its intrinsic fascination, but also because as a social species it provides a prelude to the complex organizations seen in the other ants, notably fungus farming in attine ants and carnivorous raids by army ants, that in their different ways provide a series of insights into evolutionary convergence (Chapter 8). In the weaver ants the nest is constructed by drawing together leaves that are sufficiently flexible to provide a tent-like structure.[70] While the leaf is being held in position other workers fetch larvae, the heads of which are equipped with silk glands. Gripped by the mandibles of the worker, the larva keeping itself rigid, a tiny patch of silk is laid down on the leaf surface before the larva is carried across to the other side of the leaf where the silk thread, spun out during its transport, is reattached. Not only do ants construct tents; so also do some tropical bats,[71] although the latter do not of course employ silk.[72] And the purpose of these tents? Protection of a sort, but apparently for males guarding their harems.

MATRICES AND SKELETONS

So silken webs and tents full of expectant females may be the order of the day, but wheeled animals steadily progressing across the plains of Africa while admiring a fleet of Fortean bladders seem to be excluded: it appears as if the possibilities of life have been quite thoroughly explored. That we can have some confidence in this conclusion can also be gleaned from various attempts to identify all possible combinations of a particular biological feature. In this instance the approach is to define a matrix that provides a reasonable proxy for combinatorial space, and thus allows one to see whether the same solution to an evolutionary problem has been found repeatedly as against evolving rarely, if not only once. Most interesting in this context are the combinations that 'ought' to exist but do not. A pioneering analysis in this area was by the famous botanist George Ledyard Stebbins, who looked at character combinations in flower structure.[73] He found that only about a third of the possible combinations were actually realized, but of the unoccupied majority many were either structurally impossible or evidently poorly adapted. Stebbins also noted that common solutions were due to several factors, including ease of construction and what might be called phylogenetic facilitation; that is, closely related groups of flowering plants are predisposed to arrive at certain floral arrangements. In the latter case it is important to stress that since Stebbins published his paper in 1951 there have been very extensive revisions in plant systematics, especially on the basis of molecular data. It would be an interesting exercise to see in the light of this new evidence how much of the matrix of flower structure is actually occupied by convergent forms.

A similar approach to using a matrix was adopted more recently by Roger Thomas and Wolfgang-Ernst Reif[74] in an ambitious attempt to encapsulate all types of animal skeleton, in a so-called 'skeleton space' (Fig. 6.2). This 'space' was defined by such features as shape,

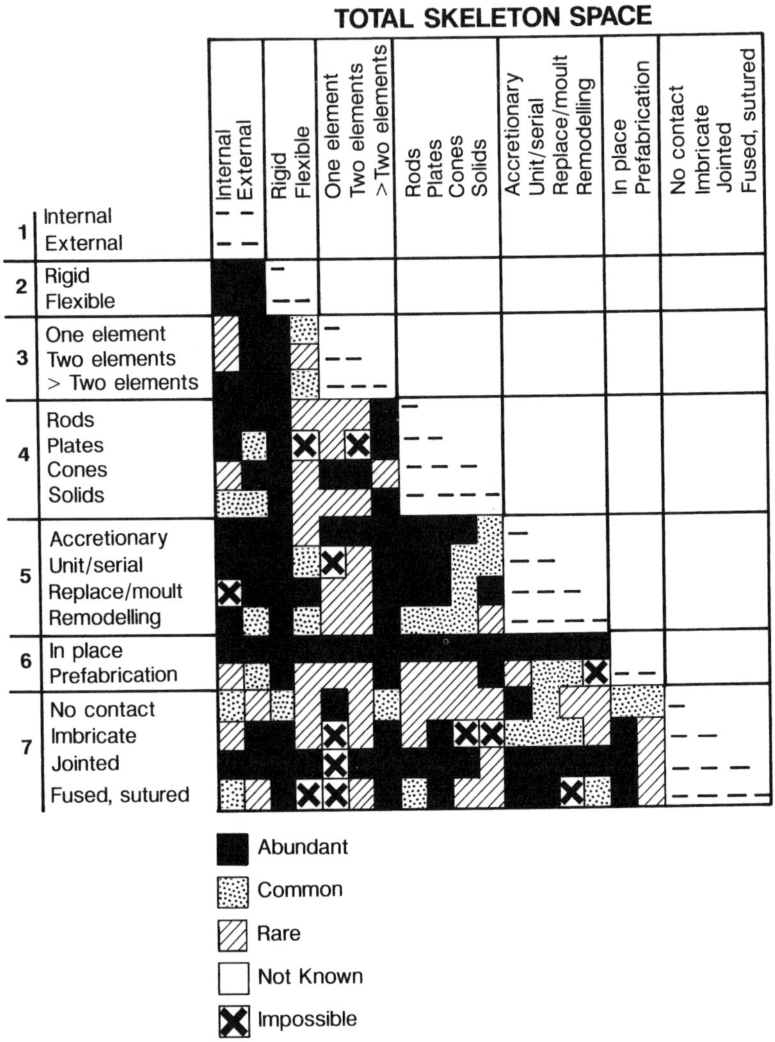

FIGURE 6.2 A plot of 'skeleton space', a matrix that divides the skeleton into seven major categories (and 21 variables), starting with whether they are internal (as in us and other vertebrates) or external (as in arthropods). This gives a total of 186 pair-combinations, of which a handful are effectively impossible. Of the remainder, the majority are either abundant or common, having evolved many times; in rare instances evolution of such an arrangement usually has occurred at least twice. (Redrawn from fig. 12 of Thomas and Reif (1993; citation is in note 74), with the permission of the Society for the Study of Evolution and the authors.)

style of growth and assembly, and degree of interplay. The matrix was defined by a total of 186 possible paired combinations, and significantly it showed that practically no combination, other than the structurally improbable, had not been invented at least once. Indeed, in many cases a position on the matrix showed multiple occupancies: in skeletons convergence is frequent. This leads to a general point, to which I shall return later (Chapter 10) concerning the link between the exploration of biological 'space' (however defined) and convergence, to the effect that barring the physically impossible and adaptationally compromised, it appears that as a general rule all evolutionary possibilities in a given 'space' will inevitably be 'discovered'.[75]

In the context of skeletons it is also worth remarking on some further instances, both specific and general. An interesting example concerns that remarkable biomineral, bone. This shows an immense versatility, particularly in its so-called trabecular form in which thin struts of bone (the trabeculae) orientate themselves according to the prevailing stress field, thereby combining strength with lightness. The precursor of bone, both embryologically, and probably in evolution, is cartilage. In such groups as the sharks and rays cartilage continues to serve as the principal skeleton. Yet clearly this material has design limitations, as is evident from the jaw of the cow-nose ray where a sort of 'trabecular cartilage' has developed. This is structurally and functionally convergent on trabecular bone, presumably evolving in response to the stresses imposed on the jaw as the creature grinds its molluscan prey.[76] Bone typifies the vertebrates, and while other groups of animals possess many and wonderfully variable skeletal configurations, it might be thought that bone is a unique biological invention. What could be more different from a vertebrate than a barnacle? These are familiar on seashore rocks and are actually sessile crustaceans, that is, close relatives of the shrimp and lobster. Barnacles encase themselves in a 'box' of calcareous plates, yet remarkably one species (*Ibla cumingi*) has evolved phosphatic plates with a distinctive and complex structure convergent on lamellate bone.[77]

Skeletons are usually thought of as tough and resistant, or at least until the snail meets the thrush. Yet biologically the concept of the skeleton can be cast much wider, to soft organs filled with fluid. Probably the most familiar example is penile tumescence when engorged with blood, a useful trick that has evolved several times.[78] Moreover, hydrostatic skeletons are an integral arrangement both of entire animals, such as the earthworm, and of specific organs, such as the eye stalks of snails and legs of spiders. In each case the pressure of the effectively incompressible fluid acts as an antagonist to the contraction of a muscular sheath. Hydrostatic skeletons are widespread and in activities such as burrowing often play the key role.[79] This is not, however, the case in the various limbless vertebrates, such as the snakes, which, as might be expected, rely on the contraction of longitudinal muscles in association with the backbone. There is, however, a very striking exception in the form of a group of limbless amphibians, relatives of the frogs and newts, which are known as the caecilians.[80] It has long been recognized that they are much more efficient burrowers than other limbless vertebrates, and it now transpires that this is because they have effectively transformed themselves into worms.[81] By allowing the backbone to move independently of the skin, as well as employing a system of helically arranged tendons, the overall style of locomotion is strongly convergent on various invertebrate burrowers: a vertebrate body plan has been remarkably transformed into a hydrostatic skeleton.

PLAY IT AGAIN!

But what of counterfactuals, the replaying of history that exerts such an eerie fascination? Just consider the 'what ifs' of European history: a successful Spanish armada? A Catholic Europe, with science emerging in a society that presupposed that the Universe was both rational and had a moral architecture?[82] A Charles the First who engaged and defeated the Scots in 1639? No English Civil War? No Lord Protector? An eighteenth century without the brilliant chemist Antoine Lavoisier? No reform of the French gunpowder industry, a vital component of the armament shipments to the American colonies in rebellion against the British Crown?[83] Napoleon Bonaparte triumphant at Waterloo and Europe's first police state fully operational by 1820, rather than three more attempts (1870, 1914, 1940) – and as for the the fourth? German hegemony and the Euro by 1916 as the British Cabinet avoids conflict by persuading the Germans not to invade 'gallant, little Belgium'? England invaded by the Nazis in 1940? The Bloomsbury group taking their cyanide, while in Cambridge a very senior figure glances at King's College chapel and then carefully adjusts his swastika armband?[84] Looking for trends in history is now deeply unfashionable, yet so far as the biological equivalent is concerned we find that here at least trends have a reality, and are not only self-evident in the history of life but are even testable in the laboratory.

The organism of choice is, unsurprisingly, a bacterium, specifically the microbiologist's equivalent of the fruit-fly and the mouse, that is, *Escherichia coli*, more usually referred to as *E. coli*. This has some substantial advantages for the study of evolution.[85] The genetics of this bacterium are well understood, and population growth is very fast: in favourable conditions the cells divide as frequently as once every 20 minutes. Moreover, even as the descendant lineages are allowed to mutate along a myriad of evolutionary trajectories the ancestral 'race' can be preserved by the simple expedient of freezing, to be reawoken when a comparison needs to be made between the starting point and a population that might have been evolving for literally thousands of generations.

And it is on this basis that Richard Lenski, Michael Travisano, and their colleagues have undertaken a series of elegant experiments to see how evolution replays itself. In one experiment these investigators set up 12 separate populations of *E. coli*.[86] Each was then allowed to diversify – effectively to evolve – independently for 2000 generations. At this stage each of these populations was divided into three, and the resultant 36 populations were then left to evolve yet further, for another 1000 generations. At this stage, 36 different evolutionary trajectories have been followed. So far, so good, but now these populations are introduced to a novel substrate, the sugar maltose. *E. coli* much prefers glucose, and maltose is biochemically challenging. For each population a new 'history', the struggle with maltose, will now unfold. The investigators specifically wanted to see what would happen with the bacteria living on the maltose substrate after 1000 generations had elapsed. The real objective of the experiment was to weigh up statistically the relative importance of three factors: those of chance (life's grand lottery), adaptation (with the implications of trends and optimization), and history (play it again!). Which is the most important? If chance played the main role then the 36 populations would be expected to evolve in all sorts of different directions, each caught in a contingent net of circumstances. Conversely, if adaptation by selection was important then the populations should converge on one (or more) stable end-points. And how on earth can we rerun history multiple times? That is the point of arranging the 36 replicates, each of which, having been allowed previously to diversify, now represents a different starting point when transferred to the 'maltose world'. Clearly enough at the beginning of the rerun, that is, transfer to the maltose substrate, history has to be highly significant, simply because each population represents a different starting condition. After the 1000 generations of maltose-based existence the historical component was strongly reduced, and correspondingly there was a strong degree of convergence; that is, different routes led to the same end-point (Fig. 6.3). Such convergence implies, of course, a corresponding increase in adaptation as E. coli 'learns' to get to grips with maltose. And the role of chance? In this evolutionary experiment, at least, it was of negligible importance. The simple conclusion is that history does not go away, but is swamped by the effects of convergence. There are many other experiments involving the rerunning of evolutionary histories of this bacterium. As noted, E. coli depends on sugars, but what happens if independent lineages are sometimes deprived of food and then glutted? To cope with this recurrent 'feast and famine' the separate lineages arrived at different genetic solutions, but even so the overall direction of evolutionary change was again strongly parallel.[87] And the tape of evolution can be rerun with respect not only to food substrates of E. coli but also to adaptation to changing temperatures: once again chance and history are not the main players.[88]

There are certainly some important qualifications to these elegant experiments,[89] but it is also important to emphasize that simply because convergence overrides history this does not mean that evolution comes to a dead end.[90] Even with a similar end result, that is convergence, the different populations of bacteria remain genomically very diverse, and also, despite adapting to the new substrate of maltose, the genomes do not 'fossilize' but remain very dynamic.[91] Hence stabilization in the form of convergence need not imply the cessation of evolution. And the roles of chance and history are not always negligible; in these experiments cell size was determined mainly by these factors.[92] Moreover, in further experiments,[93] again comparing evolution on glucose and maltose substrates, the populations left to replicate on their 'home' substrates rapidly adapted in much the same way to make the best of the food source, but when they were swapped to the other substrate their behaviour in either case was very different.[94] Travisano concluded that 'Apparently small differences in

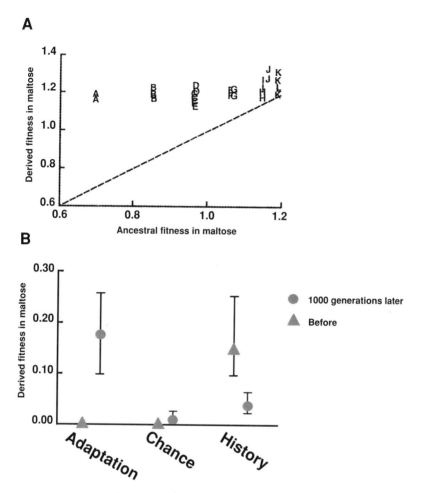

FIGURE 6.3 Rerunning the history of evolution, with the bacterium *Escherichia coli*. The upper panel shows the relative increase of fitness of 12 replicate populations (A→L) of *E. coli*, each subdivided three times, when placed on a maltose substrate. The dotted line would be the case of no increase in fitness, and clearly all 36 sub-populations have increased their relative fitness, even those that ancestrally had a high fitness to maltose. The lower panel disentangles the relative roles of adaptation, chance and history during this process. Note chance has a very limited role, that of history declines substantially, while that of adaptation increases markedly. (Reprinted (excerpted) with permission from M. Travisano, J. A. Mongold, A. F. Bennett and R. E. Lenski, *Science*, vol. 267, pp. 87–90. Fig. 2A and 2B. Copyright 1995. American Association for the Advancement of Science.)

the selective environment can greatly alter patterns of adaptation and divergence.'[95] These experiments are important, both because they can quantify to some extent the role of history, and also because despite the apparent simplicity of the system they reveal in their various ways a number of unexpected complexities. Particular responses in a given lineage of *E. coli*, when faced with a metabolic or environmental novelty, are often not predictable, presumably because of our relative ignorance of either physiological or genetic factors. Even so, these experiments also demonstrate recurrent examples of parallelism: despite different starting points, different lineages often converge on particular destinations.

A particular strength of the *E. coli* experiments was that the quantitative data could be handled statistically by analysis of the variances. Nevertheless, it is natural to ask whether what applies to a flask full of *E. coli* is of any relevance to the wider world and the likelihood of evolutionary convergence. Amitabh Joshi reminds us of some of the principal problems that might accompany any attempt to use bacteria as a guide as to how the world works, not least in the case of animal evolution. The main points he makes are: (a) the only observations made are at the beginning and end of the experiment, e.g. 1000 generations of adaptation, so 'making it difficult to study the time course of changes in the relative contributions of adaptation, history and chance'; (b) being asexual, bacteria evolve largely by mutation rather than sexual exchange of genes; (c) the relative importance of 'ancestry and past selection history' are not easy to separate; and (d) bacterial populations are usually orders of magnitude larger than most eukaryotes, so what is termed 'genetic drift' may be less likely.[96] So what happens when we attempt an analogous experiment, rerunning the tape of evolution on laboratory animals rather than *E. coli*? Almost inevitably the animal in question has to be the fruit-fly (*Drosophila*). In the context of one such set of experiments, looking at the adaptation of larval feeding rates according to the degree of crowding,[97] Joshi's concluding remarks are telling: 'Thus, the footprints of history, at least in sexually reproducing species, would, metaphorically speaking, appear to be no more than transient impressions on genomic dust'.[98]

In other cases, however, the footprints of history may not be so easily erased, especially if the clock of evolution runs backwards. This was the aim of a series of experiments, again using *Drosophila*, to see how easy it might be to reverse evolution by returning the fruit-flies to the ancestral environment from which they had diverged by artificial selection, e.g. for timing of reproduction, over many generations.[99] In most cases there was either a complete or partial convergence to the ancestral state: evolution had run backwards. In the relatively few other cases a new evolutionary trajectory was followed or, rarely, evolution effectively stood still (no change). Here, therefore, there are various histories, with particular genetic architectures predisposing some (but not all) lineages to convergence. Even so, when evolutionary reversal to the ancestral state occurs it is patently due to 'selection for adaptation to the ancestral environment'.[100] Given the ubiquity of selection so we can expect a ubiquity of convergence.

Adaptation is ubiquitous and convergence is an inevitable consequence of the constraints on life, although placing convergences into any sort of quantitative framework is still in its infancy. The work on *E. coli* and *Drosophila* certainly shows one way forward, and there are some interesting parallels to these investigations. One such, albeit on a very different scale, is Dolph Schluter's investigation of different communities of finches.[101] By studying these birds and the communities[102] they formed he was able to show that in terms of variance there were significant convergences in various traits, such as the number of species and mean size. Only in the case of mean shape, which was controlled by phylogenetic factors, could the role of history be identified as important. Much the same story emerges elsewhere. One particularly striking example of convergence in which the historical contingencies are pushed to one side concerns a group of Caribbean lizards known as the anolids. Jonathan Losos concluded that their convergence is identifiable, not only in terms of striking similarities between various species representing specific anolid eco-morphs on different islands, but also in overall community profile.[103] One important inference of his work is that the ecological structure has evolved by competitive interactions that determine how the habitats came to be occupied. This analysis of anolid convergence is justifiably well known, but it is by no means the only such example: among the mammals the rodents are similarly instructive.[104]

The investigation of these vertebrate communities and their convergences is a sort of natural experiment. Even if it were thought desirable to arrange a rerun this would probably require thousands of years, and possibly much longer. That, of course, is a particular advantage of using a rapidly evolving and relatively simple organism like the bacterium *E. coli*. A number of similar experiments have already been undertaken using other microorganisms,[105] and with advances in technology there are new possibilities for experimental investigation at even the molecular level. This topic will be discussed subsequently (Chapter 10), but here it is worth noting one particular piece of work, by Amelie Karlstrom and her colleagues, because like the *E. coli* story it is based on a rerun of 'history'.[106] Karlstrom *et al.* noted that in the field of biochemistry there are often multiple solutions to a given problem, but even so particular motifs do reoccur. This they investigated as a kind of repeated evolutionary trial, by designing a particular protein (a catalytic antibody aldolase). This showed a similarity in efficiency and mechanism to the naturally formed equivalent, but in the two trials the investigators found that a particular amino acid (lysine) at a key site (number 93 in the chain) was a recurrent necessity if the enzyme was to function properly.

ATTACKING CONVERGENCE

Such multiple evolutionary trials suggest that life is far from some sort of contingent muddle, yet the criticism might be levelled that they are either rather general, such as the example of the matrix of skeleton space, or very specific, good for an anecdote over a pint of beer in the Green Dragon but hardly of any real evolutionary significance. Some examples do indeed seem to be little more than curiosities, such as the Jurassic crustacean that is reminiscent of the trilobites,[107] or a fossil bryozoan that adopts a form of spiralling fans that as a bryozoan design had disappeared more than 200 million years earlier.[108] Occasionally these reappearances seem more like zombies.[109] Such is the case in certain fossil snails found in sediments of Triassic age, that is, the geological interval that followed the great end-Permian catastrophe. They are remarkably similar to the pre-catastrophe snails, but it is not clear whether this is because they are in fact survivors of the debacle (Lazarus taxa) or actually represent new species, but evolving along the same old tracks (Elvis taxa).[110]

What convergence is, and what it might imply, I hope will become clearer as this book unfolds. A few preliminary remarks, however, may be apposite. First and most obviously, biological convergence can mean many things and operate at many levels. As I shall argue, however, there are some common implications, despite the apparently bewildering range of examples. Second, the branching nature of evolution means that although the identification of convergence presupposes a known starting point, as often as not the reality is an almost infinite regress of investigation. Nevertheless, I believe the topic of convergence to be important for two main reasons. One is widely acknowledged, if as often subject to procrustean procedures of accommodation. It concerns phylogeny, with the obvious circularity of two questions: do we trust our phylogeny and thereby define convergence (which everyone does), or do we trust our characters to be convergent (for whatever reason) and define our phylogeny? As phylogeny depends on characters, the two questions are inseparable. In reality there are, of course, independent data sets and sometimes an historical record, so the enterprise is not crippled. Even so, no phylogeny is free of its convergences, and it is often the case that a biologist believes a phylogeny because in his or her view certain convergences would be too incredible to be true. This is mostly correct, but not always. The second reason for the importance of convergence has certainly been touched upon by others, but, I suggest, is not only neglected but might provide the nucleus of an interesting research programme (see Chapter 10, pp. 308–310). In brief, convergence offers a metaphor as to how evolution navigates the combinatorial immensities of biological 'hyperspace', a topic already touched upon in Chapter 1. Convergence occurs because of 'islands' of stability, analogous to 'attractors' in chaos theory. This view of life, however, begs two interesting questions. First, given the immense dimensions of this 'hyperspace', how is it that navigation is ever successfully achieved? Variants of this question are, of course, repeatedly proposed by opponents of evolution. Convergence gives us a clue as to how the metaphorical Easter Island is located, but just as this Pacific Island is surrounded by oceanic wastes, so too perhaps are the 'islands' of biological habitability. I suspect that not only will the bulk of biological 'space' never be occupied, but it never can be either.

It is therefore my strong belief that the topic of convergence is very far from being just another example of the frustrating imprecisions of biology where apparent rules and laws melt into exceptions and counter-examples. To echo the idea that convergence might give some surprising insights into a deeper structure of biology, there is some evidence that might at first seem to be simply anecdotal. During my time in the libraries I have been particularly struck by the adjectives that accompany descriptions of evolutionary convergence. Words like 'remarkable', 'striking', 'extraordinary', or even 'astonishing' and 'uncanny' are commonplace. It is well appreciated that seldom are the similarities precise,[111] and this in itself is as concrete a piece of evidence for the reality of evolution as can be provided. Even so, the frequency of adjectival surprise associated with descriptions of convergence suggests to me that there is almost a feeling of unease in these similarities. Indeed, I strongly suspect that some of these biologists sense the ghost of teleology looking over their shoulders. Nor is this an unworthy sentiment. The eeriness of convergence is central to how evolution navigates across the combinatorial immensities of biological 'hyperspace'. And when we look at some of these examples of convergence, it is not difficult to see why.

Take, for example, the remarkably close correspondence between the first walking leg of the praying mantis and another insect, known as *Mantispa*.[112] As is self-evident, both are modified from a generalized insect leg for grasping (Figure 6.4), yet they belong to distantly related groups within the insects, respectively the mantids and neuropterans.[113] The similarity of the raptorial appendage is striking enough, yet there is evidence that such an arrangement evolved independently a third time, in another group of insects, the possibly unfamiliar rhachiberothidids.[114] This case exemplifies a more general point because some investigators had earlier allied these insects with the mantispids, whereas a more extensive phylogenetic analysis points to convergence. Each insect therefore seems to have arrived at effectively the same raptorial solution. Note also how in the mantids and mantispids the similarities extend to the large eye and prehensile 'neck'. Incidentally the power, rapidity, and accuracy of the predatory strike, at least in the praying mantids, is quite remarkable.[115] The striking action takes two-tenths of a second, and hits home 90% of the time. A standard method of releasing a sudden surge of power is by the elastic release of stored energy, a feature found convergently in many animals and also employed in the catapult. Despite its advantages the mantids do not employ such a system, presumably because of the delays imposed with reloading the elastic. Their remarkable striking ability includes a muscle whose contraction rate equals that of the fastest mammalian muscle.

While on the topic of insect legs it is also worth mentioning that pad-like attachment structures, of which there are two alternative

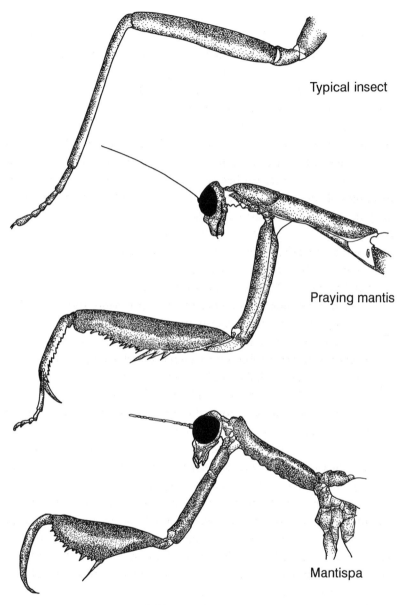

Typical insect

Praying mantis

Mantispa

FIGURE 6.4 A remarkable case of convergent evolution, between the raptorial fore-limb of the praying mantis and the neuropteran *Mantispa*, both arriving at the same solution independently from the generalized insect leg. (Redrawn from figs. 1–3 of Ulrich (1965; citation is in note 112) with the permission of the author, *Natur und Museum* and Senckenbergischen Naturforschenden Gesellschaft.)

designs, are also convergent within the flying insects.[116] Insects are a rich source of insights into many other convergences, some of which (e.g. compound eyes (Chapter 7), eusociality (Chapter 8), halteres (Chapter 7), silk (see above, p. 115), and trachea (Chapter 10) are mentioned elsewhere. So, too, are the related crustaceans. In the immediate context of attack and defence the repeated evolution of the pectinate claw is a striking example.[117] Yet more intriguing is the repeated emergence of a crab-like form[118] from among the decapod crustaceans, otherwise familiar in the form of the lobster. This example is of particular interest because, in contrast to many, if not most, examples of convergence where the arrangement has a clear functionality in response to a particular adaptive challenge, in the recurrent emergence (five times) of a crab-like form there is no apparent association with a specific mode of life or environment.[119]

Let me give you another example, again from the world of attack and mayhem. This is in the form of the independent evolution of dagger-like canines in both placental cats (the sabre-tooth felids)[120] and a group of South American marsupials known as the thylacosmilids (Figure 6.5).[121] In fact, the evidence suggests that even within the placental felids the sabre-tooth habit evolved several times.[122] The dinosaur enthusiast Bob Bakker has also suggested that an analogy to these sabre-toothed mammals can be found in the allosaurids of the Jurassic.[123] Although as a group the marsupials, best known in the guise of the kangaroo and wombat, tend to be regarded in some generalized sense as inferior to the placentals, this is too simplistic.[124] So, too, the rich, but now largely extinct, diversity of South American marsupials is widely regarded as having been competitively inferior to the onslaught of the placental mammals that surged south when the linkage to North America was secured via the newly emergent Panamanian isthmus several million years ago.[125] In fact, in the marsupial thylacosmilids the sabre showed a number of what appear to be design advantages when compared with the placental equivalents, including the possession of a protective flange (the placental sabretooth known as *Barbourofelis* shows the nearest equivalent), a self-sharpening mechanism (presumed to act against some sort of horny pad), and a deeper insertion into the skull that presumably afforded a more secure housing for the canine. It need hardly be stressed that despite this manifest convergence neither group escapes its imprint of phylogenetic history, which is marked, for example, in the specific structure of the teeth.[126] It is also worth remarking that the

FIGURE 6.5 Convergence in the sabre-tooth, between the marsupial thylacosmilid (upper) and placental cat (lower). (Copyright Marlene Hill Donnelly (the Field Museum, Chicago), with permission, reproduced from fig. 10 of L. G. Marshall's The great American interchange – an invasion induced crisis for South American mammals (pp. 133–229), in *Biotic crises in ecological and evolutionary time*, edited by M. H. Nitecki (Academic Press, Orlando, Florida).)

development of these massive canines, which it appears were more probably used to shear through the flesh of the prey rather than to engage in stiletto-like activities, should not be taken to imply a uniformity of hunting adaptations.[127]

This instance of sabre-teeth is just one of the many convergences between marsupial and placental mammals. Textbooks tend to emphasize the various similarities between the marsupials of Australia and placentals living in other regions of the world. John Kirsch (note 124) also remarks on the more general convergences between placentals and marsupials, and rightly remarks that such examples are only of real interest if the resemblance turns out to be more than superficial. In particular he draws attention to interesting examples of convergence in the brain structure of marsupials and placentals, notably those involving sensory input through stereoscopic vision and whiskers, while John Johnson draws attention to striking convergences in at least three regional brain specializations,[128] examples that are ultimately of considerable significance in the context of the rise of intelligences (Chapter 9).

It is also worth mentioning that a number of striking convergences can be found *within* the placental mammals themselves,[129] such as between the rodents and the group known as the artiodactyls (hippos, cows, and so forth), as well as among the artiodactyls themselves.[130] So, too, broader comparisons between the adaptive radiations of placental mammals that originated in two broad regions, defined as Laurasia (Asia, Europe, and North America) and Africa, define numerous similarities. Thus, in the component supergroups, referred to respectively as the Laurasiatheria and Afrotheria, the many parallels include the emergence of ungulate-like, aquatic, and semi-aquatic ('otter'), insectivore and burrowing, and even anteater (note 21) forms.[131] Yet more intriguing examples of convergence in mammals come from comparisons between the evolutionary pathways of ungulates and the predators that pursue them. The palaeontologist Bob Bakker[132] notes that in six separate evolutionary lines a whole series of anatomical features (e.g. reduction of side toes, elongation of long bones (metapodials) of paws, but shortening of fingers (phalanges)) follow 'the same morphological pathway'. He concludes that 'this striking case of iterative parallelism and convergence ... is a powerful argument that observed long-term changes in the fossil record are the result of directional natural selection, not random walk through genetic drift'.[133]

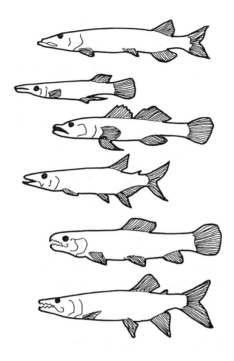

FIGURE 6.6 Convergence in the pike morphology, i.e. fusiform, sit-and-wait/stealth fish-hunters, arrived at independently (top to bottom) by the classic pike (*Esox lucius*, Esocidae: Alaska), *Belonesox belizanus* (Poecillidae: Central America), *Gobiomorus dormitor* (Eleotridae: Central America), *Acestrorhynchus microlepis* (Characidae: South America), *Hoplias malabaricus* (Erythrinidae: South America), and *Hepsetus odoe* (Hepsetidae: Africa). (Redrawn from fig. 11D of Winemiller (1991; citation is in note 134) with permission of the author and the Ecological Society of America.)

In their different ways mantids and sabre-tooths are powerful examples of convergence in the context of overpowering prey. Hunting style has, moreover, wider fields of similarity. In an assessment of diversification among freshwater fish Kirk Winemiller drew attention to recurrent patterns of what he called ecomorphological convergences.[134] One example is the repeated evolution of an eel-like morphology from phylogenetically divergent sources, as seen in the North American brook lamprey, neotropical swamp eels, and African spiny eel. In the context of hunting, and especially lunging predators, there are again striking convergences on different continents, this time towards a pike-like morphology (Fig. 6.6) and again from ancestors that are only distantly related to each other. Given these convergences among the fish, it is not surprising to see other examples of recurrent evolution. These include similar patterns of teeth, as in sea-breams,[135] or related trophic specializations within the endemic cichlids of central Africa.[136] Fish, therefore, in their various ways are a rich source of insight into evolutionary convergence, but as we shall see (Chapter 7) there are yet more remarkable examples, such as with the generation of electrical fields and endothermy, that in the wider scheme of things are arguably even more instructive.

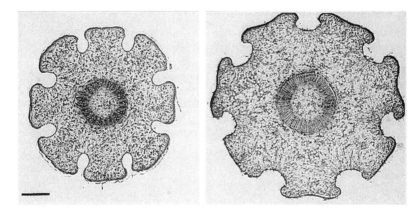

FIGURE 6.7 Convergence in desert plants, specifically the stems of the New World (Sonoran Desert) cactus *Peniocereus striatus* (left) and African (Kenya) spurge *Euphorbia cryptospinosa* (right). Scale bar is 1 mm. (Reproduced from fig. 3b, g of Felger and Henrickson (1997; citation is in note 137), with permission of the authors and Cactus and Succulent Society of America.)

CONVERGENCE: ON THE GROUND, ABOVE THE GROUND, UNDER THE GROUND

Convergence is pervasive, and moving from the animals let us take a particularly striking example from the plants. Consider two desert plants, one a cactus (*Peniocereus striatus*) that grows in the Sonoran Desert of Mexico, the other a type of spurge (*Euphorbia cryptospinosa*) from east Africa. Both are rather unprepossessing plants, straggly, and to the uninitiated, which the plant presumably hopes includes ravenous herbivores, look as if they are dead. Not only are the plants quite similar in general habit, but the stems are astonishingly convergent,[137] with a characteristic cross-section that shows flat ribs and intervening recesses (Fig. 6.7). This arrangement represents a clever adaptation, because in times of severe drought and water stress the ribs fold tightly together, thereby shielding the furrows where the pores (stomata) for gas exchange are located, but from which water can also readily escape. The interiors of the stems are also very similar, with the interior cells storing water and a central pith rich in food (starch). The similarities between these plants even extend to the reddish pigment, located just beneath the surface, which confers the moribund appearance to the plant. Of course, there are differences such as the details of the stomata, which in the cactus can be sealed with some sort of water-soluble substance, whereas in the spurge there are short spines that, it is hypothesized, restrict air circulation and thus water loss.

Cacti and spurges are only distantly related, and their common convergence is because of the rigours of living in an arid environment, although the degree of similarity is, to put it mildly, noteworthy. To a considerable extent evolutionary convergences have been placed in the context of adaptation to extreme environments. Not surprisingly, this specific and striking correspondence between the stems of a cactus and a spurge is only one of a whole swath of similarities that almost inevitably unite plants living in conditions of extreme aridity, the so-called xerophytes.[138] Similar remarks apply to the various Mediterranean-style floras in areas as remote from the Mediterranean itself as West Australia, South Africa, and Chile,[139] and extend to plant–animal interactions such as seed dispersal by ants.[140] The floras are not, it needs to be stressed, identical and the various differences also have evolutionary significance. Even so, in a survey of how plants function from the tropics to the tundra it is clear that convergence, for example in leaf structure,[141] is a pervasive phenomenon.[142]

Defying gravity and exposed to wind, not to mention such loads as a fruit crop or a passing arboreal animal, it is not surprising that the anatomy of plants reveals many structural constraints, nor is it odd that the number of solutions to a particular biological problem may be severely limited. One such example is the convergence seen among the girder-like structures, known as petioles, which support leaf systems.[143] In itself this is to be expected: buildings are not made of blotting paper and neither are plants. There are, however, more specific organizations seen in plants that would seem to be unique. What could be a better example of evolutionary uniqueness, of the sheer quirkiness of life, than a flower? Other worlds we may presume will have their equivalents to plants. Most probably they will build their cell walls with a carbohydrate similar to cellulose, and they will almost certainly power their photosynthesis by using chlorophyll (p. 110). So, too, in form they will range from ground-creeping herbs to lianas and trees.[144] All these are rather generalized predictions, based either on the universal nature of biochemistry (see p. 77) or reflecting the broad constraints of life.

On the other hand the specifics, such as flowers, let alone anything remotely similar to say a tulip or a daisy, are quite out of the question. They are surely a contingent happenstance of life on Earth, not life on Threga IX. From a historical perspective this must in one sense be true given that the flowering plants (the angiosperms) evidently evolved from a single ancestor, probably at some time in the Jurassic. Matters may, however, be rather more complicated. Whether the smiling biped on Threga IX offers us a tulip is not really relevant; remember we are in pursuit of general biological properties. Thus, in the case of the flowers, they display several specific characters. Most obvious, and appreciated, are the two whorls of leaves modified to form the often brightly coloured petals and the associated bract-like sepals. The reproductive arrangement follows the well-trodden path of male (pollen from the stamen) goes to female (via the stigma and ultimately the contained ovule), the happy conclusion of which is the fertilized seed.[145] To be sure, the structure of the flower is specialized, sometimes remarkably so as in the orchids, but the basic principle of plant sex (pollen meets ovule) is from an evolutionary perspective much more ancient than the flowering plants themselves. Yet no other group of plants has a reproductive organ exactly like those of the flowering plants. Not only is the flower highly distinctive, but its evolutionary origins are still rather mysterious: seldom can a review paper avoid parroting Darwin's remark to the effect that the origin of the flowering plants remains 'an abominable mystery'.[146]

Yet, as is well known to botanists, there exists an otherwise rather obscure assemblage of plants known as the Gnetales.[147] They in turn belong to a larger group, known as the gymnosperms, a group that is likely to be more familiar in the form of the pine tree and other conifers. As with other gymnosperms, the reproductive organs of the Gnetales are located on cones, but in this case they are remarkably flower-like.[148] Even more intriguing is a rather strange phenomenon known as double fertilization. In both the flowering plants and the Gnetales the pollen grain carries two nuclei, which having landed on the female stigma are delivered to the ovule by the agency of the pollen tube. One of the male nuclei has, of course, the heroic task of fertilization, which leads to the formation of the seed. The second nucleus, however, is not the runner-up because it fuses with another female nucleus. Hence the term double fertilization. In the case of the Gnetales naturally enough this second fertilization provides an extra, so-called supernumerary, embryro.[149] In the flowering plants, however, there is a further twist to the story because typically yet another female nucleus joins in to produce a triploid cell (i.e. three haploid sets of chromosomes combine, one set from each of the three nuclei). This cell, equivalent to the supernumerary embryo, then divides to form a mass of tissue known as the endosperm. Triple sex is not just a baroque ornamentation: the endosperm has a specific function, that is, to provide nourishment to the adjacent and growing seed. This is widely regarded as one of the key innovations to explain the overwhelming success of the flowering plants, whose evolutionary radiation has many interesting parallels to the Cambrian 'explosion' of marine animals.

So distinctive is this arrangement of double fertilization that, when it is taken together with the flower-like structures of the Gnetales, a strong supposition has arisen that the origins of the flowering plants are to be found in a gnetalean-like group that flourished perhaps about 150 Ma ago.[150] Nor are flowers and double fertilization the only similarities that exist between the two groups. Among the long list of features in common between the flowering plants and Gnetales, particular mention has been made of the similarities in vessels (that is the xylem, where the water-conducting tubes are open rather than consisting of a series of discrete cells (tracheids))[151] used to transport water through the plant, and in the genus *Gnetum* the venation of the leaf. Taken together, these similarities have been used to support a close evolutionary connection between the two groups and thereby explain the unique origin of flowers.

Yet all this turns out to be convergent. Despite some early research from molecular biology supporting a link between flowering plants and Gnetales, more extensive analyses both decisively reject this relationship and propose that Gnetales is surprisingly close to the conifers.[152] So on this basis it is argued that the cardinal features of flower-like structures and double fertilization[153] have been arrived at independently, as indeed have a number of other characters.[154] So, too, other supposed similarities, such as the structure of the xylem, are now shown to be exaggerated.[155] All this, of course, presupposes that the phylogenetic conclusions are correct. It needs to be pointed out that there are a number of extinct plants, notably a group of interest to the herbivorous dinosaurs and known as the bennettitaleans, which also have some flowering plant-like characteristics. Accordingly, given current phylogenetic uncertainties, it remains possible that some features actually evolved earlier in the history of plants and so were respectively inherited by both the Gnetales and the flowering plants.[156] Most of the latest research (note 152), however, continues to place the Gnetales very close to the pine trees and their relatives,[157] so making the argument for convergence the more likely. That such features as specialized sex organs (a.k.a. flowers) open to insect pollination,[158] not to mention the nutrition of the embryo,[159] should emerge in more than one lineage of advanced plants is surely not that surprising.[160] Perhaps flowers are an evolutionary inevitability; and, moreover, even here the options are not unlimited. Once again, convergence of floral anatomies is widespread,[161] reflecting such constraints as methods of pollination by both insects and birds.[162] So we might speculate that not only does Threga IX have plants and flowers, but perhaps something not so different from a bird pollinating a flower.

Birds themselves provide further compelling examples of evolutionary convergence. There are, for example, repeated trends: towards dark plumage in tropical seabirds;[163] in the many independent examples of long-distance migrants the development of more pointed wings[164] as well as parallel changes in the skeletal structure;[165] a cluster of characters (for example, in wing and leg structure) that converge in at least some grain- and insect-eating birds living at very high altitudes;[166] convergences in the aerially feeding insectivorous birds[167] and nectar feeders;[168] and a striking convergence between certain aquatic birds, specifically the grebe and the loon.[169] Indeed, birds as a whole are a rich source of insights into the prevalence of evolutionary convergence,[170] as well as having some striking similarities with other groups. A well-known example is the parallels between humming birds and the sphinx moths, both hovering above flowers as they draw out the nectar.[171] As we shall see in Chapter 8, the birds are even more instructive in a wider context, as when we encounter that 'honorary mammal' the kiwi, consider the physiology of warm-bloodedness, or compare their powers of vocalization: once again striking convergences emerge.

There are, moreover, other sorts of convergence that are equally intriguing. Birds form a popular part of many peoples' diet, but in New Guinea one bird, known as *Pitohui*, is avoided because of its unpleasant taste. The bird's repellent nature stems from a chemical (a type of steroidal alkaloid) that interferes with the sodium channels of the nervous system. The chemical itself is most probably sequestered from the bird's diet. Yet this chemical defence mechanism has evolved independently in the Neotropical poison-dart frogs,[172] and has also been found in another New Guinea bird, probably unrelated to *Pitohui*.[173] Nor is this the only example of chemical convergence; a repellent is also found in the crested auklet. This bird, a denizen of the subarctic Pacific and Bering Sea, has evolved a chemical convergency with a group of insects known as the heteropterans, in which it occurs in repugnatorial glands.[174] So, too, in plants themselves the extensive employment of defensive alkaloids, cyanogenic glycosides, and proteinase inhibitors also appears to be rampantly convergent.[175] Herbivores, of course, take steps to avoid noxious plants, and in the case of the lizards, which are predominantly insectivorous carnivores, the several independent shifts to plant-eating have led to a correlated evolution of chemosensory abilities that represents a type of convergence in response to a common selective pressure.[176]

Within the mammals also there are compelling examples of convergence. One of the best-documented case histories concerns the so-called fossorial species, that is, animals such as the mole which dig their way through light soils. Just as there is a parallel between the sabre-tooth cats and marsupial thylacosmilids, so the example of convergence of the Australian marsupial mole to its placental equivalent is a well-known and striking example. Somewhat less widely appreciated is that among the extinct marsupial faunas of South America (home of course to the sabre-toothed thylacosmilids) there is another striking parallel to the placental moles, this time in the form of a remarkable group known as the necrolestids.[177] But the convergences do not stop there, and among the extant burrowing mammals the similarities extend across a much wider range of groups than just the mole and its various marsupial equivalents. In total this evolutionary convergence encompasses no less than three orders, 11 separate families, and at least 150 genera (Fig. 6.8).[178] Convergences are extensive, involving not only anatomy, but extending to physiology, behaviour, and even aspects of genetics. Some of the most obvious similarities, such as powerful digging forelimbs with prominent claws,[179] rudimentary eyes and small size (including small testicles, sensibly enough for the males dragging their way along tunnels), are rather predictable. Living in burrows, where levels of CO_2 from soil respiration are often

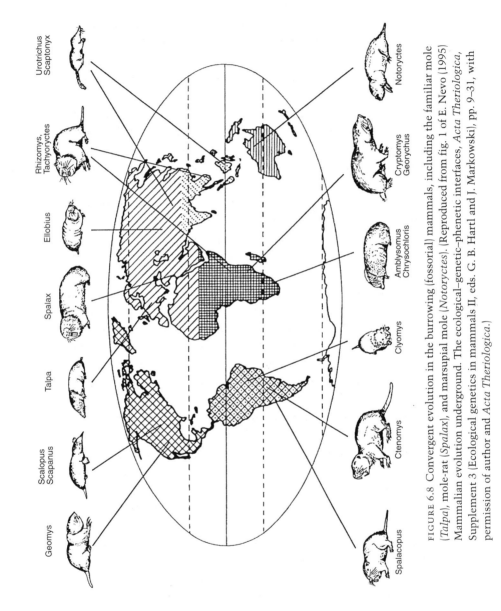

FIGURE 6.8 Convergent evolution in the burrowing (fossorial) mammals, including the familiar mole (*Talpa*), mole-rat (*Spalax*), and marsupial mole (*Notoryctes*). (Reproduced from fig. 1 of E. Nevo (1995) Mammalian evolution underground. The ecological–genetic–phenetic interfaces, *Acta Theriologica*, Supplement 3 (Ecological genetics in mammals II, eds. G. B. Hartl and J. Markowski), pp. 9–31, with permission of author and *Acta Theriologica*.)

elevated, places physiological burdens on these digging mammals, and here too we see convergences. In the following chapter I explore some of the other fascinating examples of convergence in terms of sensory systems. The subterranean mammals provide some outstanding examples, most notably in the case of hearing (pp. 190–191). Unsurprisingly, visual systems are correspondingly usually highly reduced, but as we shall see (pp. 175–177) in the case of the bizarre star-nosed mole there is another even more remarkable convergence between its nose and the eye.

As I shall argue, these convergences involving sensory systems are of considerable importance in assessing the likelihood of recurrent emergence of such biological properties as intelligence. Linked to this is the generally rather neglected evidence for convergence in behaviour (see also Chapter 10). Study of the burrowing mammals also throws up apparently quirky examples of behavioural convergence, such as their response when confronted with a carrot.[180] In a survey of more than 20 small mammals, it was found that without exception the subterranean species much preferred to start eating at the lower tip. Rather remarkably, one type, the Zambian common mole rat, can identify the polarity of the carrot even if both ends of the root have been cut off. This uniform preference for attacking the carrot at one end makes sense, of course, because a burrowing mammal is much more likely to encounter the lower end of a plant root. In the case of the burrowing animals here, too, there are other striking parallels in behaviour; typically they are solitary, communicate seismically,[181] and are generally highly aggressive.[182] Such convergences of fossorial activity are, of course, easiest to document in extant faunas, but it is also worth drawing attention to some striking convergences in fossil taxa. The controversial status of the possible Cretaceous bird *Mononykus* is returned to below, but equally remarkable are highly specialized palaeanodont mammals (*Epoicotherium* and *Xenocranium*) from the Oligocene of North America. These animals, equipped with powerful frontal claws, a prominent snout, and lacking sight, are strikingly convergent in many respects on other fossorial mammals, especially the group of insectivores known as the African golden moles (or chrysochlorids).[183]

Perhaps the most remarkable example of behavioural convergence involves the evolution of eusociality, most famously in

FIGURE 6.9 The naked mole-rat, *Heterocephalus*. (Photograph courtesy of H. Burda, University of Essen.)

Heterocephalus, better known as the naked mole-rat (Fig. 6.9). The preface of the main book[184] on naked mole rats has a telling story:

> In the mid-1970s R. D. Alexander [one of the book's editors] ... lectured on the evolution of eusociality ... In an effort to explain why vertebrates had apparently not evolved eusociality, he hypothesized a fictitious mammal that, if it existed, would be eusocial. This hypothetical creature had certain features that patterned its social evolution after that of termites (e.g., the potential for heroic acts that assisted collateral relatives, the existence of an ultrasafe but expansible nest, and an ample supply of food requiring minimal risk to obtain it). Alexander hypothesized that this mythical beast would probably be a completely subterranean rodent that fed on large tubers and lived in burrows inaccessible to most but not all predators, in a xeric [dry] tropical region with heavy clay soil (p. viii).

The mythical beast emerged as the naked mole rat.

Eusociality refers to a colonial system whereby typically only one female is reproductive, and the remaining individuals are divided into several castes, e.g. workers, that operate in a cohesive and coordinated way, notably in food collection and care of the young. Despite

FIGURE 6.10 Another social mole-rat, a bathyergid. (Photograph courtesy of H. Burda, University of Essen.)

its complexity, eusociality is patently convergent. One notable feature of this convergence is not only the social structure, but also the fact that the colony makes strenuous efforts to protect the breeding female/queen from danger. Eusociality has evolved several times in the mammals, including another group of mole-rats (the bathyergids, ·Fig. 6.10)[185] as well as some social voles.[186] The overall convergence, however, extends much further because there are striking parallels with various groups of insects (most famously in the hymenopterans (ants, bees, and wasps)), but also the termites and various other groups such as the gall thrips[187] and certain beetles.[188] In the insects, where of course it is best known, not only is eusociality rampantly convergent, but also remarkably instructive in terms of parallels with human organization and activities (agriculture, warfare, and competitive exclusion; see Chapter 8). The often striking differences between the various castes of a eusocial insect, most familiar in the contrast between the bloated and near-immobile queen versus beweaponed and agile soldiers, has another parallel in the naked mole rats because rather remarkably (for a vertebrate) the colony also shows morphological castes.[189] Nor are insects the only eusocial arthropods; this characteristic has also evolved in some crustaceans, in the form of a coral-reef shrimp (*Synalpheus*),[190] where eusociality has evolved independently three times.[191]

A recurrent theme of this book is not only the implications of convergence for evolution, but also the problems it can pose for its resolution. Not, I hasten to add, in terms of the reality of evolution. This is emphatically not in question; rather the reverse: the details of convergence actually reveal many of the twists and turns of evolutionary change as different starting points are transformed towards common solutions via a variety of well-trodden paths. Rather, convergence, if not identified, may lead to blatantly erroneous phylogenetic reconstructions. The constraints imposed by the burrowing mode of life offer a potentially telling example of this problem. There is an extraordinary fossil reptile, *Mononykus*, from the Upper Cretaceous of Mongolia. It has attracted considerable attention, because of both its strange morphology and its proposed relationships. Bipedal, and with a fore-limb equipped with a powerful claw, *Mononykus* has been interpreted as a flightless bird on the basis of various skeletal characteristics.[192] As such it is regarded as a key discovery in the evolution of early birds, even though *Mononykus* certainly remained firmly on (or in) the ground, and its age means that it post-dates a number of more advanced birds that have been unearthed in strata of Lower Cretaceous age in north China.[193] When originally described the strange arm of this animal, with its powerful claw, was recognized as similar to that of digging mammals, although other features of *Mononykus* appeared to preclude this mode of life.[194] Zhonghe Zhou, however, has suggested that the adaptation to burrowing is the correct interpretation,[195] and by neglecting this and other convergent features the role of *Mononykus* as a very strange bird needs rethinking.

Moles, other fossorial mammals, and possibly even *Mononykus* offer, therefore, an excellent example of numerous separate evolution trajectories converging towards a common solution, best adapted for occupation of this demanding habitat. It is hardly surprising, therefore, that if we turn our attention to a somewhat analogous ecology, that of burrowing frogs[196] and reptiles, especially the limbless snakes and the sand-dwelling (and sand-diving) lizards, we again find instructive examples. Among the snakes, the convergences include those of brain structure, reflecting the various adaptations to a fossorial existence, including diminished vision and the ability to communicate seismically. So far as the burrowing reptiles are concerned, once again there are certainly differences but also convergences.[197] Among the lizards such a fossorial mode of life has evidently been adopted independently at least eight times, and the mechanical problems imposed by entering and moving in the aeolian dunes have led to many parallels.[198] Even so, historical antecedents play their part. As Nick Arnold cogently remarks, with specific respect to ear reduction in iguanian lizards, 'Different lineages of organisms often evolve a number of similar traits independently, but the order in which these are assembled may often be different, even in ecological analogues, especially if the taxa concerned are not closely related . . . This [is the] phenomenon of equipotentiality, where more or less the same overall condition is reached by different routes'.[199] These comments are of key importance to the overall theme of this book: the evolutionary routes are many, but the destinations are limited. In a related context it is also worth remarking that many lizards have toes equipped with fringes, which assist not only with movement across loose sand, but also other substrates, including running across water.[200] Not surprisingly, the evolution of toe fringes is both convergent and correlated to the nature of the substrate upon which the lizard moves.[201] In a related fashion the evolutionary trend in lizards that become limbless, and thereby adopt a snake-like habit, has manifested itself several times. Significantly, despite this convergence, there are only two effective destinations represented by distinct types of ecomorph: small and short-tailed forms that burrow and larger 'grass-swimmers' with long tails.[202]

If we reconsider deserts, but this time in the context of the mammals living largely above the surface, again we find a richness of convergences.[203] This is especially true among the rodents,[204] many of which are bipedal hoppers, nocturnal and possessing enormous ears: the desert kangaroo rats are well named. Here, too, there are some interesting details. Deserts in South America, notably the Monte of Argentina, and Australia are effectively unoccupied by these specialized rodents, yet, as Michael Mares notes, we can be sufficiently confident from the general prevalence of convergence that in these deserts the same story so far as it concerns rodents will probably unfold.[205] If evolution is not inevitable, it is at the very least highly predictable. South America is of particular interest because it appears that an extinct group of marsupials, known as the argyrolagids,[206] were strongly convergent on the placental desert rodents, especially the kangaroo rats and a convergent equivalent known as the jerboas. G. G. Simpson simply remarks, 'the argyrolagids present one of the most striking known examples of evolutionary convergence'.[207] Mares[208] proposes an interesting explanation for the relative failure of the incoming placentals to diversify in this desert habitat in South America. He suggests that when they flooded south from North America as the Panamanian isthmus emerged to separate the Atlantic from the Pacific (see p. 130) this niche had already been occupied by the argyrolagids, which had then only recently become extinct. One of the immigrants, a rodent known as *Eligmodontia*, does, however, seem to be evolving in the direction of turning into something like a kangaroo rat.

To conclude: it does indeed appear to be the case that on the extremities of life the options narrow, choices diminish, and convergence is the norm. Let us imagine some suitably remote planet, inhospitable, with vast deserts, and a thin atmosphere. Life, if it existed, might be strangely reminiscent of similar regions on Earth. Tough plants, burrowing animals, and in the sky flying organisms with remarkably pointed wings: a sort of mirror image of our world when the going gets tough. But surely, so the argument continues, in the more benign areas where the mainstream of life enjoys an exuberance of diversity, matters are far less constrained. In this riot of forms, we might suppose, nearly anything is possible. Evolution is little more than a chaos of contingency, free of trends and with a myriad of equally likely destinations. Here the problems of convergence melt away. Nothing could be further from the truth.

7 Seeing convergence

If I enter a dark wood and happen to hear a woodpecker hammering at a tree, I at least have the potential satisfaction of knowing that here indeed are two unique products of evolution – human and woodpecker – or so it would appear. The example of the woodpecker is often presented as one of the best examples of evolutionary uniqueness, with its specific adaptations that enable it to make percussive attacks on trees. The corollary that follows is that the evolutionary process will lead to other unique end points:[1] on this planet, as it happens, woodpeckers and humans; somewhere else, life, but no woodpeckers, no humans. It may be centuries, if ever, before we can test the latter supposition, but this may not be necessary. Here on Earth the history of life can unfold many times, especially on isolated islands and microcontinents. In Madagascar, for example, there are no woodpeckers, but another group of birds, the vangids, is identified as a 'true substitute'.[2] Nor might this be the only example of convergence on a woodpecker. To be woodpecker-like does not automatically mean you have to be a bird, and three groups of mammals have, in a number of different ways, converged on a woodpecker-like habit.[3] To be sure, nothing precisely like a woodpecker has evolved independently, but broad biological generalities still arise. So, too, in an analogous way, can we say humans are really unique? As we shall see in Chapter 9, again much is convergent. Yet surveys of life also rightly emphasize its richness, fecundity, and beauty, if not its strangeness. Time spent in at least some woods is conducive to reflection: think of all the curiosities of adaptation leading not only to those woodpeckers, but also to acid-squirting beetles[4] or swimming animals encased in the equivalent of newspaper.[5] All very intriguing, if not odd; perhaps some might say even bizarre? Is this not evidence of an almost infinite number of evolutionary pathways?

Yet now, still in my metaphorical wood, I see a cat. On its left ear sits a mosquito. Both animals can see me: the cat through the dilated pupils of its camera-eyes, the mosquito, at least after a fashion, through its compound eyes.[6] But again, at least I know that not only have both these types of eye – camera and compound – evolved several times, but even the very mechanisms of seeing in terms of the neural architecture have also shown multiple convergences. Have I missed the right road? I give a sceptical cough; both animals can hear me, one through its triangular, furry ears and the other using the so-called Johnston's organs. What could be more different? Yet the auditory mechanisms are convergent. The cat and the mosquito can smell me, but again the differences between the nostrils of this feline mammal and the 'nose' of that blood-sucking insect (actually located on its antennae) are in reality superficial. This is because the mechanisms of olfaction are also strongly convergent. Even the way in which the cat and the mosquito walk is convergent in terms of their neural circuits, reflexes, and motor control: four legs or six, it really does not make very much difference.[7]

A BALANCING ACT

So, if convergence is going to be a guiding principle in understanding evolution, then of all the areas worth investigating one of the most interesting must surely be to look at what constraints, if any, accompany the development of sensory organs. It is here, if anywhere, that we can approach the wider problem of the evolution of nervous systems, brains, and perhaps ultimately sentience. And this in turn might give some clues as to whether indeed intelligence is some quirky end point of the evolutionary process or whether in reality it is more-or-less inevitable, an emergent property that is wired into the biosphere.

A similar question has been succinctly posed by David Sandeman.[8] He writes, 'it may be of interest to ask whether there are sensory organs that are unique and *not* convergent . . . One of the most bizarre and ingenious equilibrium systems that have arisen is without doubt the halteres of the dipteran flies' (his emphasis).[9] The halteres, it may need to be explained, are the organs for balance in the flies. They are located on the back of the fly and resemble tiny drumsticks. In flight they swing continuously to act as gyroscopes that serve to detect angular accelerations. It has long been realized that the halteres are derived from the pair of hind-wings of the fly (most insects have two pairs of wings), and are indeed, in Sandeman's terminology, 'bizarre and ingenious'. Surely these qualify as a unique biological invention? Not a bit of it. Sandeman's question was rhetorical. He reminds us that very similar structures are found in another group

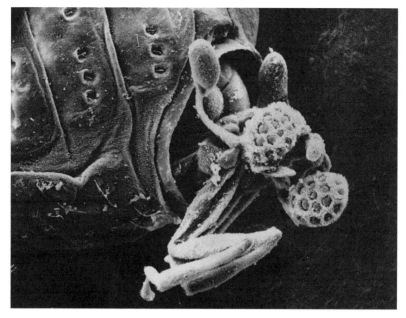

FIGURE 7.1 Electron micrograph of a strepsipteran insect, in the form of the male emerging from its host. Note the drumstick-like halteres arising from behind the head, as well as the compound eye, the arrangement of which is convergent on that of the trilobites. (Photograph courtesy of Dr J. Kathirithamby, University of Oxford.)

of insects known as the strepsipterans (Fig. 7.1), which on molecular evidence do not seem to be closely related to the dipteran flies.[10] Strepsipteran halteres are different, however, because they are derived from the pair of fore-wings. They are found only in the males, whose wings are put to use for a short mating flight that they undertake once they emerge from their host, such as a wasp, in search of the females that (with few exceptions) remain permanently within the host.[11] The parasitic strepsipterans differ, therefore, quite markedly from the dipterans, yet it is clear that their halteres act in the same gyroscopic fashion.[12] However bizarre they may seem, halteres are not a unique invention. Indeed, there is evidence for yet other convergences. One is in an African beetle that has modified its wing-cases (known as the elytra) into haltere-like organs;[13] and a fourth example can perhaps be found in a group of hemipterans known as the coccoideans.[14] The coccoideans are probably most familiar for the swollen galls the larvae form in plants, but it has been observed that in some of the adult males the hind-wings are 'reduced to haltere-like structures'.[15]

Swinging tiny drumstick-like structures in the air is one way of keeping a stable course, and man-made gyroscopes adopt a somewhat similar principle. For animals, a much more widespread method of achieving balance is by means of structures known as statocysts, which typically consist of some grains of a mineral attached to fine hairs. This arrangement is sensitive to gravity, the nervous connections to the hairs being stimulated according to how the heavier grains fall. Statocysts have evolved repeatedly, and are found in groups as diverse as the jellyfish, crabs, and cephalopods.[16] Indeed, a comparable structure is even known in the group of single-celled organisms, the ciliates.[17]

In humans and other vertebrates balance is achieved by an extension of the statocyst principle, employing mineral grains known as the otoconia. These are located within sac-like structures housed within the inner ear. Immediately adjacent to these sacs are the so-called semicircular canals, which are also involved in maintaining a sense of balance. Such a sophisticated system might at first be thought to be without a parallel. Primitively, that is in the jawless fish, the lamprey, there are two canals, but in the more advanced vertebrates there are three canals, set at right angles to each other. The canals are filled with fluid and contain sensory hairs that by registering angular accelerations transmit information, albeit in a way not fully understood, to the brain to permit effective and controlled balance. Unique? No, because a convergent arrangement is found in the swimming crabs, although here the arrangement is directly derived from the more primitive arrangement of the statocyst.[18] In comparison, however, these statocysts are strongly modified, and within the canals hair-like receptors monitor the fluid flow induced by the crab's movement. Even so, why should a crab in this respect imitate a fish? As so often when discussing convergence, the mode of life gives the necessary clue. These crabs are very agile swimmers, capable of rapid manoeuvre in pursuit of their prey. Balance is clearly at a premium, and it is not so surprising that they have arrived at a system strongly reminiscent of the one seen in the more primitive vertebrates with their two semicircular canals. As might also be expected, these canals are filled with fluid. In contrast to the vertebrates, however, this fluid is composed of sea water and is incorporated each time the crab moults its skeleton. At this juncture it has to secrete a new skeleton, including a pair of the balancing organs which, once filled with sea water, are then sealed to the outside world. In conjunction with a sophisticated visual system these balancers allow these active and fast-swimming crabs to perform their adroit manoeuvres.

The senses, five or otherwise, may be separately identifiable, but the various inputs received by the animal through the nervous system will require appropriate integration. In the case of balance in the swimming crabs an acute visual control is an unsurprising corollary. In reviewing this linked system in the crabs and comparing it with the vertebrates Sandeman remarks that they have both 'achieved the same end using similarly constructed sense organs',[19] albeit from very different origins. Sandeman continues, however, by noting that notwithstanding the sophisticated sense of balance the overriding factor in this system is the effective capture of light, and so it is to the topic of the eye that we now turn.

RETURNING THE GAZE

Yet, concerning the eyes there is an obvious difference.[20] Vertebrates, including humans, use a camera-like arrangement, whereas the crabs employ the compound eye that is typical of most arthropods. Has the principal of convergence broken down? Not at all, because it is among eyes that we find some of the most compelling examples. The existence of camera- and compound eyes reminds us, of course, that solutions to biological evolution need not be unique, but are simply very strongly constrained.[21] Even so, in at least the case of eyes it is possible to argue that the camera-eye is inherently superior to the compound arrangement: both are viable, but when it comes to association with sentience the camera-eye may prove to be the winner.

When considering convergences between camera-eyes it is almost inevitable that comparisons will be drawn between the camera-eyes of vertebrates and those of the advanced cephalopods,[22] notably the squid and octopus, which are one group of the molluscs (Fig. 7.2). This is the stuff of textbooks, but there are well-known differences. Most notable are those between the relative position and detailed structure of the light-sensitive layer, the retina, which arise as a result of the different embryologies in vertebrates and molluscs. In the molluscs the retina is derived from the outer layer (known as the ectoderm, and effectively equivalent to our skin). As it infolds to make

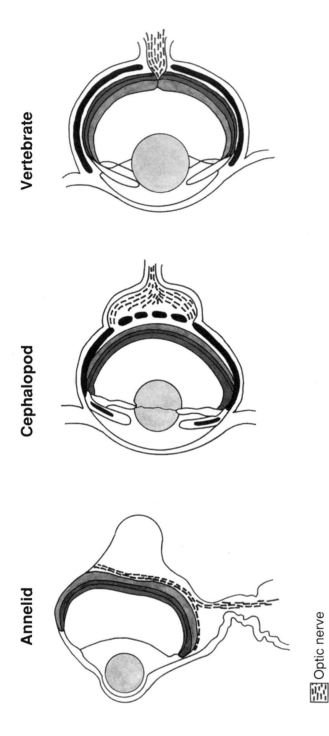

Annelid　　　　**Cephalopod**　　　　**Vertebrate**

Optic nerve

Retina

Pigmented layer

Nuclear layer

FIGURE 7.2 Convergence of the camera-eye, including the classic comparison between the octopus (cephalopod) and human (vertebrate), as well as the alciopid polychaete (annelid). (Redrawn from various sources.)

the eye-cup, the associated nerve cells extend into the body to make their connection with the brain. In contrast, the vertebrate retina is effectively an outgrowth from the central nervous system. The net result of this process is that, in contrast to the molluscs, in the vertebrates the nerve cells come to overlie the retina. The exit point of these nerves to the optic tract, which then leads to the brain, results in a 'hole' in the retina, better known as the 'blind spot'. Accordingly the arrangement of nerve cells and retina in molluscs and vertebrates is reversed, the cephalopods having arguably the more 'sensible' design of nervous layer beneath sensory retina.

The cephalopod lens is also formed in a rather different way.[23] It has a rather remarkable layered structure,[24] and being inflexible cannot change its shape to assist in the process of focusing.[25] It is still not clear exactly how the cephalopods manage to focus, but most probably it is by moving the entire lens backwards or forwards.[26] The style of construction of the lens may also explain why at least some cephalopods are able to correct for spherical aberration. (This is a phenomenon in which the light rays, on passing through the lens, fail to focus at a single point, resulting of course in a blurred image.) Even so, correction for such aberration seems to be less refined than in the fish. In the fish eye the refractive index (which is a measure of how much the light ray is bent as it passes from one medium to another) varies through the thickness of the lens. In cephalopods, however, spherical aberration may be dealt with by using non-spherical lenses, which in turn is possible because of their different mode of formation.[27] These differences perhaps represent relative disadvantages of cephalopods as compared with fish.

Control of eye movements is, of course, critical. This is achieved by the so-called extra-ocular muscles that are attached to the exterior of the eyeball. Through what is known as the oculomotor control system, routed via the brain, this arrangement is linked to the organs of balance and so provides a coordinated arrangement, referred to as the vestibulo-oculomotor reflex.[28] As already noted there are interesting convergences in the systems of balance, which in cephalopods are represented by the statocysts and in vertebrates by part of the inner ear. Given the nature of the oculomotor system it is not surprising to find convergences in the brain pathways between some cephalopods, such as the octopus, and the vertebrates.[29] The degree of similarity between these systems is, in one sense, hardly surprising, but there are differences: for example, comparison of the oculomotor reactions, which allow the eye to swivel in its socket by the contraction of the extra-ocular muscles, shows the vertebrate arrangement to be superior.[30]

All these differences are important, but it is still the case that the similarities between our eyes and those of a cephalopod are very striking. It is rather less well known that a similar camera-eye has evolved independently in several other groups. The most notable example is in a group of marine annelids, close relatives of the more familiar earthworms.[31] The group in question is known as the alciopids.[32] These eyes are strikingly similar to those of the vertebrates and cephalopods (Fig. 7.2), and because the annelids are also relatively closely related to the molluscs the retina has the same arrangement as in the latter group.[33] There is, moreover, another and striking convergence in the alciopid eye. This is in the form of the so-called accessory retinas, which are light-sensitive patches located nearer to the front of the eye. These, too, are strongly convergent on similar structures found in some deep-sea fish and cephalopods.[34] Returning to the molluscs, but this time considering the gastropods (or snails) we find that in this group a camera-like eye seems to have evolved independently at least three times.[35] Specifically these are in the pelagic heteropods,[36] the familiar shore-snail known as the winkle (Littorina),[37] and a large herbivorous tropical snail called Strombus.[38]

Nor does the list of convergently evolved camera-eyes quite end there: there are two more examples, each in their own way quite surprising. The first example comes from the much more primitive cubozoans. These are a type of jellyfish (hence the alternative name of cubomedusae) renowned both for their highly toxic stings and for their remarkable eyes. The eyes are similar in construction to other camera-eyes, with a large lens located in front of the retina[39] but separated by a layer of cells that may help in focusing.[40] Cubozoan jellyfish belong to a primitive group of animals, the cnidarians, which also includes the sea anemones and corals. While primitive eye-spots are known in other cnidarians,[41] at first sight the sophistication of the cubozoan eyes, which typically total eight arranged around the margin of the swimming bell, is quite surprising. Cubozoans, however, are active and highly agile swimmers, have obvious visual acuity,[42] and uniquely for cnidarians engage in bouts of copulation.[43] What is particularly interesting is the relative simplicity of the nervous system, which consists of a nerve net linked to a series of four pacemakers, a neural architecture that is effectively imposed by the jellyfish body plan.[44] Thus there is no brain,[45] yet complex eyes and sophisticated behaviour.

Seeing without a brain has certainly attracted notice, although as we shall see below (p. 165) other organisms have an eye that evidently can focus an image without even the benefit of a nervous system. But the case of the cubozoans deserves wider scrutiny. As one worker has remarked, 'If cubomedusae have developed unique image forming eyes, is it possible that they have also developed a unique method of processing the information?'[46] In this context it is also worth noting that nerve nets are by no means confined to primitive animals. They also occur in the echinoderms (the group represented by the starfish, brittle-stars, and sea urchins) and their relatives, the somewhat more obscure hemichordates (which, as it happens, are also quite closely related to the chordates, the group to which we belong). Interestingly, there is now evidence for unexpected complexities of function in the echinoderm nerve net.[47] Not only that, but there is now evidence that part of the brittle-star skeleton has been converted into a sort of compound eye, convergent on the arrangement seen in the trilobites (p. 159). Whether anything like an image is seen by the brittle-star is very conjectural, but here too the suspicion is that these brainless animals possess hidden levels of sophistication.

More specifically, knowing how it is that cubozoans see might even have a bearing on extraterrestrial alternatives. What is evident in these animals is that the nervous system shows levels of autonomy, so that at times it can act as an integrated network yet in other circumstances show a greater degree of independence and be capable of dealing with directional inputs. As Richard Satterlie and Thomas Nolen remark,[48] 'The presence of four pace-makers in cubomedusae, as well as their semi-independent coupling, allows significant modulation of swimming by complex sensory systems and fast, directionally accurate behavioral responses in a radially arranged, non-cephalized nervous system. Functionally, one could view [this] as a form of condensation of neural networks, analogous to cephalization in bilateral animals'.[49] What they are saying, albeit in somewhat technical language, is that coordinated and rapid activity may not automatically necessitate a brain with an attendant battery of sensory organs. Just as there are several ways of building eyes, each also convergent, so perhaps arising from a common substratum there is the possibility of several sophisticated neural arrangements evolving, and perhaps even distinct pathways to advanced intelligence marked by such features as plasticity of behaviour, memory, temperament, and even sleep?[50] These possibilities are explored in subsequent chapters, but we can at this point note that it is surely an exciting prospect to explore both the nature of the building blocks (e.g. proteins) and the levels of complexity that lead both to obvious differences and to striking similarities in the intelligences of cephalopods, insects, and vertebrates. And does an underlying neural commonality undermine the principle of convergence? Not at all. In the next chapter we shall look at the extraordinary sophistication of the social insects: here, if anywhere, is the nearest analogy to an extraterrestrial intelligence. There may well be several types of intelligence, but the number will be restricted and the pathways to each type strangely similar. And arguably there will be deeper, if not universal, similarities: in principle all will understand algebra.

So camera-eyes are remarkably diverse in their distribution, and they form an obvious contrast to the compound eyes that are so typical of the arthropods. In due course we shall see that in exceptional circumstances a compound eye can be transformed into a sort of camera-eye. Yet this is not the only route: another example of a sort of camera-eye is encountered in some of the spiders, notably those adept at net-casting, including a type known as *Dinopis*. This is a hunter of a special sort, suspending itself from a web but itself clasping a net of sticky and elastic silk. When prey is detected, the spider extends its net, rapidly drops and then pushes the net against its prey. Then, as it ascends, the net folds around the hapless victim.[51] The spider is nocturnal, and like other members of this group has eight eyes. In *Dinopis* two of the eyes, located on the posterior median part of the head, are strikingly large.[52] (This explains why these spiders are often called the ogre-faced spiders.) In cross-section each eye consists of a cup-like retina and overlying lens, the lens being embedded in an iris. Interestingly, the texture of the lens changes in consistency through its thickness, from jelly-like at the front to being harder further back. Evidently this, in combination with the retina, acts to correct for spherical aberration, in a manner similar to that of the fish lens (see above). The spectacular eye of this spider does not provide a good image; its function is rather to collect light, which is a useful attribute given the spider's crepuscular activities. A somewhat similar camera-like arrangement is found in the antero-median eyes of jumping spiders, but there the emphasis is very much on image formation,[53] to the extent that they can recognize video pictures of their prey (and mates).[54] In addition, in these spiders the retina has a quite extraordinary stacked structure, which even incorporates a telephoto system of a kind.[55] It has also been proposed that this multi-layered retina makes colour vision possible, and evidence exists for spectral sensitivity.[56] What is not in doubt is that for their size these spiders have a quite remarkable visual acuity. The occurrence of simple eyes, effectively camera-like, in these spiders is curious. This is because the majority of arthropods, the phylum to which the spiders belong, employ compound eyes, as evidently did the ancestors of the spiders. As we shall see (pp. 160–162) the nearest approach to a simple eye in the other arthropods is based on a radical reconfiguration of the standard compound eye. It remains an open question as to why this spider route was not taken more often.

The hallmark of the animals that evolved camera-eyes, be they squid, vertebrate, heteropod snail, alciopid polychaete, cubozoan jellyfish, or even spider, seems to be that all are active, mobile, and typically predatory. At first sight, the two other examples I have already briefly mentioned, those of the snails *Strombus* and *Littorina*, seem to be marked exceptions. *Strombus* is generally slow-moving and herbivorous. In remarking on the strombid eye Mike Land makes no attempt to conceal his puzzlement when he writes, 'What the eye is used for, other than simple taxes [i.e. responses to stimuli] for which it seems over-adequate, remains a mystery'.[57] Yet there is considerable evidence for electrical, and by implication visual, activity in the eye of *Strombus* (note 38). Perhaps we have underestimated this seemingly unassuming gastropod? Strombids are a very successful group in the tropics, and appear to have displaced similar forms such as the pelican-shells (aporrhaids) into colder waters.[58] Not surprisingly, members of both groups are anxious to avoid attack, especially by fish, crabs, and drilling snails, but the strombids are particularly adept. As Kaustuv Roy notes, when escaping they 'generate a series of very rapid leaps away from the predator ... [showing] some of most specialized and effective escape responses known among gastropods'.[59] Perhaps a sophisticated camera-eye is more useful to the strombids than might at first be thought?

What of the other snail, *Littorina*? There is little doubt that the winkles can see better than most other gastropods. This acuity is evident from a number of tests that show, for example, that various species of *Littorina* can orientate themselves at night, perhaps by recognizing silhouettes,[60] and are adept at navigating both by day and beneath starlight.[61] It is perhaps this snail's habitat, which is typically a gently sloping tidal flat,[62] that provides the explanation, inasmuch as there is a premium in being able to recognize particular shapes, especially plant stems. This is because as the tide rises, and the predators move in, *Littorina* can climb the stems to relative safety.[63] Such optical acuity is probably an important component in explaining the immense success of the humble winkle.

EYES OF AN ALIEN?

It is beginning to look, therefore, as if active, fast-moving, and at least in some cases intelligent animals opt for the camera-eye.[64] So, too, we need to remind ourselves that commonalities of eye structure may lead to convergences in brain structure and the evolution of visual centres, which is at least the case within the mammals.[65] Yet, so far as visiting aliens go the story more often looks to the immense cylinder, embedded in the cricket pitch, the hatch slowly opening, a hideous scrabbling before the bug-eyed monster levers itself out and begins its obligatory, laser-cannon-powered destruction of the pavilion. The inspiration for this abrupt interruption of the smooth running of the Home Counties is presumably the alien-like appearance of insects, robotic limbs, and empathy-free compound eyes. Clearly, given the immense success of the arthropods, the compound eye has its advantages. For example, the eyes of some deep-sea crustaceans are remarkably adept at collecting the minute quantities of light available at depth, and do so by employing a sort of fibre-optic system.[66] This specialized arrangement has arisen independently in two pelagic groups, specifically a galatheid crab and the euphausiids.[67] Not only that, but it now seems likely that the compound eye itself is convergent in the arthropods, evolving independently in a group of crustaceans, the ostracods.[68] It is interesting also to note convergent evolution of the compound eye in the sabellid annelids[69] and bivalve molluscs[70] (Fig. 7.3). In the sabellid annelids this type of eye evidently has evolved independently several times.[71] In both these groups the shared convergence seems to reflect, however, a rather different adaptation from

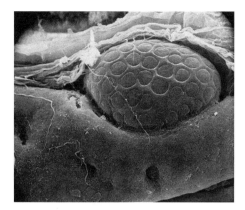

FIGURE 7.3 Convergence in the compound eye, in the case of the mantle eye of the bivalve *Arca*. (Photograph courtesy of Dr T. R. Waller, Department of Paleobiology, National Museum of Natural History, Smithsonian Institution.)

that found in the arthropods; here the compound eyes appear to act as a sophisticated optical alarm.[72]

Yet it is in the arthropods that the compound eyes are most familiar, ranging as they do from the familiar example of the insects to the calcite lens of the extinct trilobites.[73] Trilobite eyes are often claimed to be unique, but in fact they show two interesting convergences. The first is in terms of general organization, and entails a comparison with the eyes of the strepsipteran insects,[74] a group that we encountered earlier in the context of their convergently derived gyroscopic halteres. The overall resemblance to trilobite eyes, especially the more advanced type known as schizochroal (see below) is quite striking, but strepsipteran eyes (Fig. 7.1) are not calcitic. An equivalent arrangement, albeit on a smaller scale, is found, however, in a group of echinoderms mentioned above, the brittle-stars. Here, on the upper side of the five arms, there are the characteristic calcitic ossicles that make up the flexible skeleton, but evidently modified for the purposes of vision.[75] As in the trilobite lens, the optic axis is parallel to the line of sight, an important feature because the mineral calcite shows strong double refraction. This means that light passing through the crystal at an angle to the optic axis is split into two rays, resulting in the formation of a double image. Along the optic axis of the crystal, however, the rays remain together, so that the transmitted image remains single.[76] This convergence in calcitic eyes may extend further because each brittle-star ossicle is composed of a series of lenses, each forming a dome-like structure, which is strongly reminiscent of a trilobite compound eye. Just what the brittle-star 'sees' is rather conjectural because the nervous system is a diffuse net without a central brain (see note 47), and is therefore reminiscent of the cubozoan jellyfish (see above). Even if images are not perceived, the brittle-star's sensitivity to external stimuli is evident in its ability to avoid predators rapidly and in the striking changes in its colour on a daily cycle.[77]

Trilobite eyes have also attracted considerable attention because the internal structure of the more advanced type of lens, which is referred to as schizochroal,[78] appears to have a double component. Such a configuration, it is claimed, acts to minimize the spherical aberration of the perceived image. Accordingly this lens structure evidently anticipated, by hundreds of millions of years, human technology with its grinding of lenses with the same aim of minimizing such optical distortion of the image.[79] Interestingly, a somewhat similar arrangement, again to correct spherical aberration, is found in the calcitic lenses of the brittle-star (note 75). So far as the trilobites are concerned, not everyone, however, is entirely convinced by this elegant story. At least in some cases post-mortem changes, such as recrystallization in the eye structure, may provide artefacts of structure that were not present in the living trilobite.[80]

The magic of being able still to see through the lens of a trilobite[81] does not mean that one can necessarily avoid a distorted vision of these extinct animals, but let us return to the relative merits of camera- and compound eyes. Will the alien monsters be bug-eyed, or should we prepare ourselves to return the steady gaze of sentience through an eye like ours? As it happens, more probably the latter. This is because, despite its versatility, the compound eye simply does not match the light-collecting powers of the camera-eye. Apart from the ogre-faced and jumping spiders discussed earlier (pp. 156–157) there seems to be only one exception to the rule that arthropods typically have compound eyes (see also note 73). This concerns the unique eyes of a shrimp belonging to a group known as the mysids.[82] As in arthropods, the principal eyes are compound, but embedded in the posterior region of each eye there is an extraordinary addition in the form of a giant lens (Fig. 7.4). In effect the shrimp has evolved a simple eye, consisting of a single enormous lens that overlies a giant crystalline cone. This serves to channel the light to a retina composed of more than a hundred of the individual optical units that are referred to as the rhabdoms. Effectively these rhabdoms, which in the normal compound eye

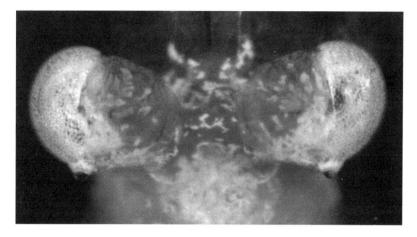

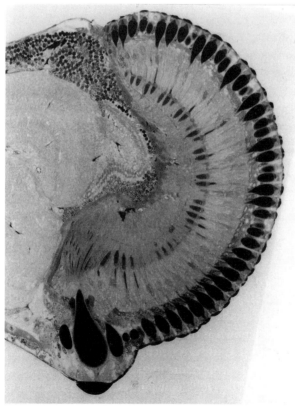

FIGURE 7.4 The compound eye of a mysid shrimp. Above, entire eye with simple eye located on posterior side. Below, transverse section of eyes. Simple eye is on lower side. (Photographs courtesy of Professor D-E. Nilsson, University of Lund.)

each underlie a separate lens, have been diverted to supply the giant lens. In life the eyes of the shrimp can move freely, and used together the two giant lenses evidently act as a binocular. Looking at this animal changes our perception; as Dan Nilsson and Richard Modlin remark, 'Viewing the live animal from behind ... gives an almost uncanny feeling of being observed'.[83]

So here in one particular arthropod we have a design that approaches that of our own eyes; yet from the point of view of evolution it appears to be a dead end. As the investigators remark,

> The ultimate development is ... a large single facet and crystalline cone supplying light to a common retina of numerous rhabdoms. There can be no further improvement along this line and [the giant eye of this shrimp] has reached the ideal design for a compound eye. Ironically, the ideal compound eye is not really a compound eye at all. It has been turned into a simple eye in order to exploit the advantages of imaging through a single aperture. There is thus little reason for surprise over the unique design in [this shrimp]. It would be more appropriate to wonder why [it] does not share its ingenious solution with other crustaceans or with insects.[84]

The answer may lie in the difficulties in reorganizing the neural circuitry and, as it happens, with the notable and already noted exception of some spiders (see p. 157), nearly all other arthropods remain stuck with their compound eyes.

In the final analysis camera-eyes are not only different from compound eyes; they are better. This is because in the compound eye the visual array can be increased both by expanding the overall area of the eye and also by the increasingly dense packing of the lens. There are, however, limits to both, and in particular as the lenses shrink in size so their ability to collect enough light to function is compromised. Calculations suggest that if humans were to rely on a compound eye to achieve equivalent vision it would have to be at least a metre across, and more realistically up to 12 metres wide (Fig. 7.5).[85] On distant planets there might (or might not) be bug-eyed aliens, but any astronomers are almost certainly going to be equipped with camera-eyes. Perhaps, too, they will have binocular vision, a reasonable enough assumption given its convergent acquisition in birds and mammals.[86] Yet, it is also worth remarking here that in a certain

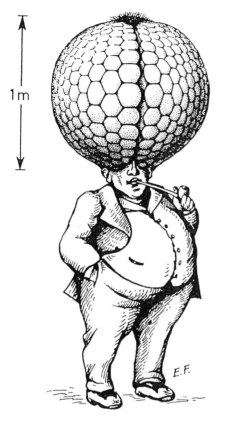

1 m

FIGURE 7.5 Why we don't have compound eyes. A conservative estimate of the size of eye a human would require to provide even a near-equivalent of the camera-eye. (Reproduced with permisssion from *Neural principles in vision*, edited by F. Zettler and R. Weiler, in the chapter (pp. 354–370) by K. Kirschfeld, fig. 7c; 1996, copyright Springer-Verlag and also with the permission of the author.)

sense the differences between compound and camera-eyes are, at one level, rather superficial. For example, the process of visual tracking in which an animal's coordination of vision and locomotion ensures successful interception of prey (or mate) shows that between humans and such insects as the mantids and hoverflies[87] there are 'some remarkable similarities'.[88] Not surprisingly, given these similarities, binocular vision has also evolved independently in a number of insects.[89] As we shall see later, there are also convergences in the molecular architecture of vision, notably among the proteins known as crystallins and rhodopsins, both of which are essential for any animal eye to work. Moreover, the convergences discussed below in such sensory modalities as olfaction and hearing make it less surprising that there may be parallels in the neural processing of, say, insect and vertebrate eyes.[90]

The limitations of biological form and the reality of the adaptive framework provide a realistic guide to what there will be 'out there'. Even let us suppose that our extraterrestrial has the inevitable camera-eyes, but these are mounted in little turrets and able to swivel independently. Again we can find a striking convergence here on Earth. Such an arrangement is well known in the chameleon lizards. It entails some remarkable changes to the basic camera-eye. Among the most striking of these is the suppression of the lens and its replacement for refractive purposes by the cornea, which is duly equipped with muscles to effect the changes in shape necessary for focusing. How odd, but this rearrangement is strongly convergent on the eyes of the sand-lances, a group of fish.[91] And despite the different media, air as against water, there is a common adaptive explanation. Both chameleons and sand-lances are capable of exceedingly accurate and rapid strikes at their respective prey: in effect the lunging of the entire fish is equivalent to the darting of the chameleon tongue.[92]

In discussing eyes a number of qualifications are important. First, it is agreed that eyes have evolved independently very many times. In their classic paper Luitfried v. Salvini-Plawen and Ernst Mayr[93] write, 'Summing up the different and convergent sequences towards eye perfection in general, there are about 20 or even more independent lines of differentiation including at least 15 cases of independent attainments of photoreceptors with a distinct lens.'[94] Although I have emphasized the importance of compound and camera-eyes and their convergences, unique optical configurations do occur. Perhaps the best example is the remarkable eye (or eyes, given that an individual has a whole series distributed along the shell margin) of the scallop, which although camera-like has a system of internal mirrors.[95] As Michael Land aptly remarked,[96] this represents 'an almost unique example of a type of eye that *should* exist, but which had not been found until relatively recently'.[97] Even so, the reflectivity of the eye depends on the mirror-like properties of the material guanine.[98] This compound has already been encountered in the rather different context of providing thin, flexible sheets that would help to gas-proof a Fortean bladder (pp. 112–113). In an optical context, however, not only is guanine found in scallop eyes, but it also forms the reflective structures, known as the tapetum, that are found in the eyes of certain vertebrates, notably deep-sea fish, where they act to concentrate light.[99] Nor does its 'optical' use stop there, because the silvery guanine is also widely used as a reflector to channel the light generated by the luminescent organs known as photophores.[100]

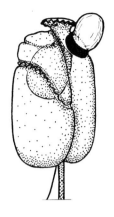 

FIGURE 7.6 The optical apparatus of two dinoflagellates, *Erythropsidinium* (left) and *Warnowia* (right), magnified respectively ×460 and ×380. Note the bulbous lens and underlying cup. (Reproduced from fig. 1 of Greuet (1968; citation is in note 106), with permission of Urban and Fischer Verlag. (*European Journal of Protistology.*)

Second, and even more interestingly, eyes are by no means restricted to animals. Most familiar are the various sorts of eye-spot (more strictly light antenna or photosensory apparatus) in the protistan eukaryotes such as the single-celled *Chlamydomonas* or globular colonial *Volvox*.[101] Here, too, independent evolution is widespread,[102] and it also turns out that the visual protein (rhodopsin) employed in these eye-spots may be convergent on the equivalent protein found in our eyes and those of other animals (see pp. 171–172). Typically the protistan eye-spot has a reddish or orange colour; this is due to a layer of carotenoids that serve to reflect the incoming light back against the photoreceptor.[103] Even so, among the examples of protistan eye-spots, the optical system in the group known as the dinoflagellates is particularly remarkable.[104] In this group a variety of eye-spots occur,[105] but in certain taxa (e.g. *Erythropsidinium* and *Nematodinium*), the organism, which, remember, is a single cell, bears a protruding bulb-like structure. In appearance, this is almost telescope-like (Fig. 7.6). Its structure is complex, but in essence it consists of a pigmented cup (melanosome) surmounted by a crystalline lens (or hyalosome)[106] which can refract light so that it focuses on the underlying sensitive layer.[107] Despite the absence of a nervous system, this tiny optical protuberance, less than a tenth of a millimetre long, is strikingly convergent on the animal eye,[108] although most probably it originates from a chloroplast.[109] Somewhat nervously, in describing the behaviour of these dinoflagellates, Ester Piccinni and Pietro Omodeo[110] remark, 'It is unthinkable that an apparatus of this sort, in spite of the sophistication of its design, can function like an image-forming eye',[111] but they speculate that these organisms may detect their prey and then entangle it with explosively discharged thread-like structures,[112] which are also convergent.[113]

CLARITY AND COLOUR VISION

The existence of these primitive eye-spots exemplifies once again the important, but I believe relatively neglected, principle of inherency, whereby the basic building blocks of complex structures are available long before they are recruited for new and more sophisticated tasks. This observation is important for two reasons. First, if such building blocks, necessary for the assembly of complex structures, are readily available – as seems to be the case – then, unless there is some neglected and at present mysterious principle of evolution that imposes an almost insuperable barrier of some sort, there seems no reason why, in a world subject to adaptive scrutiny and competition, that sophisticated forms, such as eyes, should fail to evolve repeatedly. Second, and as a consequence of the first point, it is apparent that evolutionary novelty is often only skin-deep. This is because its emergence relies more on co-option and redeployment than invention.

Thus, in the case of the eye, what, in some ways, could be more remarkable than having a window into the head, that is transparent tissues that not only allow transmission of light but remain functional for decades? The solution is provided by a number of specific proteins that serve to confer transparency to the lens. For this reason they are known as the crystallins, although it is important to stress that they are not specifically crystalline in the way, for example, of a mineral like quartz. As is now well known, these crystallins clearly must have been co-opted from serving some previous function. This is because originally these proteins evolved for completely different purposes, most usually associated with providing the organism with resistance to physiological stress.[114] Of these stress resistors, the best examples are the heat-shock proteins that are widespread and found even in bacteria: in this sense crystallins are inherent in evolution from billions of years ago. Nor is their recruitment difficult to understand. Long-term stability in the face of environmental insult is clearly highly desirable, especially when repair mechanisms that entail a blood system would compromise optical transparency. In addition, the proteins recruited for this purpose are typically rather small in size. This favours close packing in regular arrays that, together with a water-rich environment, confers the necessary transparency.[115] And however strange the construction of the eye may be, such as is seen in the remarkable mirror optics of the scallop eye (which actually has a direct parallel to some telescopes[95]), the transparency is still conferred by the agency of crystallins.[116]

The importance of the crystallin proteins in conferring transparency to tissues is nicely shown in what might be called a 'reverse eye'. In some squid, for example, the body bears bioluminescent organs (the photophores) that emit flashes of light. The tissue of this 'reverse eye' is derived from the musculature, yet not only are there sometimes lenses[117] but once again the transparency is conferred by recruitment of a crystallin.[118] More generally, the wide range of co-opted proteins that provide transparency, be it in eyes or 'reverse eyes', is another striking example, not only of convergence, but also of biochemical convergence.[119] Just as I speculated that alien astronomers (if there are any) would be searching the skies using camera-eyes, so it would also seem more than likely that inside their eyes the lenses would be packed full of the direct equivalents of the crystallin proteins.

The other essential component of any eye is a mechanism for converting the incoming photons into electrical signals that, somehow, the brain can interpret as an image. In detail this entails a very complicated sequence, but the key element is a group of proteins known as the opsins. Of these the protein rhodopsin is probably the most familiar. Its basic structure is of some interest in that we shall encounter the arrangement of seven α-helices, each of course made of a string of amino acids, spanning two sides of a lipid membrane in another sensory context (pp. 180–181). Embedded within the rhodopsin molecule, and attached to two of the α-helices, is a molecular unit known as a chromophore. This is usually a derivative of vitamin A, known as retinal[120] and hence explains why such a vitamin disorder, sometimes genetic, can lead to difficulties with sight.[121]

Rhodopsin is, therefore, as integral to vision as are the crystallins. This protein also offers further insights into the nature of convergence in at least two ways. The first is rather specific, and concerns the – to us – familiar property of colour vision. The second is more general and also controversial: how many times has rhodopsin evolved? To humans the glory of a rainbow is appreciated because our colour vision is trichromatic (blue, green, red), a relatively unusual condition among the animals. In mammals, for example, it is otherwise found only in the apes and their close relatives, the Old

World monkeys, and convergently in the howler monkeys of the New World[122] (as well as possibly in some of the more primitive prosimians[123]) and some Australian marsupials.[124] That we are able to see in colour is generally thought to be because our tree-dwelling ancestors had strong incentives to locate either appropriately coloured fruit[125] or leaves.[126] In some of the New World monkeys, such as the marmoset, there is a particularly intriguing mixture of colour discrimination whereby all the males see dichromatically, whereas the females are either trichromatic or dichromatic according to whether the individual is respectively heterozygous or homozygous.[127] Here, too, there is some evidence to suggest that this peculiar arrangement might have evolved independently among the New World monkeys,[128] and might in fact confer an adaptive advantage in the co-operative searching for fruit by males and females, given that each is attuned to different visual signals.[129] Some support for this notion might come from humans who happen to be dichromatic. Their apparent deficiency is offset by a remarkable ability to penetrate the disguises of camouflage, be it military[130] or natural, as in detecting otherwise cryptic moths.[131] Many other mammals[132] and some invertebrates are dichromatic, while other animals can see beyond our range of visual sensibility into the region of the ultraviolet.[133] As is well known, on account of the relative commonness of red–green colour deficiency in male humans, colour vision has a genetic basis,[134] and it is also evident that the origins of trichromacy are related to episodes of gene duplication.

The changes in the structure of the rhodopsin molecule that make possible the absorption of particular wavelengths of light, and so the perception of colour, depend on substitutions of one amino acid by another at key sites along the length of the protein.[135] In red–green vision the replacement typically involves the insertion of a hydroxyl-bearing (i.e. OH) amino acid in place of a non-polar one (i.e. one that 'dislikes' water), although changes that involve the substitution of a different-sized amino acid can also play a role.[136] Quite why a particular substitution should alter the wavelength of light absorbed has only recently become clear.[137] The sites of substitution are highly specific and are limited to a mere handful, of which even fewer are really important. This has led to the identification of the so-called 'site rules', which for red–green vision involves five sites. It is not so surprising to learn, therefore, that not only has red–green vision evolved independently, but at least in some cases its molecular basis is strongly convergent. One well-known example is the convergence in red vision between a fish and mammals where two, and possibly three, sites show identical substitutions.[138] Indeed, within the vertebrates the evidence continues to accumulate that the 'five-site rule' is very widespread,[139] and although there are some variations these are much more minor.[140] So, too, with the registering of blue wavelengths of light, where in a number of vertebrates there are a number of specific substitutions of the rhodopsin molecule.[141] Rather remarkably, in at least some toothed whales and seals there has been a loss of blue-sensitive visual pigments, an absence that is evidently convergent in these aquatic mammals.[142] For animals living in deep water a shift in the opposite direction, towards the red end of the spectrum, is to be expected, and here, too, at least in a series of distantly related teleost fish, the spectral tuning is again controlled by a series of specific substitutions.[143] For ultraviolet (and violet) vision in various vertebrates, the story of site substitutions is slightly more complicated. In the ancestral form, ultraviolet vision evidently depended on amino acid replacements at about eight specific sites, and for the most part only limited substitution is possible. The birds, however, regained sensitivity to these short wavelengths by the replacement of an amino acid at a completely different site.[144]

Rather less is known about the molecular controls on spectral absorption, and by implication colour perception, among the invertebrates, but broadly the same picture emerges.[145] The case of the cephalopods, such as the squid, is particularly important because of the striking convergence in terms of the overall arrangement of the camera-like eye (Fig. 7.2). The octopus is evidently colour-blind, but in one species of squid, where the rhodopsin[146] shows a spectral shift, there is a key substitution at one site that is evidently the equivalent of one of the mammalian sites involved with colour vision.[147] What of those animals, notably the arthropods, with compound eyes? They certainly have colour vision,[148] and in at least some instances there are interesting convergences with the vertebrates. Thus in some butterflies both blue- and red-shifts have exact parallels with mammals, while among the butterflies themselves other convergences exist.[149] So, too, among the crayfish, in those regions of the rhodopsin molecule that are similar to those for the vertebrates several sites may be involved with a shift in spectral absorption.[150] As a group, insects have a wide range of spectral sensitivities. In the fruit-fly

(*Drosophila*), for example, the principal window of vision is in the blue and ultraviolet range, and this too appears to depend on strongly constrained amino acid sequences. Certain other site substitutions, however, have parallels in the vertebrates.[151]

It should also be noted that many animals, such as fish and birds, incorporate colour filters in their retinas by employing oil droplets to absorb particular parts of the light spectrum.[152] This also appears to be convergent. So, too, many animals have independently developed the capability of detecting polarized light, perhaps most famously the honeybee, which uses differences across the sky to assist it to navigate. In certain instances the sensitivity to polarized light is such that it is analogous to colour vision, and hence is referred to as polarization vision. This is known to occur in the stomatopods,[153] a group of crustaceans long known for their extraordinary eyes[154] and a lifestyle that encompasses effective hunting and communication. It is perhaps less surprising that polarization vision has evolved convergently in the cephalopods such as the octopus and squid.[155]

UNIVERSAL RHODOPSIN?

As it happens, a number of other proteins can act as light receptors, and they too provide some illuminating cases of convergence. Yet the rhodopsin molecule seems to offer various advantages: as Russell Fernald[156] has remarked, in the context of vision although an opsin 'is not the only way to detect light . . . it has proven irresistible for use in eyes'.[157] Rhodopsin is remarkably widespread, and apart from its near-universal adoption in the various visual systems this molecule has long been known to occur in the bacteria. These organisms exist as two kingdoms, the Archaea and the Eubacteria, of which a typical representative is *E. coli*. The most notable examples of rhodopsin are in those Archaea referred to as halophiles. As their name indicates, they live in salt-ponds, where their immense numbers frequently colour the lagoons red or purple, which is the colour of the pigment associated with their rhodopsin. The molecular architecture of this so-called bacteriorhodopsin,[158] with seven α-helices spanning a lipid membrane, is the same as the molecule in our retina. So, too, is its basic dependence on light, but in the bacterial variety the energy of the Sun's photons is used to drive a proton pump[159] to transfer hydrogen ions (i.e. protons, H^+) as the basis of the cell's energy system, specifically by the synthesis of ATP.

Bacteriorhodopsin has received a vast amount of attention. This is principally because of its now-classic role as an ion pump and the desire to elucidate its detailed structure as a protein.[160] Yet a remarkable fact and an intriguing inference concerning bacteriorhodopsin have emerged. The former concerns its widespread distribution. Its occurrence in archaeal bacteria is well known, but only recently has it been appreciated that in oceanic ecosystems this molecule occurs in the eubacteria (in which organisms it is referred to as proteorhodopsin), where it appears to play a key role in the harvesting of the sunlight in a way analogous to photosynthesis.[161] Nor does the distribution of bacteriorhodopsin stop there. As we have seen, various single-celled protistans, including some green algae, have eye-spots. It is hardly surprising, therefore, to find rhodopsin in the single-celled *Chlamydomonas* and colonial *Volvox*.[162] This story has had its convolutions, especially with the identification of a membrane-bound protein, referred to as chlamyrhodopsin, in the eye-spot. This, however, has a molecular structure that is markedly different from that of genuine rhodopsin.[163] It is now clear that not only does rhodopsin itself occur, but it evidently has a bacteriorhodopsin-like structure.[164]

Bacteriorhodopsin, however, is even more widespread, and it also occurs in the fungi. One instance is in the parasitic chytridiomycete fungi. Their reproductive cells, known as zoospores, are equipped with a rhodopsin-bearing photoreceptor that guides this swimming stage towards the light.[165] In many, and perhaps all, these fungi the rhodopsin employed is equivalent to the bacteriorhodopsin. The intriguing question, therefore, is whether this type of rhodopsin is strictly the same as that found in our eyes, or is in fact convergent. The latter seems to be more likely. First, it has long been appreciated that there is no discernible similarity in the amino acid sequences when the two types of rhodopsin are compared.[166] Furthermore, despite the overall similarities of molecular architecture there are some specific differences.[167] In their review of rhodopsins, John Spudich and colleagues[168] conclude,

> the genome sequence data presently available and the three-dimensional structures of the molecules themselves argue that nature discovered retinal [i.e. the vitamin A derivative] twice, and both times found it useful, when solvated with seven helices, for photosensory signalling as well as other phototransduction

functions. The two-progenitor hypothesis would require that archaeal sensory rhodopsins and mammalian rhodopsins have converged on remarkably similar mechanisms of receptor photoactivation. Such closely similar mechanisms could result from 'likely reinvention' determined by the inherent properties of retinal as a chromophore.[169]

The remarkable similarity of the molecular architecture of the two rhodopsins may, therefore, well be convergent. If this is the case it is perhaps not surprising that bacteriorhodopsin is also very sensitive to substitutions at key sites. Thus, in a way analogous to substitution at key sites controlling colour vision, so meddling with site 85, for example, can completely destroy the ability of the bacteriorhodopsin to pump protons.[170] And the parallels may be even closer, because it appears that some types of bacterial rhodopsin are also capable of a colour discrimination,[171] suggesting once again the possibility of a universal property. Wherever life sees light it will be through the agency of a rhodopsin, and where it sees red it will depend on key amino acid substitutions. So, too, the sentient inhabitants of Threga IX will also enjoy those spectacular red sunsets, by camera-eyes, through transparent crystallins, and by means of rhodopsin.

Light reception by organisms is therefore very widespread. Rhodopsin may have evolved twice, but other proteins can also act as light-receptors. Opsins may well be best, but the other examples are also instructive in the context of both light sensitivity and also convergence. For us, the role of light is effectively synonymous with vision, yet, as is widely appreciated, there are also perceived cycles of light and dark. Most notable, perhaps, are the circadian rhythms of the 24-hour solar cycles, of day and night.[172] Also important is the longer-term seasonality, with the most familiar effects in higher latitudes of autumnal colours in plants, as well as hibernation or migration in animals. Such rhythms and changes are very widespread and are monitored by light-sensitive molecules. Of central importance are the so-called cryptochromes, cryptic neither in being difficult to find nor in establishing a photoreceptor function, but because until quite recently their true structure was problematic. Cryptochromes are proteins with a flavin component (and linked to a vitamin B_2-based pigment) that are sensitive to blue light.[173] They occur, for instance, in the retinas of mammal eyes, where they play an important role in the setting of the circadian clock, effectively the discrimination between night and day.[174] The importance of this clock is familiar from the disorientating effects of jet lag; more seriously, disruption of the circadian clock is implicated in major malfunctions during the night shifts of nuclear reactors and chemical plants.[175] Danger might also lie elsewhere. For example, disruption of circadian rhythms by the widespread use of artificial light may also be implicated in the increased prevalence of breast cancers.[176]

Given their role in light sensitivity, it is unsurprising to learn that cryptochromes occur also in algae[177] and plants.[178] What is more interesting is that the plant cryptochromes appear to be very ancient, whereas those of animals are more recent and an independent invention.[179] Such a convergence is less surprising in one sense, because in both cases the precursors of these cryptochromes lie with a group of proteins known as the photolyases whose function is to repair DNA damaged by light, especially by ultraviolet radiation.[180] Curiously enough, although humans have cryptochromes most probably we lack photolyases, thereby increasing our vulnerability to ultraviolet radiation and, by implication, skin cancers.[181]

SMELLING CONVERGENCE

Given that key molecules required for vision, such as rhodopsin and the crystallin proteins, evolved in single-celled organisms this suggests that given time[182] and the adaptive value of light discrimination then the evolution of the eye seems to be a near inevitability. And so fascinating is the story of the evolution of the eye, not to mention our particular dependence on vision, that we tend to forget other types of sensory world, even when walking the dog at night with the air full of bats: to us other senses, such as those of olfaction and echolocation, are almost closed books. Yet the principle of evolutionary convergences should encourage us to think, not only of recurrent re-emergences of given senses, but also about deeper similarities of sensory input, such as those of hearing, generation of electrical fields, and perhaps even the sense of smell. In this last case there have been some curious experiments, such as those by the remarkable Victorian polymath Francis Galton. Today he is probably best remembered for his interest in hereditary principles and the related application of eugenics,[183] the monstrous nature of which was articulated with characteristic prescience by G. K. Chesterton.[184] Galton's autobiography[185] is still entertaining, even if it reveals an individual to whom self-doubt was an almost complete stranger.[186] Perhaps that may explain some of the less appealing aspects of Galton. Thus, in his absorbing account of the recognition of the importance of fingerprints in forensic investigation, Colin Beavan[187] argues that for all his contributions to this area, driven in part by the hope that fingerprints would yield eugenic insights, Galton was less than fair to the other pioneers, not least to one Henry Faulds. Indeed, Beavan shows that behind the façade something unpleasant, but not so rare, lurked: 'Other people's success aroused a venomous jealousy in Galton.'[188] Well, well.

In his memoirs Galton mentions a paper he had written, entitled 'Arithmetic by smell'.[189] Galton's interest was evidently sparked by what is called non-verbal representation, crudely whether we (and by implication other species) can think and handle abstract concepts without recourse to words. So Galton decided to investigate one of the further reaches of mental representation. Here is what he wrote:

> Leaving aside Colour, Touch and Taste, I determined to try Smells. The scents chiefly used were peppermint, camphor, carbolic acid, ammonia, and aniseed. Each scent was poured profusely on cotton wool loosely packed in a brass tube, with a nozzle at one end ... A squeeze of the tube caused a whiff of scented air to pass through the nozzle. When the squeeze was relaxed, fresh air was sucked in and became scented by the way. I taught myself to associate two whiffs of peppermint with one of camphor, three of peppermint with one of carbolic acid, and so on. Next, I practised small sums in addition with the scents themselves, afterwards with the mere imagination of them. I banished without difficulty all visual and auditory associations, and finally succeeded perfectly. Thus I fully convinced myself of the possibility of doing sums in simple addition with considerable speed and accuracy, solely by imagined scents. I did not care to give further time to this, as I only wanted to prove a possibility, but did make a few experiments with Taste, that promised equally well, using salt, sugar, quinine, and citric acid.[190]

It is easy to smile at Galton's equivalent to an olfactory abacus,[191] yet these convergences of sensory input may have more subtle evolutionary implications. Consider, for example, the star-nosed mole (Fig. 7.7).

FIGURE 7.7 The star-nose mole, a typical mole in terms of its powerful fore-limbs, but remarkable for its tentacular nose, with its nostrils clearly visible. (Photograph courtesy of Professor K. C. Catania (Vanderbilt University, Tennessee).)

We have already met the moles as a splendid example of fossorial convergence among the burrowing mammals. The star-nosed variety is unusual in constructing its burrows in wet and boggy ground, and it also spends a considerable amount of time swimming.[192] This mole inhabits the eastern part of North America, but it may be a recent immigrant because fossils have been found in Poland.[193] It is almost blind, but by way of compensation has an extraordinary nose consisting of 22 mobile and fleshy appendages, somewhat like a star (Fig. 7.8). This nose, however, is not used for smelling, nor do the tentacles serve to capture food directly. Rather, as Ken Catania has so elegantly shown, this remarkable structure retains its sensory powers for the detection of prey, but is modified in a most interesting fashion.[194] The surface of the nose is densely studded with button-like sensory structures, known as Eimer's organ (Fig. 7.8).[195] It is estimated that the nose carries about 25 000 of these mechanoreceptors. Not surprisingly, the nose is exceedingly sensitive, and although it is only about a centimetre across it is supplied with five times as many nerves as run into the human hand. But the real surprise is that not only is this nose ultra-sensitive, but more importantly it shows some remarkable parallels to the eye. In particular, one region of this nose towards the

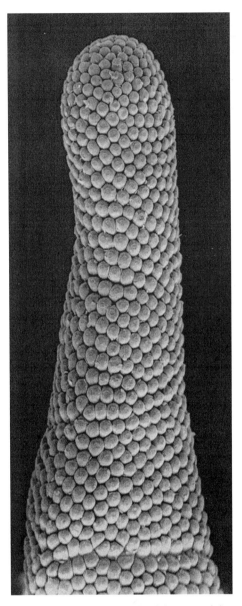

FIGURE 7.8 Details of the nose of the star-nosed mole. Above, detail of a tentacle, studded with the sensory Eimer organs. Opposite, electron micrograph of the nose. The analogue of the fovea is represented by the two central lobes beneath the nostrils. (Photographs courtesy of Professor K. C. Catania (Vanderbilt University, Tennessee).)

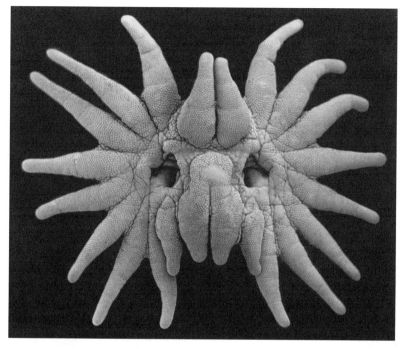

FIGURE 7.8 (cont.)

central area is especially well supplied with nerve endings. As such it can be compared directly to that region of the eye known as the fovea, where the retina possesses exceptional sensitivity. Both this part of the mole's nose and the fovea show a comparable degree of nervous acuity.[196] In making comparisons between nose and eye Catania is unequivocal about their similarities when he writes, 'The many surprising parallels between the somatosensory system of the mole, and the visual systems of other mammals, suggest a convergent and perhaps common organization for highly developed sensory systems.'[197] Interestingly, in another underground mammal, the naked mole-rat, which we met earlier (Chapter 6, p. 142) in the context of the convergent evolution of eusociality, nearly a third of that part of the brain allocated to touch perception (the somatosensory cortex) is rededicated to the prominent incisor teeth.[198] Effectively these teeth take the place of the usual rat whiskers, and given that the naked mole-rat is for all intents and purposes blind, it is not surprising that the area of the brain normally employed in vision appears to have been almost entirely taken over by the somatosensory cortex. So, in a way analogous to that of the star-nosed mole, these naked mole-rats probably 'see' with their teeth.

'Seeing' with a nose (or even teeth) allows for some intriguing speculations. Imagine a planet ejected early in its history from its parent solar system, which as we saw earlier may be an all-too-common occurrence (Chapter 5, p. 86). A position whirling through the Stygian gloom of interstellar space might seem to make such a lonely world an improbable abode for life,[199] especially at an advanced level. Probably so, but let us imagine also an abundance of radioactive elements in the planetary crust, whose continuing breakdown by decay serves to maintain subterranean ecosystems at viable temperatures. Perhaps in caverns, which on Earth extend at least 1.5 km beneath the surface, any animals would be 'seeing' in much the same way as the terrestrial star-nosed moles and naked mole rats. Some hint of how adept such hypothetical animals might be also comes from studies of how the blind mole-rats navigate their complex burrow systems in a way that would do justice to a rat in a maze.[200] Sensory cues may possibly come from the Earth's magnetic field, but it also seems conceivable that the mole-rat is able to build a cognitive map: one more hint of mentality in 'dumb' animals (see Chapter 8).[201] There are even instances at the surface of the Earth in which eyes are lost, or more strictly eaten, without deleterious effects. One particular example concerns a population of tiger snakes living on Carnac Island, off the western coast of Australia. Silver gulls attack these snakes and remove the eyes, but the snakes are in a sense 'over-designed' and they are able to survive[202] using other sensory mechanisms.[203] It is also worth noting that while snakes cannot re-grow eyes, where regeneration is possible, as in arthropods, amputation of a sensory appendage may lead to its replacement by other appendages with their own neurological (and sensory) characteristics.[204]

The remarkable proboscis of the star-nosed mole provides a number of other useful and relevant insights into the tension between novelty and constraint in evolution. In one sense the origins of this mechanosensory nose are not so peculiar; the related talpids, that is the other true moles, also have highly sensory noses, again studded with Eimer's organs. More specifically, comparisons with moles such as Townsend's mole (a species of *Scapanus*) give a good idea of how the star-nose might have arisen.[205] Even so, in its evolved form the star very much represents a biological novelty. Many animals have

fleshy appendages of various kinds, which routinely employ a particular developmental pathway that depends on specific developmental
genes (notably one called *distal-less*). In the star-nosed mole, however, the appendages develop in a unique manner, without parallel
elsewhere.[206] Why this mole should pursue such a novel developmental route when the genetic template for appendage formation is
readily available, and indeed presumably coded for in the limbs of
the mole, may seem rather odd. There are, however, particular constraints. Catania and his colleagues suggest that the need to keep the
neural connections, without which of course the nose is functionless,
means that the standard method of 'let's build an appendage' simply
is not available. Finally, remembering the topic of convergence, it is
worth mentioning that the development of the elaborate nose is not
without consequences for the architecture of the skull. In particular,
the chewing apparatus is quite modified (e.g. weaker dentition and
more fragile mandible), and as such is convergent on other animals
that are either vermivores or termite eaters and thus tend to swallow
their prey.[207]

As noted above, the star-nosed mole does not employ its spectacular proboscis for olfaction. But here, too, those animals that do have
a sense of smell show some striking similarities. So it is that despite
major anatomical differences the basic mechanism of olfaction seen
in the insects (usually located in the third segment of each antenna)
and vertebrates (the nose, of course) depends on nervous structures
known as the glomeruli.[208] These act to connect and integrate the
olfactory messages that are then conducted as electrical signals via
the nervous system to the brain, where the discrimination of odours
takes place. To be sure, the proteins (and thereby genes) that serve
to bind the odours, as the first step in olfaction, are different,[209] but
as John Hildebrand and his colleagues have emphasized, despite the
molecular genetics being quite different, our sense of smell and that
of the insects (and other arthropods) works on the same principles,[210]
and so encompasses such specific examples as the sex pheromones.[211]
Not only do the properties of the insect glomeruli 'have precise counterparts in vertebrate olfactory bulbs',[212] but Nick Strausfeld and John
Hildebrand go on to remark, 'Basic types of olfactory information
coding are remarkably similar in both vertebrates and insects' with
various excitatory responses to odours strongly indicating 'common
strategies for processing odor information at the level of glomerular

relay neurons'. These writers are fully aware of the implications when they continue, 'The primary olfactory centers in the brains of vertebrates and insects thus appear to share canonical cell arrangements and common physiological properties ... Furthermore, vertebrates and insects show remarkable parallels in events underlying the development of glomeruli ... Do such commonalities suggest that the glomerulus, as an olfactory functional unit ... originated in a common bilaterian ancestor [i.e. to insects and vertebrates]? Or, is common design and physiology the consequence of convergent evolution? Is there a uniquely logical response to common selective pressures such that to construct a glomerulus requires the same set of rules?'[213]

A definitive answer is not yet possible, but given the differences in the molecular genetics of olfactory receptor coding, the evidence points strongly to convergence. This is because the predecessors of the vertebrates, the amphioxus animal (see Chapter 1), and the insects, that is the aquatic crustaceans,[214] do not possess such glomeruli, which indicates that the method of effectively identical olfaction arose independently.

Once again it seems that the options are limited. There may be only a few ways to smell, and, as we have seen in the star-nosed mole, if a nose is not used for olfaction it can end up as the sensory equivalent of an eye. And again there may be deeper similarities pointing towards the inevitability of evolutionary end results. The processing of the olfactory signals, and especially the recognition of particular odours, involves a synchronized neuronal activity that leads to an oscillatory firing of the neurons. The associated networks are not only similar in both the insects and the vertebrates, but they in turn may also find parallels to what occurs in the processing of visual stimuli in the mammals, at least.[215] Further evidence for more profound similarities between the processes of olfaction and vision also comes from an investigation of proteins known as arrestins which, as their name suggests, are involved with the termination of a physiological process, such as the quenching of phototransduction in rhodopsin.[216] Given that the proteins involved with olfaction have a basic similarity to such visual proteins as rhodopsin,[217] it would not in principle be surprising to find arrestins also involved with the olfactory transduction pathway. What, therefore, is the more remarkable is evidence for a specific family of arrestins that is involved in both visual (eye) and olfactory (antenna) transduction in insects.[218] This is not the only example of such overlap in molecular transduction in the two systems,[219] and it is further evidence for a deeper, and largely unappreciated, commonality of these sensory systems.

THE ECHO OF CONVERGENCE

The richness and versatility of the sensory worlds does not end here. Many animals, in addition to the star-nosed mole, inhabit gloomy and crepuscular worlds, but generate elaborate and sophisticated fields of perception that arguably are as refined and sensitive as any visual or olfactory system. Most familiar in this regard, perhaps, are the extraordinary powers of echolocation possessed by the bats,[220] a group that offers other insights into convergence.[221] The basic principles of echolocation are fairly self-evident, and are of course employed in a somewhat similar way by many people who are blind.[222] So, too, with cats.[223] In the case of the fast-moving bats, however, what is particularly remarkable is that even when the animal is very close to its insect prey, where the neural responses can no longer keep track of the returning echoes, the bat still keeps a lock on its prey. Evidently the bat must employ a filtering mechanism of some sort, but how this signal is actually interpreted by the brain is still mysterious.[224] Another intriguing feature is that not only will the bat navigate through crowded vegetation in pursuit of its prey, but the hunted moths have acute hearing and are capable of evasive action.[225] More extraordinary are those moths that also produce ultrasonic clicks, the purpose of which appears to be to jam the bat's sonar.[226]

Echolocation is not confined to the nocturnal sonar of the bats. It is also well known in various marine mammals, such as the dolphin,[227] but in the context of considering convergence we shall postpone a visit to these wonderful animals because of their involvement in the yet more intriguing area of intelligence (Chapter 10). Concerning the techniques of echolocation there is, however, another fascinating example of convergence that is perhaps rather less known. This involves those birds that inhabit the deep, dark recesses of caves, where they often share their domiciles with the bats. Notable in this regard are the South American oilbirds and the Asian swiftlets, whose saliva-bound nests are eagerly sought for those who find bird's nest soup exceptionally tasty. As their name suggests, the oilbirds have extensive fat deposits, which if rendered provide an oil of exceptional quality and purity. These birds attracted the notice of the great traveller and naturalist Alexander von Humboldt, and subsequently were engagingly described by Donald Griffin.[228] Oilbirds are nocturnal. They are also noisy animals, but can they echolocate when travelling to or from their perches in the caves? Griffin listened to them as they poured out of their cave at twilight, remarking, 'Perhaps the clicks were call notes, perhaps they were something analogous to profanity, or perhaps they were symptoms of some other avian emotion whose nature we could not guess.'[229] His subsequent experiments, however, showed that the ability to echolocate was genuine, with the series of sharp clicks clearly audible to humans. Continuing work[230] has shown that the resolution of oilbirds' echolocation is rather crude, at least when it comes to avoiding discs deliberately suspended in their flight path.

Interestingly, despite having well-developed eyes the oilbirds appear to lack binocular vision,[231] which is otherwise a general characteristic of birds and convergent on other vertebrates (p. 162). In some ways this makes the behaviour of the Asian swiftlets all the more extraordinary, since their echolocatory abilities enable them to avoid much smaller objects.[232] Although their echolocation does not rival that of bats,[233] nevertheless in their natural habitat the pitch-dark caves are filled with flying swiftlets, seldom if ever colliding and readily finding their respective nests. Bats, dolphins, and even some birds, therefore, have all entered a sensory world that is apparently almost completely unknown to us (note 227). It is fascinating to speculate what sort of echolocatory image these animals 'see'. Although it is now time to turn to the even stranger world of electrical perception, it is worth remembering that as the commonalities of convergence continue to emerge so we may find that apparently alien perceptions, and even mentalities, may not be as remote as has sometimes been imagined. Even to enter the mind of a bat may not be as difficult as is sometimes supposed. We return to this topic in Chapter 9 (pp. 265–266).

SHOCKING CONVERGENCE

Of the sensory systems it is perhaps those involving the generation of electrical fields that are the most remote from human experience. Perhaps it is not surprising that it has been suggested that the exquisitely sensitive nose of the star-nosed mole possesses an electrical sensitivity, but this proposal is viewed with considerable caution.[234] Even so, such sensitivity is certainly possessed by some other mammals, notably the primitive monotremes.[235] It is among the fish, however, that we find not only some of the most remarkable examples of electric generation (and perception), but also – and by this stage are we really surprised? – yet more striking examples of convergence. Thus, electrogeneration has evolved independently at least six times and in each case entails the modification of muscle cells, albeit from a variety of locations.[236] The phenomenon of electrical generation by fish must have been shockingly apparent to incautious waders and fishermen for a very long time, and certainly at least as far back as classical times.[237] A famous physician, close to the Imperial family of the first century AD, one Scribonius Largus, wrote of the use of the electrical discharges of the torpedo ray, a familiar denizen of the Mediterranean and Atlantic coasts (although it is found in all oceans), for the relief of intractable headaches.[238] Such a therapy was also endorsed by Dioscorides, a celebrated medical authority active at about the same time as Scribonius Largus, who, despite a reputation for travelling, spent at least some time in Tarsus.[239] He also knew of this treatment, but more alarmingly employed the torpedo's electrical capacity for the treatment of the prolapsed anus.[240] Interestingly, the greatest of the ancient physicians, Galen, disputed the efficacy of this galvanic treatment not only for headaches but also for more fundamental matters.[241] Nor was familiarity with the electrical properties of certain fish restricted to the classical world. African tribes, for example, were also familiar with the therapeutic powers of such fish, and according to reports provided by Jesuit missionaries and other travellers the inhabitants of Ethiopia employed the discharges for a variety of purposes, ranging from the expulsion of evil spirits to the control of fevers.

The properties of electrical fish also attracted the attention of both doctors and scientists in nineteenth-century Europe. Among the former were those intrigued by the surge of enthusiasm associated with mesmerism and other examples of what was termed 'animal magnetism'.[242] Of the scientists, the most famous was the physicist Michael Faraday, who undertook an extensive series of experiments in the late 1830s. These, not surprisingly, focused on the measurements of the electrical current but also included observations on the killing of prey and its ingestion by the fish. Faraday was a thoroughly hands-on, or in this case, hands-in, scientist. In one such experiment he and two colleagues had their hands in the tank when 'suddenly [the fish] gave a shock which startled us all and was perfectly satisfactory as to

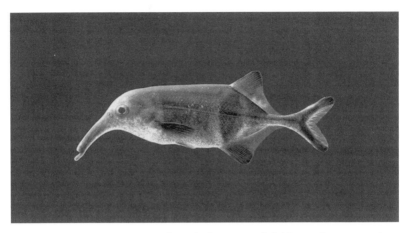

FIGURE 7.9 A mormyrid, or elephant-nose, fish (*Campylomormyrus*).
(Photograph courtesy of Professor C. D. Hopkins (Cornell University).)

the generality of the discharge. Mr Gassiot evidently felt it least', but Faraday dryly continues, '... I daresay Mr Bradley [the] most.'[243]

The fish Faraday employed was an electric eel from South America,[244] which belongs to a freshwater group known as the gymnotids. They show a series of quite remarkable convergences with an effectively unrelated group, the mormyrid fish (Fig. 7.9),[245] which inhabits the lakes and rivers of Africa.[246] In both gymnotids and mormyrids the twin processes of electrogeneration and electroreception are strongly convergent, although it should be noted that the underlying neurological mechanisms are certainly different.[247] These fish have been the subject of intensive experimentation, and the functional significance of their electric signalling and responsiveness has given glimpses into an extraordinary electrical world, which might, however, have analogies with our more familiar sensory modalities. Experiments typically involve the recording of electrical activity and observation of behaviour, but while an important strand of investigation concerns studying how individuals react, it is also possible to 'fool' the fish with an artificial equivalent, consisting of an electric dipole connected to a computer that can respond to the electrical signalling of the fish.[248]

Almost without exception, the electric organ is derived from muscle, not from nervous tissue,[249] but unsurprisingly the cells (or electrocytes) are specialized and commonly form a structure equivalent to a stack of gelatinous discs with a rather remarkable series of interpenetrating stalks.[250] As might also be expected, the electric organ itself has evolved. Primitive and advanced arrangements can thus be recognized, but there is evidence in this organ for both reversion as well as convergence.[251] Control of the electric organ is from the brain, via particular nerves known as the electromotor nerve axons. The resultant electric organ discharge (or EOD) that is transmitted into the water varies remarkably in duration, frequency, number of peaks, and polarity according to the species concerned.[252] There is, moreover, one important difference between the gymnotids and mormyrids inasmuch as with one exception the mormyrids produce the electrical signal as discrete pulses, whereas the gymnotids produce an effectively continuous signal as a wave form. In the case of the gymnotids there is also some evidence that the signals associated with communication, especially important in the sexual context, may be generated in a different region of the fish from those concerned with the imaging of the surrounding environment.[253]

An interesting possibility is that one of the driving forces towards an increasing complexity of the signals may have arisen as a defence against those predators that are capable of detecting the electrical impulses. Thus, by shifting the frequency of part of the signal the fish become more difficult to detect,[254] although it certainly does not render them immune to attack.[255] But this is not the end of the matter, because the enhanced electrical signal can then be utilized in sexual communication.[256] The gymnotid and mormyrid fish therefore live in what is effectively an electrical world in which the signals are received by the electroreceptors and then transmitted to the brain, which allows the animal to perceive inanimate objects,[257] especially when encountered in a novel context,[258] and to recognize or communicate with other inhabitants of the murk, be they friends or foe. It is also interesting that the two types of electroreceptor found in the skin, to detect direct and alternating currents respectively, are again very similar in both gymnotids and mormyrids.[259]

Because the environment in which these fish swim is effectively one of electrical uproar arising from the competing signals of other fish, it is hardly surprising that each species produces a highly specific electric signal. Such call signals, analogous to those issued by individual aeroplanes, are doubly important because the general electrical racket produced by dozens, if not hundreds, of electric fish nearby is further augmented by the frequent tropical thunderstorms with the associated lightning, adding to the 'electrical cacophony [and] ... background roar'.[260] Not only is there distinction between species, but in some cases either sex or the juveniles of a particular species also produce a characteristic and distinct waveform. For the most part, however, the signals of individuals of a given species appear to be identical. This may, however, be a simplification because in one species of gymnotid the signatures of individuals are discernibly different, a factor of considerable importance in social contexts such as defence of home territory.[261] But there are exceptions, effectively a convergence, where up to four species produce a similar electrical signal. It comes as little surprise to learn that in such cases the fish aggregate in schools.[262]

In this electrical Babel separation of signals is, therefore, essential. What happens, however, if two fish of the same species produce their respective signals simultaneously? Potentially the result would be problematic, for the two signals would lead to destructive interference, and the whole point of signalling would be lost. To circumvent this problem the fish have evolved a mechanism that is referred to as the jamming avoidance response (JAR) in which the fish changes its signal frequency.[263] Sensible enough, but two things are remarkable about this avoidance of jamming. First, the fish are astonishingly sensitive to potential interference. They will respond to signal modulations with an amplitude difference of 0.1% and a timing disparity of 400 nanoseconds or less, each fish shifting its signal pattern in a few microseconds.[264] Second, and even more remarkably, the algorithm used by the gymnotids and mormyrids to shift the signal has, of course, evolved independently but it is identical. Nor do the convergences end there, because computationally similar neural algorithms also occur in the owl, where acuity of sensory perception is acoustic rather than electric.[265] In the owl,[266] not surprisingly, neither the neural mechanisms nor the implementations are identical to those of the electric fish, but this does not prevent the emergence of deeper similarities. So when we look at the general arrangement in gymnotids and mormyrids, it is clear that even though different parts of their respective brains are employed for the interpretation of the electrical signals there are also a number of significant similarities in the neural circuitry.

Given the degree of convergence between the gymnotids and mormyrids[267] in terms of electrical activity it comes as less of a surprise that they occupy similar environments, living in waters that are either muddy or otherwise opaque 'blackwater', and typically showing nocturnal activity.[268] In such a setting the powers of vision are decidedly restricted[269] and, at least so far as the mormyrids are concerned, an early investigator, M. R. McEwan, remarked, that 'The sense of sight of these fishes is not at all keen.'[270] This observation was consistent with her study of the mormyrid retina, which demonstrated a number of peculiar structures that probably represent adaptations to the low levels of light.[271] Aspects of their life cycles and reproduction are also strikingly similar, including the embryological stage.[272] Although the body form of each group is rather different, in terms of their ecology, and especially feeding, there are important parallels in these fish. Thus there is a recurrent tendency to evolve elongate and tubular mouths (Fig. 7.9) with weak jaws and a feeble dentition. In arriving at this arrangement, these fish have opened a new resource in the form of a rich bottom-dwelling fauna of worms and larvae that more orthodox arrangements of mouth and jaw would find difficult to exploit.[273] Moreover, in at least some instances the prey is detected by the generation of the electrical field.[274] Indeed, Tyson Roberts identified an even closer link between feeding and electrical production when he wrote, 'I would go even further, and suggest that the interrelation between electrical faculties and feeding played a decisive role in the initial divergence of the gymnotoids and mormyroids from non-electrically specialized ancestors,'[275] as they entered the twilight zone of the Congo and Orinoco. Nor should this mud-grubbing life lead us to underestimate the extraordinary nature of these electric fish. As already indicated, the jamming avoidance mechanism is clearly a complex response involving a sophisticated set of adaptations. By now it should be clear that these fish occupy a rich world of electrical signals and social communication and, relative to at least some other fish, one of enhanced perception. This is consistent with a neural plasticity that in filtering out the predictable indicates that these fish occupy a contextural environment in which learning and memory are the norm.[276]

So it appears that, in a way analogous to the star-nosed mole, these fascinating fish 'see', not tactilely but in an electrical world.[277] So, too, it seems reasonable to think of other electrically sensitive animals, such as the sharks, also constructing an electrosensory landscape.[278] Nor need this parallel, however unfamiliar it is to humans, surprise us, because when we look at the brain activity

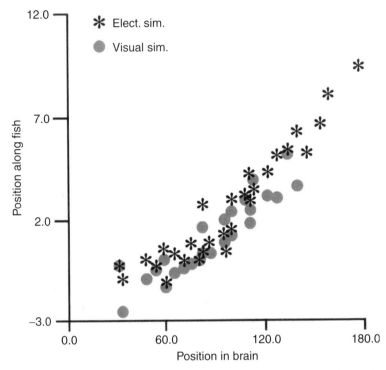

FIGURE 7.10 Seeing in the mormyrid fish, in terms of visual stimulation (dots) and electrical stimulation (asterisks) with respect to the response of a moving image or electrical stimulation along the body length versus the corresponding response position in the brain, specifically tailward of the anterior margin of the optic tectum. The overlap and parallelism of either stimulation suggests that the sensory modalities are in one way equivalent. (Redrawn from fig. 7A of W. Heiligenberg and J. Bastian (1984). The electric sense of weakly electric fish. *Annual Review of Physiology*, vol. 46, pp. 561–83, 1984. With permission, from *Annual Review of Physiology*, vol. 46, © 1984 by Annual Reviews, www. annualreviews.org, and the authors)

associated with both the electrical and visual systems of a mormyrid there are obvious similarities (Fig. 7.10). Despite their weak vision (note 271) these fish are not actually blind, and in at least some of the gymnotids there is evidence that the electrical and visual signals are integrated.[279] Indeed, some have argued that the parallels between electroreception and sight can be pursued to the point where electroreception yields some sort of analogy to colour.[280]

It is a fascinating, and largely untouched, problem to begin to speculate just how different are the sensory modalities of animals. A relatively familiar example of this is the ability of some people to use one sense perception to appreciate or understand another, such as seeing music in terms of colours, or more rarely as taste or shape.[281] So, too, Galton's experiments (note 189) using olfactory stimuli for mathematical calculations indicate the existence of equivalent modalities. So, too, experimental procedures in which animals are either deprived of one modality, e.g. sight,[282] or subject to transplants,[283] reveal evidence for neural plasticity. Thus, a part of the brain that in normal circumstances would be involved with vision is now able to process auditory information. So, too, transplantation, involving transsexual exchange of appendages (specifically feeding claws) in fiddler crabs, induces new sensory modalities that linked to the central nervous system invoke novel behaviours. More specifically in the case of the electric fish, it has been suggested that the organization of nervous centres involved with electrosensory perception (the electrosensory line lobe) has intriguing similarities to mammalian hearing (in the dorsal cochlear nucleus).[284] What is important in this context is that the mammalian system is an evolutionary innovation, since equivalents to this neuronal arrangement do not exist in the reptiles or amphibians. Similarities of neuronal arrangement may not guarantee, of course, equivalence of perception. There is more than a hint, however, not only of emergent systems in advanced organisms (yes, evolutionary progress is real), but also of an underlying commonality that might presuppose that extraterrestrials with nervous systems will hear, see, and smell in very much the same way as we do, and if that is so will also possibly have similar mental processes.

It should be pointed out that electric fish have not abandoned other sensory mechanisms. Mormyrids, for example, also have an acute sense of hearing.[285] These remarkable fish, however, show another peculiar feature that is probably instrumental in sound reception. In a number of fish the sense of hearing is evidently improved by connecting the swim-bladder to the skull through a series of bones (the weberian ossicles), which presumably serve to amplify the sound. This arrangement is found in various fish, such as the minnows, as well as the gymnotids. The mormyrids, in a sense, go one better because their embryological development leads to parts of the swim-bladder coming to lie, as isolated sacs, against each ear region, and these sacs are probably instrumental in their exceptionally acute hearing.[286] It is hardly surprising to learn that at least some mormyrids are adept at producing sounds, which arise from the action of a drumming muscle connected to the swim-bladder. In the mormyrid *Pollimyrus* the courtship sounds, described as grunts, moans, and growls, are broadcast by the males and evidently form an important complement to their electrical signalling.[287] It appears, therefore, that even if an animal experiences a rich sensory world, be it visual or electric, other modalities are by no means neglected and may show intriguing parallels. So, too, with hearing, to which we now turn.

HEARING CONVERGENCE

Hearing, of course, depends on the transmission of vibrations, and their transduction into electrical signals that in ways analogous to sight and smell are interpreted as sound. For most of us hearing is almost as important as sight, and since the invasion of land by the tetrapod vertebrates, effectively beginning with an amphibian-like creature,[288] this sensory modality to detect and track airborne sounds has been extensively developed.[289] Yet the evolutionary evidence suggests that the evolution of the tetrapod ear, with its tympanic membrane and associated features, such as pressure-relief mechanisms and differential ossification to provide an otic capsule that helps to refine the auditory mechanism, has evolved independently several times among the vertebrates.[290] More specific convergences are also found. For example, in the vertebrates impinging sound waves are ingeniously amplified via the hairs located within the spiral-shaped part of the inner ear known as the cochlea.[291] This structure contains specialized hair cells that are involved with auditory perception. While the differences between the arrangement found in birds as against mammals have long been appreciated, it is also evident that there are important convergences.[292] So, too, within the mammals themselves there are some interesting examples of convergence, perhaps most notably among the many subterranean species that, as already noted (pp. 139–141), provide many other insights into convergence. In their fossorial environment the sensitivities of hearing, not surprisingly, are shifted towards the lower sound frequencies. Even so, vocalizations and hearing carry only for about five metres.[293] Across longer distances communication is by seismic pulses, e.g. head-banging or foot-drumming, and does not appear to utilize the ears.[294] The auditory systems of the underground mammals nevertheless show some striking convergences linked to the shared mode of life. Unsurprisingly, the external ear lobes (or pinnae) are greatly reduced, if not absent. Within the middle ear there are a number of parallel changes to meet the need for detecting low-frequency sounds; far from this part of the ear being vestigial it is actually highly adapted.[295] So, too, within the inner ear there are further convergences, such as those concerning the arrangement of the semicircular canals that probably play a significant role in navigation through the maze-like world of these mammals.[296]

Other animals hear, but in groups such as the insects this has long been assumed to be a passive process, lacking the sophisticated physiological features found in mammals, in which acuity of hearing is sensitive to the animal's metabolism and also shows a feature known as autonomous vibration, which is important in the process of sound amplification. This now appears to be incorrect. In such dipteran insects as the mosquito the equivalent of the ear is a complex antennal structure, including the so-called Johnston's organ. This structure is richly endowed with nerve cells, is correspondingly very sensitive to sounds,[297] and now it transpires that the hearing of mosquitoes is much more similar to that of mammals than was realized. As Martin Göpfert and Daniel Robert[298] remark,

> Like vertebrates, mosquitoes actively improve hearing by
> enhancing sound-induced vibrations in a frequency- and
> intensity-dependent way. Considering the vastly different
> anatomies of mosquito hearing organs and vertebrate ears, such
> similarity in the mechanical processing of sound is surprising . . .
> it seems reasonable to assume that active auditory mechanics in
> vertebrates and insects evolved independently and in parallel.[299]

Nor is the versatility of the Johnston's organ confined to land. In water it is capable of detecting pressure variations, as in the whirligig beetles. When swimming through water these insects are evidently capable of sensing pressure waves produced during their swimming that are reflected off adjacent objects. Not only does this help to prevent collisions, but it has also been noted that even in the densest flocks of gyrating whirligigs the individuals never crash.[300]

Other insects have hearing but typically this depends on a tympanal arrangement.[301] This is analogous to our eardrum, being composed of a thin sheet of cuticle. Interior to this is an air-filled cavity and a complex structure known as the chordotonal organ.[302] Such an arrangement is found in many insects, such as the moths,[303] as well as the group (the orthopterans) that includes the grasshoppers and crickets. The songs of the orthopterans are familiar to us, and serve them for sexual communication. For the crickets and their relatives, however, such broadcasts carry risks, not least the unwanted attention of parasitoids. These include a particular group of dipteran flies known as the ormiinids (e.g. *Ormia*), the females of which seek out hosts to lay maggots on or near the luckless cricket.[304] Now, as we have seen, the standard dipteran 'ear' is in the form of the Johnston's organ. This is, however, ill-equipped to register the intense and high-frequency sounds generated by the crickets. One can almost guess the solution, because the ormiinids have evolved a tympanic ear, located behind the head, which is strongly convergent on that of the crickets.[305] As the investigators aptly remarked, 'for a fly to act like a cricket, it must hear like one.'[306]

This example is also interesting in several other respects. First, it is a rare example of acoustical sexual dimorphism; after all, it is the female carrying her maggots that needs the more acute hearing. Why the males also have good audition is less obvious, but may result from the common need to avoid predatory bats.[307] In addition, despite their tiny size and relative simplicity, the acuity of these ears and their directional ability are little short of astounding.[308] The ormiinids belong to a larger group of dipterans, the tachinids. There are thousands of related species in this group, and all are parasitoids. In contrast to the ormiinids, however, they rely on visual and olfactory cues in search of their hosts. Why then did the orminiids evolve this ear? Presumably it was because the nocturnally singing crickets would otherwise be impossible to locate; so their pinpoint accuracy in hearing opens up a whole new resource, made possible by the 'invention' of an ear. So far as the orminiids are concerned, this is an evolutionary novelty, but it is necessarily convergent on its host. Nor is this some fluke-like event. As we have already seen, the tympanal ear of insects is blatantly convergent and is an evolutionary inevitability given the co-option of sensory stretch receptors that are inherent to this sort of body plan, that is one consisting of an articulated exoskeleton. More specifically, the orminiid 'solution', which is how to hear its host, has evolved in another group of parasitic flies known as the sarcophagids, who home in to the sound of cicadas.[309]

These examples of convergence, interesting as they are, are of more general importance because they lead to a consideration of deeper similarities of sensory perceptions. In particular, the process known as transduction, whereby a mechanical force (e.g. sound) is converted into an electrical signal, is evidently the same in insects and vertebrates.[310] This, in turn, however, begs the question of how many molecular alternatives, in this case the ion channels, necessary to generate the electrical signal are really available, and the extent to which such channels and the component proteins that form them have been recruited for this purpose from some pre-existing function. Similarly, as Martin Göpfert and Daniel Robert remark (note 298), the genes (notably *Math 1* and *atonal*)[311] involved in vertebrate hair-cells and the insect Johnston's organ are the same. This commonality, however, may be less informative than might first appear. Thus, the *atonal* gene is involved both with the development of stretch receptors (i.e. the chordotonal organ) from which the insect tympanal ear is derived (notes 302 and 305) and photoreceptors,[312] as well as the olfactory nervous system.[313] A parallel, and more famous example, is the control of eye development by the *Pax-6* gene.[314] It would be a mistake to think that this gene 'makes' an eye, any more than *atonal* 'makes' ears: they are really switches, and in the case of *Pax-6* its primitive function may have been involved with the differentiation and patterning of nervous systems.[315] This would explain its role, not only in eye formation, but also its involvement in olfaction,[316] the brain,[317] and the pituitary.[318] As for many developmental genes, a role in the nervous system need not preclude co-option, and this might explain its appearance even in the pancreas (see also p. 240, Chapter 9).[319] To be sure, the compound eyes in arthropods (and almost certainly their convergent equivalents in annelids and molluscs) and camera-eyes of vertebrates and cephalopods (and almost certainly their convergent equivalents in heteropod gastropods and alciopid polychaetes) are all controlled[320] in one way or another by *Pax-6*, but the eye structures themselves, in terms of macroscopic arrangements, nevertheless, have evolved by convergence. So, too, with ears.

I argued above that any alien astronomers would use a camera-eye, and when they ask for a drink we can be pretty sure the mechanism of hearing will be strangely familiar. So, too, as they sniff the gin and tonic they will register the same olfactory signal of juniper berries, while convergence of taste should lead them to close momentarily those camera-eyes. And as already indicated, there will be yet deeper similarities, not only in transduction mechanisms using recurrent protein designs, but also the way in which meaning is conveyed and recognized. Human speech, and thus hearing, is well known for what is referred to as 'categorical perception'. That is, despite a continuum of sound we divide the signal into discrete categories. So, too, with other vertebrates, such as birds,[321] and also insects.[322] As Robert Wyttenbach and his colleagues remark, 'categorical perception may be a ubiquitous feature of perceptual systems in many animals, invertebrate as well as vertebrate.'[323] Star-nosed moles and stridulating crickets, echolocating bats, and electric mormyrids are all reminders, not only of a rich sensory universe, but also of one in which emerging mentalities may find fertile ground for evolutionary innovations but ultimately shared experience.

THINKING CONVERGENCE?

When we reconsider this range of animals, embedded in their respective environments of complex sensory perceptions, it seems safe to conclude that, despite what to us are largely alien means of perception, the many analogies that exist with our sense of vision are a powerful argument for the importance of convergence in evolution. But is this so important? At one level the conclusion might seem rather trivial. After all, it only reminds us that the world is subject to constraints; nervous systems interpret the sensory input, be they from noses, eyes, ears, or electric organs, and they all have a fundamental similarity of operation. Hence the trajectories towards any style of perception in any possible evolutionary history are indeed greatly restricted. Convergences are inevitable, but again, so what?

There is, however, something else worth knowing. Let us return to the mormyrid fish, but this time to look at their brain.[324] Not only does it have a strange, almost crystalline, internal structure, but also a well-developed series of lobes (Fig. 7.11). Most remarkable in the latter context is the enormous enlargement of the cerebellum (notably the region known as the valvula cerebelli), although a number of the other regions (e.g. the acoustico-lateral area, mesencephalon, and corpus cerebelli) also contribute to its extraordinary size. Consider for example the remarks by R. Nieuwenhuys and C. Nicholson when they write, 'Opened the skull of a mormyrid fish reveals, however, a totally different picture [from the standard teleost brain, say of a trout]. It is immediately apparent that the brain is of a relatively enormous size and that none of the structures mentioned above [in the basic brain]

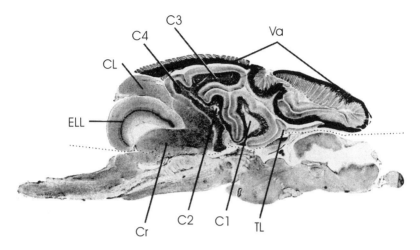

FIGURE 7.11 Longitudinal section through the brain of a mormyrid fish. Note, in particular, the immense size of the cerebellum, which comprises all parts of the brain above the dotted line. The various parts of this hypertrophied structure include the remarkable valvula cerebelli (Va), the central cerebellar lobes (C1 to C4), the crista cerebellaris (Cr), the electrosensory lateral line lobe (ELL), the transitional cerebellar lobe (TL), and the caudal cerebellar lobe (CL). Total length of the brain is about 12 mm. (Reproduced from fig. 7 of Nieuwenhuys and Nicholson (1969; citation is in note 324), with permission of R. Nieuwenhuys. Photograph kindly made available by Professor C. C. Bell, Oregon Health Sciences University.)

can be readily seen.'[325] In Fig. 7.11, note, in particular, how the cerebellum 'flows' over much of the rest of the brain, in a manner reminiscent of the massive neocortical expansion of the human cerebral lobes. Not only are these lobes enormously enlarged, but in addition to this hypertrophy the total area of the brain is further increased by both a series of deep folds and a series of fine ridges. It is all the more intriguing to note that in relation to its body size the mormyrid brain is comparable to the disproportionately enormous brain within our skulls. Not only that, but more than half the consumption of body oxygen in the mormyrid is accounted for by the brain, a figure three times greater than for humans.[326] And this is hardly surprising; as already noted, these fish live in a rich electrical environment, requiring sophisticated neural processing involving recognition of other fish and navigation in near-darkness through a cluttered and potentially lethal environment. And the gymnotids, with their convergent electrical system; what of their brain sizes? Certainly their enlargement is less dramatic than in the mormyrids, but even so they too have – in comparison with most other fish – larger than average brains.[327] The enlargement of the gymnotid brain is located in a region known as the torus semicircularis,[328] and this appears to be the principal area of electroreception. Even so, the overall similarities of the cerebellum-like structures in both gymnotids and mormyrids are strikingly equivalent in terms of sensory processing and thus reinforce the degree of convergence in these two groups of electrosensory fish.[329]

If brains can get big independently and provide a neural machine capable of handling a highly complex environment, then perhaps there are other parallels, other convergences that drive some groups towards complexity. Could the story of sensory perception be one clue that, given time, evolution will inevitably lead not only to the emergence of such properties as intelligence, but also to other complexities, such as, say, agriculture and culture, that we tend to regard as the prerogative of the human? We may be unique, but paradoxically those properties that define our uniqueness can still be inherent in the evolutionary process. In other words, if we humans had not evolved then something more-or-less identical would have emerged sooner or later. Such an idea is dismissed out of hand by nearly all evolutionary biologists, but that is exactly what the next two chapters aim to show.

8   Alien convergences?

I have just woken from a hideous nightmare, the sort of dream that haunts you for days afterwards. I have visited, or been sent to, a terrible world, and to make things worse it had strange parallels to ours, only it was alien and mechanical. I am standing on a wide highway, with robot-like workers rushing in either direction. On either side of this stream miniature versions scurry everywhere, patrolling the margins, ceaselessly on guard. Such is the flow of traffic that the earth has long since been beaten flat; not a shred of vegetation obstructs the road. In one direction the army of helots marches empty-handed, but towards a truly immense forest which I can see only dimly as there is something not quite right with my vision. My sense of smell is, however, overwhelmingly acute, and wave after wave of olfactory messages urge me to obey the twin commands of unceasing work and obedience. Still looking towards the distant forest, I see that from this direction the army returns, each worker scurrying forward with a parcel of food. New olfactory orders are issued, and I am impelled to follow the returning stream. After walking for hours I am driven into a set of huge subterranean chambers. Here the activity is at a frenzy, yet there is also an underlying order. My senses reel: from the ceilings hang huge clumps of what look like some sort of fungus, but it is being tended with the utmost care and attention. Now I can see workers busy with weeding; others seem to be applying herbicides, and yet more are either collecting pieces of the fungus or extending the zones of cultivation. Yet for all its strangeness this world is a little too familiar for comfort. The workers rushing past me are indeed literally bug-eyed monsters, sinister antennae constantly in motion, with hardened carapaces, and in some cases carrying miniature versions of themselves. But it is also clearly an organized society, a vast underground city entirely dependent for its well-being on the ceaselessly maintained fungus farms. As I awake from this dry and soulless world, I realize suddenly that I have not been among the extraterrestrials, but have been mingling with the leaf-cutting ants who quite independently of humans have invented agriculture.

FIGURE 8.1 Attine ants at work. Left: with mandibles poised above a leaf. Right: with a minima. (Photographs courtesy of Cameron Currie, University of Kansas, Lawrence.)

DOWN IN THE FARM

What, at first sight, could be more specific than agriculture? To become a farmer entails a series of familiar processes, from maintenance of gardens, transport, weeding, application of herbicides, manuring, cropping, to the exchange of cultures. That is effectively how we pursue our agriculture. So, too, and convergently, do the leaf-cutting ants (*Acromyrmex* and *Atta*)[1] that flourish in Central and South America.[2] It was they who were the subject of my *Vorstellung* at the beginning of this chapter. As their common name suggests, these remarkable insects (Fig. 8.1) use their sharp mandibles to excise pieces of leaf[3] that are then transported back to the communal nest. Although typically associated with forests, these leaf-cutting ants have managed to extend into desert environments.[4] In some species there is a remarkable division of labour, in which a rather small number of worker ants leave the nest early, climb the trees, cut away the leaf pieces, and then drop them to the ground. There another group, arriving slightly later, find the pieces, cut them smaller, and carry them to the 'road', where a third group of ants transports the vegetation back to the nest.[5] Another example of such division is either the 'bucket brigade' whereby some species of ant hand over plant material at junctions of the 'road' system[6] or the formation of caches when supply exceeds the rate of processing the leaves within the nest.[7] Travel is often via well-defined and more-or-less permanent paths,[8] which may stretch for considerable distances, sometimes 100 metres. On the main thoroughfares the ground may be cleared to form a smooth highway that in turn maximizes walking speeds. In these social insects three distinct castes are recognized, and in addition a miniaturized variety (the minims) may also occur. Some of these minims spend time working in the gardens, as well as riding on the larger ants, where they engage in grooming and cleaning activities. Another important role, only recently appreciated, is to patrol the edges of the trails along which the foragers march, on the lookout for danger and when appropriate issuing an alarm.[9] Yet others 'hitch-hike' on the leaf fragments as they are being carried back to the nest, in part probably to defend the leaf-carriers from attack by parasitoid flies (known as phorids) and possibly also to begin the preparation of the leaf prior to its arrival at the nest.[10] In the nest itself, the soldier caste is vigorous in defence.

The collected leaves are not eaten directly[11] but are used to provide a mulch for fungus gardens that are located within the nest. The leaves, of course, are fresh and the initial preparation, which as already noted may start during transport, includes a stripping away of the outer waxy layer. This process, achieved by a sort of licking, also appears to inhibit the activity of associated microorganisms, the control of which is a central necessity to the health of the fungus farm. Thereafter the leaves are shredded and pulped, and at this stage are ready for fungal innoculation.[12] The saprophytic activities of the fungi break down the plant material, especially the resistant cellulose, and so provide an edible crop for the ants.[13] The gardens are subject to careful and ceaseless maintenance. Weeding,[14] especially of infected areas, is undertaken principally by tiny ants (the minims). In weeding several minima typically loosen the offending item before it is removed by larger ants. In addition to weeding, the minima also engage in grooming, that is the removal of alien spores. If necessary the ants will also transfer the fungi to parts of the nest with more suitable humidity[15] and temperature.[16] In addition to weeding there is also the activity of pruning, again to encourage the harvest.[17] The application of fertilizers is in the form of a manure, an excrement rich in nitrogen as well as an enzyme supplement.[18] The harvest is ready and, in the more advanced types of cultivation, cropping involves the cutting off of the knob-like ends of the fungus, which are rich in protein, sugars, and other compounds.[19]

As on our own farms there is, however, a recurrent danger of invasion by pathogens. For the ants' fungal gardens the principal risk comes from a virulent parasite, also a fungus, *Escovopsis*. If it is not controlled, the garden is soon converted into a mouldering and blackened ruin.[20] How do the ants avert disaster? They apply the equivalent of a herbicide, specifically in the form of an antibiotic.[21] This is derived from streptomycetean bacteria, the filaments of which grow attached to the bodies of the ants.[22] In addition to these fungicides it is likely that other chemicals, including again antibiotics, secreted by the ants themselves, also help to inhibit the growth of unwanted bacteria and fungi.[23] Both provide defences against infestation, but interestingly the ants show a trade-off whereby newly imported leaves are protected by the ants (mostly minims) using their own secretions. This helps to provide an all-round microbial defence. However, in the older parts of the fungus garden, where the fungal biomass is higher and presumably more vulnerable to attack by the parasitic *Escovopsis*, the larger ants responsible for this area of maintenance rely on the bacterial defences.[24] Nor are these the only risks faced by the ant-gardeners: on occasion a specialized predator aggressively sweeps in, expels the attine ants, and usurps the garden.[25] These so-called 'agro-predators' are a particular species of myrmicine ant, unrelated to the leaf-cutters and related attine ants. In addition to usurping the gardens these myrmicines have been observed to place their larvae adjacent to those abandoned by the vanquished attines, where most probably the latter larvae act as a convenient source of protein. Yet more terrible fates await other attine nests when their citadel falls, after stiff resistance and appalling casualties on both sides, to army ants. Here part of the resistance entails plugging nest entrances and building barricades with leaf fragments.[26]

MILITARY CONVERGENCE

In parentheses, one should note that the army-ant syndrome has evolved independently several times.[27] This particular example of convergence is important for two reasons. The first is simply connected to the intrinsic fascination of a sophisticated biological organization, replete with militaristic metaphors and an associated folklore of marauding columns of invincible ants sweeping all before them. In fact, in some parts of the world an invasion of a house is welcomed as it sweeps out pests and other unwelcome visitors, and for the most part the army ants offer no danger to humans.[28] Arguably of greater importance is the emergence of a complex system that encapsulates a remarkable social system and thereby defines a different sort of collective intelligence.[29] As such the army ant adaptive syndrome has several noteworthy and recurrent features. The most celebrated aspect is the relentless mobile columns, capable among other things of building bridges constructed of their own bodies (see also note 32), and self-organized into an advancing front that can span as much as 20 metres of tropical forest floor. Animals flee,[30] but those either trapped or overwhelmed are rapidly dismembered, the pieces being despatched back to the nest by exceptionally efficient teamwork.[31] Such transport of the dismembered pieces of prey is referred to as super-efficient, inasmuch as the total transported to the nest typically involves cooperative efforts and exceeds in bulk what the individual ants could achieve, however the piece was divided. Such super-efficient transport is one of the convergent features of army-ant behaviour. It is also important to note that recruitment to the team is highly dynamic, with individuals 'choosing' involvement. The nest itself provides protection for the young, yet it is built of the living bodies of the ants.[32] Interestingly, in these bivouacs (Fig. 8.2), which are temporary because the colony is highly nomadic, there is a careful thermoregulation of nest temperature.[33]

In the context of this book, however, the principal topic is not convergence *per se*, but the attempt to establish the likelihood of the repeated emergence of complex biological systems. Thus it comes as little surprise to learn that population sizes of some leaf-cutter ant colonies are enormous, with the inference that such numbers can be sustained only by the ants' careful system of fungal agriculture. Yet, these 'cities' are overtopped in numbers by some species of army ant, notably *Dorylus*, where the population of a single colony can reach 20 million workers. Such enormous numbers seem to be a result of both growth and aggression, and as such emerge as a result of an arms race.[34] Even though genuine arms races in natural biological systems are probably rather uncommon,[35] such examples as the army ants awake uneasy echoes of human organization.

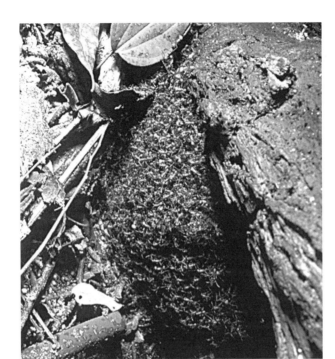

FIGURE 8.2 The bivouac of the army ant *Eciton*. (Photograph courtesy of David Kistner, California State University, Chico.)

Not only is this true for the army ants, but study of other social insects may reveal other interesting evolutionary parallels. Consider, for example, the advanced eusocial bees,[36] with their familiar division of labour, notably the queen and pollen-gathering workers. These animals have fascinated humans for millennia. In England it is still the custom, when the beekeeper has died, for a friend to go to the hive and tell the news to his or her bees. In a different way, in past centuries the industry and organization of the hive invited obvious political analogies. Now this fascination extends to evolution with the realization that bees have cognitive capabilities and a plasticity of behaviours that otherwise are known only in the vertebrates,[37] and, no less surprisingly, other features such as exhibiting sleep-like states.[38] Or is it surprising? Sleep may have several functions, but one widely agreed purpose is the need to consolidate memories.[39]

In any event, the complexity of bees not only has some intriguing parallels to vertebrate mental processes, but has been arrived at

by an independent evolutionary trajectory that may have an unnerv-
ing parallel in our hominid history. This is because there is evidence
that a once-wide diversity of eusocial bee groups has been whittled
down through competition to only two groups (the apinid honey[40] and
meliponinid stingless bees[41]) over geological time.[42] Interestingly, although the evolution of advanced eusociality in these two groups of
bees has generally been regarded as stemming from a common ancestor, recent work suggests that they arrived at this arrangement by
convergence.[43] Moreover, Michael Engel remarks that it appears that
during the

> geological expanses of time, aggressively foraging advanced-
> eusocial species had a negative impact, not only on non-social
> species[44] but also and especially on other advanced eusocial
> lineages. A similar situation is reported from another highly
> competitive, social species (albeit not eusocial) for which the
> fossil record indicates the aggressive exclusion of related social
> genera and species that may have caused their extinction – namely
> *Homo sapiens* ... the only survivor of a prior hominid
> radiation.[45]

'Big, fierce societies'[46] may lead not only to the emergence of collective intelligences but to ones that are, from our point of view, so
alien as to be useful in envisaging extraterrestrial societies. In this
context some remarks by Nigel Franks, a leading specialist on army
ants, are intriguing, if also in places self-confessedly speculative. In
reviewing the extraordinary social structure of these insects he observes how it is among the army ant colonies 'we see the emergence
of flexible problem solving far exceeding the capacity of the individual', a capacity that finds no rival in other animals, apart from
humans. This, as Franks stresses, presupposes effective communication, so that 'intelligence, natural or artificial, is an emergent property
of collective communication ... This is exactly what happens when
army ants pass information from individual to individual through the
"writing" and "reading" of symbols, often in the form of chemical
messengers or trail pheromones, which act as stimuli for changing
behavior patterns.' So it is that we can see a transformation of a system from being hard-wired to flexible via 'increasingly sophisticated
patterns of communication'.[47] Such a pheromone-based society[48] may

be one of the few genuine alternatives to the emergence of human-like intelligence, which as discussed in the next chapter is itself almost certainly convergent. In genuinely speculative mode Franks draws attention first to eye reduction (each eye to a single facet in *Eciton*), which poses the question as to how the colony can navigate, given the vestigial eyesight of each worker. He continues,

> In my wildest dreams, I imagine that the whole swarm behaves like a huge compound eye, with each of the ants in the raid zone front contributing two lenses to a 10- or 20 m wide "eye" with hundreds of thousands of facets. Each ant has possibly only the slightest directional preference, but through tactile signals and trail pheromones these preferences might be collated and amplified across the swarm. In this way the army ants could comprise a parallel-processing computer of intriguing yet awesome simplicity.[49]

Since writing this, Franks has continued his exploration of the convergent organization of army ant societies, and has also presented an absorbing analysis of the self-organization of the swarm raids. As he points out, the rules for a raiding front can in themselves be quite simple, even though the net result is uncannily 'intelligent' with an 'active architecture'. His comments on biological self-organization are of particular interest. It is, he explains, an assumption that such self-organization is 'order for free ... By this they [other authors] seem to suggest that self-organizing biological patterns occur so automatically that natural selection has had little influence over the evolution of their structure. I believe that such a viewpoint is extremely misleading'. So when we consider how army ants organize their swarm raids Franks argues that the processes of natural selection remains in force, for instance, by

> rather precisely [tuning] both the behavioural rules of thumb of the ants and the chemical properties (e.g. volatility) of their pheromones ... Self-organization theory does not suggest that natural selection has had *no role* in the creation of certain patterns in biology – rather it suggests that natural selection has *rather less to do* than one might expect given the complexity of the global structure. As illustrated by the swarm raids of Old World and New

World army ants, natural selection may have had to select for a surprisingly small and simple set of rules to generate swarm-raiding patterns. Thus self-organization theory may help to explain why we observe such a high level of convergent evolution in certain biological structures.[50]

CONVERGENT COMPLEXITIES

The behaviour of complex social insect societies, and specifically the army ants, is more than a digression. This is because the emergence of such biological complexity may be much more constrained than is sometimes imagined.[51] At the least it is a reminder that carnivorous species can be more successful, at least in terms of colony size, than the farming ants (and, as we shall see below, termites). Success, as even the most jaded (if not jealous) scientist well knows, is relative. Size, we are reassured from the most reliable sources, is not all. In their relative adaptive contexts social wasps and bees have taken over their respective worlds. So, too, the attine ants have evolved a remarkable system of fungal cultivation. In each of these examples we have evidence of extensive appropriation of resources, capable of maintaining large and complex societies. Life beyond these 'cities' and 'armies' remains, of course, diverse and marvellous: the biological world is by no means reduced to a monochrome. Nevertheless, the repeated rise of such societies, and at least evidence of the displacement and ultimate extinction of less successful equivalents, suggests that such an arrangement is a biological inevitability.

So, too, in any specific case if a group 'decides' to adopt a particular strategy, say agriculture, then the routes to success will be very limited. It is time, therefore, to return to the parallels between ant and human farming, and thereby consider some implications of such a commonality. As in our mushroom farms, the activities are carried out underground, since, unlike plant crops, the fungi have no need for sunlight. The farms are located in elaborate nests, equipped with ventilation shafts[52] and in certain species also dump-pits – some large enough to house a man[53] – which are used for waste disposal (Fig. 8.3). Just as in human societies, control of the waste is very important for the health of both the colony and the fungal colonies, and it now appears that the risky business of waste management[54] may be 'allocated' to the older workers, nearing the ends of their lives and less valuable to the colony. It seems that once assigned to dump management[55]

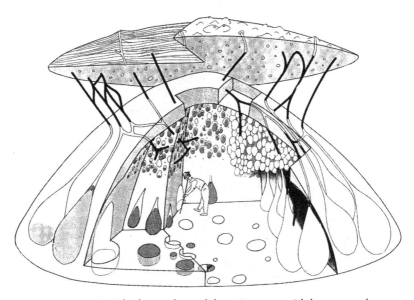

FIGURE 8.3 The fungus farm of the attine ants, with large central chamber containing pendant clumps of the fungus and one worker (!) Note the tunnel system leading to the outside world, and the capacious dump-chambers. (Reproduced from fig. 29 of Jonkman (1980; citation is in note 53), with permission of Blackwell Wissenschafts Verlag GmbH and the Jonkman family.)

the workers seldom leave, not least because of an aggressive response by other occupants of the nest.

Human farming has evolved independently a number of times, and so, too, has agriculture among the attine ants. Present evidence indicates that the domestication of fungus was first achieved approximately 50 Ma ago, but there have been repeated episodes (at least five) of domestication of various groups of fungi as distinct cultivars.[56] What originally might have triggered this innovation? Here, too, there could be parallels to the invention of human agriculture, where it has been suggested that in some instances, as in ancient Mesoamerica, its development was spurred by significant climatic changes.[57] So, too, a comparable set of upheavals, perhaps even the ecological mayhem that succeeded the K/T impact event, could have been the crucial factor in shifting these ants to agriculture.[58] This ability to cultivate fungus is more than just an evolutionary curiosity. Not only is it strongly convergent on our agricultural activities, but as Neal Weber has remarked, 'the evolution of a unique skill in fungus culturing has freed these ants from the usual food limitations and has enabled them to expand in colony size. In the process a few species have developed some of the largest ant colonies known,'[59] with populations estimated to reach at least seven million,[60] and typically exceeded only by the aggressive army ants (p. 201). Gigantic populations dependent on highly organized societies: does it sound familiar? And, as with the eusocial bees, there may even be lessons for us. The fungal gardens are usually monocultures. But as humans know (or should know), monocultures have their risks, being notoriously liable to invasion by viruses, fungi, and other pests. The use of the streptomycene antibiotics (note 22) has, of course, a direct parallel in human medicine. What seems quite remarkable is that the resistance conferred by the antibiotics seems to have been maintained by these ants for literally millions of years, in comparison to our experience where in only a few decades we have seen the seemingly inexorable rise of 'superbugs', resistant to nearly all treatments. In reality the system of ant farming and its crop protection is presumably much more dynamic. It appears that the more virulent strains of pathogenic fungi occur in the more advanced species of attine ants, which in turn are more reliant on monocultures. It is hardly surprising that there is evidence for exchange (and stealing) of fungal cultures between colonies.[61]

The agricultural activities of the attine ants are surely one acme of arthropod organization, and further parallels with human agriculture may well emerge. As a leading worker, Ulrich Mueller, has remarked: 'The more I study [the fungal gardens] the more analogies I find between human agriculture and ant agriculture.'[62] Given the nutritional value of fungi and the ease of propagation, it is not so surprising that other insects have also learnt the art of agriculture. Thus the ambrosia beetles, a type of weevil, have evolved a mutualistic relationship with a group of ascomycetes known as the ophiostomatoid fungi.[63] The general arrangement is certainly rather different from that of the attine ants because these beetles bore into tree trunks, so producing a complex series of galleries. The beetles are responsible for introducing the fungi. These are carried in containers (mycangia)[64] that consist of cuticular pockets with associated secretory glands that evidently 'control the growth and form' of the fungi.[65] The mycangia share, therefore, a basic function, but their form is extremely varied, and in different species they occur on many parts of the carapace. As ever in evolution, 'needs must': who cares whether the container (in this case the fungal-ferrying mycangia) comes from, so long as it works? So far as beetle and fungus are concerned, both sides are winners. Evidently the advantage to the fungi is to be smuggled into new tree hosts, incidentally making these beetles a serious economic pest.[66] The beetles benefit because the fungi help to block plant defences, such as the secretion of resins,[67] and also provide a source of food. Yet in other respects this fungal agriculture is reminiscent of that pursued by the ants. For example, it has become an obligate association, shows vertical transmission from generation to generation, and provides the main food source, and interestingly in these beetles appears to have 'evolved at least seven times'.[68]

It is also possible that a convergent situation has arisen in the termites, which although often referred to as 'white ants', are actually relatives of the cockroaches, whereas ants are closer to the wasps. In one group, the macrotermitinids, there is a striking and close symbiotic union between fungus and termite.[69] In contrast with the attine ants, which are restricted to the New World, this association is an Old World one, probably originating in Africa and then spreading to the Middle East and southern Asia. Termite nests are among the more spectacular of animal constructions, with an elaborate system of chambers and tunnels.[70] The tunnels are also used for foraging runs outwards from the nest, for up to 50 m; one calculation gave for a single nest a total tunnel length of almost 6 km. The foraging tunnels repeatedly branch but crossroads are very rare. The walls of tunnels are smoothly plastered and sharp bends are cambered. These tunnels enable excursions to be made in relative safety; the termites ascend to the surface by short access tunnels that slope upwards from the main tunnels. These access tunnels are less well built, and may be sealed when not in use. Their diameter is such that they can also be blocked by the heads of the large soldier termites. The nest itself is engineered to allow both the circulation of air and the venting of waste gases. Also present are storage areas and, at least in experimental set-ups, latrines. As is also the case in other colonies of social insects (and their analogues), there are other inhabitants, some beneficial, some not. Some termite colonies, for example, play host to a group of flies known as the phorids, and here, too, convergences emerge.[71]

Growing within the termite nest are spectacular combs of the fungus, aptly named *Termitomyces*. It belongs to a group known as the basidomycetes[72] and thus is related to the fungi cultivated by the attine ants, as well as to the more familiar mushrooms and toadstools.

On occasion, *Termitomyces* emerges from the nest to form on its outside a mushroom-like fruiting body (which is edible to humans), but evidently this fungus cannot survive outside the termite colony. Thus, if the termites are removed the combs are soon invaded by other sorts of fungi and bacteria. Maintenance of the combs may involve salivary secretions and weeding, yet earlier workers such as Roger Heim[73] were dismissive of the term 'cultivation'.[74] Since then views have changed, and as Johanna Darlington briskly notes: 'Heim's views of the adaptable fungus forcing itself upon the reluctant termite is at odds with more recent experience of *Termitomyces* as a delicate and finnicky organism requiring care and cosseting. It seems best merely to take note of Heim's arguments and pass on.'[75] Yet, while the association seems to be agricultural, the precise nature of this symbiosis is still somewhat elusive. Again to quote Darlington,

> The fungus is a passive partner, absolutely dependent on the termites for its survival and success. The termites, on the other hand, invest a lot of time and energy in caring for the fungus. They gain substantially from the relationship, although it is proving difficult to discover exactly how. The evidence favors a nutritional role, with the fungus comb acting as an external stomach,[76] breaking down less digestible components of the forage and so making them available to the termites ... An advantage of the system is that the digestive work of the symbiont is carried on outside the termites' bodies, unlike the gut symbionts that do a similar job for the lower termites. Lower termites have to carry fermentation tanks around with them, while the Macrotermitinae leave them parked at home.[77]

Evidently this agriculture is different from that of the attine ants, just as human agricultures vary.

Other convergences also emerge. Communication among ants is famous for its pheromone-based system, but in the termites seismic communication is achieved in the soldier caste by drumming the head against the substrate,[78] a method also used by subterranean mammals (Chapter 6, p. 141). Seismic communication is also used by the leaf-cutter ants.[79] In these ants, although the sound is produced as a stridulation, it is transmitted through the ground. Typically stridulations are produced only when the individual attine ant is immobilized, such as occurs by collapse of part of the earthworks. These cries of distress elicit rapid digging by other workers, who can detect sounds through 5 cm of earth and will begin rescue work if the thickness is no more than 3 cm. In the termites this alarm reaction can evidently be triggered either by air currents or vibrations. The more advanced fungus-growing termites are particularly sensitive in this respect. Rather remarkably, the signal can be propagated by chains of soldiers, each re-amplifying the signal in a way analogous to the now-redundant methods of human communication using smoke and drums,[80] as well as the periodic re-amplifications observed in the transmission of signals along the nervous system.

Although they are in certain respects less sophisticated, it is appropriate to mention in passing the so-called ant-gardens, which are arboreal earthen structures known as 'cartons'. In this arrangement certain species of epiphytic plant are encouraged to grow on the nests, from seeds that are collected by the ants and planted in the nest wall. In due course these plants germinate, grow (and thereby provide a source of extra-floral nectar), extend roots that probably help to strengthen the nest, eventually flower, and so produce seeds, the fruit of which is eaten before they are planted … and so the cycle continues.[81] In a few instances fungi are employed in the 'cartons', probably to help bind the structure and possibly also to release antibacterial chemicals. Interestingly, in some cases the fungus is effectively a monoculture, apparently maintained by weeding and feeding. So far as can be told, however, the fungal products themselves are not directly cropped.[82] Other activities of these garden ants include the pruning of surrounding vegetation, probably to create a 'fire zone' to reduce the risk of invasion by other ants,[83] the collection of vertebrate faeces, presumably as fertilizer, and perhaps the choice of plants that, like those cartons with fungi, release chemicals that help to keep pathogens at bay.[84]

This section on the ants (and termites) has included more than its fair share of digressions, but I hope that the common thread of exploring convergences still runs through the narrative. Consider again the army ants: living in the midst of their mobile and aggressive column has its self-evident risks, yet various species have managed to insinuate themselves into the ants' social system. Most striking in this instance are probably the various staphylinid beetles. These have effectively transformed their bodies into an ant shape that readily deceives the actual ants.[85] Hiding yourself in an aggressive raiding column may yield various benefits, not least an uncontested share of the

FIGURE 8.4 A staphylinid beetle in association with the army ant
Eciton. (Photograph courtesy of David Kistner, California State
University, Chico.)

booty and also some protection from predators. Typically the beetles
travel in the centre of the column, and are not found in the vanguard.
In times of excitement, such as during an episode of emigration, the
beetles may ride on the ants themselves and despite continuous antennal interrogation never seem to be recognized as interlopers. The
way in which these beetles have transformed themselves into ant-
mimics is very striking (Fig. 8.4), and the selective pressure to survive in a jostling mass of aggressive workers can explain how such
a transformation in the staphylinids has occurred independently on
multiple occasions.[86] In at least one instance the convergence extends
to a colour matching, whereby a geographical variation in the ant coloration is matched by the beetle.[87] As David Kistner (Fig. 8.5) points
out, given that the ants are effectively blind, this mimicry supports
the idea that it serves to dupe 'educable predators which sit by the
raiding columns to pick up insects which are stirred up by the raid'.[88]
Nor is this the only example of a convergence between a mimic and
an army ant. A common association is the attachment of mites to
the exterior of insects (and other animals such as birds and mammals,
clinging respectively to feathers and hair; pass me the nit comb), but
in the case of the mite *Planodiscus* 'the sculpture of the mite and

FIGURE 8.5 David Kistner and his wife in the field in Ecuador, collecting
staphylinid beetles and army ants. (Photograph courtesy of David
Kistner, California State University, Chico.)

the ant's leg is nearly identical. Also the arrangement and number of
setae on the mite approximates the arrangement and number of setae
on the leg. Thus when the ant grooms its leg, the tactile stimulation
[as it passes over the mite] will be similar to that of the leg itself.'[89]

In Chapter 6 (pp. 116–117) I briefly introduced the tent-building
capacities of the aptly named weaver ants. Here, too, there are some
striking cases of mimicry, which here involve weaver ants and a crab
spider, *Amyciaea forticeps*.[90] As it happens, the weaver ant in question, found in India and known as the Indian Red Ant (*Oecophylla
smaragdina*), is also associated with another spider (*Myrmarachne
plataleoides*) that in turn is 'a perfect mimic of the red ant; so perfect is this mimicry that even experienced biologists may pass it by
as an ant, in the field.'[91] Despite its close association with the Indian
Red Ants, which can be very aggressive, *M. plataleoides* takes good
care to avoid the ants. Possibly it employs its mimicry to fool other

ants from whom it steals their young, as well as to avoid predation by being mistaken for its genuinely aggressive counterpart. The other spider, *Amyciaea*, has only a generalized similarity to the Indian Red Ant, but this is compensated for by a rather remarkable behavioural convergence (see also p. 285, Chapter 10) which seems to entail its pretending to resemble an ant in distress. Thus, when a nearby ant adopts an alarm attitude, thereby disturbing the other ants. This then gives the spider an opportunity to attack. Once seized, the meal is concluded as the spider suspends itself from a silk thread, safe from the other ants.[92]

The convergence of mimicry of insects and spiders to an ant morphology has 'evolved at least 70 times',[93] and as such gives a series of fascinating insights into both the malleability of biological systems and the likelihood of establishing adaptive explanations. The transformation of staphylinid beetles to an ant-like form has already been noted, and given that ants are also insects it is relatively easy to see how other insects might come to mimic them. Even so, there is a subtlety inasmuch as the closeness of mimicry might be compromised as the insect changes in size and shape. The solution, of course, is for the mimic to resemble successively, in a series of moults, two or more species of ant.[94] The case of the spiders, the other principal ant mimics, is more remarkable, given the more obvious differences in body plan, and here, too, illusions connected to segmentation, colour, appendages, eyes, and surface texture have all been arrived at by a series of ingenious modifications. Not surprisingly there are degrees of mimesis, but in at least some instances there is evidence that the phylogenetically most derived are also the most ant-like,[95] and as such will define evolutionary trends. As already noted for one spider, these may be supplemented (or complemented) by various behavioural modifications. Not all insects need look like ants to insinuate themselves into ant colonies, and many closely associated species evidently rely on compatible and thereby convergent chemistries or textures.

In passing one should note that the vast area of biological mimicry, where one species comes to resemble another, is a perfectly good example of convergence. Its classic manifestation is in Batesian mimicry, where an animal, usually an insect, comes to resemble closely an unrelated but noxious species and thereby escapes predation. Mullerian mimicry, on the other hand, is where two species, both unpalatable, converge to resemble each other. So well known are these mimicries and their various manifestations,[96] that for the purposes of this book they can, I hope, largely be taken for granted.[97] The complexity of both ant and termite colonies, not to mention their various guests and symbionts, also provide a rich field within the topic of convergence, and one that can be analysed at several levels. Passing comment has already been made on such built structures as trails and tunnels, farms, and cartons, and it is clear that much remains to be discovered about these various constructions in an adaptive context.[98]

These insects, moreover, are by no means the only arthropods to have evolved constructional abilities. Take, for example, the beach-dwelling orypodid crabs. These crustaceans are highly territorial and protect their domains from intrusion by a variety of complex behaviours that include the building of mud fences and also the plugging of neighbouring burrows. Sometimes the occupant is evicted, but in other cases it is entombed. As might be expected, at least some of these behavioural repertoires are convergently acquired within this group.[99] Nor is this the only example of such sophistication of behaviour. The fiddler crab is famous for its signalling, achieved by movement of its hypertrophied claw. It now transpires that the signalling for mates, in a process known as 'lekking' in which numerous associated males signal simultaneously to choosy females (a well-known characteristic of some birds), also occurs in these crabs.[100]

HEARTS AND MINDS

Not only do certain invertebrates provide some intriguing parallels to the behavioural sophistications of the vertebrates, but they also approach them in terms of activity and intelligence. Earlier I emphasized the convergent similarities between the camera-eyes of cephalopods and vertebrates (p. 151).[101] In addition, despite the rather different arrangement of the cephalopod brain,[102] the size of this organ relative to the body exceeds that of many fish and reptiles.[103] In addition, in at least the cuttlefish, the blood–brain barrier,[104] essential for controlling the chemistry of the brain and thereby a prerequisite for high level integration, 'is as tight as that of mammals'.[105] The cephalopods also appear to share another characteristic, the possession of molecules similar to those known in vertebrates as neurotrophins, a type of protein growth factor. This, as has been suggested, may be a prerequisite for the emergence of complex brains.[106] Cephalopods are fascinating animals and, notably in the cuttlefish, they can show an astonishing range of chromatic signals that are under direct nervous control. To see the colours flowing across a cuttlefish is a quite extraordinary sight, and although the colour changes are well attested with respect to camouflage and predator avoidance, it is difficult to believe that it is not in some way 'emotional'.[107] In the case of the octopus, there is an intelligence that includes a flexibility in behavioural repertoires and the ability to learn and remember.[108] Evidently the octopus, far from being a rule-bound machine, is capable of acting in an autonomous fashion.[109] In their natural habitat octopuses have a sound grasp of the seascape: as one report[110] commented, the octopuses 'often zigzagged through multiple substrata and depths, with many obstacles obscuring their visibility of the horizon. The divers [who were tracking them] were often surprised when a forage ended and they realized the octopus had arrived back at its den while the divers themselves were still disoriented.'[111] Some workers have gone so far as to talk about the individual temperament, if not personalities, of the octopus.[112]

Given these characteristics, it is scarcely surprising that the octopus and the squid are very active animals, with maximal power outbursts that compare favourably with those of human athletes. This type of activity presupposes an effective circulatory system. As such it finds strong parallels to ourselves in two ways: the structure of the aorta and the pattern of circulation. Although rather different, both are related to the necessity of supplying oxygen at a fast enough rate. In such circumstances high blood pressures are essential. Blood leaves the heart via the arteries, and to deal with the repeated fluctuations in pressure it is hardly surprising that the arterial walls are rich in specific proteins that confer an elasticity to the throbbing tube.[113] Nor is it really surprising that the structure of the aorta wall in squid and human is strongly convergent.[114] The other parallel concerns the general arrangement of the circulatory system. Elevated blood pressure is essential for an active (and intelligent) animal, but if directed to the respiratory organs (lungs or gills) there is a danger that it will rupture the delicate membranes across which the gases are exchanged. The solution is simple and to some extent convergent. Only in the cephalopods[115] and vertebrates is the circulatory system completely enclosed in vessels and so capable of operating at high blood pressures. In both there is effectively a dual system, whereby the blood is first pumped to the lung/gill in order to collect the oxygen (and dispose of carbon dioxide). The blood is then returned to a second set of chambers in the heart where it is dispatched at full force into these elastic arteries. Even so, despite the shared principle there are significant differences. The branchial heart in cephalopods, which as the name indicates is responsible for feeding blood to the gills, is separate from the main pump. Furthermore, it is not very muscular, and probably has additional excretory functions.[116] The main systematic heart is powerful and muscular, but its structure only approximates to the vertebrate arrangement.[117]

Despite their manifest differences, the convergences between the cephalopods and vertebrates have attracted the attention of many biologists. Of greatest significance surely are those that pertain to the camera-like eye (Chapter 7) and brain (note 108). Others are probably less significant, but still intriguing. Earlier, in discussing the hypothetical Fortean bladders I remarked on the convergence between the relatively familiar fish swim-bladder and that of an octopus (note 35, Chapter 6). Another interesting convergence with the vertebrates is the development of a cartilage-like tissue in the head.[118] In fact, the roster of convergences is still not complete. Consider, for example, that part of the cephalopod sensory system associated with the skin. This is strongly analogous to the lateral-line system found in the fish and aquatic amphibians,[119] and it too is sensitive to pressure waves travelling through the water.[120]

Having reviewed earlier (Chapter 7) not only convergences in particular sensory systems but also possible underlying commonalities (which are equally important), it is not surprising that, at least so far as the fish are concerned, there is evidence that the lateral line can help the animal to form a 'hydrodynamic image'.[121] What sort of 'map' or 'image' this might form in the brain is still a matter for speculation. It is likely, however, that the input from the lateral line produces a 'pressure world', analogous both to the 'electrical world' of the mormyrids and other electrosensitive fish and to one that is integrated with other sensory inputs, such as vision.

Self-evidently, as an aquatic adaptation the lateral line of the fish (and amphibians) was lost as the tetrapods clambered on to land. In the case of the manatees (or dugongs), which are secondarily aquatic mammals, the arrangement and anatomy of the post-cranial hairs are something of an evolutionary novelty. Their sensitivity to changes

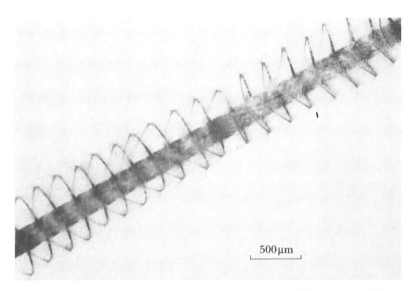

500μm

FIGURE 8.6 A surprising convergence on the lateral line system of fish and cephalopods, as manifested in the trailing antenna of a penaeid crustacean. The structure defines a tunnel with spaced bunches of sensory hairs, and so is a direct analogue of the pressure-sensitive system of the other two groups. (Reproduced from Pl. 1a of Denton and Gray (1985; citation is in note 123) with the permission of the authors and the Royal Society.)

in water pressure, however, offers an intriguing analogy to the lateral line,[122] and is consistent both with the poor eyesight of these creatures and with the murky waters they inhabit. Because I regard the convergences among sensory systems as being of particular importance, in the context of lateral lines it is surely necessary to remark on an even more striking example: that between the lateral-line system of the fish and the modified antennae of some swimming crustaceans known as the penaeids.[123] These pelagic arthropods have a pair of flagella-like antennae (Fig. 8.6), which project from the head at right angles before bending sharply and trailing parallel to the animal as it swims through the water. As in many other arthropods, the antennae have hair-like setae, but these are recurved and so define a tunnel-like structure on the floor of which are located sensory setae. Despite their quite different origin these antennae function in effectively the same way as the lateral line of a fish; in each case the basic structure is a tube with sensory hairs sensitive to changes in the water pressure.

HONORARY MAMMALS

So building complex circulatory systems, peering at the world through a camera-eye, and employing intelligence with a large brain have all evolved convergently. We are, however, in pursuit of the humanoid, and I shall assume that the galactic equivalent is in some sense mammal-like. This may seem too bold a claim, but let us see. So far as the Earth is concerned, there is a simple natural experiment. What we need to do is find a landmass which the mammals have failed to colonize. There is an excellent example, and it is called New Zealand. Indeed, Jared Diamond[124] went so far as to say 'New Zealand is as close as we will get to the opportunity to study life on another planet.'[125] For at least 85 million years[126] these islands have remained isolated in the Pacific Ocean, too remote to be colonized by any of the terrestrial mammals, other than by the bats and much later the boat-travelling Polynesians who arrived about AD 1000. New Zealand, however, had plenty of other inhabitants. Particularly extraordinary are the giant wingless crickets, known as wetas, relatives of the grasshopper and locusts. These are ecologically convergent on mice and rats, 'resembling rodents not only in their biomass, but also in nocturnal foraging and diets, use of diurnal shelters, polygamy and even their droppings',[127] and when surprised leaping across the forest floor.[128]

Nor are these the only animals in New Zealand to approach mousedom. Diamond remarks that the 'Stephens Island wren ... was ... the world's only known flightless songbird and functioned as an avian mouse.'[129] So, too, one of the endemic bats (*Mystacina*) is also convergent on a mouse-like habit, being partially terrestrial and when walking on the ground folding its wings to protect the delicate membranes.[130] Presumably like the bats the ancestors of these birds had originally flown there, but perhaps because of the absence of other ground-dwelling competitors many of them became flightless. These include an extraordinary nocturnal parrot, known as the kakapo, which grows as big as a turkey.[131] The most famous are the kiwi and the much larger moa.[132] These birds are not closely related and, arrived at flightlessness convergently.[133] The moas have vanished, and their extinction was almost certainly by way of the stew-pots, roasts, and fricassees of the Maoris.[134] The much smaller kiwis, however, avoided the category of a walking larder. If, as for the moa, we had only the bones, it is unlikely that we would ever have realized that in a land without mammals, the kiwi (Fig. 8.7)

FIGURE 8.7 The kiwi, a flightless bird, emblem of New Zealand, but also an 'honorary mammal'. (Reproduced from the figure on p. 103 of Calder (1978; citation is in note 135), with the permission of the artist, Alan D. Iselin.)

has converged strongly in this direction.[135] The feathers are rather atypical in being shaggy, in fact almost fur-like. In part this must be because these flightless birds live in burrows. Like many of the equivalent-sized mammals, kiwis are also nocturnal, are strongly dependent on a sense of smell, and unusually have nostrils at the tip of the beak. Moreover, the set of feathers around the mouth is modified into whisker-like structures.[136]

It is in its reproduction, however, that the kiwi (Fig. 8.7) is the most mammal-like. Not so far as giving birth to live young; like the monotreme mammals, such as the duck-billed platypus, the kiwi still lays eggs. Given the widespread occurrences of ovoviviparity, that is, where the egg is retained in the female who ultimately gives birth to live young (and as discussed below is strongly convergent), it is something of a mystery why, with one dubious exception concerning a budgerigar living in Dorking,[137] no bird has managed to become capable of live births. There has been no shortage of suggestions, including potential problems of immunology (leading to rejection of the embryo), the type of sex determination (which in birds depends on the female, with a ZZ–ZW system of chromosomes as against our XX–XY male determination), and perhaps most plausibly the difficulty in dispensing with a shelled, calcareous egg.[138] Even so, it is possible to predict that if ovoviviparity were ever to evolve in the birds the species concerned would show the following characteristics: flightlessness; a lower than average body temperature (one reason why internal incubation may be difficult for birds is their elevated body heat, at c. 40 °C); and a single egg rather than a clutch. And the closest candidate to this list? Yes, the kiwi.[139]

So, even though the kiwi is not ovoviviparous, it still remains the case that its reproduction is otherwise strongly mammal-like and unlike that of other birds. Thus, both ovaries remain fully functional, with the ovum released in the alternate fashion. The incubation period of the egg is exceptionally protracted, equivalent in time to a mammal of the same body weight. In reviewing this flightless denizen of New Zealand, William Calder III was in no doubt that this convergence was more than superficial when he wrote,

> When one adds to this list [that is reproduction] the kiwi's burrow habitat, its furlike body feathers and its nocturnal foraging, highly dependent on its sense of smell, the evidence for convergence [with mammals] seems overpowering. Only half jokingly I would add to the list the kiwi's aggressive behavior ... When I intruded on his domain at night, [the male kiwi] would run up to me snarling like a fighting cat ... and drive his claws repeatedly into my ankles until I went away. For this behavior and the many other reasons [given] ... I award this remarkable bird the status of an honorary mammal.[140]

GIVING BIRTH TO CONVERGENCE

The case of the kiwi is a useful reminder that despite sticking to eggs, many other non-mammalian vertebrates have adopted the system of ovoviviparity, in which the egg is retained in the mother prior to a live birth. The squamate reptiles (lizards and snakes) are particular masters of ovoviviparity: Richard Shine estimates that in this group this type of reproduction has evolved independently approximately 100 times,[141] while Daniel Blackburn suggests that overall viviparity has evolved independently approximately 132 times,[142] with examples also in the anuran amphibians[143] and fish.[144] Not surprisingly, the retention of the embryo in the female reproductive tract has led to other convergences, most notably the emergence of a placenta.[145] Particularly remarkable in this respect is the similarity between the mammalian placenta and an ovoviviparous Brazilian lizard (*Mabuya heathi*) which, as Blackburn and his colleagues remark,[146] 'has strongly converged upon a reproductive pattern long believed to be unique to eutherian [i.e. placental, as against the metatherian marsupials] mammals'.[147] The relevant features, in addition to the placenta and its nutrient supply, include a remarkably small egg (*c.* 1 mm), comparable in size to a typical mammalian ovum. Not surprisingly the period of gestation, for a reptile, is prolonged. This has the curious side-effect that females can be impregnated while juvenile and give birth to their live young when adults. Other species of the lizard *Mabuya* show similar features,[148] and it is possible that such viviparity originated at least four times in this group.[149] Other viviparous lizards show interesting features convergent with mammals. In a European species of *Chalcides* (*C. chalcides*) the placentome, a specialized area involved with absorption, develops to such an extent that at the time of birth it can be cast off, and thus uniquely for the reptiles is equivalent to the mammalian afterbirth.[150] A curious parallel also exists in the spade-nose shark. Here, too, the ovum is minute (*c.* 1 mm), and quickly implants in the uterine wall where it receives nourishment through a placenta-like structure. The combination of tiny egg and major maternal input again leads to a massive increase in the weight of the developing young.[151]

In reviewing the occurrences in the lizards, Daniel Blackburn emphasized that 'In several important respects, matrotrophic *Mabuya* and *Chalcides* have converged strongly with mammals. Features that have evolved independently among eutherian mammals and in one or both of these two reptilian groups [total at least seven],'[152] and he concludes that

> Classic textbook examples of convergence, such as the evolution of flight, and fusiform body shapes in aquatic vertebrates, pale beside those relating to viviparity and matrotrophy. Among viviparous vertebrates evolutionary convergence has operated at virtually every level – not only with respect to viviparity itself, which has originated over 130 times, but at the level of the organ, the tissue, and the cell. Such convergences are all the more striking when we consider that the historical sequences through which viviparity and matrotrophy have evolved apparently differ from lineage to lineage. Moreover, the independently-derived

specializations for fetal nutrition refute theriocentric [i.e. mammal-centred] views of mammalian reproduction and placentation as uniquely specialized.[153]

Blackburn's analysis epitomizes many of the important features of convergence, especially the degrees of similarity and the fact that the end-result may be arrived at by various historical pathways. Another important facet of convergence is emphasized in a study of sharks and rays, where viviparity (see also note 151) is estimated to have evolved independently at least nine times. In at least this group there is a very strong polarity towards viviparity, with only a few counter-examples.[154] It is worth noting, however, that the strong phylogenetic trend to viviparity is not matched by the evolution of matrotrophy (nutrition provided by the mother in addition to the yolk), which evolutionarily is much more labile. This has evolved independently four or five times, but also with multiple reversals.

In the context of non-mammalian ovoviviparity, there is one other item worth mentioning. Earlier (Chapter 5, p. 96) I suggested that as a contingent episode the asteroid that hit the Earth 65 Ma ago, and so led to the demise of the great reptiles, which thus facilitated the subsequent rise of mammals, may be misconstrued. Not as a historical episode: there seems little doubt that the mammalian radiations are a direct consequence of the disappearance of the Mesozoic reptiles. That, however, does not address the question that concerns the likelihood, if not the inevitability, of the emergence of such biological properties as 'mammal-ness'. Viewed in this light a different picture emerges, and for two reasons. First, as the Tertiary planet cooled, culminating of course in the ice ages, the mammals (which, recall, appeared about 220 Ma ago, at about the same time as the dinosaurs) would have been at an advantage anyway, well adapted to temperate and even polar environments. This suggests that the rise of active, agile, and arboreal ape-like mammals, and ultimately a hominid-like form, would have been postponed, not cancelled. My guess is that without the end-Cretaceous asteroid impact and in due course progressive planetary refrigeration, the appearance of the hominids would have been delayed by approximately 30 Ma. But even that is a local history. It is interesting that, at least in the lizards, the evolution of viviparity is associated with colder climates.[155] It is not the only factor, but recalling also that 'honorary mammal' the kiwi, it too suggests that 'mammal-ness' will not be the exclusive property of one evolutionary group. If 'we' had not emerged, then (as this and the next chapter aim to show), rest assured that a viviparous, warm-blooded, vocalizing, and intelligent species would have done so.

Finally, as if it needed any emphasis, it is important to stress, as Nicholas Dulvy and John Reynolds specifically remark with respect to shark and ray viviparity 'Our data do not support a linear, irreversible progression toward a 'pinnacle' of maximum maternal input.'[156] Quite so; evolution *is* labile, it *does* show reversals, but the point still remains that the emergence of various biological properties is in response to adaptation and is governed by selection. Nor does this mean that the selective pressures operating in an ocean need be the same as on the savannah, although as we shall see later (p. 250) here, too, there are some surprising convergences, specifically between sperm whales and elephants. Convergence simply tells us that the evolution of various biological properties is certainly highly probable, and in many cases highly predictable.

WARMING TO CONVERGENCE, SINGING OF CONVERGENCE, CHEWING CONVERGENCE

The stories of the kiwi and non-mammalian ovoviviparity are two reminders that being a mammal and evolving towards 'mammal-ness' need not be the same thing. There are at least two other features that typify mammals but are in fact also strikingly convergent: warm-bloodedness and vocalization. Concerning the first, endothermy (that is, the phenomenon of having a body temperature elevated in comparison with the ambient environment), is rampantly convergent and occurs not only in the birds and mammals, but also in various fish (teleosts and sharks) and even in insects.[157] The example of the fish might seem somewhat removed from the path of humanoids, but may in fact have some relevance. In the fish endothermy is clearly convergent, and has probably evolved on at least five separate occasions. Examples include the remarkable modification of the eye muscle in some scombrid fish, to form a heater to warm the brain and retina (see note 118, Chapter 7). Whole-body endothermy in fish is perhaps best known in some tuna, famous for their fast swimming and wide ranges. Barbara Block and her colleagues[158] stress that this is a sophisticated adaptation, and entails a concatenation of changes. These include the development of countercurrent heat exchangers at three locations in the body, the movement of red muscles towards the interior of the animal, combined with the body becoming larger and thicker, and a style of locomotion known as thunniform, in which swimming is achieved by oscillation of the tail rather than sinuous deformation of the entire body. Yet this complex arrangement was convergently arrived at in the lamnid[159] and alopiid sharks, where again it probably evolved independently in each group.[160] There are, moreover, more specific convergences[161] such as the 'Striking parallels in both structure and function [that] exist between the shark orbital rete [a fine network of blood vessels] and the mammalian carotid rete which serves as a brain cooling system in mammals'.[162]

The adoption of this type of endothermy is significant because it evidently allowed an expansion of a habitable zone, and in these fish it made possible the invasion of a vast new realm, specifically the cooler (and deeper) parts of the ocean. As Barbara Block and her co-workers note, the multiple origins of endothermy indicate 'Strong selection for this energetically costly metabolic strategy'.[163] In other words, endothermy is expensive, but well worth the cost by being highly adaptive and accompanied by success. Costs, as ever, have risks, and in the context of endothermy there is another interesting physiological convergence. This involves a pathological condition known as malignant hyperthermia, in which the muscles produce excessive heat that if unchecked leads to severe tissue damage and death. In mammals it has a genetic basis, and may manifest itself in anaesthetized humans undergoing surgery, as well as in genetically disposed pigs that panic in the slaughterhouse.[164] A strikingly similar syndrome is found in highly stressed tuna, especially those caught on a hook and line. This leads to so-called 'tuna burn', badly damaged musculature and subsequent rejection by the fastidious *Sashimi* enthusiast.[165]

What then is the significance of the convergent evolution of endothermy in the birds and mammals? C. G. Farmer has argued that among the welter of possibilities the key factor is parental care.[166] Of central importance is the need for sustained exertion, to enable the parent(s) to collect sufficient food not only to maintain their ravenous fledglings or litter, but also for them to grow as quickly as possible and thereby decrease their vulnerability. Even in those birds and mammals that characteristically save energy by entering a state of torpor and letting their body temperature decline, incubating birds and pregnant mammals typically retain a higher body temperature. Farmer suggests that the original trigger for bird and mammal warm-bloodedness was connected to reproduction and the increased production of hormones, e.g. from the thyroid, which in turn are also important determinants of metabolic rates. She continues by noting that

> The convergent evolution of ... parasagittal limb posture ... a large pulmonary diffusion capacity ... high oxygen-carrying capacity of the blood ... a large tissue diffusion capacity ... a completely divided cardiac ventricle ... an extensive coronary circulation and compact myocardium ... high systematic blood pressure ... and a single aorta ... may be due, at least in part, to selection for an ability to sustain vigorous exercise which is requisite for birds and mammals to provision their young.[167]

Parental care is, of course, a particular hallmark of the birds and mammals,[168] and Farmer concludes that this in turn may have promoted both 'complex social structures [and] ... vocal signaling'.[169] And that might seem to be the furthest these convergences can be pushed. Birds are warm-blooded, indeed generally showing significantly higher body temperatures, and of course they can also sing. Is the ability to sing only a vague similarity? Not a bit of it; now it is time to see just how far vocalizations are also convergently arrived at.

The ability of birds to vocalize needs, of course, little emphasis: twittering, booming, squawks, hooting, and melodious song are all familiar, as are the mimics, such as some parrots and the mynah birds. One African Grey I knew could produce a remarkable repertoire of sounds, including the Greenwich time pips with eerie accuracy and now, her companion Caroline Pond tells me, is well up with mobile phones and car alarms. And there are more poignant instances. Darwin, for example, in a section on the extinction of races and tribes remarks that 'Humboldt saw in South America a parrot which was the sole living creature that could speak a word of the language of a lost tribe.'[170] Even so, we tend to regard bird mimics as simply amusing, and their songs as only incidentally beautiful and in reality vocal expressions of territorial demarcation and the urgent priorities of mate attraction. Perhaps so, but it transpires that in reality there are under-appreciated similarities between our vocalizations and those of the birds, both in terms of songs and musical output[171] (and even drumming with a stick[172]) and neurology. Thus, with respect to song, both birds and humans share such features as 'interval inversions, simple harmonic relations, and retention of melody with change of key', while our 'simple melodic canon ... is reminiscent of the matched countersinging of many bird species.'[173] So, too, claims of human uniqueness with reference to such matters as the voluntary control of the supralaryngeal tract, a prerequisite for human language, in reality finds parallels in African Grey parrots.[174]

So far as neurology is concerned, the behavioural biologists Allison Doupe and Patricia Kuhl remark,[175] concerning birdsong and human speech, that 'there are striking similarities in how sensory experience is internalized and used to shape vocal outputs, and how learning is enhanced during a critical period of development', and they continue, 'similar neural mechanisms may therefore be involved'.[176] Significantly, Doupe and Kuhl use the phrase 'strikingly similar' a number of times. Thus, in referring to the vocal control systems in songbirds and humans, they comment how continuing study 'reveals numerous anatomical and functional similarities in the organization of neural pathways for vocal production and processing ... both fields [of enquiry] grapple with strikingly similar questions about how sensory and motor processes interact in vocal learning and production,'[177] while concerning the learning and self-organization of sounds they remark that the 'perceptual patterns stored in memory serve as guides for production [and are] strikingly similar'[178] in birdsong and human speech.

In some ways the similarities of these various processes are all the more surprising, given both the various differences in avian and mammalian brain structure, e.g. the absence of a multi-layered cortex in the avian brain, and in some species of bird strong sexual dimorphism of song production. Yet the similarities, striking or otherwise, still emerge. Even so, Doupe and Kuhl are careful to qualify these remarks, noting that although the parallels are striking there are also a number of obvious differences, most notably the human possession of a grammar (but see Chapter 9). Yet what they rightly call the 'numerous parallels'[179] between my remarking to my companion on the beauty of a bird's song and the song itself, suggest that not only warm-bloodedness and viviparity but also at least some mechanisms of both vocalization and song may be widespread across the Galaxy (see note 132, Chapter 9). So, too, given the recurrent emphasis on evolutionary convergence, it is not surprising to learn that in the birds the powers of song and vocalization have evolved at least twice.[180]

Song, the birth of live young, and warm-bloodedness, not to mention that honorary mammal, the New Zealand kiwi, all provide compelling examples of convergence that show how features we associate with the mammals are more widespread and may, therefore, reasonably be expected to emerge elsewhere. Nor do the examples end here. Mammalian dentition is well known for its complexity. Earlier, we encountered the convergent evolution of the massive canines in the sabre-toothed cats and marsupial thylacosmilids. In addition to the canines, most mammals have nibbling teeth (incisors) and the familiar battery of molars for grinding or shearing.[181] That this dentition makes sophisticated food processing possible has no doubt contributed to the evolutionary success of the mammals. In contrast, the dentition of the reptiles is much simpler. Typically it consists of a sharp array of pointed teeth, and if there is any difference along the length of the jaw it is usually only one of relative size. Exceptions, however, are known, and they show striking convergences with the dentition of mammals. Not only that, but this type of dental convergence has arisen at least twice,[182] in two distantly related groups of reptiles, both of which lived in the Cretaceous. The first concerns a teiid lizard, from sediments close to the Cretaceous–Tertiary (K/T) boundary in Montana. Here the posterior teeth on both the lower (dentary) and upper (maxillary) jaws have become molariform, and resemble the so-called tribosphenic arrangement of mammalian molars with its series of interlocking cusps.[183] In addition, the method of regular tooth replacement, whereby old and worn teeth are discarded and new ones take their place, which is the norm in reptiles, is suppressed.[184] As in the mammals, after the loss of the deciduous (or 'baby') teeth, the next set is for life. The second example concerns some crocodilans from China and Malawi, whose dentition has again ground towards the molariform solution. In these examples chewing appears to have been somewhat less effective in its action, without direct occlusion. Unlike the teiid lizard, which probably ate insects, these crocodiles were vegetarians.[185] While considering the somewhat improbable topic of plant-eating crocodiles, it is surely worth mentioning in passing another example from the fossil record of a group of crocodiles (the ziphodonts) that became effectively fully terrestrial and in doing so developed something rather like hooves.[186] And are crocodiles unique? Well, not really, or at least so far as the skull is concerned. The spinosaurs,[187] hitherto regarded as a rather enigmatic group of theropod dinosaurs, have a skull that is interpreted as a crocodile mimic.[188]

By now, I hope it will be clear that many of the evolutionary features that help to define the human are convergent. If such features as warm-bloodedness, vocalization, and even agriculture can evolve independently, then so, too, on any suitable planet the same will emerge. Yet at this stage there is surely a dimension missing. To be sure these and many other convergent features serve to delineate complex biological systems, but the scope is still very wide-ranging and encompasses animals as disparate as ants, tuna, and kiwi. Even if we grant that 'mammal-ness' is a biological property rather than a historical contingency, we still seem to be far removed from anything specifically human with such hallmarks as bipedality, tool-making, culture, and intelligence. Let warm-bloodedness, vocalization, and even agriculture be convergent, but surely the hallmarks of the human are simply the quirky results of contingent happenstance. On Threga IX there will be much that is reminiscent of Earth, but – so it is widely believed – any consciousness will be submerged in the inarticulate and any music will be little more than harmonic babbling. The opposite turns out to be the case.

9   The non-prevalence
of humanoids?

What we know of the social insects, and especially the extraordinary organization of agriculture and warfare among the ants, is striking both in terms of their convergence and in the almost alien nature of these complex societies. Certainly the jointed skeletons, the compound eyes, the miniaturized clones, and apparently robotic social organization are a familiar staple of science fiction. Suppose that there are advanced extraterrestrials: will they be like us, at least vaguely humanoid, or so alien as to defy belief and perhaps even recognition, let alone communication? The majority certainly tends towards the latter opinion. It probably owes as much as anything to George Gaylord Simpson, one of the last century's great evolutionary biologists.[1] He was a prolific writer, and among the 16 papers he published in 1964 was one baldly entitled 'The nonprevalence of humanoids'.[2]

Simpson's article, presented with characteristic force and intelligence, was a sustained protest against what he saw as unwarranted extrapolation and conjecture. He argued that the history of life as revealed on Earth could not possibly be taken as a useful guide to biological events anywhere else in the Universe. Simpson presented his argument carefully, acknowledging that planets suitable for habitation would probably be in fairly short supply, and he conceded further that the likelihood of life itself arising was even lower. But no matter: as Simpson remarked, the Universe is a big place, so however uncommon life was, the total number of planets with life must be quite large. But, he concluded, whatever happened on these remote worlds in terms of evolution would bear little resemblance to what we see on Earth. Not surprisingly, given Simpson's life's work,[3] his argument was strongly neo-Darwinian: while no stranger to the importance of convergence[4] he was adamant that the emergence of anything like a human would be an evolutionary fluke.[5] As Irven DeVore exclaims, 'Extraterrestrial intelligence? Not likely.'[6] As such, our presence on this planet was *a priori* unpredictable because the evolutionary trajectory that led to us was only possible according to the 'precise conditions of our actual history.'[7]

INTERSTELLAR NERVOUS SYSTEMS?

Nor was Simpson the first to argue along such lines. George Beadle,[8] for example, also acknowledged the probability of extraterrestrial life, but dismissed the likelihood of any such organism being hominid-like, remarking that 'the *a priori* probability of evolving man must have been extremely small – for there were an almost infinite number of other possibilities.' He pessimistically continued, 'Even the probability of an organism evolving with a nervous system like ours was, I think, extremely small because of the enormous number of alternatives.'[9] Perhaps so, but what are the constraints? Central to an effective nervous system is the sodium channel, which by pumping sodium ions across a membrane provides the electrical activity necessary for the rapid propagation of action potentials (i.e. changes in voltage) and hence transmission of the signal along a nerve. This arrangement is sometimes said to be unique to the animals, evolving first in the jellyfish and their relatives.[10] Even if its primitive function was linked to the evolution of the nervous system, it is worth noting that the sodium pump has also been adopted for a wide variety of other physiological processes.[11] Sponges, which lack (and perhaps never had) a nervous system, are regarded as the most primitive group of animals and show no evidence for electrical activity in the form of action potentials.[12] Even so, there is a sophisticated single-celled organism, the heliozoan known as *Actinocoryne* that consists of a 'head' with radiating filaments and a stalk that attaches the cell to the substrate. If disturbed, the stalk and filaments are capable of remarkably rapid contraction, and the action potential that triggers this response is mainly dependent on sodium.[13]

Sodium transport[14] is also found in the bacteria, where it has various roles, such as in respiration and the movement of the rotary flagellum.[15] Rather remarkably it now transpires that in at least one bacterium this transport, which can occur in a number of different ways, is specifically by a sodium channel.[16] The evolution of sodium channels therefore does not seem to involve some wildly improbable process. It should also be noted that ion channels dedicated to the movement of potassium and calcium ions, which are fundamental in controlling cell physiology and excitability, are effectively universal.[17] It seems very likely that the sodium channel evolved from a pre-existing calcium channel.[18] Nor was this necessarily very difficult to achieve, given that single mutations can change the behaviour of the sodium channel so that it acquires the characteristics of a calcium channel.[19] So why not carry on using calcium, since it was already available? One problem is that calcium has a number of key roles within the cell, and sudden fluxes, necessary to generate the electrical signal associated with nervous activity, would lead to intracellular mayhem.[20] Safer by far to design a specific mechanism to handle fast action potentials. The fact that the sodium channel evolved independently of the animals in a single-celled heliozoan, capable of very rapid contractile movements, strongly suggests that at least from this point of view Beadle's scepticism about a nervous system being able to evolve may be over-pessimistic. Not only that, but Milton Saier[21] points out that the various transport proteins, including those involved with ion transport,

> apparently exhibit a common pattern of six tightly clustered transmembrane helical segments with both the amino and carboxy termini localized to the cytoplasmic membrane surface. It may be that a three-dimensional transmembrane structure consisting of six nearly parallel α-helices is particularly well suited to transmembrane channel formation. Such constraints could account both for the retention of this structural motif during evolutionary divergence of homologous transport proteins and for convergence of evolutionarily unrelated transport proteins.[22]

Thus, as we saw earlier, not only are there good arguments that aliens would see and smell using very similar proteins to those we use, but their electrical conductivity would again converge towards the same solution.

THE CONCEPTUALIZING PANCAKE

It is worth remembering that the scepticism of such people as Simpson and Beadle was set in the context of the beginnings of the exploration of outer space. In addition, it coincided with the growing interest in what came to be known as the SETI (Search for Extraterrestrial Intelligence) projects, galvanized by Frank Drake and others with the formulation of the now-famous Drake equation (which attempted to estimate the number of extraterrestrial civilizations[23]) and the OZMA Project (which was a preliminary search for signals using a radio-telescope in Green Bank, West Virginia).[24] Since then the question of the non-prevalence (or otherwise) of humanoids has remained an important ingredient in the SETI programmes, not least because of its potential implications for communication. Leonard Ornstein,[25] for example, took a view broadly similar to that already set out by Beadle and Simpson. His remarks echo the view of Temple Smith and Harold Morowitz,[26] which I introduced earlier (Chapter 1) in the form of the metaphor of navigating to Easter Island: that is, finding functional states, be they proteins or brains, in a vast 'universe' of alternatives, most of which are probably entirely maladaptive, is problematic. At first sight the alternatives are so remote from each other that any end-product, such as the one that we label as humanoid, would seem to be as fortuitous an outcome as anything else. Thus, Ornstein wrote: 'The full set of messages [that is the end-products of the Darwinian process arrived at by mutation] tested by selection, from the beginning of life, constitute only a minute and probably unrepresentative sample of different possible messages from which the sample has been "drawn" ... Therefore, no matter how prevalent *life* might turn out to be, biological evolution on earth can easily have generated many "inventions", perhaps including intelligence, which are unique in the universe.'[27] Despite this, Ornstein reminds us of how it is that by convergence 'evolution [can] easily "rediscover" certain useful classes of "technical solutions" to problems of survival, more or less independently of evolutionary starting points.'[28] As a possible proxy for intelligence he considers that favourite topic in convergence, the similarity between the eyes of advanced cephalopods and vertebrates. This, of course, was explored in some detail earlier (p. 151). But having got that far, Ornstein suddenly takes a blind alley, so to speak, by suggesting that in fact these eyes are not really independent at all, so saving his argument of potential uniqueness.[29]

So, should we accept the arguments against the prevalence of humanoids put forward by Beadle, Ornstein, and Simpson? I think not. In the same year that Simpson published his review, another worker, Robert Bieri,[30] took effectively the diametrically opposite view. He reminded us of the 'severe limitations [which] are imposed on the number of routes available to evolving forms. The number of alternative possibilities is by no means infinite; on the contrary, the number is quite limited. This limited number of available routes has led to the innumerable cases of convergent evolution in plants and animals,'[31] a few of which I have reviewed in the previous chapters. Bieri concluded by saying: 'If we ever succeed in communicating with conceptualizing beings in outer space, they won't be spheres, pyramids, cubes, or pancakes. In all probability they will look an awful lot like us.'[32] Similar sentiments are offered by the physicist Philip Morrison in a discussion[33] of extraterrestrial life when he writes, 'Evolution does not repeat itself; its paths across the badlands of the adaptive landscape are too complex. But all the same it frequently converges upon states that seem quite similar (attractors?) along several distinct paths,[34] He continues with pertinent comments related to the emergence of such evolutionary oddities as complex social systems and tool manufacture, both as we shall see being convergent.

So are we, as humanoids, in some sense either very probable, or perhaps even inevitable? Given that much of this book is about convergences, no bets need be placed. With a sample of one, that is the Earth's biosphere and its four billion years of history, it can always be argued that a descendant of Bieri, an astronaut, will land on a remote planet and strike up a conversation with one of those conceptualizing pancakes, but even at the most basic levels it now seems increasingly likely that whatever alternatives there might be they are going to be highly restricted. How many avenues are available even for the origin of life? Given our lack of success in this area (Chapter 4), could there be only one? Similarly, as I explained in Chapter 6, George Wald has argued that wherever there are planets whose biosphere is dependant on the light of a star, there, too, there will always be chlorophyll (and the associated antenna proteins) harvesting the radiant energy.[35] What of DNA? In a brilliant research campaign Albert Eschenmoser and his team have been constructing what they call an aetiology of nucleic acids, a library of alternative possibilities in the realm of DNA.[36] Some of the alternatives may equal the versatility of DNA; some, perhaps may even exceed it. Nevertheless, the various peculiarities of DNA, touched upon in Chapter 2, are undeniable: it, too, might be a molecule uniquely suitable for biological processes. Even the strange optimization of the genetic code[37] indicates that although there must be alternatives, they are not going to be abundant.

If life is universal, it also seems likely that it has a universal basis. The aim of this chapter, however, is to extend the argument for pervasive convergence and thereby question the thesis concerning the non-prevalence of humanoids. It may need no further emphasis, but what matters is the emergence of particular evolutionary properties. It is presumably immaterial which specific lineages develop the necessary complexities of, let us say, camera-eyes and intelligence, so long as some do. On Earth we can test this by reference to convergence. Nor need this have very much to do with SETI: as I have noted, life may well be a universal principle, but that does not prevent our being alone. So, too, the potential for humanoids may be universal, but that does not rule out the Earth being unique.

THE BRICKS AND MORTAR OF LIFE

In tackling Simpson's challenge there are obvious questions: at what level of organization should we begin? At what stage do humanoids become inevitable? Perhaps, as just mentioned, we can allow ourselves some basics, such as proteins and DNA. But then where? So far as humanoids are concerned, perhaps our starting point has to be the great apes? But here there are plenty of convergences, ranging from the tools of some New World monkeys to the social organization of dolphins. Perhaps we should start with the mammals? But what exactly is a mammal? Given that other groups of vertebrates converge in various ways on the mammals, we may wish to distinguish between the evolutionary lineage we call mammals and some wider property of 'mammal-ness'. Perhaps the potential for humanoids lurks somewhat deeper in evolution, maybe among the fish? Here surely the *sine qua non* of terrestrial existence for vertebrates, the four legs, is a fluke-like development from the pre-existing fins of the aquatic fish? But in fact the development of legs seems highly probable, given a series of parallel trajectories towards limbs in the Devonian fish.[38] Still not deep enough? Perhaps some sort of worm? But here too there is plenty of evidence for vermiform convergence, even among the vertebrates.[39] To be sure, man is not a worm; the point is simply that at each and every stage convergence emerges.

Ape, mammal, fish, worm, or even single cell? There is no simple answer to the question about the stage at which something like ourselves becomes overwhelmingly probable. To provide a focus, however, much of this chapter will concentrate on those elements that we might regard as the hallmarks of the humanoid – large brain, intelligence, tools, and culture – all of which, needless to say, show examples of convergence. Despite this emphasis it is worth re-emphasizing the notion of inherency. By this I mean that there is a more fundamental level, inasmuch as many of the molecules (or for that matter, genes) that we rightly regard as essential for our existence were in fact already in place at much earlier stages in the history of life. Indeed, in some respects this basic architecture (we might call it a sort of biochemical scaffolding) is as central to the discussion of evolutionary inevitabilities as the veneers of life, such as sentience.

A good example is the iron-bearing protein haemoglobin. For us, and other animals, it has a vital role in the transport of oxygen in the red blood corpuscles. Haemoglobin, however, is much more widespread. In addition to its occurrence in some plants, most famously in the root nodules of such leguminous plants as the clover, haemoglobin is also found in more primitive organisms from single-celled protistans, such as the ciliates, to bacteria, including the cyanobacteria.[40] To a first approximation, irrespective of whether the haemoglobin is in the blood of a cow, the clover it is cropping, or in the cyanobacteria forming the scum of a nearby pond, the protein is doing the same thing, taking care of oxygen. But there the similarities cease. In the leguminous clovers and the cyanobacteria nitrogenous compounds are synthesized, but their formation critically depends on the exclusion of oxygen. Hence a role for haemoglobin. This protein is highly effective; David Goodsell has referred to it, perhaps incautiously, as 'perfectly designed for oxygen transport.'[41] Haemoglobin is not the only protein involved with oxygen transport, but as we shall see (Chapter 10) in the case of the related myoglobin and the copper-bearing haemocyanin, as well as that of haemoglobin itself, there is again evidence for convergent origins.

The proteins involved in oxygen transport (and storage) are important because, however ancient they are, and from whatever predecessor they were recruited, they are a vital ingredient in the success of large, metabolically active animals, including humanoids. Muscles, and even more so brains, require copious quantities of oxygen. But that is only one component. Another key element in the emergence of at least animal complexity is those chemicals suitable for nervous activity. A key molecule in the context of nervous transmission is acetylcholine. This compound plays a key role at the junction where a nerve fibre (carrying the electrical information from the brain) abuts against muscle fibre (the contraction of which will lead to a desired action, e.g. running away from a charging ungulate). At this meeting point the signal has to be transmitted across a narrow junction known as the synapse. All in all it is a sophisticated system, yet this molecule is very ancient.[42] Acetylcholine is found in bacteria and plants, and quite clearly has been co-opted in the animals for neural functions. We know this also because molecules characteristic of the process of nervous transmission and activity are found in single-celled protistans, notably the ciliates.[43] In these tiny organisms there are no nerves, let alone brains; so what are these molecules employed for? Ciliates show a sexual process known as conjugation in which individual cells must first confirm their compatibility (recognition) before adhering in order to allow exchange of genetic material. All this depends on the molecule acetylcholine and a number of other key receptors.

The molecular architecture of the nervous system is complex in other ways. For example, certain hormones, the neuropeptides, are also important. At first sight it is quite surprising that hormones such as corticotropin, which is important in the pituitary gland, as well as β-endorphin-like molecules and dopamine, have also been found in the ciliates.[44] Given that these organisms lack any sort of nervous system, the function of these neuropeptides is somewhat enigmatic. Nevertheless in their own way the ciliates are sophisticated organisms and use of messenger molecules is to be expected.[45] George Mackie[46] has aptly referred to these molecules occurring in protistans as 'prophetic', and they underline the likelihood of more complex structures, say a brain, emerging from the unicellular substrate.

Other types of complexity in both bacteria and various eukaryotic microbes have been comparatively well known for many years, not least the remarkable propensity for certain bacteria (the myxobacteria) and eukaryotic slime moulds (Dictyostelium) to aggregate into quasi-multicellular organisms, some of which display a mobile slug-like behaviour. There is, however, newly emerging information, across a wide front of enquiry, that is demonstrating hitherto unappreciated levels of complexity that correspond in a number of interesting ways to the social behaviour of animals and higher organisms.[47] Thus, aspects of sociality such as foraging and cooperative hunting, specialized dispersal forms, genetic altruism, and (perhaps most interestingly) communication, using various chemical signals,[48] have now been identified. As Bernard Crespi remarks, 'The social phenomena uncovered so far allow the first direct comparisons between microorganisms and macroorganisms [and in the text he lists seven social attributes], which reveal convergences in behavior that are clearly suggestive of adaptation.'[49] Not only that but, as pointed out by various authors,[50] the social interactions, synchronized activity, and communication between microbes confer on them multicellularity of a sort. This makes one wonder whether even on planets that, in my view implausibly, were to remain stuck in a microbial cul-de-sac,[51] parallels to the complexity of animals and plants might not still emerge.

GENES AND NETWORKS

There is one escape route from this apparent dilemma of the simple being equipped with molecules essential for the complex. In the fine print of life's contract there is a clause allowing the lateral transfer of genetic material between distantly related organisms. Could, let us say, animals 'invent' neuropeptides, and then export the code to all and sundry? There are certainly now a number of well-attested examples of such transfers, but for the most part they seem to entail exchange between more primitive organisms.[52] At this stage it seems much more likely that in a way analogous to the crystallin proteins of our eyes, so, too, long before there were nervous systems, at least some of the key molecules were there 'ready and waiting'. So the principle of inherency, widely recognized, but more seldom articulated, appears to hold.

There is, however, a more general problem of assessing how the sort of complexity we associate with humanoids emerges from life's molecular substrate. This is because of both the immediate question as to when and where on any evolutionary timetable the emergence of complex forms becomes inevitable, and to anticipate in Chapter 11 the extent to which genetic fundamentalism has any bearing on explaining both the structure of life and its potentialities. Thus, despite their iconic status, it is not clear that DNA and genes are in much of a position to provide any sort of helpful metric when it comes to deciding either what is inherent or what is complex. First, it has long been appreciated that the overall quantity of DNA an organism possesses provides effectively no guide to its relative complexity: single-celled organisms may contain far more DNA than a human.[53] The enormous variation in the size of genomes has led most biologists to regard the 'excess' DNA as a sort of molecular 'junk', surplus to requirements.

Another problem, and equally serious, is the news emerging from the genome projects in which the DNA is mapped and the total number of genes tallied. For a complex animal, such as a fruit-fly

(*Drosophila*) or nematode worm (*Caenorhabditis*) to function, thousands of genes are required, significantly more than for the bacteria, for which a total of about 4000 is fairly typical. Thus, in the worm *Caenorhabditis* the estimated total is just over 18 000. So far, so good, but in its own way it is a relatively simple animal. As an adult it has a fixed number of cells, which is one of the reasons why it is particularly suitable as an experimental species. This is because each lineage of dividing cells can be followed exactly from the fertilized egg until the final number in the adult is reached. The cuticle is certainly rather complex, but the general organization, e.g. gonads, gut, of *Caenorhabitis* is fairly straightforward. No eyes either, and the worm has a relatively simple brain. So perhaps 18 000 genes are about right. Now consider the fruit-fly. A sophisticated flier with its gyroscopic halteres, a complex brain with capability for memory and courtship, compound eyes, and a well-differentiated body. All in all a complex animal, and how many genes? If the worm needs 18 000 what about 30 000 for the fly? Not a bit of it: the fly actually has substantially *fewer* genes, totalling 13 600.[54] So what about 'the pinnacle of creation': us? Until recently the estimates of our gene total were in the order of 100 000, but they too are now being revised downwards; some people are suggesting as few as 30 000. But perhaps we should not be so surprised. Claims for the primacy of the gene have distorted the whole of biology, and as I shall discuss in Chapter 11 other views are perhaps worth entertaining.

Neuropeptides in single-celled organisms without a nervous system, and more complex animals built with fewer genes than simple animals, might in their different ways be thought to pose a problem in biology. How much of a complex organism, say a humanoid, has evolved at a much earlier stage, especially in terms of molecular architecture? In other words, how much of us is inherent in a single-celled eukaryote, or even a bacterium? Conversely, we are patently more than microbes, so how many genuinely evolutionary novelties can we identify that make us what we are? It has long been recognized that evolution is a past master at co-option and jury-rigging: redeploying existing structures and cobbling them together in sometimes quite surprising ways. Indeed, in many ways that *is* evolution. New insights that confirm this view are now becoming available from the rapid development in our understanding of developmental biology and the associated genes. The examples most often cited concern features of animal architecture such as the body axis,[55] dorso-ventrality,[56] wing/limb and other appendages,[57] eyes (camera versus compound),[58] and heart.[59] Most often this is expressed in terms of comparisons between fly (arthropods) and mouse (vertebrates). The conservation of genetic instructions between these rather distantly related animals originally came as a considerable surprise, but is now realized to be part of a much wider pattern. Even so, there is still a world of difference between the human advancing with raised newspaper in hand, camera-eyes glaring versus the fly through its compound eyes gazing at the *London Review of Books* descending rapidly. Splat? Missed again? Tut! Tut!

Clearly other genes, and more importantly the respective networks of developmental processes, in a fly ascending to safety and the frustrated human are different. The tension that has arisen from the discovery of widespread genomic similarity of organisms that are otherwise widely disparate in terms of anatomy and behaviour is only now being explored. Differences there must be, but at a molecular level these may be trivial, sometimes even a matter of a handful of amino acid substitutions. Alternatively, they may arise from already well-known evolutionary processes, notably gene duplication. In the case of the newspaper-waving human and the other vertebrates this latter process, notably the double duplication of the *Hox* genes, is widely interpreted as offering the fish and all their descendants new evolutionary opportunities.[60]

The details of what will be discovered along the way, as the various gene networks and cascades are documented, will in their various ways offer fascinating insights into the structure of life. Why, for example, do practically all mammals have a fixed number of neck (cervical) vertebrae? In giraffes and moles, for example, the lengths of the respective necks could hardly be more different, but in both the number of cervical vertebrae is seven. In contrast, in the other vertebrates this total is much more variable. All things being equal, it would be more 'sensible' for the giraffe to multiply the number of neck vertebrae, rather than being 'forced' to elongate each of the seven it has. Why then the constraint? An intriguing suggestion, made by Frietson Galis,[61] is that in the mammals a presumably fortuitous coupling has arisen from the involvement of key developmental genes (especially *Hox* genes) in both the laying down of the axial skeleton, including of course the cervical vertebrae, and the process of cell proliferation.

If, owing to some developmental abnormality,[62] the patterning of the axial skeleton is upset so, too, there is a tendency to develop childhood cancers. These are examples of uncontrolled cell proliferations, which in this case originate in the developing embryo. For mammals, departure from seven spells lethality. Moreover, in mammals some cancers may owe their initiation to the production of highly reactive molecules (known as free radicals). In the mammals, at least, the free radicals are an unavoidable by-product of an active metabolism. It may be no coincidence that the few exceptions to the rule of seven in neck vertebrae are in the metabolically sluggish animals, such as the torpid sloth. In this sense the rule of seven in mammalian necks is a good example of stabilizing selection, and may be the 'price' to pay in ensuring the effective development of very complex organisms. Such a constraint has, therefore, its costs, but when we see the diversity of mammals it seems that a restriction to seven cervical vertebra in animals as diverse as bats and camels has been more than offset in other respects.

The difficulties in making simple equations between genetic architecture and the body form of either a worm or a humanoid, let alone something more primitive, are considerable. Given the widespread conservation of genes among the animals, it is hardly surprising that some must date back to at least the dawn of their history.[63] A further problem is the evidence for extensive redeployment of genes so that they end up with multiple functions. There are now many examples, but the famous *Pax-6* will serve well because its activity is sometimes regarded as almost synonymous with the origin of eyes. True up to a point, but as already mentioned in Chapter 7 in the mammals *Pax-6* is also implicated in the development of other organs, including the nose,[64] brain,[65] pituitary gland,[66] gut,[67] and pancreas.[68] To be sure, in these animals *Pax-6* is involved with the formation of eyes, but in the eyeless nematodes *Pax-6* is still expressed.[69] There is a further complication because in at least one group of animals (the flatworms) *Pax-6* is, as expected, involved in eye formation, but if it is 'knocked out' during the process of regeneration (for which flatworms are famous) then eyes still form.[70] As already noted in Chapter 7 (see note 315), it is possible that *Pax-6* was originally involved with differentiation of nervous tissue and thereby became involved with sensory structures at the anterior of primitive animals which had at most rudimentary eyes. Indeed, the fact that *Pax-6* is also involved with crystallins[71] and rhodopsin,[72] which as proteins far pre-date the eye, is probably a result of co-option for effective optical development.

There are many more examples of multiple functions and re-deployment of developmental genes.[73] Nor is this at all surprising; these genes are effectively switches that instruct, presage, or order other genes essential for the formation of particular tissues. This has led to a protracted, if not agonizing, discussion about what really is the same in evolution. If *Pax-6* makes all eyes, then despite the differences between camera- and compound eyes are they not really equivalent? Not in any useful sense, because despite the remarkable genetic conservation[74] the genetic instructions that encode for the fly eye differ in other ways from those for the vertebrate eye.[75] Rising above a molecular perspective is essential if we are going to understand evolution. Indeed, there is already an uneasy feeling that patterns of genetic expression may tell us more about the constraints of building organic architecture than providing insights into particular origins.[76] Co-option of genes for new functions is hardly unexpected, but the questions of how? when? and why? are taking more time to sort out. If new recruitments are commonplace, perhaps even ubiquitous, this opens the possibility that, at least in certain instances, if evolution wants to build something it may not have many choices. Lisa Nagy,[77] for example, remarks 'Should vertebrate and insect limbs be considered homologous [that is, descended from a common ancestor] because they are patterned by similar gene networks? Or is the similarity an example of molecular convergence, representing not an extreme conservation of limb construction throughout metazoa, but merely a consequence of a limited number of molecular tools that an organism has available to change its form?'[78]

In a similar vein, a study of a developmental gene known as *distal-less* points to repeated co-option in convergently derived structures. As its name suggests, this gene is typically expressed at the tips of embryonic appendages, and it has been extensively studied in the arthropods. We now find, however, that *distal-less* is employed much more widely than just in limb development during the early developmental stages of various arthropods. As with *Pax-6* the original function of this gene is not certain, but some evidence suggests that its primary role was linked to the development in the embryo of the nervous system, and especially the sensory organs. Now it so happens that in arthropods many of the sensory organs are located on the appendages, and accordingly when there was a need for improved sensory perception so parts of the body protruded to extend the spatial range of the sensory cells. Only later were such outgrowths on occasion employed for such purposes as locomotion. The widespread expression of the gene *distal-less* is, therefore, effectively a reflection of the recurrent and independent evolution of such limbs: in a sense *distal-less* hitchhikes as a sensory protrusion and is subsequently transformed to allow an additional function such as a leg or an antenna.[79] So limbs, like eyes, may be underpinned by a similar genetic architecture, but the end-product is still convergent. Nowhere may this principle be more important than in the origin of advanced nervous systems and the role of a gene known as *otx* (the vertebrate homologue of the *otd* (*orthodenticle*) gene found in the fruit-fly).[80] In vertebrates *otx* plays an important role in the early stages of brain development. So, too, does *otx* in insects, but the expression patterns are not identical. Most probably the gene is conserved, but echoing Nagy the researchers Nic Williams and Peter Holland[81] also remark that while conservation is the most plausible explanation, the alternative of 'convergent evolution cannot be ruled out by current molecular evidence'.[82]

JACK, THE RAILWAY BABOON

So, genes and developmental pathways are a very important part of the story, but in themselves they cannot provide the entire picture. Differences there patently are, but despite their sometimes momentous consequences it may be that from a molecular viewpoint the rearrangement of a gene or protein is trivial.[83] Indeed, in some ways we hardly needed years of frenetic activity in this area of genetics and developmental biology to learn this. At a rather early stage of this research programme the then surprising fact emerged that genetically the difference between humans and chimps is negligible,[84] even though there is a vast difference between living in the trees and boarding the much-delayed 08.06 train from Cambridge – or is there?

What is inherent in more primitive mentalities? The evidence tends to be rather anecdotal, but consider the following account told by Euan Nisbet.[85] It concerns the baboon, Jack, who lived near to Port Elizabeth in South Africa, shortly before the time of the Boer Wars. The baboon had been adopted, when young, by a Mr James Edwin Wide who worked on the railway. In an earlier railway accident Mr Wide had lost both his lower legs, and subsequently he trained Jack in some rather surprising directions. To get to his post at the signal box Mr Wide travelled on a special trolley, and Jack's job was to put the vehicle on the rails and push Mr Wide to work (Fig. 9.1). On the down-hill sections of the track both human and baboon enjoyed the ride. Nor was the trolley Jack's only duty. Other responsibilities included pumping and carrying water, gardening, and locking doors and, when necessary, handing over a key to passing train drivers who needed to unlock the points giving access to the coal yard. On one occasion, when Mr Wide had injured his arm in a fall, Jack, who had already been trained to work the levers for the signals, following commands from his human friend, took over the actual signalling. As Euan Nisbet continues, on the basis of reliable witnesses: 'Jack knew every one of the various signals and which lever to pull – not unnaturally the railway passengers objected initially but the baboon never failed during his many years of work.' It is scarcely surprising to learn that the emotional bond between ape and man was very close, and found expression by mutual grooming. On one occasion offensive remarks made to Mr Wide by another railwayman led to Jack jostling him off the platform. One wonders what Jack would have made of the much-delayed 08.06. After nine years of service, Jack died of tuberculosis. Mr Wide was broken-hearted.

This story, which is certainly true, and strangely moving, is meant to be more than a diversion, because in its own way it is another example of evolutionary inherency. So we return to the question of the prevalence, or otherwise, of humanoids. With so many of the building blocks of life in place, even at the time of the Cambrian 'explosion' more than half-a-billion years ago, when does the merely likely become the almost inevitable? Here, in our tracking of the humanoid, let us grant, for the sake of the argument, that we have organisms that are able to move freely, have well-developed sense-organs and a nervous system capable of interpreting all manner of signals. On Earth we call them animals. What then among all the millions of species is special to us? One feature that might come to mind, so to speak, is intelligence, but as we shall now see we are not the only players.

GIANT BRAINS

The many convergences documented in the last chapter, from the agriculture of ants to the vocalization of birds, are strong evidence that the evolutionary emergence of many complex systems is highly probable, if not inevitable. Yet the sceptic will still pause in thought.

FIGURE 9.1 Jack standing against the trolley with Mr Wide; note the lever frame in the background. (Photograph courtesy of Euan Nisbet, Royal Holloway, University of London.)

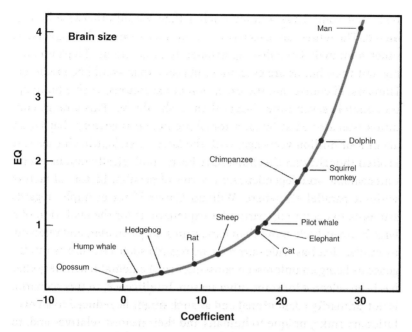

FIGURE 9.2 Relative size, or encephalization quotient (EQ), of
mammalian brains plotted against a coefficient that relates total volume
of the brain against the surface area and volume of the cerebral cortex.
This comparison is less familiar than the EQ calculated on the basis of
brain and body volumes, with an expected exponent of 0.67. (Redrawn
from fig. 2 of M.A. Hofman (1982), Encephalization in mammals in
relation to the size of the cerebral cortex. *Brain, Behavior and Evolution,*
vol. 20, pp. 84–96, with the permission of the author and S. Karger AG.)

Such similarities are indeed intriguing but do they really address the
central question as to whether or not humans as a biological property are inevitable? This is because as well as having a complex social
system, agriculture, placentas and live birth, warm-bloodedness, and
vocalization, we have something else. Even if some reptiles give birth
to live young, and others chewed the equivalent of celery, their mental
powers neither were nor are conducive to rumination. Whatever may
be said in favour of reptiles, their brain size is distinctly disappointing.
Bigger brains are largely the prerogative of the birds and mammals, although as we saw earlier the electrosensory mormyrid fish also weigh
in with a hefty brain. To a first approximation the size of the brain
scales to the body mass (Fig. 9.2).[86] Most mammals, the group upon
which I shall now concentrate, have a brain whose size matches the

body, but some have a brain smaller than would otherwise be predicted. The tenrec, an insectivore, is one such example.[87] Conversely, other mammals have disproportionately large brains. Elephants are big, but their brains are even more massive than would be predicted. Humans, of course, are the exception of exceptions, with a brain approximately seven times bigger than it 'should' be. This strange condition was arrived at by an astonishing neural trajectory that began about four million years ago, with the later australopithecine apes. It is often thought that this must have been a biologically unique event, unrepeatable and dependent on a series of peculiar historical factors without parallel elsewhere. William Calvin,[88] for example, regards attempts to conjure up particular explanations for the evolution of a large brain as little more than caricatures. In particular, and referring to an idea that has wide currency, he regards such a feature as intelligence as being an unforeseen consequence of a neural machinery that has been selected for some other reason. Intelligence, in this scenario, is not primarily adaptational and as such might be reduced to an evolutionary fluke, unique to humans and their nearest relatives and, to echo G. G. Simpson (see note 2), unlikely to be found elsewhere in the Galaxy.

In fact the evidence suggests otherwise, at least on this planet. To start with, there is strong evidence that among the primates those with bigger brains show more innovatory behaviours, social learning, and tool use, while among the birds those with greater behavioural flexibilities and adventurousness (or fewer neophobias) again have larger brains.[89] This accords with the fact, returned to at various points below, that some monkeys, parrots, and crows are all markedly intelligent. So, too, of course, are the great apes, but at least in this case the fact that chimps, for example, have important parallels to human mentality is hardly surprising. To find striking similarities to human intelligence that might persuade the disinterested reader that such represents a general biological property likely to emerge on any suitable planet, we need to turn to the toothed whales (the odontocetes) and especially the dolphins.

It has long been recognized that for their size some of the toothed whales, of which the dolphins are one group, have large brains.[90] A straightforward comparison with other groups of mammals, of which the humans and related anthropoid primates are the most relevant, nevertheless runs into some difficulties. Rather self-evidently the size

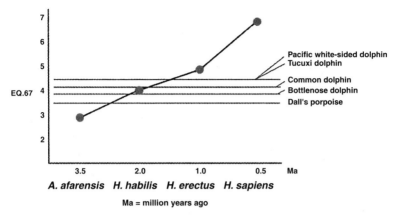

FIGURE 9.3 A comparison of encephalization quotients, calculated as the ratio between brain and body sizes to the expected component of 0.67, a porpoise (c. 3.5), four species of dolphin (c. 4.0–4.5), and the increasing values in the hominid lineage from the australopithecines to modern humans. Note hominid EQ only pulls past that of the biggest-brained dolphins about 1.5 Ma ago. (Redrawn from fig. 3 of L. Marino (1996; citation is in note 92) with the permission of the author.)

range of the bigger whales far outstrips the great apes: a typical killer-whale weighs about seventeen times as much as a gorilla. Scaling of body and brain masses across these size differences is not necessarily easy. More importantly, the buoyancy of sea water means that a large body size can be achieved without the imposition of a crushing gravitational burden, and in addition a large part of the body tissue consists of blubber. As its primary function is that of insulation, it is neurologically inert and accordingly requires no investment by the brain tissue. The net result is that toothed whales have a rather different ratio of body weight to brain mass when compared to the primates. Appropriate corrections therefore have to be made, but once this is done it can be shown that not only do several types of dolphin (including the Pacific white-sided, Tucuxi, common, and bottlenose) have large brains, but in proportion to the brains of our three nearest relatives (chimp, gorilla, and orang-utan) they are significantly larger.[91] Indeed, as Lori Marino has shown, until about 1.5 Ma ago, these dolphins were the biggest-brained creatures on the planet. Only then, at about the time of *Homo erectus*, did the brain size of hominids overtake that of the dolphins (Fig. 9.3).[92]

The questions then arise as to how, when, and why did dolphin brains get so large, and to what extent are the convergences with the brains of the great apes actually informative?[93] So far as the origins of the story are concerned, it is not surprising to learn that well-preserved fossil skulls of toothed whales are only moderately common. Nor has it been easy to measure the so-called endocranial volume, which is a fair guide to brain size, at least until the advent of non-invasive medical techniques, such as computer tomography (CT). In any event, the fairly limited evidence suggests that the early whales had brains of unremarkable size, some of which were, indeed below average.[94] The transition by whale ancestors from a terrestrial habitat to an aquatic existence, the elucidation of which has been one of the triumphs of palaeontological investigation,[95] was evidently not in itself a spur to bigger brains. A dramatic increase in brain size is apparent when present-day dolphins are compared with these early ancestors, but the exact history is still not resolved. The increase probably started in the Oligocene, about 30 Ma ago, followed perhaps by a further jump in the Miocene. Nor, too, is it yet clear whether these spurts of encephalization were geologically sudden or more gradual.[96] Nevertheless, it appears that by the Miocene (if not before) brain size had increased suddenly, especially in the ancestors of the porpoises and dolphins. It is surely significant that this vast organ, which metabolically is ruinously expensive,[97] has been maintained for a protracted period, probably far in excess of that for hominids, and possibly for as long as 20 Ma.[98] Evidently big brains may be, in at least some circumstances, adaptively useful, and are not just fickle blips of happenstance that in due course sink back into the chaotic welter of the evolutionary crucible.[99]

What, however, might have initiated one or more upsurges in dolphin brain size? At present perhaps the best evidence is that the trigger was environmental, specifically the dramatic cooling of the Southern Ocean.[100] Among the consequences of this event, which was the harbinger of global refrigeration that culminated in the present ice age, was increased oceanic productivity as marine upwelling intensified.[101] As Australia pulled away from Antarctica, marking the final break-up of the once immense supercontinent known as Gondwana, so the circum-Antarctic current became firmly established, the ice caps began to spread down from the mountains of the Antarctic continent, and in the adjacent oceans the baleen-sievers and predatory toothed whales both diversified, the deep waters filling with their clicks, whistles, and other vocalizations. This environmental trigger has an interesting parallel, perhaps, with the hominid story because it has been suggested that the major increases in brain size in such forms as *Homo erectus* may have been encouraged by increasing aridity in Africa, imposing new stringencies but also spurring new possibilities.[102]

Large brains may therefore be favoured when the environment offers a special challenge. Their persistence for many millions of years requires, however, additional explanations. Of these it may be that the emergence of sophisticated social organizations (with an emotional dimension) and the necessary corollary of advanced vocalizations may be especially significant. The social structure of dolphin groups, especially in the well-studied bottlenoses, is directly relevant to the theme of evolutionary convergence.[103] At first sight it might seem rather remarkable that the bottlenose dolphins show 'striking parallels in social organization and complexity with that of the chimpanzee',[104] as well as parallels to the ateline or spider-monkeys,[105] a group to which I return below on account of a series of other instructive convergences. Such associations fall into a category known as fission–fusion societies, which as the name suggests are rather fluid in their composition, with alliances and coalitions of varying durations.[106] Societies of this type are, as Marino notes, 'extremely complex because they represent a constantly dynamic social situation involving the movement of different individuals into and out of groups at various times'.[107] Even so, some individuals may stay together for protracted intervals, and a common feature is a stable alliance of two to three male dolphins. These males form temporary consortships with females. Mating is probably promiscuous, but so far as the females are concerned it is also highly coercive.[108]

The similarity with chimp societies is another fine example of convergence within social systems. Nor is this resemblance likely to be accidental. On the contrary, it is much more likely to be adaptive and arises because despite the radical ecological differences there is a deeper constraint imposed by the patchiness in space and time of food resources in both ocean and jungle.[109] The parallels are not exact; why should they be? Moreover, as studies of the dolphins continue, so further complexities in the structure of their societies are emerging. They indicate that, in some ways, despite having a fission–fusion society, the dolphins have advanced further than our closest relatives, the chimps. It is now apparent that their societal structure is more complex than hitherto realized, and that with dolphin groups form so-called 'super-alliances'. This clearly implies that their intelligence is well suited to handling an extended social network comprising at least a hundred individuals.[110] Such intelligence is also well attested in the examples of cooperation between dolphins and humans, notably in the dolphins' assistance with fishing by providing clues as to where to cast the net or in herding the fish.[111]

This example of convergence in a sophisticated social context finds another very interesting parallel, but this time between two larger-than-average mammals, the sperm whales and the elephants.[112] Linda Weilgart and her colleagues comment that despite their self-evident differences in 'a remarkable number of ways, including life history and ranging behavior, sperm whales and elephants resemble each other more than they do other animals – even ones that share similar ancestries, diets, environments and predators. The closest resemblance is found in their complex and unusual, but comparable, social organization.'[113] In both these gigantic mammals the females and young form highly social units, highly communicative with various long-distance vocalizations.[114] Socialization is intense, and in the sperm whales, for example, there seems to be a form of 'babysitting' in which the vulnerable young are cared for by other adults when their mothers are engaged in deep dives in pursuit of food.[115] The males, in contrast, are solitary and wide-ranging, and return to the mating game only when they are not merely sexually mature but big enough to win contests. Social complexity, communal care of the young, intelligence, and memory, as well as longevity, seem to be the key ingredients in driving this remarkable convergence.

In the sophisticated and dynamic milieu of many cetacean societies, it is not surprising that vocalizations are complex and varied, especially in the dolphins. As has been repeatedly pointed out, the relative opacity of water means that facial expressions are of very limited use, and so sound production and acoustics have largely taken priority. In the specific case of the dolphins their fission–fusion societies are probably highly dependent on the production of recognizable whistles and other noises.[116] Indeed, there is evidence that the vocalizations are far from a cacophony. Dolphins are accomplished mimics, and the speed and accuracy with which they can imitate given sounds is highly impressive.[117] Evidently they can make contact across quite substantial distances, and one group of researchers have identified what they term 'whistle matching', in which the receiving dolphin returns an effectively identical whistle to the sender.[118] This, however, is a very controversial notion, and it is much more likely that the whistles are simply contact calls. In particular, Brenda McCowan and Diana Reiss[119] present evidence that groups of about 12 dolphins have a shared whistle type, but embedded in this there are subtle variations that in part are individualistic. The extent to which individuals do or can recognize each other remains moot, but all would agree that dolphin whistles are by no means fully understood; interestingly the type of dolphin known as *Cephalorhynchus* does not produce whistles. Its vocalizations, however, are convergent on those of the phocoenid porpoises.[120] It has been remarked that given their ability for mimicry and their frequent contact with humans it is rather surprising that dolphins' powers of imitation have not extended to our speech.[121] As it happens, certain other marine mammals are not so restricted. A famous example concerns the captive Harbour seal, one Hoover. Perhaps because of a bout of illness, followed by attention-seeking, the seven-year-old Hoover started to mimic English words.[122] As the investigators reported, 'One observer wrote in the files, "he says 'Hoover' in plain English." I have witnesses'.[123] Perhaps because he lived in Maine, Hoover spoke with a Boston accent, although the tendency to slur words made him sound inebriated. A somewhat similar case concerned a beluga whale, who also repeated his name and evidently enjoyed socializing with humans.[124] Lori Marino also tells me that in the New York aquarium the belugas imitate the sound of the elevated train that runs near by.

The failure of dolphins to imitate human speech should not, however, distract from more intriguing similarities. In their case what is especially significant is the way the infant dolphins learn to vocalize.[125] As has been pointed out by various investigators, this process has strong parallels with humans: there is a whole range of sounds that evidently represent babbling, over-production, and finally an attrition to a more standard repertoire.[126] Not only that, but notwithstanding the debate concerning 'signature whistles' (note 118), it is known that dolphin vocalizations are considerably more complex than was once thought.[127] It is tantalizing to think that such vocalizations, and their parallels in birds (note 175, Chapter 8) and some primates,[128] might provide an analogy as to how humans learnt to speak. In this context it is also worth recalling some remarks by Marc Hauser and Peter Marler, who commented that 'no one would, of course, claim homology between [bird] subsong and [human] babbling.' Rather, similarities would be viewed as convergences, illustrating 'a basic set of strategies that any species would be likely to employ if it embarks on the development of a system of communication based on learned signals'.[129] So, too, Diana Reiss remarks on the discovery of a 'surprising complexity and plasticity in the communication, orientation, and navigation systems of many species ... diverse species either use or can learn to use, to different degrees, symbolic or referential communication for intraspecific or interspecific exchanges. This suggests that there may be a convergence or continuity in the communication and cognitive abilities in animals from different evolutionary paths.'[130] These comments on the commonalities of vocalization and thereby the transmission of information are potentially of universal significance, as it is now being realized that dolphins might provide a model of how to achieve communication with extraterrestrials. Nor need this be the only clue. Readers of Mary Doria Russell's *The Sparrow* will recall how the aliens were first detected in a transmission of their singing; so, too, the song of the humpback whales[131] may be a contribution to a universal music.[132]

Nor should these comparisons in communication be taken as some sort of fanciful whimsy: it is possible to teach dolphins to understand sentences and take the appropriate action.[133] The actual experiments involved two dolphins, each one of which was taught an artificial language based respectively on computer-generated sounds (approximating to, but sensibly enough not identical to, their whistles) and on signs given by the hand and arm movements of a human. What emerged was a degree of comprehension that not only associated 'words' as symbols for given objects, e.g. a ball or hoop, but mastered both syntax (word order) and semantics (meaning), even when the 'words' were presented in a novel order. Such instructions provided a short cut to learning new tasks, that otherwise would have involved laborious routines and training. Indeed, this work has been taken further and there is now good evidence that dolphins can understand human gestures that refer to different parts of their bodies.[134] Thus a gestural symbol achieves, for the dolphin, a semantic meaning. It is difficult to avoid the conclusion that dolphins are capable of abstract thought via mental representations that entail symbolic referents.

Beyond humans such an ability is rare indeed, as Marino notes: 'The fact that artificial "language" studies can only be meaningfully attempted with these very few species [dolphins, bonobo chimps, and perhaps African Grey parrots] suggests that these few species have converged toward a level of cognitive complexity that allows them to understand a simple but symbolic and rule-based system of communication.'[135] This is not to imply, incidentally, that in the specific case of dolphins, they may one day develop a language with grammar[136], but their syntactical competence seems to be firmly established.[137] Human language may, on this planet, be unique, but waiting in the wings of the theatre of consciousness are other minds stirring, poised on the threshold of articulation.

In this sense the wide divergences of opinion as to the extent to which animal communications are indicative of the origins of human language, let alone to the possession of sentience by non-human species, are less important than the realization that what we call language is an evolutionary inevitability. Most of the cognitive substratum is already firmly in place. Nor should we be fooled by the routine criticism that because only *some* primates, cetaceans, or parrots achieve syntactic competence then the rest somehow have failed. That competence may in fact be more widespread than we realize.[138] But even if it is not, then the explanation for its evolutionary emergence is contextual, notably in terms of highly complex social worlds where communication and memory are adaptations to a constantly changing world.[139] It would actually be more surprising if convergences did not occur in such complex contexts. Of course, the belief still persists that the emergence of human language was a contingent fluke, but everything else we know about evolutionary convergence and the exploration of functional 'spaces' makes this seem increasingly improbable. And what of grammar and syntax, how the world, as well as the word, achieves meaning? So far as humans are concerned, a universal grammar may have evolved as a result of natural selection that optimizes the exploration of 'language space' in terms of rule-based systems.[140] These themes resonate, of course, with much of this book, that is, to address the question as to how life 'navigates' to particular functional solutions. This agenda has the potential, if not the promise, of revealing both a deeper structure to biological organizations and a predictability to the process.

Notwithstanding decades of interest, the more we learn about dolphins the more remarkable they seem to be. Not only are they intensely vocal, but they show other convergences. For example, their powers of memory, at least so far as lists are concerned, are strongly reminiscent of the memory processes seen in humans,[141] and it seems likely that at least short-term memory is retained as internal representations.[142] So, too, there are similarities in their response to uncertainty, the dither factor. Broadly speaking, when faced with a problem, humans decide either to avoid it (escape) or to collect more information. The difficulty is that in times of high uncertainty a decision to collect more information may lead to a catastrophic error. In comparing the uncertainty responses of humans and dolphins an investigation by David Smith and his colleagues concluded that 'The dolphin performed nearly identically ... Human and dolphin uncertain responses seem to be interesting cognitive analogs.'[143]

Another observation seems at first just a curiosity. Dolphins sleep, but they also need to swim continuously lest they drown by failing to come to the surface. What do they do? Effectively either the left or the right side of the brain falls asleep, while the other side remains fully conscious with the corresponding eye open. Interestingly, birds have convergently arrived at this arrangement.[144] It seems possible that at least some birds can engage in such unihemispheric sleep while flying, but its principal function seems to be as a protection against attack. For dolphins, however, having only half a brain asleep is apparently to keep in contact with the rest of the school.[145] Dolphin brains therefore show a high degree of what is referred to as interhemispheric independence. This is presumably a precondition for laterality, which is a well-known feature of the human brain, and may possibly occur also in the dolphins. As remarked above, dolphin brains (Fig. 9.4) are large, and the hominids outstripped them only about 1.5 Ma ago. So far as the degree of folding of the neocortex is concerned, it seems that in this respect the dolphin brain is the more convoluted.[146] The same applies to the cerebellum, at least in the bottlenose and common dolphins, in which the cerebellum is significantly larger than in any primate, including humans.[147] Marino and colleagues suggest that this enlargement of the cerebellum is not only linked to coordination of movement, but has broader functions

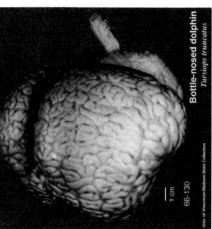

Human *Homo sapiens*

Univ. of Wisconsin-Madison Brain Collection

69-314

5 cm

Less convoluted
Hemispheres not as independent
No paralimbic lobe
Slower auditory processing
No echolocation

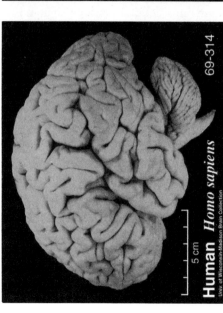

Bottle-nosed dolphin
Tursiops truncatus

Univ. of Wisconsin-Madison Brain Collection

66-130

1 cm

More convoluted
More hemispheric independence
Unique paralimbic lobe
Faster auditory processing
Echolocation

FIGURE 9.4 A comparison of the human (left) and dolphin (right) brains, with some of the principal differences listed. (From http://brainmuseum.org/specimens/cetacea/dolphin/index.html, courtesy of L. Marino, Emory University.)

connected with the complexities of sensory processing, memory, and cognition.

These differences, in neocortical folding and cerebellum, underscore a key fact. This is to the effect that the large brain of the dolphins may have arisen from a common mammalian substrate,[148] but it is very far from being identical to that of the human.[149] The contrasts include a marked and very different development of particular areas and lobes, with the temporal and parietal regions emphasized in dolphins as against the frontal and occipital in humans (and other primates). Particularly noticeable is the development in the dolphins (and other cetaceans) of a unique paralimbic lobe, unknown in other mammals, which evidently has a special role in a variety of functions, including vision, touch, and motor activities. As Marino notes, 'The segregation of the limbic and supralimbic regions by an interposed paralimbic lobe is a radical departure from the typical terrestrial mammalian pattern of cortical evolution;' she continues, 'The fundamental topography of cortical sensory projection regions in a cetacean brain also stands in stark contrast to that of the primates.'[150] Nor are these the only differences: the neocortex is thinner, the internal laminae are less defined, and the cellular structure is generally simpler.[151] To quote Marino again, 'The differences in gross morphology and topological arrangement of cortical sensory zones between cetacean and primate brains persist at an even deeper level in the form of profound differences in cortical cytoarchitecture.'[152] Yet despite all these differences the degree of cognitive convergence is striking.

The presence of an enlarged cerebellum has already been alluded to, and it is probably no accident that it is similarly enlarged in the microchiropteran bats, which also echolocate.[153] We should also recall, at this juncture, the immense cerebellum of the mormyrid fish with their highly developed capacities for electrogeneration and electroreception (Chapter 7, p. 182). All these animals are masters of sensory processing, possessing in their different ways remarkable acuities. In the case of dolphins and humans it will be apparent that despite both are mammals their brains, as Lori Marino stresses, 'represent two fundamentally different cortical organizational themes ... [and thus] compels the conclusion that any similar complex cognitive processes between primates and cetaceans are convergent.'[154] The evolution of the brain cannot therefore be divorced from its adaptational requirements and a corresponding moulding of function to those needs.[155]

FIGURE 9.5 Dolphins with their bubble rings. (Photograph courtesy of Brenda McCowan, University of California, Davis and Lori Marino, Emory University.)

This is important, not only because it confers a predictability to the evolution of brain structure (see pp. 266–267), but because it also suggests that while the routes to adaptive success, of which one is a higher intelligence, may be quite strikingly different (as in dolphins and humans) the end-points converge again and again.

It remains the case that in some ways the dolphin brain is a curious amalgam of a rather archaic brain, reminiscent of the relatively primitive condition seen in such mammals as the hedgehog and the bat,[156] combined with the massive enlargement of the deeply convoluted lobes that gives it an uncanny similarity to our own brains (Fig. 9.4). Even so, the complexity of the dolphins' social life, sophisticated and specific communication, and ability to mimic and learn clearly indicate that complex intelligence is not the unique preserve of humans, and that in some respects the convergence seen in dolphins is more compelling than the classical comparisons with the apes. To this list should, of course, be added the playfulness of dolphins, perhaps best exemplified by the blowing of bubble-rings (Fig. 9.5) where both some degree of forethought in their production and an ability to manipulate the ascending circle of air are again consistent with some degree of cognition.[157] What is often regarded as a key ability, that of self-recognition (usually tested by using mirrors), has long been suspected in dolphins,[158] and has now been unequivocally demonstrated.[159]

This finding is of particular importance because it helps to refute the notion that something like a human can emerge only by some quirky path of evolutionary happenstance. As the investigators, Diana Reiss and Lori Marino, remind us,

> 'Bottlenose dolphins, great apes, and humans all possess high degrees of encephalization and neocortical expansion ... Yet the brains of dolphins are markedly different from those of primates on many levels ... reflecting the fact that the cetacean ... and primate ancestral lines diverged at least 65–70 million years ago. The present findings imply that the emergence of self-recognition is not a byproduct of factors specific to great apes and humans but instead may be attributable to more general characteristics such as a high degree of encephalization and cognitive ability ... More generally, these results represent a striking case of cognitive convergence in the face of profound differences in neuroanatomical characteristics and evolutionary history.'[160]

The evidence therefore seems to be increasingly consistent not only with a cognitive capacity, but with one arrived at independently and convergently with humans. These abilities, it seems, can encompass a comprehension both of symmetry and of signs,[161] thereby indicating that, as I have already suggested, the notion that dolphins might well be capable of abstract thought is not in itself intrinsically absurd.

In the wider context of the toothed whales (such as the killer whales) and other marine mammals (such as the elephant seals and the Weddell seals) it has long been recognized that their vocalizations are not necessarily uniform across their geographical ranges, but in some instances can be divided into recognizable variants, that is, dialects.[162] Some investigators are now going further, and speak of the cultural transmission of behaviours (that is, a state in which transmission of information is no longer entirely reliant on the gene). And what is patently true of humans most probably applies also to various marine mammals, and perhaps not surprisingly to some birds.[163] For various whales such transmission involves not only vocalization but other activities, such as feeding behaviours.[164] In the light of such work it is difficult to escape two conclusions: first, that the emergence of cultural capabilities represents a continuum, and second that convergences are inevitable. This is not to deny that humans have gone further; they have what has been termed a 'hyperculture', but it does not rule out such a phenomenon evolving elsewhere, whether on Earth or on Threga IX. That cultural attributes are manifested in such features as vocalization should not make us forget that their contexts are ones that are firmly embedded in a social matrix. This in turn has intriguing ramifications, such as, for example, the convergent development of the female menopause. Far from it being a puzzle why evolution 'permits' post-reproductive females, there are in fact probably sound adaptive reasons why it should in some complex societies.[165] More than an echo of this has also been found in the African elephants, where the oldest of the females also tends to be the wisest, the best able to remember elephants from other groups, and thus able to transmit the requisite social knowledge to her own 'family'.[166] As already emphasized (note 112), elephant and sperm-whale social structures are strongly convergent,[167] and it has been repeatedly pointed out that to slaughter the largest and oldest in such matrilineal clans is a stupid course of action if the result is to deprive the group of the accumulated knowledge that enables it to prosper. It is a small irony that as the Darwinian synthesis was emerging in mid-Victorian Britain the rooms were lit by the burning of sperm whale oil, and the soothing music from the piano arose as the fingers moved across the elephant ivory keys.[168]

In the case of whale vocalizations it is also evident that a number of discrete categories are recognizable, but in the context of their possible cultural transmission what is significant is that only certain call types change with time, whereas others remain invariant. As certain call types change in a particular social group (which has a matrilineal social structure; see note 165), they will typically diverge in structure from the sounds made by another social group, although it is also possible that on occasion calls will in fact come to resemble one another. In either case there is a clear implication that within a social group a component of this change must be by learning, that is, by a cultural instruction. In their study of a pod of killer whales and their vocalizations Patrick Miller and David Bain[169] inclined to a view that the changes arose by cultural transmission mediated by learning, and they were not afraid to suggest that such learning might occur 'as a consequence of exposure to the sounds of tutors'. They concluded that the 'Vocal similarity [of the pod] seems to be correlated with the multilayered structure of killer whale society, and ultimately to the social interactions that contribute to the stability of the social structure.'[170] Shifts in sound production might certainly, as the various investigators are careful to point out, be underpinned by a genetic change, but it seems just as likely that it is by imitation, improvisation, and experimentation. So, too, in humpback whales the arrival of some 'foreigners' led to a rapid adoption of the new song, literally a 'cultural revolution'.[171] And should we be so surprised? Like humans, these marine mammals are long-lived, show a high degree of cognition, demonstrate prolonged parental care, and form cohesive societies.

In their various ways a number of marine mammals, and especially the dolphins and some of the other toothed whales, are intriguing in their social complexity and cognitive abilities. Much is convergent with other intelligent mammals, including humans, and it is a clear enough indication that at least in this biosphere if we had not emerged as the cerebral species then at some point, and probably sooner rather than later, someone else would. In this sense, humans, as a biological property, were inherent from at least the Cambrian period, if not before. The life of a dolphin appears to be rich and complex, but in evolutionary terms these large-brained cetaceans seem to have reached an impasse. Because they are highly adapted for swimming and living in the sea, they would seem to be unlikely candidates for the emergence of an advanced technology. Even so, despite their aquatic milieu, when it comes to tool use then, once again, the flippered dolphins surprise us. At least one group, principally more solitary females, has learnt to root up conical sponges and stick them on to their anterior beak, a structure known as the rostrum.[172] Several alternative explanations can be entertained. Perhaps it is just that dolphins enjoy fooling around. Another possibility is that the sponges have medicinal properties. The most likely explanation, however, is that the sponge acts as a natural glove and as the dolphin rootles around in the sea bed, so it receives some protection from an alarming array of venomous animals, such as the stone-fish, scorpion-fish, sting-rays, sea-snakes, and the occasional blue-ringed octopus. Nor is this the only example of tool use by cetaceans. We saw earlier evidence of cognitive capacities in the formation of bubble rings by dolphins, but the humpback whales produce nets and curtains of bubbles to trap their prey.[173] Perhaps on planets that are entirely oceanic (see p. 92) all advanced cognition resides in cetacean-like societies. There, as Michael Denton reminds us,[174] access to fire and the skills of smelting and metallurgy would for ever be denied.

GRASPING CONVERGENCE

If the path to the humanoid is characterized by any feature other than the carrying of a large brain rich in mentalities, it is the tools dropped on the way. Chosen and then discarded, tools epitomize intelligence and the purposeful. To be sure, those constructed by animals may seem primitive in the extreme, but it is clear enough that human technologies ultimately are based on the same antecedents. More importantly it is also now obvious that such an evolutionary ability may be rare, but it has emerged independently a number of times: inherent in evolution is not only intelligence and cognitive sophistication, but also technology. Tool use, *per se*, is of course well documented in a wide variety of birds,[175] but a particularly famous example is provided by the New Caledonian crows.[176] This example has excited interest because of evidence for the manufacture of both particular tool types, including hooks,[177] and their standardization, features that had been thought to be effectively restricted to the advanced primates. In addition, Gavin Hunt documents lateralization in tool use and a rule-based method of construction.[178] He also stresses that in a number of respects the more famous chimpanzee tool cultures do not match that of these crows, and concludes:

> At the least, crows provide an extant species for learning about the neuropsychology associated with ... tool-making, such as handedness, hook use and the shaping of tools to rule systems, including an opportunity to see whether left-hemisphere specialization of the brain for the organization of sequential, manipulatory behaviours in tool-making might indeed be phylogenetically very ancient. If crows' tool behaviour involves cultural transmission, they also offer the opportunity for studying tool-making by pre-modern humans where cognitive, behavioural and social processes may have resulted in largely repetitive rather than innovative tool manufacture, and symbolism and language were rudimentary or absent.[179]

Apart from the parrots, already mentioned, and for which considerable evidence for cognitive sophistication exists,[180] many ornithologists have remarked on the general intelligence of crows. Recent observations on tool modification and goal-directed activities in the American crow bear this out.[181]

Tool use in chimps has, not surprisingly, received extensive attention. Most of the work focuses, reasonably enough, on the use of sticks and stones.[182] The closely related bonobo is also known to engage in tool-making. Remarking on one particular case, Nicholas Toth and his colleagues[183] noted that the bonobo in question, one Kanzi, showed 'exceptional progress to date [but] his skill in flaking stone still contrasts sharply to that of Oldowan hominids'.[184] They were unsure as to whether this was more a reflection of manipulative abilities as against cognitive constraints. Interestingly, however, chimps also use twigs for dental care in activities that include cleaning the teeth and helping to yank out deciduous ones.[185] Much of this activity is self-directed, but examples are also known of one individual performing elementary dentistry on another chimp.

Nor is tool use among non-human primates confined to chimps. Significantly in some of the New World monkeys, whose evolutionary history has been separated from that of the Old World monkeys (and their descendants, the great Apes) for about 30 Ma, have convergently acquired tools. In particular, capuchin monkeys (*Cebus*), and also some of the callitrichids (specifically the golden lion tamarins[186]) show extensive tool use for a variety of purposes, including a primitive lithic technology when provided with suitable materials in captivity.[187] Suzanne Chevalier-Skolnikoff[188] also emphasizes that tool use is unlikely to have arisen by trial and error (fiddling around if you will) but is a direct product of advanced sensorimotor abilities: that is, capability combined with motivation. It is difficult to escape the conclusion that once such abilities are in place, tool use becomes an inevitability. In reviewing the earlier literature on tool use by capuchins, Gregory Westergaard and Stephen Suomi provide an effective counterpoint to Chevalier-Skolnikoff's comments.[189] They remark, 'We suggest that the ability to make and use simple stone tools is a primitive behavioral capacity that may have been "discovered" numerous times and utilized by more than one hominid genus and species'.[190] This idea echoes an earlier suggestion by Sue Parker and

Kathleen Gibson of parallel developments in the capuchin monkeys and great apes.[191] This theme was returned to by G. C. Westergaard and his colleagues[192] in summarizing a discussion of handedness, locomotion, and tool use. They remark, 'We do not hold that parallels between capuchins, on the one hand, and apes and hominids, on the other, resulted from homologous processes. Instead, we speculate that an array of behavioral similarities, including food-sharing and tool-use, evolved through convergent processes in *Cebus* and the common ancestor of great apes and modern humans.'[193] In parallel to the chimps, however, the capuchin tools do not rival the most primitive of hominid technologies, known as the Oldowan. Nor is it entirely clear why the capuchin seems relatively adept at tool use in captivity, but there are few reports of such activity in the wild.[194] As Elisabetta Visalberghi[195] notes, this may be due to lack of observation or to their tree habitat, whereas primate tool use is very much associated with ground activities.[196] It is also worth remarking that in capuchin tool use the manipulation has a pronounced bias to right-hand employment.[197] It is significant that this discussion is put in a selective context by the intriguing possibility of a separation in the brain 'between language and object manipulation'.[198] In parenthesis, it is important to stress again that the point of this book is to explore the likelihood of the emergence of certain biological properties. It is not my intention to persuade you that capuchins will one day evolve into humans; they won't. To start with, their mentality has strengths, but also limitations.[199] Nor need this tool use necessarily be an appropriate guide to early hominid activities.[200] The simple point is rather that tool use is far from being some sort of quirky by-product of evolution, and when tools are first employed other things may become far more probable, at least in some circumstances.

The examples of the capuchin monkey and especially the New Caledonian crow demonstrate that tool use is patently convergent. For us tools and hands are almost synonymous, yet the birds (and even dolphins) are a reminder of valid alternatives. Yet, by and large, tool use would seem to be a prerogative of the vertebrates. So far as there are rivals in a terrestrial context, for advanced complexity we would most probably turn to the insects, some of whose marvels have already been explored. Consider, however, this account by Samuel Wendell Williston of the behaviour of a parasitic wasp.[201] It is quite a long quotation, but if at the end you don't rub your eyes, obviously you are more immune to surprise than I am. Thus Williston writes:

> An insect, alighting, ran about on the smooth, hard surface till it had found a suitable spot to begin its excavation, which [when finished] was ... nearly vertical, and carried to a depth of about four inches ... The earth, as removed, was formed into a rounded pellet and carefully carried to the neighboring grass and dropped ... When the excavation had been carried to the required depth, the wasp, after a survey of the premises, flying away, soon returned with a large pebble in its mandibles, which it carefully deposited within the opening; then, standing over the entrance upon her four posterior feet [so leaving the front two free], she ... rapidly and most amusingly scraped the dust with her two front feet, 'hand over hand,' back beneath her, till she had filled the hole above the stone to the top. The operation so far was remarkable enough, but the next procedure was more so. When she had heaped up the dirt to her satisfaction, she again flew away and immediately returned with a smaller pebble ... and then standing more nearly erect, with the front feet folded beneath her, she pressed down the dust all over and about the opening, smoothing off the surface, and accompanying the action with a peculiar rasping sound. After all this was done ... she laid aside the little pebble and flew away ... Soon, however, she comes back [with] ... the soft green larva [which] ... is laid upon the ground, a little to one side, when, going to the spot where she had industriously labored, by a few rapid strokes she throws out the dust and withdraws the stone cover, laying it aside. Next, the larva is dragged down the hole, where the wasp remains for a few minutes, afterwards returning and closing up the entrance precisely as before [an action that is repeated four or five times] ... The things that struck us as most remarkable was the unerring judgment in the selection of a pebble of precisely the right size to fit the entrance, and the use of the small pebble in smoothing down and packing the soil over the opening.[202]

CONVERGING ON THE HUMANOID

This example of stone tool use in wasps underlines my repeated emphasis that it is not my intention to suggest that there is only a single type of intelligence. Nor do I intend to imply that technologies are necessarily dependent on a unique configuration of the brain: dolphins themselves hint at the opposite possibility. Even within the mammals it can be shown that there are several trends in terms of the structural organization of the brain whereby different regions, such as the hippocampus, neocortex,[203] or olfactory bulbs, are variously developed or suffer relative atrophy. One such study, by Willem de Winter and Charles Oxnard,[204] has many points of interest. As its basis it took three major groups of mammals, specifically the bats, insectivores, and primates. To a first approximation each group has gone its own way: bats into the night sky in pursuit of moths, insectivores rootling around for worms and suchlike during nocturnal forays, and primates swinging through the jungle. Each inhabits a very different world, and not surprisingly in terms of neural architecture each of the three groups occupies a largely separate area of 'brain space', as defined on the basis of the relative importance of the different regions of the brain.

That different mammals emphasize different sensory modalities has long been known, and writers attempting to enter the mind of, say, a dog will typically try to transpose the wealth of olfactory 'images' into something more amenable to human comprehension. So, too, we can ask to what extent equivalences might exist between sensory assimilations dependent, say, on olfaction, electroreception, or echolocation. The last case is of particular interest because in a famous essay Thomas Nagel argued, as part of a consideration of what it is to be conscious, that the mind of a bat was effectively unknowable to us.[205] Nagel's article has attracted an enormous amount of attention, but it may be that he underestimated the underlying degrees of similarity in sensory perceptions. Thus, in referring to bat echolocation he remarked that 'bat sonar, while clearly a form of perception, is not similar in its operation to any sense that we possess, and there is no reason to suppose that it is subjectively like anything we can experience or imagine'.[206] Nagel continued that even for such experiences as we might share with a bat, for example 'pain, fear, hunger, and lust ... [they] have in each case a specific subjective character, which it is beyond our ability to conceive. And if there is conscious life elsewhere in the universe, it is likely that some of it will not be describable even in the most general terms available to us.'[207] But is this too pessimistic a view? The many examples of convergences in sensory modalities already addressed, and exemplified by the star-nose mole that 'sees' with its nose, suggest that there are deep commonalities in both neural systems and the associated transduction proteins. These may enable wildly different sensory systems, in effectively unrelated animals, to build similar cognitive maps.[208] If that is so, then arguably the root problem of consciousness, the qualia (e.g. the redness of the sunset on Threga IX), have deep-seated similarities that require an interpreter rather than blank incomprehension. Incidentally, this is not necessarily to deny another point made by Nagel, to the effect that our neural architecture may for ever preclude the comprehension of certain realities. Thus he writes 'one might also believe that there are facts which could not ever be represented or comprehended by human beings, even if the species lasted forever, simply because our structure does not permit us to operate with concepts of the requisite type.'[209] An interesting thought, especially if all intelligences have a universal neuronal basis.

Whatever truth there may be in these suppositions, the study by de Winter and Oxnard (note 204) of how these three groups of mammals occupied the various zones of 'brain space' leads to some very interesting, if perhaps by now unsurprising, conclusions. As has been long known, within each of the three groups there are recurrent convergences in terms of life, habit, and behaviour. For example, among the bats carnivory, by which is meant the ability to capture live vertebrate prey, has evolved several times.[210] So, too, among the insectivores there have been multiple convergences associated with the transition from the terrestrial environment either to the aquatic realm or to a burrowing mode of life; and in the primates leaf-eating has evolved several times.[211] Such convergences are, of course, a principal theme of this book, but the work of de Winter and Oxnard is of particular importance because they find that the convergences that lead to the adoption of similar life-styles are mirrored in the brain structure. Neurology, and by implication mentality, overrides phylogeny. One significant implication of this analysis is that it points to the pervasive influence of evolutionary selection: how else can we explain such similarities? The net result, as Winter and Oxnard note, is that one can 'infer important ecological and behavioural attributes of a mammal from ... knowledge of its brain proportions alone.'[212] So confident are these researchers that in one case, those bats that visit plants, for example, to feed on nectar, they predicted that certain of the Old World bats would show this ecology even though at

the time the habit of plant visiting was believed to be restricted to the
New World bats. Subsequent observations on the Old World murinine
bats confirmed this prediction.[213] Indeed, these and other analyses[214]
indicate that the general rules of brain organization confer a wider predictability to a number of evolutionary processes. As Leah Krubitzer
has remarked on the evolution of cortical structure,

> it is hypothesized that future changes [in cortical-field evolution]
> will be shaped by similar mechanisms. Indeed, while the product
> in a given lineage of several million years of further evolution
> cannot be predicted exactly, which features are likely to be
> retained, the types of modification that are likely to occur, and
> what will not happen can be predicted with some certainty.[215]

Research on the convergence and lability of brain structure, and
its correlations with function and ecology, also has an immediate bearing on the following section, which specifically addresses hominid
origins. One key step in this process is evidently locomotor adaptations. As it happens, primates possess a very distinctive quadrupedal
gait, and this may well owe its origin to the art of climbing along
and between fine branches. Very specialized, and no doubt unique?
No, the same gait has evolved independently in the marsupials, again
meeting the adaptive challenge of living near the tops of trees.[216] Associated with this is the separation in function between the principally propulsive hind-limbs versus the use of the fore-limbs (arms)
for grasping and feeding.[217] Such an arrangement is regarded as 'the
primary hominoid adaptation',[218] and it represents, of course, a prerequisite for such features as mobile hands, arm-swinging, and in some
cases ultimately bipedality. Yet it is also known that this locomotory
style so characteristic of the apes is strongly convergent on the New
World monkeys,[219] specifically the atelines, or more familiarly the
spider-monkeys.[220]

It would hardly be surprising if such locomotion and especially
arboreal agility sowed the seeds for a particular type of intelligence.
And in a way analogous to the bats we see that the occupation of
'brain-space', as defined by de Winter and Oxnard (note 204), shows
close parallels between the Old World apes and their New World analogues. This is strongly echoed in the increases in the respective mass
of the neocortex among these (and other) primates. The increase defines a clear trend, parallel and independent, that is consistent with

selective pressure. Moreover, despite the evolutionary trend that can be defined simplistically as 'lemurs to humans,'[221] the details are more complicated. In particular, New and Old World monkeys show overlap because both have complex social structures that in various respects show convergences.[222] The case of the spider monkeys is particularly interesting because their social organization is 'remarkably similar' to that of the chimps,[223] whose parallels to the fission–fusion society of dolphins were addressed earlier.

Remembering that Old World monkeys (known as the cercopithecoids) and hominoids (including the apes) are a sister-group, and recalling the convergences of locomotion and brain structure between Old and New World primates, it seems entirely plausible that by one route or another, be it as an island population (see below) or one spreading across open prairie or pampas, a terrestrial (as against arboreal) New World monkey would have adopted a bipedal stance following the same evolutionary path as the transformation of ape to australopithecine. After all, preadaptations of locomotory ability, parallels in brain structures, and complex social systems all point in this direction. Are we so surprised when we read what Martin Moynihan has written about *Cebus*, the capuchin monkey:

> Captive *capucinus* and *apella* [two of the species of *Cebus*] tend to explore and play with all the objects they can lay their hands on even when they are not hungry. They are very prone to take things apart. They seem to have a passionate desire to discover what is inside or behind anything that can be pulled or plucked or dismembered. They can also put familiar and strange objects together, sometimes in quite surprising ways, when the spirit moves them ... they are more easily bored than any other primates with which I am acquainted ... With all their limitations, however, capuchins can cope with some classes of stimuli with remarkable ingenuity, almost unparalleled below the level of man himself.[224]

If we hadn't walked out of Africa then probably sooner, rather than later, our analogues would have strolled[225] out of South America, holding tools (see notes 187–189), and probably enjoying the taste of meat.[226]

Carnivorous, with tools, and a restless intelligence? Perhaps, but bipedal? In the sense of true bipedality humans are certainly very unusual.[227] To be sure birds, and some of their relatives, the extinct dinosaurs, are bipedal; so, too, are the kangaroos. Apes are typically tree-dwellers, and on the ground proceed by what is known as knuckle-walking, which is forced upon them by having a hand designed principally for grasping. Why the lineage leading to humans became bipedal is controversial, but the fossil record – although fragmentary – indicates that it happened shortly after the divergence with the common ancestor, presumably a somewhat chimp-like beast. Of the various hypotheses explaining bipedality and the concomitant freeing of the hands, those connected with thermoregulation and the necessity of keeping cool during times of activity in open, unforested areas are attractive.[228] This is, however, only one idea and not surprisingly discussion on the origins of bipedality remains lively, especially with growing evidence that the first hominids may have lived in forest environments rather than open savannah.[229] As our knowledge of the hominid 'bush' of diversification rapidly expands, there is, moreover, another dimension to this problem. Thus, in a commentary on one sensational new find, that of the c. 6-Ma-old *Sahelanthropus* from Chad (see note 245), Bernard Wood[230] remarks that the continuing accumulation of data 'predicts that because of the independent acquisition of similar shared characters (homoplasy), key hominid adaptations such as bipedalism, manual dexterity and a large brain are likely to have evolved more than once.'[231]

There is also evidence that bipedality evolved independently, not only among the hominids, but also in the apes. In what is now Tuscany, but about 7 million years ago,[232] was a series of islands in the Mediterranean, where *Oreopithecus*[233] flourished, most probably having migrated out of Africa. For about 3 Ma this ape enjoyed a period of geographical isolation,[234] and then something both remarkable and convergent happened: bipedality emerged. Even though reconstructions of this Miocene ape sometimes portray it as swinging through the trees, detailed analysis[235] of the Tuscan fossils strongly indicates that 'bipedal activity made up a significant part of the positional behavior of this primate. The mosaic pattern of its postcranial morphology is to some degree convergent with that of *Australopithecus* [the genus that preceded *Homo*] and functionally intermediate between apes and early hominids.'[236] The arguments for *Oreopithecus* being bipedal depend on the analysis of the hip[237] and foot bones. Nor do the convergences stop there: it seems that this ape also independently evolved a hand with a type of precision grip apparently similar to the arrangement seen in the more primitive australopithecine humans.[238] This, of course, is a vital prerequisite for the use of tools. But why did *Oreopithecus* become bipedal? The reasons suggested for getting up on the hind feet are particularly interesting. The argument centres on the advantages of insularity.[239] Living on islands[240] meant that the large predators were no longer a menace, and in addition food could be readily collected from low-hanging branches.[241] Things were fine until reconnections were made to the outside world and the carnivores returned (note 234). *Oreopithecus* vanished, and the world had to wait a couple more millions of years before the next adventure to involve bipeds with a precision grip got under way.

It is also worth noting that, at least so far as the vertebrates are concerned, the ability to use the fore-limbs for skilled manipulation is remarkably widespread and is not restricted to the higher primates, as is sometimes thought. Examples can be found among the frogs and many mammals, including some marsupials, lemurs, and rodents. It seems, therefore, that when bipedality arose hands were very much ready for use. As it happens, these repeated occurrences of skilled fore-limb manipulations are generally thought to have arisen convergently, although others argue that they are basic to the tetrapod vertebrates.[242]

As we imagine the bipedal *Oreopithecus* wandering around the islands of future Tuscany, it is tempting to think that some such crucial step in the evolution of early humans also took place on islands, protected from the terrors of the jungle and savannah.[243] But, now, as the only surviving species of *Homo*, should we not count our luck, admiring ourselves in the mirror as a unique and otherwise irreplaceable end-product of the evolutionary process? If evolutionary history had gone otherwise, it might have led to a planet filled with grunts, whistles, howls, and other yatterings, but never the music of *Don Giovanni* or *Parsifal*. The evidence suggests otherwise. It certainly needs to be acknowledged that the study of human evolution is riven with controversy. What is generally agreed, however, is that even though you and I are, by definition, lineal descendants of a bunch of African apes, hominid phylogeny can no longer be construed as an evolutionary railway: from heavily browed and small-brained tree-climber to smooth-headed bicyclist. Despite the limitations of the fossil material and the taxonomic squabbles, it is clear that we are sole survivors of quite a substantial 'bush' of hominid diversification. It probably needs no emphasis, but spectacular as have been the advances in understanding the diversification of hominids, there is no doubt that further discoveries will be made. Consider, for example, the report of a new form, *Kenyanthropus*, a contemporary of the African australopithecines and dating from about 3.5 Ma.[244] Even more remarkable are substantially older discoveries, notably of a hominid 6 Ma old (*Sahelanthropus*) from Chad,[245] and considerably more controversially what appears to be a bipedal hominid (*Orrorin*) from coeval sediments in the Tugen Hills in Kenya.[246] Now, however, *Homo sapiens* is alone, having lost (or killed?) its sister-species *H. neanderthalensis* about 30 000 years ago. Much is made of this dramatic reduction in hominid diversity, a seemingly perilous shrinkage to just one surviving lineage. Implicit in this view is the notion of human uniqueness and our vulnerability: when we go, by nuclear incineration or social mayhem, that's it. Evolution gave us the chance, and we blew it. Not quite.

CONVERGING ON THE ULTIMATE

Giant brains, tool use, bipedality, and even a precision grip are not, therefore, specific to humans. To be sure, in our species they are each developed to a high order, but their independent emergence suggests that in principle there is no reason why they could not be similarly honed. Still, that does not rule out the possibility that the human lineage possesses unique features, which if lost by some chance would never be able to evolve again in the contingent turmoil of continuing evolution. The facts, however, point in precisely the opposite direction. When we peer into that 'bush' of hominid diversity (note 244), encompassing perhaps as many as seven species of *Homo* and a similar number of the more primitive australopithecines (including *Australopithecus*, *Paranthropus*, and *Ardipithecus*), and the more remote *Sahelanthropus* (and perhaps *Orrorin*), then a familiar tune is heard. Much has been made of this sprawling bush of hominid diversification from, among others, those anxious to demolish any sense of linearity in our history, especially if the implication is that this evolutionary trajectory might even be progressive. This discussion, however, misses the point. Recall that we are interested only in the emergence of particular biological properties: some will be of principal interest to a restricted group; for example, teeth to dentists. Others might be of universal interest; for example, intelligence. In either case, however, the convergences emerge.[247]

Concerning dentistry, one of the most spectacular examples involves an offshoot of the australopithecines. They are characterized by huge grinding teeth (appropriately called megadontic) and the associated development of a massive jaw musculature. This evolutionary excursion, into what is generally agreed to be serious vegetarianism, actually arose independently at least twice, and perhaps three times.[248] These robust australopithecines are instructive because of what they tell about the realities of evolution and the rules of convergence. The first time these powerful teeth were developed (in *Paranthropus aethiopicus*) is correlated with a massive accentuation of two sets of jaw muscles, specifically depending on the deployment of the posterior temporalis and anterior masseter. On the second occasion when the robust habit emerged (in *Paranthropus robustus* and *P. boisei*), one of these muscles had already become reduced so that the grinding process largely depended on the action of the masseter muscle.

As Rob Foley[249] noted, in reviewing this case of convergence, 'Clearly, in evolution there are more ways than one to crack a nut.' He then went on to remark that 'these anatomical differences [between the robust australopithecines] reflect alternative ways of solving the problems of heavy chewing, and which one evolved depended upon what had already happened in the lineage'.[250] Quite so, yet Foley then concluded, 'More interesting though, is the whole issue of convergence.' It is to this topic that he returns in a short but stimulating chapter published in *Structure and contingency*.[251] A central insight made by Foley, echoed frequently, but equally widely neglected, is his remark that 'Convergence is perhaps the strongest evidence for adaptation.'[252] Foley is in no doubt that key features in hominid evolution, such as 'bipedalism or increased meat-eating, characteristics that underlay the hominid adaptive radiations, are adaptive, functional, and the product of natural selection.'[253] He does not dèny the roles of contingency and other stochastic factors – why should he? – but his concluding remarks are judicious and tally with what we know about the realities of evolution. Thus he writes, 'while contingency plays a part in the timing and location of evolutionary events, the way these events are played out, the final biological outcomes, are strongly influenced by selection, adaptation and function,' and he concludes, 'evolution is the outcome of both stochastic and deterministic processes. As such, should the tape of life be replayed, undoubtedly, there would be many differences, but there would also be a very significant number of similarities.'[254]

Hominids evolve, and unsurprisingly given its ubiquity, convergence is inevitable. The independent development of nut-cracking is a dramatic example, but there are also more subtle instances. In becoming more gracile and showing a reduction in tooth size our own species *Homo sapiens* marks a departure from the general hominid robustness and actually approaches once again the more primitive condition found in our pre-australopithecine ancestors. The differences between us and the australopithecines are, of course, much more obvious. In some ways the most significant is the general absence in the australopithecines of any evidence of a stone tool culture. The earliest such occurrences, the Oldowan Industrial Complex, are from Gona in Ethiopia and are dated at about 2.5 Ma.[255] An array of slightly younger (c. 2.3 Ma) tools from Kenya[256] is important because reassembly of the fragments (cores and flakes) shows that tool-preparation involved a high degree of dexterity. Although most are evidently the product of early *Homo* it is almost certain that the late-stage australopithecines, including *Paranthropus*, were also tool-makers.[257]

Most probably the nascent skill in tool use was already present in the common ancestors of *Homo* and *Paranthropus*. What is, however, of particular interest is the evidence during geological time for brain-size increase in the *Paranthropus* lineage, independent of but parallel to that seen in *Homo*.[258] Why it was that the brain sizes increased has, of course, been the subject of protracted debate, but the convergence in *Paranthropus* might help to constrain these possibilities. Thus, Sarah Elton and her colleagues reinforce one current idea when they write, 'The manufacture and use of stone tools as a means to access meat eating might have been a crucial factor in hominin encephalization. However, it is also possible that increased meat eating and exploitation of patchy resources occurred because the existing cognitive ability and skills of hominids enabled them to make tools, and therefore facilitated meat eating.'[259]

The likelihood that lithic technologies would evolve more than once has already been touched upon, and so it seems reasonable that at some time, sooner or later, one or more lineages would rampantly encephalize. Thus, in a discussion of early hominid evolutionary homologies and homoplasies Daniel Lieberman and his colleagues[260] remark, 'An additional question raised [by our study] . . . is whether enlarged brains evolved more than once in hominid evolution . . . there is in fact no theoretical reason why, if large brains are advantageous, that this should not have occurred independently among different Pliocene hominid lineages.' Acknowledging the many difficulties in obtaining enough reliable data, Lieberman and his colleagues comment that their analysis raises 'the interesting possibility that increased encephalization evolved independently in more than one clade of hominids.'[261] In a similar vein, Gerrit van Vark[262] has argued, on the basis of a multivariate statistical analysis of cranial material dating back to *c.* 1 Ma, that two separate lineages show independent trends towards human form. As he remarks, '"Hominisation" is not as unique a process as many may think.'[263] Nor does van Vark think this process of hominization is yet complete: in this case it may well be legitimate to extrapolate existing trends into the future.

While these ideas will certainly be controversial, it is beyond dispute that much of the hominid evolutionary 'bush' became associated with stone tool cultures. A striking and much remarked-upon feature is the conservatism of these cultures,[264] with vast epochs of time elapsing before there is a noticeable shift in technology. Stone tools, of course, may give a rather one-sided view of these cultures. Thus, the discovery of wooden spears 400 000 years old,[265] evidently designed to act as javelins rather than to be employed in thrusting, are timely reminders, not only of the cognitive range of extinct hominid species (in this case *Homo erectus* or a near-relative), but also of how much of these cultures may have literally rotted away.

Other hints of increasing sophistication are the controlled use of fire from approximately 400 000 years ago[266] (earlier records are controversial[267]), and pathological conditions that imply protracted care of the sick. These include what appears to be a case in *Homo erectus* of hypervitaminosis, an agonizing condition brought on by the unwise consumption of carnivore livers containing vast and potentially lethal quantities of vitamin A, with the implication that the afflicted individual received long-term care rather than being left to the hyenas.[268] Despite these hints, and remembering that much evidence has been lost, it still seems that in general the pace of cultural change was, from our technophilic perspective, torpid in the extreme. Much has been made, and justifiably so, of the so-called Upper Palaeolithic Revolution.[269] This refers to the dramatic appearance of sophisticated tool-kits,[270] ornaments, e.g. beads, and perhaps musical instruments[271] beginning about 50 000 years ago. On the last item, Patricia Gray and her colleagues comment, 'Remarkably, many different types of scales can be played on reconstructed prehistoric flutes, and the sounds are pure and haunting.'[272] This culture is, of course, a product of our own species and unsurprisingly their anatomy is indistinguishable from ours. Yet the so-called anatomically modern humans appeared substantially earlier, about 125 000 years ago, although prior to the Upper Palaeolithic Revolution for the most part they churned out a markedly different set of stone tools. These are referred to as the Mousterian culture. While these implements are perfectly adequate they are decidedly less complex, and once again strangely conservative, showing little change over aeons of time.

What may have been a rather abrupt transition from Mousterian to the more advanced Aurignacian tool-kits in Europe was possibly heralded by substantially earlier developments, such as those reported from central Africa.[273] Indeed, the very concept of an Upper Palaeolithic revolution is now in question.[274] Nor should this surprise us: perhaps what happened in ice-age Europe was a telescoping of events that elsewhere was a more protracted process. Surely the important point is that over a relatively brief time we see a clear emergence of advanced cognitive abilities. Nor should we automatically assume that the emergence of sophisticated cultures can occur only once: evidence for multiple origins for agriculture and domestication of animals points in exactly the opposite direction. Nevertheless, the earliest occurrences of such cultures are highly sporadic,[275] and there does seem to be a real shift in the mental landscape when, about 55 000 years ago, the first compelling evidence for symbolic composition is found.[276] Here, too, there are earlier harbingers such as a fragment of bone 70 000 years old from the Blombos cave in South Africa, with a set of largely parallel cut marks,[277] and even more spectacularly, and from the same site, rare pieces of engraved ochre.[278] Here arguably is the key benchmark, in what was to lead to a cultural effloresence made famous by the cave paintings, figurines, and other sculptures that were in production from about 35 000 years ago. If humans were inevitable from the Cambrian period, a visit to the Moon was on the cards when the Palaeolithic painters surveyed the bare cave walls of Les Chauvet.

So here, at long last, we have found an example of evolutionary uniqueness. Other species of hominid might warm their bodies beside the fire, hunt with spears, and care for their sick, but were they really capable of abstract thought and symbolic representations?

Who are those figures in shadows, tracking our own history? Step forth the Neanderthals, a much-researched and on occasion much-abused group.[279] Either way their study is accompanied by continuous controversy. Nevertheless, the majority opinion is that they represent a separate species, *Homo neanderthalensis*, a view established on differences in skeletal structure[280] and development[281] and more recently on the basis of the recovery of ancient DNA.[282] Their massive and powerful construction, and details such as huge noses, are evidently adaptations to living in hostile, near-tundra conditions, of dealing with bouts of intense cold, and an erratic food supply that led to the adoption of hunting patterns rather different from those of *Homo sapiens*.[283] Our and Neanderthal brain sizes are equivalent, and it is beyond all reasonable doubt that they controlled fire and employed the red pigment ochre, possibly for body decoration. Unequivocal evidence for cannibalism also exists,[284] but the discovery of skeletons with severe impairments indicates that the infirm and crippled were at least sometimes tended to. Famous examples come from the Shanidar caves in Iraq.[285] Of the series of Neanderthal skeletons, a majority shows evidence of various traumatic injuries. One of the individuals (Shanidar I) had particularly dramatic injuries, perhaps caused by a rock fall, that included cranial damage, a fractured foot, and an arm that had withered. This individual lived, for a Neanderthal, to a relatively advanced age, and the extent of his disabilities indicates a society capable, for whatever reason, of care and compassion.[286] Even so, Neanderthal life was evidently robust and demanding. A general survey of Neanderthal trauma[287] confirms the prevalence of injuries, as well as the very high incidence of neck and head damage. Thomas Berger and Erik Trinkhaus remind us that although the old and infirm might receive care, the general 'dearth of older Neanderthals' suggests that 'these hominids did not sacrifice the survival of the social group as a whole when it was threatened by an immobile individual'. Despite the likely employment of spears in hunting, these may have been used principally for thrusting rather than throwing. As Berger and Trinkhaus dryly note, 'Given the tendency of ungulates to react strongly to being impaled, the frequency of head and neck, as well as upper limb, injuries seen in the Neanderthals should not be surprising.'[288] And if the Neanderthal escaped the charging ungulate, then there was also the risk of attempted murder.[289] What is especially intriguing is the evidence for deliberate burial (Fig. 9.6). This too

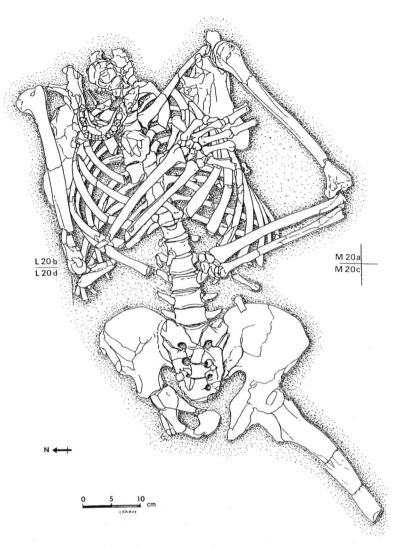

FIGURE 9.6 The Kebara Neanderthal burial. Inhumation was probably deliberate and may have involved decapitation and removal of the lower jaw. (Reproduced from the figure on p. 229 of B. Arensburg *et al.* (1985), Une sépulture néadertalienne dans la Grotte de Kébara (Israel). *Comptes Rendu des séances de l'Academie des Sciences, Paris*, ser. 2, vol. 300, pp. 227–30) with permission of the Academie des Sciences, Paris.)

has attracted scepticism; one paper is entitled *Grave shortcomings.*[290] Certainly the evidence for the laying of flowers with the corpse seems circumstantial,[291] nor is there convincing evidence for grave goods, at least in the eyes of most investigators.[292] Nevertheless, it seems almost certain that at least in certain circumstances the Neanderthals took care of their dead;[293] perhaps for hygiene, perhaps for some other reason?

And yet the consensus is that a vital spark was missing. It seems inconceivable that Neanderthals with their use of fire and burial of the dead did not have some kind of language; who knows, perhaps they sang? However, the only tangible evidence, which depends specifically on the position of a small bone (the hyoid) in the throat, is only indicative of some sort of language ability.[294] Differences between the vocal tracts of Neanderthals and modern humans do, however, exist, and while there is no doubt that the Neanderthals had a facility for language, there is still a fierce debate about whether their fluency approached that of modern humans.[295] The general consensus, however, seems to be that to a first approximation Neanderthals and humans were equally fluent. Richard Kay[296] and his colleagues, for example, use the size of the hypoglossal canal, which supplies the nerves to the tongue, and which is effectively the same size, to argue that vocal capabilities of Neanderthal and human were equivalent. A similar conclusion[297] is based on the evidence for, and origins of, thoracic nervous innervation and its links to sound control. Language is also consistent with the production of sophisticated stone tools, but one has to note that these tools are functional and conservative. To be subjective, there is little sense of beauty in their form.[298] In fact, the so-called Mousterian tool-kits, made both by Neanderthals and by humans, are in a European context similar in range of types and skills in execution, until about 50 000 years ago when we begin to diverge in technological complexity and usher in the next cultural stage, the Aurignacian.[299]

In contrast, the Neanderthals seem to have plodded on. Judged from about 40 000 years ago, but with foreknowledge of their coming doom, one might have predicted the last Neanderthal to have gone to the grave[300] still clutching in his or her massive hand a Mousterian tool. This, however, did not happen, because about 35 000 years ago there was a rather extraordinary cultural effloresence. Exemplified by what is referred to as the Châtelperronian,[301] it is marked by

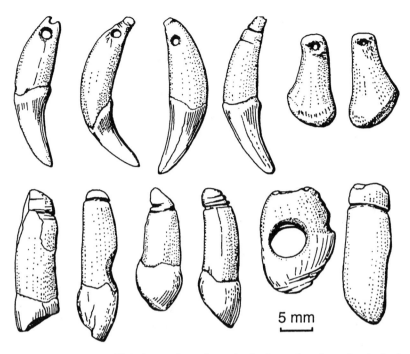

5 mm

FIGURE 9.7 Châtelperronian culture, in the form of perforated or incised teeth and bones, presumably to be worn as ornaments. (Reproduced from fig. 13 of Zilhâo and d'Errico (1999; citation is in note 308), with permission of the authors and Kluwer Academic/Plenum Publishers.)

the Neanderthals not only making more advanced tools but, even more remarkably, by shaping bones and teeth to form artefacts. Notable finds from the excavations at Arcy-sur-Cure[302] include necklace pieces (Fig. 9.7) composed of canine teeth and fossils. This breakthrough, which for obscure reasons did not spread to Iberia,[303] took place after tens of thousands of years of seeming cultural stagnation, and has been interpreted as largely imitative on the Aurignacian cultures of the incoming *H. sapiens*, if not acquired by trading or even scavenging abandoned sites.[304] A useful analogy, suggested by Paul Mellars, is with the cargo cults of such islands as New Guinea and the New Hebrides.[305] On some of the islands, such as Tanna in the New Hebrides where the John Frum cult still exists, the indigenes construct elaborate cult centres which include model aeroplanes, upon which it is hoped the desired objects will arrive.[306] If these artefacts had been found in an archaeological context, it would not be sensible, as Mellars reminds us, to conclude that the islanders had any practical knowledge of aerodynamics. So too, perhaps, with the Neanderthals: 'Very good, Arthur; not bad at all, now if we can just hold the flint a bit higher ... well, never mind, I expect we can use it for something; now if you would like to bring a flint from that pile over there, we'll continue the lesson ...'[307]

A radically different interpretation, however, is now being put forward that denies that the Châtelperronian culture is imitative. Instead the evidence is used to indicate that this breakthrough, where a necklace had meaning, was independent from *H. sapiens* and thereby convergent.[308] To say that this proposal is controversial is putting it mildly: battle-lines are being drawn.[309] The disagreements revolve around the dating, which depends on radiocarbon technology, and the stratigraphic order of the various cave deposits. The cave stratigraphy is notorious for the complexity of the infill histories and for the problems of subsequent disturbance, by hominids, animals, or even careless excavators. Yet the evidence is intriguing. In one key locality, the famous Grotte du Renne at Arcy-sur-Cure in Yonne, central France, exceptional Châtelperronian skeletal remains and artefacts have been excavated. Interestingly, despite the finesse of the artefacts, the skeletal morphology of these Neanderthals remained highly robust, although there are some hints of minor changes in the loading patterns of the forearm and shoulder.[310] In any event this Neanderthal occupation patently underlies and is separate from the subsequent Auriginacian influx that marks the arrival of *H. sapiens*. These Châtelperronian tools were not imported from some trading post, but were obviously manufactured on site.

In addition, the Châtelperronian itself was not completely homogenous. Some argue that it is divisible into a series of separate and advanced cultures (with names like Szeletian (a central and east European culture) and Uluzzian (in Italy), all doing approximately the same thing at the same time. Interestingly, Neanderthal culture has its own characteristics,[311] and among the material recovered is a bear penis-bone (baculum) with carved circular markings.[312] This find, from Croatia, is probably from an upper level of Neanderthal occupation, but it is possibly derived from a deeper horizon, dated at *c.* 42 000 years BP. This is not to say, incidentally, that cultural contacts did not exist between some Neanderthals and modern humans, although the evidence for this is conjectural (see note 312). Certainly the idea that the two species (if they were genuinely separate) may have interbred is intensely controversial.[313] But if they did at least have cultural contacts, it is also worth remembering that even if the Châtelperronian cultures arose independently (which seems likely) then if there was any contact it would almost certainly be a two-way process, and *Homo sapiens* would also have picked up useful tips.[314] Concerning this ongoing controversy Steven Churchill and Fred Smith (note 299) take a measured view, and although noting that 'the last Neanderthals seem to have held to a Mousterian way of life to the bitter end [see note 303] ... [others] appear to have been full participants in the evolving Upper Paleolithic ... [and] suggest that Neanderthals had cultural capacities on a par with those of early modern humans.'[315]

Translating the stratigraphy of the caves and their contained artefacts into the daily life of a band of humans or Neanderthals, each with the daily pressures of avoiding hunger and charging ungulates, surviving illness, and on occasion encountering other groups, is bound to be contentious. Tools were made, but are now silent. When does innovation and novelty stop being imitation? This problem is not confined to the origin of the Châtelperronian cultures. For example, it has been suggested that similarities in the tools made in the Upper Palaeolithic of Europe, and specifically the Solutrean culture (c. 20000 years ago, effectively the people who painted Lascaux), and the somewhat younger North American Clovis culture can be explained only by transAtlantic crossings thousands of years in advance of either the Vikings or the Spanish. The majority of researchers are, however, unconvinced, and interpret such resemblances as there are between Solutrean and Clovis are simply convergent.[316]

Still, it is sensible to keep an open mind. What, for example, are we to make of a terracotta head, apparently about 2000 years old, that if displayed in a museum would be put in the Roman section? The only problem is that the head is from Mexico, and was found in a sequence that is unequivocally pre-Columbian.[317] A Roman ship blown off-course from Spain, crossing the Atlantic?[318] Let us further imagine that two of the passengers were a friend of Paul the Apostle and a relative of Pliny the Younger. The ship moves slowly across the Caribbean: one of them looking at the New World with radiant hope, while the pagan sinks into gloomy resignation.[319]

To conclude: it is emphatically not my purpose to argue that agricultural ants, New Zealand kiwis, playful dolphins, or tool-wielding capuchin monkeys should have inherited the world as the most sentient species. Rather it is the simple observation that the ubiquity of convergence will lead inevitably to the emergence of recurrent biological properties that define the fabric of the biosphere. Rerun the tape of life as often as you like, and the end result will be much the same. On Earth it happens to be humans, just as the author of this book happens to be an academic in Cambridge. So what? Self-evidently we humans are now utterly different. We have new concerns, new priorities and questions, and, most important of all, new possibilities.

10 Evolution bound: the ubiquity of convergence

It was a cold spring day, and the little boy had been out for too long, playing in the pond. The other members of the family were not paying much attention, and it was only with a start that the father first realized that the boy was no longer to be seen; then with a surge of anguished alarm he saw that not far from where the boy had been wading something was now floating. All attempts at resuscitation were futile. With great sadness they left the ambulance as the metal doors closed. Swiftly they made their way home, already late to prepare for the Sabbath. Young Adolf was dead.

To judge from the recurrent, if not increasing interest in counterfactual histories, be it the untimely demise of either tyrant or genius, the what-ifs of history seem to loom ever larger as the tapestry of events weaves folly compounded with stupidity, the witless leading the insane, to the soft applause of the flaccid and the time-server: 'All loyal, my lord'. So many missed opportunities, such corruption and malice, laced with vaunting pride. All seems to be dependent on the twists and turns of fate, epitomized by the loose nail of the horse's shoe by which a kingdom is lost.

So, too, it is now widely thought that the history of life is little more than a contingent muddle punctuated by disastrous mass extinctions that in spelling the doom of one group so open the doors of opportunity to some other mob of lucky-chancers. The innumerable accidents of history and the endless concatenation of whirling circumstances make any attempt to find a pattern to the evolutionary process a ludicrous exercise. Rerun the tape of the history of life, as S. J. Gould would have us believe, and the end result will be an utterly different biosphere. Most notably there will be nothing remotely like a human, so reinforcing the notion that any other biosphere, across the galaxy and beyond, must be as different as any other: perhaps things slithering across crepuscular mudflats, but certainly never the prospect of music, nor sounds of laughter. Yet, what we know of evolution suggests the exact reverse: convergence[1] is ubiquitous and the constraints of life make the emergence of the various biological properties very probable, if not inevitable. Arguments that the equivalent of *Homo sapiens* cannot appear on some distant planet miss the point: what is at issue is not the precise pathway by which we evolved, but the various and successive likelihoods of the evolutionary steps that culminated in our humanness. To remind ourselves of what Robert Bieri noted: 'If we ever succeed in communicating with conceptualizing beings in outer space, they won't be spheres, pyramids, cubes, or pancakes. In all probability they will look an awful lot like us.'[2]

UBIQUITOUS CONVERGENCE

Before moving to some examples specifically in the areas of behaviour and molecular biology, where convergence *a priori* would be surprising, let me try to show you how ubiquitous is the phenomenon of convergence. Science fiction is replete with examples of the 'insectoid', vaguely modelled on the apparently robotic scrabblings of a terrestrial counterpart. Evolutionary orthodoxy, of course, is that such a creature is a contingent accident, assembled by chance histories and circumstances. Insects are interesting; insects are monophyletic; but in the final analysis that is all there is to say. If, however, we consider 'insectoids' as a biological property then perhaps something more general emerges. So what is the design specification? Among the defining features of the insects are the following: an articulated exoskeleton arising from the process of arthropodization; compound eyes; a hexapod gait whereby three of the six walking legs are always on the ground and thereby define a triangle (two legs on one side, one leg on the other) that keeps the animal stable; respiratory tubes known as tracheae that serve to bring oxygen into the interior animal via special openings (spiracles) on the side of the body; and, to complete this list of evolutionary peculiarities, the development of complex eusocial colonies, as in the honey bees. All pretty strange, all one-offs in the great lottery of life? On the contrary, *all* are convergent. Arthropodization may have evolved as many as four times independently;[3] the convergent evolution of compound eyes was addressed in Chapter 7; a hexapod gait is inferred in eurypterids (an extinct group related to the scorpions) such as *Hibbertopterus* that moved out of water on to land;[4] tracheae evolved independently at least four times[5] (as has the method of gas exchange[6]); and a eusocial organization has not only evolved independently in many groups of insects (ants, bees, termites, wasps), but is also found in the shrimps and various mammals, most famously the naked mole-rats (Chapter 8).

Yet despite the textbook examples of convergence, such as the camera-eye of vertebrates and cephalopod molluscs, along with the seemingly arcane, for example, birds that can echolocate (p. 181) or reptiles with a dentition similar to that of mammals (p. 227), my thesis is that both the extent and the importance of convergence have been consistently underestimated. By this I do not mean so much the innumerable examples. These are mostly known only to the specialists, and even among them the convergences tend to be treated as simple curiosities, with the customary adjectives of 'remarkable' and 'surprising' (see p. 128). Rather it is my suggestion that by understanding convergence we can constrain what sort of biosphere we might expect, especially in an extraterrestrial context. To be sure, for the most part any discussion concerning convergence tends to revolve around the immediate realities of the physical world. In this sense the repeated evolution of features as distinct as stabbing canines (p. 130) or electroreception in fish (p. 182) are, on a closer analysis, hardly surprising. There are, however, other more complex convergences, such as the cognitive functions of dolphins (p. 258). Nor in this context should we be surprised to see the repeated re-emergence of advanced and sophisticated social organizations. The comparison between the elephant and the sperm-whale (p. 250) is a fine example, as are the similarities between the fission–fusion societies of chimp and dolphin (p. 249). Because of the general emphasis on physical constraints in evolution, these examples of societal convergence surely deserve to be more widely known.

Indeed, I rather suspect that the role of convergence in the behavioural realm will produce quite a few more surprises. Earlier I referred to the way in which electric fish and moths, generating respectively electrical and sound signals, show a convergence in the way they confuse potential predators. At present, however, the number of examples of convergences in the behavioural repertoire is rather limited. Indeed, at first sight such convergences would seem to be rather unexpected given the complexity and range of responses. Nevertheless the examples now available, including the courtship behaviour of houseflies,[7] lacewings,[8] crickets,[9] and bowerbirds,[10] give some indication of the many other examples of behavioural convergence that probably await recognition. The case of the bowerbirds is, of course, of particular interest because of the complexity of the decorated bowers constructed by some species.[11] Mike Hansell (note 11) also discusses whether the bowerbirds have any sense of beauty, a view made popular by Jared Diamond.[12] Hansell is clearly sympathetic to this idea, alluding to evidence of male apprenticeships and the possibility of cultural transmission. So it is that once again the birds, this time in the context of complex behaviours, hint at biological universals.

In this context the famous displays by the fiddler crabs are also relevant, both because there is indeed evidence for convergent acquisition of complex behaviour, and because of the likelihood that these are linked to the transition towards a more terrestrial life. In entering this new adaptive zone, key features, such as rapid locomotion and excellent vision, are linked to new behavioural capabilities.[13] Nor should this be a surprise. What we already know of convergence in terms of cognitive sophistication (Chapter 9, p. 258) and sensory perceptions (Chapter 7) indicates that behavioural similarities are almost inevitable. A nice example of this is in acridid grasshoppers, whose song is usually produced the rubbing of the hind-leg. In one species, however, the same song is produced by rubbing together the mandibles.[14] Even so, the examples of behavioural convergence tend to elicit the customary cries of surprise. How can we blame the investigators? Until one appreciates the background these convergences do seem to be remarkable.

It is, however, important not to adopt too uncritical a view of such convergences. Identifying examples of behavioural convergence, or for that matter any other type of convergence, presupposes, of course, a reliable phylogeny. A recurrent feature of phylogenetic reconstruction is that one set of data, say morphology, produces a phylogeny that is at odds with another set, such as behaviour.[15] Somebody has got to be wrong, and in being wrong by implication will have failed to identify certain convergences, either in the morphology or in the behaviour. What is needed, of course, is an independent data set such as molecular sequences that may enable the more reliable phylogeny to be identified. To a first approximation this works, but even molecular databases usually produce contradictory results. Phylogenies therefore remain in a constant state of flux, and this means, of course, that in some cases identification of a given convergence is provisional. So far as the thesis of this book is concerned, however, this is not a problem, inasmuch as the important comparisons involve distantly related groups whose common ancestor lived long ago in the geological past, and which very probably did not possess the relevant character we identify as convergent.

RESPIRATORY CONVERGENCE

Despite the apparent ubiquity of convergence, there are some areas of evolution where the role of convergence would be expected to be minimal. Of these molecular biology is surely the prime candidate. It is not difficult to see why. As was discussed earlier, the combinatorial 'universe' of protein space is unimaginably vast, which suggests that the likelihood of navigating to the same 'destination' is exceedingly remote. In itself this may not be terribly important. The success of proteins depends on their overall structure and, to a first approximation, the precise sequence of amino acids may not make that much difference. To be sure, the structure at particular sites may be exceedingly sensitive to the substitution of one amino acid by another, and as we have already seen in such examples as trichromatic colour vision (p. 167), these substitutions also give insights into a particular sort of molecular convergence. Such substitutions, however, presuppose an existing protein that is already functional. However, we also discover that, despite the combinatorial immensity of protein 'hyperspace', in other ways the landscape of possibilities is again somewhat limited.[16] This is simply because proteins are constructed from a limited number of 'building blocks'. Furthermore, in at least the case of certain structures such as the so-called α/β barrels[17], there is evidence that the arrangement may have been arrived at convergently. It is useful, therefore it evolves. Even so, it must be acknowledged that there is considerable tension, metaphorically speaking, in the study of protein evolution in deciding whether a structure has evolved because of a common ancestry, even when no sequence similarity remains, or because of general constraints on protein structure and function.[18]

As such, this suggests that molecular convergence at the level of proteins need not be unexpected. For example, the general view is that the haemoglobins found in organisms as different as humans and cyanobacteria are similar because they all derive from a common ancestral protein. Yet it is also possible that the haemoglobins found in the protistan ciliates and cyanobacteria actually have a separate origin,[19] albeit with the complication that the cyanobacteria may have acquired them from the ciliates by a lateral transfer.[20] In the context of these distinctive haemoglobins the possibility of 'an independent evolutionary origin' remains a real possibility.[21] So perhaps those extraterrestrials that inhabit oxygen-rich planets will also employ a remarkably similar respiratory protein. And if not haemoglobin, what else? On Earth the number of proteins is rather limited. The next most important, after haemoglobin, is the copper-bearing protein known as haemocyanin. This respiratory compound is found in many molluscs and in the arthropods,[22] but the molecular structure of the haemocyanin in either group is very different. Quite clearly haemocyanin is convergent.[23]

Before leaving the exhilarating topic of the respiratory proteins it is well worth drawing attention to yet two other examples of molecular convergence. Of these, the second is to my mind even more remarkable, because it shows how proteins can be different, but end up with effectively identical functions. In each case the story concerns the molecule myoglobin.[24] It is related to haemoglobin, but is somewhat simpler. It, too, has its bacterial antecedents,[25] but as with haemoglobin there is a strong possibility that the myoglobin that occurs in the cyanobacteria (and the protistan ciliates) evolved independently of that found in mammals (and bacteria).[26] In any event, myoglobin is best known for its role in animals as the molecule used to store oxygen, especially in the muscles. It is no coincidence that whale muscle is also conspicuously red in colour and that here the abundance of myoglobin is connected to the need to store substantial quantities of oxygen during deep dives. So, too, we see convergently high concentrations of myoglobin in other animals that experience oxygen stress, notably the burrowing mammals.[27] Even more interestingly there is evidence for a more specific molecular convergence in the myoglobin of diving and burrowing mammals in terms of an increased content of the amino acid arginine.[28]

The second example of myoglobin convergence concerns the gastropods, or snails. In a number of these animals the buccal mass, which helps to ingest food, is not only very muscular but is by any measure exceedingly rich in myoglobin. But it is not the usual myoglobin. It is a substantially larger protein, and is evidently derived from an enzyme (indoleamine 2,3-dioxygenase (in short IDO)), which is involved with the degradation of tryptophan.[29] So how did it end up as a respiratory protein, convergent on true myoglobin? There are several clues. A key fact is that the enzyme carries iron,[30] and as might be expected this structural characteristic means that it has an absorption spectrum very similar to those of other proteins, including myoglobin.[31] But in itself such a structural similarity is not enough. This is because at the stage where the oxygen is being carried, which forms part of the catalytic process, this intermediary form is very unstable. If the protein is to act in any sort of respiratory mode a key necessity, therefore, is that it must have been able to mutate that part of the protein structure close to the iron and so confer a new stability. Only then can the protein serve as a respiratory protein. And this is evidently what has happened in the abalone (*Sulculus*), in which this convergence was first noted,[32] and in a number of other snails,[33] all of which belong to the group known as the archaeogastropods. Other members of this group, however, possess normal myoglobin.

How and in what circumstances, one wonders, did the ancestor of these particular snails come to lose its ability to synthesize myoglobin? A clue of some sort might come from animals rather far removed from the snails. In the heart of a bullfrog there is no myoglobin, but an evidently altered and very small haemoglobin has taken over this essential role.[34] But what counts as essential? Rather oddly, in genetic experiments mice can also be deliberately deprived of myogloblin, but seemingly without obvious deleterious effects.[35] Such was evidently not the case for the snails mentioned above. Presumably once the original myoglobin was lost (for whatever reason), the alternative copper-bearing haemocyanin (the usual molluscan respiratory pigment) was not available as a substitute, and some other protein which happened to contain iron had to be recruited. That presumably had to be a rapid process, and in any event once again convergence emerges.

So, too, with the arthropod silks. As noted in Chapter 6, they are blatantly convergent. There is, however, a deeper convergence in the type of molecular structure that characterizes these silk proteins. As John Gatesy and his colleagues[36] point out, there are a variety of high-strength proteins. These include not only the arthropod silks, but also certain proteins found in bivalve molluscs (such as the edible mussel and oyster), where they have such structural roles as conferring toughness to the shell or the byssus threads that in some instances anchor the animal to the seafloor. All share the same motifs, suggesting that similar proteins on Thega IX will be less alien than might be expected. These examples, if anything, exemplify convergence:

whether the journey is by bicycle or airship is less important than the common destination. One suspects, for example, that if elephants and sperm-whales could somehow communicate about their strikingly convergent social systems (p. 250) it would lead to an animated conversation about how much they had in common. And in an analogous way molecular convergences might also point to deeper patterns in evolution. It is to this controversial topic that we now turn.

FREEZING CONVERGENCE, PHOTOSYNTHETIC CONVERGENCE

Animals faced with the threat of freezing can meet this challenge in various ways. A common strategy is to alter the osmotic balance of the internal fluids to produce a natural equivalent of an antifreeze: adding glycerol or glucose to the blood or equivalent is a standard biological remedy.[37] Another route is to synthesize particular proteins that act to prevent the formation of ice crystals, the growth of which would rupture the cells and so lead to death. A variety of proteins are able to confer protection against ice growth, but there is at least one striking example of convergence. It involves two groups of fish, the Arctic cod and (in freezing waters around Antarctica) a group known as the notothenioids.[38] These groups are not closely related, but both produce effectively identical protein antifreezes that are based on a sequence of tri-peptide repeats. Several lines of evidence show, however, that the genes responsible for the production of this protein are completely different.[39] In the notothenioid fish the gene[40] is evidently closely connected to trypsin production[41] in the pancreas (so representing co-option), whereas although not yet specifically identified the equivalent antifreeze gene in the Arctic cod is evidently entirely unrelated to any pancreatic trypsinogens. It will be interesting to learn the derivation of this gene in the Arctic cod, and to see whether convergences exist among some of the other antifreeze proteins. Thus fish have developed a variety of antifreeze proteins, but their distribution makes little phylogenetic sense and probably reflects an *ad hoc* recruitment by the various groups in the face of global cooling.[42] So, too, proteins have been recruited by plants and insects to prevent ice formation, and in the insects, at least, there is evidence for convergence.[43] Perhaps less surprisingly, given the nature of the convergence, at least some of these antifreeze proteins have been under positive Darwinian selection,[44] indicating that if ice is universal so, too, will be the adaptations to combat its growth in tissues.

The reason for the polar fish having such proteins is, as noted, to combat the problem of inhabiting near-freezing waters, especially adjacent to the Antarctica ice sheets. Despite their relative familiarity, if only from films showing the antics of penguins or the cavorting of polar bears, from a geological perspective such icy waters (and therefore polar ice caps) are rather unusual. For millions upon millions of years the weather forecast for most of the planet has been 'Dry and sunny, with afternoons in the low 40s, with thunderstorms after sunset'. Periodically, however, the planet refrigerates. Then once again the ice sheets smother the poles and so construct huge ice caps. That the Earth now possesses approximately 30 million km^3 of ice is because of a long period of global cooling. This began about 20 Ma ago, with the initiation in the Southern Hemisphere of the first major ice caps. As we saw earlier (p. 248), in the Southern Ocean this cooling event may have been a trigger for the diversification of the ocean-going whales, one consequence of which was the rise of dolphin brain power. Despite the growth of the Antarctic ice sheets, the world only entered the fully fledged ice age about 2 Ma ago, when the northern continents and polar ocean became respectively glaciated and covered with sea ice.

The net result is our present climatic predicament, because although we live in a warm interval there is no reason to think that the ice age is over. What we regard as 'normal' is but an interlude of moderate climatic stability (if not safety) in a longer-term pattern in which the global climate not only oscillates wildly but may plunge from one state to another in only a few decades. These oscillations are of widely varying duration – think for example of the so-called 'little ice ages' – but are most obviously expressed in times of major glacial advance alternating with warmer interglacial periods. Self-evidently, we live in one such climatic respite. Nor is the present day the warmest interval. About 400 000 years ago hippopotamuses wallowed a few miles from my office in Cambridge, and were no doubt the occasional object of scrutiny by the equivalent of Boxgrove Man,[45] members of *Homo erectus* (or a similar species). By 40 000 years ago, however, my view would have been replaced by bleak tundra, occasionally traversed by hunting parties of Neanderthals tracking the spoor of the woolly mammoth.

Knowing that where I work was once home to the near-naked Boxgrove Man or fur-covered Neanderthal depends, of course, on many lines of evidence. Records of ancient pollen and the study of such animals as the beetles are particularly important, as are the much more occasional stone tools left by the hominids. But some of the clearest evidence for these climatic fluctuations comes from the drilling of cores through the ice caps. In this way it is possible to read the history of glaciation to a remarkable degree of accuracy. As the terms *glacial* and *interglacial* imply, the glaciers sometimes advanced and then subsequently retreated as the world swung into the next warmer interval; but in the polar regions, and especially on the great ice caps of Antarctica and Greenland, the snow continued to fall, and it is here that the critical climatic evidence is encoded in the ice caps.[46] In these polar regions, as the snow turned into ice, air bubbles were trapped and retained tiny samples of the atmosphere at that time. Not surprisingly, during any given interval the proportions of the major gases, such as nitrogen and oxygen, remain constant. There are, however, other so-called greenhouse gases, notably carbon dioxide and methane, which show major fluctuations in abundance through the thickness of accumulated ice. Specifically, as the amounts of these gases increase so the planet warms and the ice begins to melt. Although much talk of global warming revolves around carbon dioxide, methane is probably also of critical importance. At present huge quantities are locked up in ice and similar hydrates, especially in the tundra and within deep-sea sediments. Once released, the methane is quite rapidly oxidized. This means that it has a short residence time in the atmosphere. However, the substantial release that is expected as a result of already rising global temperatures will in its turn accelerate the process of global warming. It is interesting that the geological record has compelling evidence for the release of massive volumes of methane resulting from the destabilization of buried hydrates, notably during the so-called late Paleocene thermal maximum (LPTM).[47] So the world is no stranger to an event of this sort, which is, of course, exactly what we are producing today by burning fossil fuels.

Present-day global warming, at least in part the result of the addition of massive amounts of carbon dioxide to the atmosphere, is likely to have perilous consequences. However, from the point of view of the last ten million years or so of geological history, the carbon dioxide being released from belching chimneys and fuming car exhausts is actually reversing a long-term decline in its atmospheric abundance. That decline, although natural, is also potentially perilous, at least for the plants. The reason is rather straightforward. Photosynthesis, of course, requires carbon dioxide and the fact is that for the plants the pre-industrial levels of this gas in the atmosphere had reached dangerously low levels. This predicament is made the more serious because, as we have already seen, the process of photosynthesis depends on the RuBisCO enzyme, which itself shows serious deficiencies, not least its inefficiency in the (inevitable) presence of oxygen (p. 108).

These difficulties are effectively unavoidable, but to escape the predicament posed by declining levels of atmospheric carbon dioxide a number of plants have found an escape route. They have evolved a new photosynthetic pathway. Here it is necessary to explain that the usual method is called C_3 photosynthesis, because the first compound (a substance known as phosphoglycerate) to be formed in the pathway contains three carbon atoms. In contrast, the new pathway, which arises as a response to desperately low concentrations of carbon dioxide, first produces oxaloacetate, which has four carbon atoms, hence C_4.[48] A significant group of C_4 plants are the grasses, and the development of grassland and savannah[49] had a profound effect on mammalian evolution, notably for such herbivores as the horses, and, many have argued, ultimately for the hominids. One should also in passing note that declining levels of CO_2 may not be the whole story in explaining the rise of C_4 photosynthesis; aridity and salinity also appear to be significant factors.[50]

The C_4 photosynthetic pathway is biochemically very complex, so complex in fact that it might be thought to have evolved in only one group of plants which fortuitously stumbled on this solution. Not a bit of it: C_4 photosynthesis is rampantly convergent and is believed to have arisen independently at least 31 times.[51] In essence the C_4 process aims to protect, so far as possible, the RuBisCO activity, and within the plant to increase the amount of carbon dioxide. For the most part this convergence depends on three critical factors. These are a close proximity between types of leaf tissue involved with photosynthesis and the initial transport of the sugars (respectively the mesophyll and bundle-sheath cells, forming the so-called Kranz anatomy); up-regulation of PEP carboxylase; and down-regulation of RuBisCO in the mesophyll. The enzyme PEP carboxylase plays a key role because it helps in the release of the vital carbon dioxide. Yet, as is so often the case, this protein has been co-opted from an earlier role; in this particular instance from functions that include control of the stomata and assimilation of nitrogen. Until recently it was also thought that an essential aspect of C_4 photosynthesis was the possession of a Kranz anatomy, but so far as a C_4 chenopodiacean plant (*Borszczowia*) is concerned its distinctive leaf anatomy does not depend on the dual-cell arrangement seen in classical Kranz anatomy.[52]

There are some interesting nuances in this story. First, although described as C_4 photosynthesis, the pathways in the respective plants show various differences and degrees of completion in the transition from the C_3 to C_4 photosynthesis.[53] As is effectively inevitable in evolutionary convergence, there are multiple pathways to the same destination. Even so, there are a number of peculiarities in this evolutionary process. One of the most obvious is that C_4 photosynthesis is not generally known among the trees, although some shrubs with secondary woody growth are C_4 photosynthesizers, and in Hawaii a rare tree (*Chamaesyce forbesii*) also has this type of photosynthesis.[54] Despite this restriction in land plants, what appears to be an independent acquisition of the C_4 pathway is known in some algae, specifically the diatoms.[55] Second, although the story of the rise of C_4 plants focuses on events beginning about ten million years ago (see note 49), there may be a deeper history. There are at least hints of a precipitous drop in carbon dioxide about 90 Ma ago and a corresponding appearance of C_4 photosynthesis.[56] More tenuous is the possibility that a much earlier drop in atmospheric CO_2, linked to the late Palaeozoic ice ages (late Carboniferous – early Permian, *c.* 280 Ma ago) also saw an independent development of the C_4 pathways.[57]

Once again evolutionary inevitabilities emerge. If George Wald is correct and chlorophyll is universal (Chapter 6), so, too, is photosynthesis. On planets where the levels of atmospheric carbon dioxide happen to decline, and given that only on tectonically active planets where water, plate tectonics, and mountain-building interact will atmospheric carbon dioxide fluctuate, the C_4 photosynthetic pathway will inevitably follow. Without C_4 plants, humans would face a rather different world. Some of the most important groups of C_4 photosynthesizers are the grasses, including wheat and rice and (originating in the New World) maize. Even within the grasses it appears that the C_4 process was arrived at independently at least four times.[58]

Now, of course, humans face the prospect of yet another world, but this time of their own making. As carbon dioxide from the burning of fossil fuels continues to pour into the atmosphere, in principle plants will benefit, but this is little cause for comfort because in reality the effects are highly variable. Plants need more than carbon dioxide, and other factors such as water stress and increasing temperatures can work in the opposite direction and so largely negate the apparent advantage of boosting atmospheric carbon dioxide. Not only that, but increasing levels of carbon dioxide, not to mention the accelerating release of the yet-more-dangerous greenhouse gas methane (note 47), will lead to many other climatic effects that, from a human perspective, will probably be distinctly unwelcome. In parentheses one might mention that much could be done to alleviate global warming today, but dithering, self-interest, and, most importantly, corporate hostility are – well – discouraging. Perhaps my children's generation will be entitled to ask why we did nothing?

THE MOLECULES CONVERGE

The convergence of antifreeze proteins and the multiple paths to a C_4 photosynthesis can therefore be added to the other examples of molecular convergence mentioned earlier, such as the 'five-site rule' for vertebrate colour vision (Chapter 7) and probably also rhodopsin itself. As already noted, the combinatorial vastness of protein 'space' would, *a priori*, suggest that examples of molecular convergence would be very rare indeed: there are, after all, so many alternatives. As it happens, however, there is a growing list of such examples. These include the proteases[59] and peptidases,[60] aminoacyl-tRNA synthetases,[61] cytokinases,[62] proteins associated with malaria,[63] NADH dehydrogenase 1,[64] lactate dehydrogenases,[65] nicotine oxidases,[66] the evolution of polysaccharide lyases,[67] light-harvesting proteins,[68] proteins associated with cartilage (lamprins) and various elastic proteins,[69] chitin-binding proteins,[70] HIV-protease,[71] antigen receptors in sharks,[72] as well as biochemical processes such as those involved with nucleotide binding by proteins,[73] possibly DM domain factors involved with sexual determination,[74] and steroid signalling.[75]

Some of these convergences are of more than academic interest, notably the structural mimicry that can arise when microbial pathogens insinuate themselves into the cellular processes of their hosts.[76] These are particularly important, for several reasons. First, the structural mimicry adopted by the pathogens has clearly come about by natural selection. In addition, this mimicry requires a molecular matching, but such a correspondence need not entail any similarity to the amino acid sequence of the host protein. This is significant in the context of navigating through the combinatorial vastness of protein 'hyperspace': common structures will emerge notwithstanding the immensity of alternative sequences. As ever, the biological solutions are extraordinarily specific. Consider, for example, the remarks by Erec Stebbins and Jorge Galán with reference to one example. This concerns a bacterial pathogen that in its invasion uses one of its proteins to take over those signal transduction pathways that in the host cell rely on such proteins as fibronectin.[77] This procedure allows the bacterium to bind to the host cell, which accordingly fails to recognize that it is under attack. In conclusion, Stebbins and Galán write, 'The fact that two completely different protein structures ... present the same residues [two aspartic acids, one arginine] in the same positions for binding attests to the power of convergent evolution.'[78] Such examples are probably very much more widespread than is at present realized, and they emphasize both the sophistication and the subtlety of these molecular processes. As Stebbins and Galán conclude, in the context of such a pathogen attack, 'convergent evolution contributes significantly to the dynamics of the evolutionary process.'[79]

That Stebbins and Galán are not wrong is now becoming clear with other examples of molecular mimicry that go to the very heart of the structure of life. What, for instance, are we to make of evidence for molecular mimicry between three different proteins (respectively known as EF-G (a translocase), eRF1 (human release factor), and RRF (ribosome recycling factor)) and the RNA which is involved in that part of the replication process known as translation.[80] In other words, despite having markedly different structures and compositions, these proteins (made of amino acids) mimic structurally the RNA (made of nucleotides). The investigator, Yoshikazu Nakamura, also points out that the ability for such a protein to act in the process of replication, in place of the usual RNA, might possibly have some bearing on the origin of life, or at least on how the transfer from the hypothesized 'RNA world' (see note 3, Chapter 4) to the universal DNA replication-protein system was achieved. This is a tempting and possibly believable goal. We may need to remember, however, the cautionary remarks of Andrew Ellington (note 41, Chapter 4) to the effect that the billions of years of molecular evolution may irrevocably sunder us from establishing the first stages in the origination of life, at least through an examination of modern biochemistry. Moreover, if convergent evolution is an 'eternal return' to the 'attractors' of functionality, then we cannot be surprised if history repeats itself.

One hint that this could be so comes from work on a type of RNA known as the hammerhead ribozyme, whose simplicity and ability to show self-cleavage makes it a tempting candidate for inclusion in the 'RNA world'. Yet, as Kouroh Salehi-Ashtiani and Jack Szostak show, this ribozyme has evolved several times.[81] Their remarks, which strongly echo those of Kenneth James and Andrew Ellington (note 86, Chapter 4), are particularly pertinent: 'Our results show that, despite the dominance of contingency (historical accident) in some recent discussions of evolutionary mechanisms [here they cite S. J. Gould's *Wonderful life*], purely chemical constraints (that is, the ability of only certain sequences to carry out particular functions) can lead to the repeated evolution of the same macromolecular structures.'[82] This book aims, if nothing else, to refute the notion of the 'dominance of contingency', and these examples of molecular convergence perhaps explain why some scientific attitudes to the reality of convergence are perhaps ambiguous. I have already suggested (p. 128) that the associated adjectival battery of surprise may have less to do with evolution than an uneasy sense of teleology. Self-evidently, if we cannot be sure whether a feature is genuinely primitive as against having evolving several times, then clearly the study of evolution, which presupposes that organisms are related because they are similar, would be severely compromised. This is the specific problem of homoplasy, viewed by most cladists as both profoundly disruptive to their neat schemes and evolutionarily of marginal interest. But, of course, evolution can be studied: the routes are many, but the destinations few, and the landscapes across which all organisms must travel are adaptive.

Caution, however, is always needed.[83] First, and this is a point that presupposes the identification of any convergence, one needs a reliable phylogeny. Even here, whether or not the end results are judged to be convergent may depend on the sort of question asked. Take, for example, one of the classic cases of molecular convergence, that of the enzyme lysozyme. This is widely employed, and acts by breaking down the walls of bacterial cells. It thereby confers protection, and is found, for example, in exposed surfaces, such as eyes and the nose, where it serves to protect these exposed surfaces from infection. Lysozyme, however, has another role. This is to assist with the digestion of the tough plant material scooped up by cows and other ruminants. The basic principle is to employ bacteria in a reaction chamber, the rumen. The role of the bacteria is to attack the plant cellulose, and so allow fermentation of the swallowed vegetation. The resultant slurry is then processed in the stomach, using the enzyme lysozyme. This arrangement of rumen and stomach with lysozyme activity, familiar from the cow, has evolved convergently in one group of primates, the colobine Old World monkeys, e.g. the langur,[84] and also a rather odd bird known as the hoatzin.[85] This animal, which lives in the Amazon forests and eats a diet made up largely of leaves,[86] is probably best known because of its supposed parallels with the bird–dinosaur *Archaeopteryx*.[87] The hoatzin has shifted its diet from the softer fruits to tougher leaves, and here too lysozyme is employed. Molecular convergence between the lysozymes of these monkeys and the ruminants has significantly also been identified,[88] and so, too, with the hoatzin birds.[89] But is it really a convergence? When the overall sequences of the lysozymes in primates and ruminants are compared, the phylogeny is as expected, whereas if the convergence were unrecognized it could be used mistakenly as a sign of common ancestry and so would 'artificially' bring the two groups together.[90] This, however, is not quite the point because it is the specific similarities within the lysozyme molecule, at five or so amino acid sites, which excite our interest,[91] because of both the molecular convergence and the implication that observations of this kind have for molecular adaptation.[92] It would be misleading, of course, to suggest that these examples of molecular convergences are particularly frequent, but examples in some key enzymes suggest that as with DNA (p. 31) and chlorophyll (p. 110) there may be some effectively universal constraints.

CONVERGENCE AND EVOLUTION

Evolutionary convergence shows that we live in a constrained world, where all may not be possible. It does not, to be sure, rule out unique solutions, nor does it guarantee the emergence of the identical. It shows, however, that at many levels evolution is seeded with probabilities, if not inevitabilities. As yet there is no overall metric that might establish a scale of evolutionary likelihoods, but, as I have repeatedly emphasized, the many biological parallels to those features that define the emergence of humans suggest that something similar will emerge elsewhere.

There are, however, several important provisos. Arguably they should have been introduced at the very beginning of the discussion. They have, indeed, already received passing mention. Their overall importance is, however, clearer once the field has been reviewed. The first, and most obvious, point is that identification of convergence presupposes a reliable phylogeny. Indeed, assumptions about the reliability of the evidence and decisions on which characters are evolutionary relevant introduces a constant risk of circularity in the argument. Is a particular character the same because it evolved from a common ancestor, or is it convergent? This point is far from trivial: those who employ the methods of cladistics are constantly aware of what they call homoplasy, arising from characters that carry no phylogenetic information because they arose independently. A related question is whether a particular character is actually primitive, but by being lost (or transmuted beyond recognition) in some evolutionary branches appears to be convergent because of its sporadic distribution across the tree of life. In many of the examples I have given these alternatives need to be taken seriously, and because of the fluidity of many phylogenetic schemes the ensuing questions will not necessarily be easy to answer. It is also worth noting that, whatever the phylogenetic scheme adopted, practically never can it escape convergence. Although this is a headache for those wrestling with local phylogenies – is the character in question the same because it descends from a common ancestor or because adaptive constraints have led to convergence? – for the most part the examples discussed in this book are from distantly related groups where the evidence for separate and independent originations is proportionally more secure.

A more significant point concerns what has been termed 'concerted convergence', which as the name implies is when a whole series of characters change in tandem in response to an adaptive pressure. Striking examples are available in groups as disparate as fig-breeding fruit-flies[93] and lilies.[94] The likelihood that a set of characters will be linked during evolution has attracted attention because of its phylogenetic implications. Clearly, if a number of characters evolve in concert then they cannot be treated as heaps of nuts and bolts subject to the atomistic scrutiny of cladists. Concerted convergence in complex systems is probably much more widespread than is presently realized, and it has two important implications and one qualification. First, the interlocking inherent in concerted convergence suggests that at least in some instances the likely outcomes of particular evolutionary trends may be even more constrained than is at first apparent. Second, this concept is consistent with the idea of one or more key adaptations whereby evolution of one feature may effectively promote, if not guarantee, a cascade of parallel changes. The qualification is that this does not mean that concerted convergence can be divorced from its phylogenetic context. Rather, as Janice Lord and her colleagues[95] have stressed, the roles of adaptation and phylogeny are complementary. While this will tend to confer a broad predictability to the evolutionary process by what is termed 'niche conservation', it emphatically does not debar evolutionary excursions into new adaptive territories when the environment (however defined) happens to shift. In this sense the much-vaunted concern about evolutionary constraints is effectively bypassed.

The difficulties encountered in encapsulating convergence are also exemplified by the problem of deciding how many really important turning points there are in the history of life and, if they had been either different or had not occurred, how distinctive would the course of life have been? This problem may not be intractable, but the problem of facing an indefinite regress of questions (e.g. is the evolution of mitochondria essential for intelligence?) and the unresolved difficulties of evolutionary inherency, especially at the molecular levels (e.g. if rhodopsin, then inevitably eyes?) make it very difficult ever to identify a starting point. If DNA and proteins are universal, then this might provide the ultimate bedrock of biological probabilities; in other words if given molecules can operate only in given ways, then perhaps this largely predetermines the emergence and nature of complex structures, such as eyes, brains, and behaviours.

A final point also needs to be addressed. It might be claimed that convergence is too elusive a concept to have any real validity: what after all does 'similar' really mean in a biological context? One response is to reject any overarching scheme, and deal with matters on a case-by-case basis. Much of this book can, of course, be read in just such a fashion. My overall approach, however, is to see in the recurrent emergence of biological properties such as intelligence, memory, and self-recognition – all, to be sure, based on prior evolutionary events but themselves convergent – a programme that is more interested in the definition and probability of complex states than the precise history that led to any particular example. Not that histories are without interest; very much the reverse, because these trajectories are a proof of evolution as a process. They do not, however, address directly the likelihood of the rise or maintenance of complex states in an adaptive framework. In this sense the study of convergence is a return to a nomothetic biology.

What then are the implications of convergence? First and foremost is that the various examples I have given will provide no comfort for the 'creation scientists', because in their various ways they provide compelling examples of the reality of organic evolution. Thus we see that the same ends may be arrived at along various, and sometime wildly different, routes. Correspondingly, very seldom is the convergence so exact as to make the organism or structure indistinguishable. Discerning the nuances of difference and the paths followed are in themselves highly informative about the evolutionary process. As I have already emphasized, however, our fascination with historical contingencies – the Caliph of London?, the Jesuit Centre for Astronomy and Mathematics in Kyoto?, the coronation of an Aztec king at Westminster in 1412? – are almost irrelevant in organic evolution. Convergence shows that we can provide first-order predictions of the emergence of important biological properties on Earth, and by inference elsewhere.

The realities of convergence have four implications for evolution, specifically the inescapable need to consider the themes of adaptation, trends, progress, and (a topic that has been surprisingly neglected) whether evolution can exhaust, at least locally, its potential. All should be unremarkable, yet for what seem to be effectively ideological reasons have been subject to sustained attack, notably by S. J. Gould. In the case of adaptation, for example, the notion of spandrels[96] has had an entirely disproportionate influence. It would be strange if organic 'architecture' lacked its equivalents to the spandrels (or pendentives) of a building, but are they of anything more than passing interest? As has been repeatedly pointed out, a failure to devise an adequate test for adaptation may tell us more about the shortcomings of the investigative method than the reality or otherwise of adaptation.

This is why convergence matters, because, as a number of workers have already emphasized, its very ubiquity confirms, as if there were ever any doubt, the reality of evolutionary adaptation.[97] For example, writing on the topic of ecomorphological diversification in freshwater fish, Kirk Winemiller[98] remarked, 'The concept of ecological convergence and independent evolution of equivalent ecomorphotypes deserves special attention because of its implications for general ecological and evolutionary theory,' and he continued, 'the remarkable qualities exhibited by several convergent pairings makes the phenomenon difficult to dismiss as arising from stochastic processes.'[99] Rob Foley[100] is more trenchant: 'Convergence is perhaps the strongest evidence for adaptation, and indicates that despite the contingent factors that lie at the base of evolutionary episodes, the rules of survivorship and reproductive success govern the final outcome.'[101] Earlier I introduced the many convergences in the subterranean mammals, and it is no accident that Eviator Nevo's masterly overview[102] of this topic is equally a renewed prolegomenon on behalf of the importance of the adaptational explanation. Using multiple examples, he proclaims that 'the adaptationist programme is as vital as ever.'[103]

Much as I disagree with the world picture of Richard Dawkins, with its questionable genetic reductionism and etiolated secular pieties, his explanations and enthusiasm for the reality of adaptation are of great value.[104] This is the way the world is, and in one way the only interesting question is how different could it be, in terms of both mechanism and end results? So far as the former is concerned, the much-vaunted plurality of alternative evolutionary mechanisms appears, at best, to be of marginal relevance. Consider, for example, the notion of 'historical burden', the constraints of past 'decisions' that guide, restrict, and perhaps even stymie a phylogenetic 'career'. That such constraints exist is undeniable, but what is far more interesting is the way in which organisms repeatedly 'get round' these problems, which is why convergences are ubiquitous. This is neither to suppose that everything is possible, nor to deny the reality of contingent happenstance in biological history.[105] Even to acknowledge the realities of convergence is not to imagine that every organism is 'trying' to evolve into a human. The Earth's biosphere is patently a product of divergences, but as this book emphasizes there is an underlying structure that imposes limits and delineates probabilities of outcomes. Some evolutionary trajectories point to increasing complexity, but simplification can be just as prevalent. Thus, there are a number of excellent examples from among the animals of what appears to our eyes as regressive evolution. These include the flatworms, whose relative simplicity was until recently thought to represent a rather primitive arrangement from which more advanced phyla emerged.[106] More extreme reduction is seen in the curious dicyemids, parasitic occupants of the kidneys of cephalopods,[107] and what is effectively a regression to a protistan state has occurred in the myxozoans.[108]

The organic world is a plenitude and a marvel, but it still has a rational structure. Simplification will arise in its own adaptive circumstances, but so, too, will complexity. And of course there are different solutions to the same problems, as we have seen in the case of the eye. When we learn, however, how this optical convergence extends to the nose of the star-nosed mole, or that the olfactory systems of insects and vertebrates are effectively the same, then we get some inkling as to how wonderfully varied our world is, but how underpinned it is by deeper commonalities. It is a truism that the general acceptance of the Darwinian explanation has not lessened the fear of many that our animal origins erode, if not deny, our human uniqueness. Yet, to anticipate the next chapter, there is another view that is more optimistic. In this view of life all share the terrestrial Creation, but we need to acknowledge that not only does our unique knowledge reveal a transcendence in wholly remarkable ways, but it also enables us to understand how the emergence of sentience is imprinted in the evolutionary process. The implications of how we choose to use the natural world, not least the other sentient species, will be obvious.

That the motor of evolution is adaptation seems incontestable, yet I also wonder if we appreciate the full range of the adaptive framework, especially with respect to molecular biology. There is, for example, the perhaps surprising sensitivity of so many molecules to change, so that even a single substitution may make the molecule effectively non-functional.[109] Even trivial differences can be disastrous. In a similar way, proteins with markedly different functions may differ only in very minor ways. Conversely, there are certain types of molecule that exist as a wide variety of variants, known as isoforms. Do they each have more or less the same function, the differences between them being little else than a rococo embellishment? Seemingly not. In the case of one protein family a total of 153 isoforms has been detected, each of which seems to have its own functional specificity.[110] The various examples of molecular convergence discussed above (p. 295) are also relevant in providing examples of adaptation to function, yet it is seldom that the investigator or commentator can resist expressing surprise. This is not because the molecules lack a function; far from it. Rather it is the difficulty of determining how a molecule evolved to its present precise configuration. It is partly for this reason that evidence for selection at the molecular level continues to attract particular interest.[111] The importance of this is, of course, not lost on those trying to bio-engineer molecules, and it is also of interest because again it allows a rerunning of the evolutionary tape, with predictable results.

CONVERGING TRENDS

This ubiquity of convergence and the likelihood that the great majority of the examples are almost certainly the result of selective processes operating in the context of adaptation has the obvious corollary that common evolutionary destinations presuppose specific, and restricted, trajectories. To mainstream evolutionary biology the existence of trends is entirely unremarkable.[112] For example, Christine Janis and John Damuth in a wide-ranging review of mammalian evolution and convergence remark, 'We regard the widespread occurrence of detailed ecomorphic convergences to be prima facie evidence for sustained, adaptive, phyletic trends at lower hierarchical levels,'[113] and in their conclusions go on to say, 'It should be clear at this point that we feel that the adaptive component has been dominant in historical trends observed in the Mammalia.'[114] To be sure, like adaptation, an apparent failure to demonstrate a trend may tell us more about the method employed than about the non-existence of the trend. Yet trends imply directionality, and perhaps progress (note 112). It is hardly surprising then that in his decades-long campaign against adaptation, S. J. Gould sought to demolish such ideas, arguing in particular that we would do better to look at the changes in variance as lineages evolve,[115] with the corollary that the trend is no more than a diffusion away from a restraining boundary.

So are trends with a directional component that might arise from a persistent pressure, such as natural selection, not only real but sufficiently common to be important? The ubiquity of convergence would argue so, but in this context I am more interested in directionality in evolution *per se*. Take, for example, what is probably the best-known evolutionary trend, that of size increase. This is known as Cope's rule. Yet, for the most part the reality of this rule has been

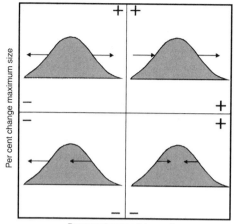

FIGURE 10.1 A graphical depiction of evolutionary changes in body size during geological time. Upper right represents the strict case of Cope's Rule, with a persistent trend towards larger size. The opposite case is shown in the lower left, while increases and decreases in the variance of body size are shown respectively in the upper left and lower right. (Redrawn with permission from fig. 10.5 of 'Body size and macroevolution' by D. Jablonski in *Evolutionary paleobiology: in honor of James W. Valentine*, edited by D. Jablonski, D. H. Erwin, and J. H. Lipps (The University of Chicago Press, Chicago and London) © 1996 by The University of Chicago, and with the permission of the author.)

a vague articulation rather than a specific set of hypotheses: after all, if it occurs all over the place then what is at issue? David Jablonski, however, injected rigour into this discussion, drawing upon his extensive knowledge of fossil marine molluscs from the Cretaceous.[116] Concerning changes in size through geological time, there are four broad possibilities (Fig. 10.1), to which can be added the character of no change.[117] Unsurprisingly, in Jablonski's analysis all four examples of body-size change occur, of which overall size increase, i.e. Cope's rule, is found in about 30% of instances, while an increase in variance (which will include size increases) accounts for another 28%. Interestingly, only a small proportion (c. 4%) shows no change, while size decreases also occur (c. 28%). So far as these molluscs are concerned, Cope's rule is not universal – why should it be? – but it is highly significant.

So, too, in a terrestrial context, when the overall history of Tertiary mammals is examined there are striking increases in body sizes, consistent with Cope's rule.[118] As John Alroy remarks in this instance, 'The basic pattern is overwhelming ... Newly appearing species are on average ... 9.1% ... larger than older congeneric species, a highly significant difference ... The only clear-cut hypothesis that predicts such a pattern is the most narrow and deterministic interpretation of Cope's rule; namely, that there are directional trends within lineages.'[119] And the birds? Here an investigation[120] of whether the evolution of body size was non-random concluded that there are 'many more instances where a large taxon putatively originated from a smaller one than vice-versa.'[121] These results, concerning the applicability of Cope's rule to mammals and birds, had already been anticipated in various ways. For example, Brian Maurer and his colleagues[122] emphasized how the evolution of body size and its tendency to increase arises as a result of the interplay between the processes of natural selection and differences in the relative rates of speciation as against extinction. This, in itself, is highly significant because it suggests that microevolution and macroevolution are not decoupled; but more importantly in this context Maurer and his colleagues robustly conclude, 'We suggest, however, that Gould's ... emphasis on randomness be replaced with an emphasis on deterministic outcomes that result largely from the role of ecological processes in speciation and extinction.'[123]

Evolutionary convergence is, therefore, consistent with what most biologists find entirely unremarkable: the Darwinian world of adaptation and trends. What has, in my view surprisingly, received rather less attention is the extent to which any particular 'design space' becomes saturated. Broadly, given the exploratory nature of Darwinian mechanisms, this could be expected, and such examples as that of 'skeleton space' (p. 117, Chapter 6) are consistent with a saturation (or exhaustion) of possibilities. But is this generally true? In an outstanding survey Peter Wagner[124] demonstrates that this does indeed appear to be a general principle. Effectively, as a cohort of evolving species (the clade) traverses geological time, so convergences come to predominate. This obviously has important consequences for the ability to recognize ancestral forms (unless the fossil record is taken into account) and the reliability of phylogenies if, indeed, convergence is rampant.

There is, however, a more general principle because the exhaustion of evolutionary possibilities might be because of the particular biological constraints of a given form (metaphorically, the clade has run out of things to 'do'), but it might equally reflect 'persistent selection favoring similar general forms.'[125] Wagner's study, which encompasses a wide variety of fossil clades, was concerned with their specific histories, which led consistently to the exhaustion of potential: in each case the tape of life was not only rerun, but rerun repeatedly. In a wider context, and as has already been emphasized, this does not negate the self-evident reality of continuing diversification. It might be objected that this undermines the principle of predictability as inferred from the ubiquity of convergence. What evolution cannot do is see into future diversification so far as the envelope of possibilities is concerned, although it can be equally sure that a great deal of what does one day evolve will have emerged in parallel circumstances, in other times and places. What we can also say is that whenever the known edge of the evolutionary envelope is reached, be it in terms of intelligence or agriculture, then it will be explored independently several times.

In their own ways the realities of adaptation and evolutionary trends (and the exhaustion of possibilities) excite little comment in the mainstream of biology; why should they? Not everything need be adaptive, nor are random walks by any means impossible, but are they effectively peripheral to the argument? And what about evolutionary progress, that term that S. J. Gould[126] gently refers to as 'noxious'.[127] Simply because evolution has delivered us to a point where only now can the word 'progress' make any sense, need not mean that it either has no relevance to the human condition or that it lacks an evolutionary reality. That the bacteria are still with us, and that without them the planet would soon grind to a halt in the absence of their recycling abilities, misses the point. Neither is progress a question of the sheer number of species, nor the supposed number of body plans. What we do see through geological time is the emergence of more complex worlds. Nor is this a limiting view. It might be premature to suppose that even the bacteria of today are some sort of 'honorary fossils', unchanged relics from the Archaean pond-scum. Nor need we imagine that the appearance of humans is the culmination of all evolutionary history. Yet, when within the animals we see the emergence of larger and more complex brains, sophisticated vocalizations, echolocation, electrical perception, advanced social systems including eusociality, viviparity, warm-bloodedness, and agriculture – all of which are convergent – then to me that sounds like progress.

A POSSIBLE RESEARCH PROGRAMME

Perhaps I am too optimistic, but I suspect that there is already a sea change under way with respect to the realities of convergence[128] in the context of adaptational explanations as against appeals either intrinsic to processes (e.g. phylogenetic constraint) or to random processes. What may be less appreciated is that the ubiquities of convergence are clear enough evidence that life has a peculiar propensity to 'navigate' to rather precise solutions in response to adaptive challenges. I would suggest that one such solution is manifested in a biological property that we choose to call 'mammal-ness'. So, too, within this 'zone' there are more localized solutions, one of which is 'ape-ness'. Mammals and apes (or any other biological entity) have, of course, arisen by way of specific historical trajectories, but in these (and many other) instances the various convergences on mammals and apes that are documented above indicate that although any history is necessarily unique, the resultant complex end form is not simply the contingent upshot of local and effectively random processes. On any other suitable planet there will I suggest be animals very like mammals, and mammals much like apes. Not identical, but similar, perhaps surprisingly similar.

This view of life is, of course, by no means the first attempt to arrive at a general biological theory that might have potentially universal applicability.[129] Nevertheless, it offers some intriguing clues that could mark a first step towards the definition of galactic-wide niches.[130] In essence the aim will be first to define a series of hyperdimensional 'boxes', based on the combinatorial immensity of the relevant variables that together encompass all the alternative possibilities. Such 'boxes' range from protein 'hyperspace' to intelligence and even societal 'hyperspaces'. With respect to the last example,[131] Lee Cronk reminds us not only of the universalities shared by all human societies, together a long and impressive list, but also that in terms of combinatorial 'hyperspace' it is possible to envisage $c.10^{53}$ alternatives, of which less than 10^3 are known. Cronk goes on to remark that the figure of 10^{53} can be reduced by collapsing the number of variables. He writes, 'If we recalculate the total number of possible combinations using only these variables, there are still over one hundred million possible combinations. Although this is a lot less than $[10^{53}]$ ... it still suggests that only a tiny fraction of the points in ethnographic hyperspace are occupied.'[132] Such an example makes it clear that the known zones of occupation must be an infinitesimally tiny fraction of the total available 'space'. The current consensus is that any such zone of occupation has been arrived at by chance, and there is a very large number of alternatives, probably well in excess of the number of habitable planets in the galaxy. Accordingly, on this basis, it is very unlikely that the occupants of one planet will look remotely like those of another. The phenomenon of evolutionary convergence indicates that, to the contrary, the number of alternatives is strictly limited, with the interesting implication that the vast bulk of any given 'hyperspace' not only never will be visited during evolutionary exploration but it never can be. These are the howling wildernesses of the maladaptive, the 99.9% recurring of biological space where things don't work, the Empty Quarters of biological non-existence.

If this is correct then it suggests that an exploration of how evolution 'navigates' to particular functional solutions may provide the basis for a more general theory of biology.[133] In essence, this approach posits the existence of something analogous to 'attractors',[134] by which evolutionary trajectories are channelled towards stable nodes of functionality.[135] This, of course, begs the question as to how the vastness of a given biological 'hyperspace' is actually navigated. This is because the number of potential 'blind alleys' is so enormous that in principle all the time since the beginning of the Universe would be insufficient to find the one in a trillion-trillion solutions that actually work. This echoes the common criticism, exemplified by Fred Hoyle's disbelief that a tornado passing through a junkyard could construct a jumbo jet, that there is anything like enough time to arrive at biological functionality. Life, it is claimed, is simply too complex to be assembled on any believable timescale. The stock response is to invoke a million monkeys typing alternatives, with the invisible hand nudging the myriad of efforts towards the correct Shakespeare sonnet (or whatever). This really misses the point, first because it presupposes that the correct version is known all along, and secondly because it fails to tackle the problem of the almost illimitable size of biological 'hyperspaces'. The metaphor of navigation, however, suggests a more fertile research programme than the mass employment of monkeys. This is because it will attempt to explain, not only the preferred trajectories to the optimum, but also evolution's uncanny ability to find the short cuts across the multidimensional 'hyperspace' of biological reality. It is my suspicion that such a research programme might reveal a deeper fabric to biology in which Darwinian evolution remains central as the agency, but the nodes of occupation are effectively pre-determined from the Big Bang.

One such node is, of course, that of the humanoid, and from the present evolutionary perspective we are undeniably unique. Yet, as I have already argued, if we had not arrived at sentience and called ourselves human, then probably sooner rather than later some other group would have done so, perhaps from within the primates, perhaps from further afield, even much further afield. Yet it is our prerogative to explore the natural world, and thereby document evolution, and in the footsteps of Darwin marvel at its diversity. These discoveries have not been without their cost. Not only are our animal origins plain to see, but so, too, it is now supposed that with our origins revealed this must banish any religious instinct: what was almost universally believed is now to be seen as an immense delusion. Now is the time, it is proclaimed, to adopt wholeheartedly the naturalistic view of life, not only to expel the mysteries, but to deride the ignorant and mould the world as we will. In the next, penultimate, chapter we shall see that things are not necessarily so simple.

11 Towards a theology of evolution?

Let us suppose that I am an immensely successful biochemist, and happen to be engaged in experiments involving gene manipulation. A couple of years ago I was attending a conference – keynote speaker, naturally – when I fell into conversation with a curious individual, who for some reason seemed much older than he actually looked. As we talked, it seemed we were walking across a plain of infinite dreariness, but his voice, his demeanour, how can I describe it? He knew all about our work, and as conversation progressed, gave me remarkable hints as to some avenues of research we had somehow overlooked. I was enthralled, and as we parted he remarked, 'I am sure we will meet again.' The next day, however, he had vanished, and checking at the registration desk, I was puzzled to find no record of my companion. It was all a little eerie, but the hints were sufficient. Now my team and I have managed to reconfigure a gene that will allow animals, and the poorer humans, to digest cellulose directly. Before long, in Bangladesh and Somalia, the main meal will be recycled newspapers. There is, unfortunately, just one small side effect, and that is it is very likely (I can explain the details if you have time) to induce childhood cancers in about one per cent of the population, especially if the individual happens to live in a deprived environment. Of course, the gene is patented, and in strictest confidence I can reveal to you alone that the biotech company, OmegaPoint, has the product ready for immediate marketing. Great riches beckon; surely I am to be congratulated?

Do I detect a sense of unease,[1] but pray why? Surely it is most regrettable that a vast proportion of the world's biomass is tied up as indigestible cellulose,[2] and at present the only simple route to the hamburger is via the bacteria living within the rumen of the millions of cattle contentedly grazing on thin grassland, once rain forest. It is, of course, only these bacteria which are capable of breaking down the refractory plant cellulose and so release compounds that elsewhere in the cow will ultimately end up as beef mince. Nothing, of course, will go to waste: spinal cord and the rectum have their uses, while anything else, in the best traditions of agrobusiness, will be ground up and returned to the cattle themselves. And if this hypothetical gene[3] has side effects, what matter? The vast bulk of scientific research undertaken by the biotechnology companies is subject neither to peer review – the accepted norm anywhere else in science – nor available for publication. That link to childhood cancers will take years to 'prove', and in reality most probably never will be, given both the medical inadequacies in much of the Third World and – well – facts that are deemed 'commercially sensitive'. And does not the race go to the best and fastest, which in the West we equate with the accumulation of stupendous amounts of money? So, am I not using the talent bestowed by evolution for the best, feeding the hungry – yes, yes, on newspaper – and making myself very, very rich?

AN EVOLUTIONARY EMBEDMENT

It is self-evident that whatever our peculiarities as humans, we are embedded in the natural world, and just as clearly we are one product of an evolutionary process that began about four billion years ago. In itself, the mechanism of evolutionary change is so unexceptional as to be almost trivial. As Martin Carrier[4] has written, 'Darwinian theory plays a role in evolutionary biology that is analogous to the one Newtonian theory plays in celestial mechanics. It provides the mechanism of change; it specifies the law-governed processes that determine how species develop and adapt in a possibly changing environment.'[5] Despite this simple process most biologists will freely acknowledge that both the routes and the products of evolution are profoundly fascinating. I have already alluded to a few of these, such as the problem of the origin of life itself. Probably of equal moment is to discover how it is that proteins fold so effectively and quickly. Evolution also presents what Denis Duboule and Adam Wilkins[6] have termed 'bricolage', that is the surprising co-option and redeployment of biological material for unexpected uses. An excellent example is the crystallin proteins of the eye lens, which as discussed in Chapter 7 (p. 166) are routinely and independently co-opted from heat-shock and other stress-related proteins that evolved thousands of millions of years before any eye could see.

Biologists also have, in the true Darwinian spirit, immense admiration for the jury-rigging of biological design, whereby co-option and modification lead to the functioning whole. And, if they are honest, they may feel a sense of unease about the fluidity and grace of adaptation. It has an almost uncanny sense of precision and balance, which humans achieve only rarely in technology or art. Not only does one admire the pervasiveness and ubiquity of integrated organic complexity, but one cannot help but be impressed by its sensitivity, in some circumstances, to minute changes so that sometimes even a single amino acid substitution in a protein can lead to a major change of functionality, if not catastrophe. And for some – probably many – when we review the broad sweep of evolution and greet first the flickering and then the full emergence of consciousness, that is all there is to say. Science must work in a naturalistic framework, and again many appear content to live with this arrangement. Roll on that cellulose-digesting gene, and let us treat the world as open to unlimited manipulation.

Yet, there are nagging doubts. Yes, it may all be due to a few misfiring neurons, perhaps an extra dollop of neuropeptide or whatever, but the fact remains that humans have an overwhelming sense of purpose. As a species we are strangely comfortable to find ourselves embedded in a teleological matrix. So the intention of this chapter is to begin to see whether the idea of a *telos*[7] is redundant, to ask if some of our predecessors who saw their religious faith either ebb or haemorrhage were both misinformed and over-pessimistic, and to enquire whether some common ground can be regained.

One clue is surely our admiration for moral greatness. Rather, however, than argue or defend any particular individual, although there are many such men and women, let us recall the cosmic view of G. K. Chesterton.

> Reason and justice grip the remotest and the loneliest star. Look at those stars. Don't they look as if they were single diamonds and sapphires? Well, you can imagine any mad botany or geology you please. Think of forests of adamant with leaves of brilliants. Think the moon is a blue moon, a single elephantine sapphire. But don't fancy that all that frantic astronomy would make the smallest difference to the reason and justice of conduct. On plains of opal, under cliffs cut out of pearl, you would still find a notice-board, 'Thou shalt not steal.'[8]

Let me conclude this introductory section with two other remarks. The first is to the effect that contrary to received wisdom I certainly do not consider religions and metaphysics to be aberrations of the superstitious, delusions of those still enmeshed in medieval credulity, unwilling to face the boundlessly happy future; a future that, strange to say, is always just around the next corner. More specifically, although the more we explore this world the better is our understanding, this is also accompanied by a growing sense of its extraordinary strangeness and beauty. Darwin at least was no stranger to these feelings, but, as we shall see, in growing older he retreated into a gloomy agnosticism, and his world became stretched and thin. And Darwin, of course, was (and even more so today, is) not alone in this growing sense of futility. And yet just as many are possessed of a strong teleological instinct, and in the words of Arthur Peacocke[9], 'Somehow, biology has produced a being of infinite restlessness, and this certainly raises the question of whether human beings have properly conceived of what their true "environment" is.'[10] More often these pangs of want turn to a dissatisfaction that in turn seems to lead only to the futile and sterile, but songs and stories tell us this need not be so.

DARWIN'S PRIESTHOOD

As I have already indicated, such views as those expressed by Peacocke and others may strike a chord in places, but I suspect that they will also be widely regarded as quaint and antiquated. Indeed these views are under full-scale assault by a group that allows a unique priority to Darwinian mechanisms and, in most cases, the primacy of the gene. These ultra-Darwinists are highly prominent, not least because the shoulders on which they stand include Huxley, Simpson, and Mayr, all giants in their various ways within the Darwinian synthesis. Let me make some initial observations, which in today's climate of accommodating pluralism and relativism will, I suppose, seem deeply unfair. I am driven to observe of the ultra-Darwinists the following features as symptomatic. First, to my eyes, is their almost unbelievable self-assurance, their breezy self-confidence.[11] Second, and far more serious, are particular examples of a sophistry and sleight of hand in the misuse of metaphor, and more importantly a distortion of metaphysics in support of an evolutionary programme. Consider how ultra-Darwinists, having erected a naturalistic system that cannot by itself possess any ultimate purpose, still allow a sense of meaning mysteriously to slip back in. Thus, the philosopher of science John Greene[12] remarks,

> Not all of the champions of the modern synthesis have been as open as [Julian] Huxley in acknowledging the religious aspect of their devotion to evolutionary biology, but most of them, especially those who reject religious and philosophical approaches to the problem of human duty and destiny, manage to smuggle in by way of simile and metaphor the elements of meaning and value that their formal philosophy of nature and natural science excludes from consideration.[13]

Despite this, such scientists have no foundation for their reaction against pointlessness other than the not unworthy and intuitive sense that the world should be built as it is; embedded in the Universe are not only neutrons but such edicts as, to echo Chesterton, 'Thou shalt not steal.' Greene has other trenchant comments that are also surely apposite. As he remarks, in the field of evolution the term 'progress' must be value-free and can only mean 'survival'. He continues, 'One would like to feel optimistic about the scientific mythology that has grown up around the theory of evolution, but it is hard to do so.' And he reminds us that in the hands of some practitioners the analysis of evolution is more like that of a myth, and one that is intellectually dishonest, 'employing teleological and vitalistic figures of speech to describe processes that are advertised as "mechanistic" and pretending to derive from evolutionary biology values that stem from classical, Judaeo-Christian, and Enlightenment sources. It deifies science, denigrates philosophy and religion, and panders to Western culture's penchant for regarding science and technology as the guarantors of indefinite progress toward some hazy but glorious future paradise.' And Greene pointedly continues, 'Worse yet, it fosters dreams of genetic manipulation and control designed to reshape imperfect human nature according to some scientistic ideal.'[14]

Third, as has often been noted, the pronouncements of the ultra-Darwinists can shake with a religious fervour. Richard Dawkins is arguably England's most pious atheist. Their texts ring with high-minded rhetoric and dire warnings – not least of the unmitigated evils of religion – all to reveal the path of simplicity and straight thinking. More than one commentator has noted that ultra-Darwinism has pretensions to a secular religion, but it may be noted that, however heartfelt the practitioners' feelings, it is also without religious or metaphysical foundations. Notwithstanding the quasi-religious enthusiasms of ultra-Darwinists, their own understanding of theology is a combination of ignorance and derision, philosophically limp, drawing on clichés, and happily fuelled by the idiocies of the so-called scientific creationists. It seldom seems to strike the ultra-Darwinists that theology might have its own richness and subtleties, and might – strange thought – actually tell us things about the world that are not only to our real advantage, but will never be revealed by science. In depicting the religious instinct as a mixture of irrational fundamentalism and wish-fulfilment they seem to be simply unaware that theology is not the domain of pop-eyed flat-earthers.

So does this matter? The world ticks along, and someone, somewhere is presumably busy trying to manufacture that cellulose-digesting gene. As a day-to-day activity science is highly pragmatic, but it also makes much wider claims to describe the world as it is. But to assume that science itself can produce or verify the truths upon which it depends is, as many have pointed out, simply circular. On this basis and as a human activity science is ultimately imperilled. The discoveries of science, as is also widely acknowledged, are not short of ethical and moral implications. Nowhere is this more true than in biology, and it is pertinent to make some reference to some of the real giants of the field as the implications of evolution sank in: Darwin, the Huxleys (Thomas Henry and Julian), and more uneasily Haeckel. Here I offer only a series of snapshots. Concerning Darwin and Huxley, there is, of course, a huge literature, but as above I draw especially on the work of John Greene.[15] This is mostly because his admiration for these individuals is also set in a metaphysical framework that declines, courteously, to be browbeaten by naturalistic dogmas. Darwin, the chief architect of evolution, whose genius is not in doubt, had a strongly theistic background, and as a young man had an admiration for Paley's arguments for biological design reflecting the hand of the Deity. Yet, as is equally well known, when the implications of his theory on evolution began to sink in, he descended (and I use that word advisedly) into a sort of pantheism, dogged by half-articulated fears, if not terrors. For him belief seems to have slowly ebbed, almost against his will.

Darwin was, to put it mildly, a retiring figure, and for a more public, if not melodramatic, view of another trek away from Damascus then it is inevitable that we turn to his staunch supporter, Thomas Henry Huxley. Interestingly, for Huxley, the adoption of the Darwinian world-picture also took on more than a tinge of theology. Thus the religious imagery that accompanies Adrian Desmond's *Huxley: From devil's disciple to evolution's high priest*[16] is revealing,[17] and while it acknowledges Huxley's secular enthusiasm, perhaps it also distorts the sort of man he really was. It seems that, at first, Huxley was content that any sense of an underlying meaning could be eroded from the fields of biology and evolution. These regions of science fell into a realm of blind chance and unconscious agony, but one that at least allowed a deistic notion that conceded, however feebly, a cryptic arrangement of forces and laws. But in the end this, too, was insufficient: Huxley had to become something for which he had also to coin a name, that is, an agnostic. This was not the decorous scepticism with which the term is usually associated: for Huxley his position reflected genuine and unresolved intellectual doubt. Even so, there is little evidence that he was willing to entertain, let alone discuss, serious theological statements, and he had an ill-disguised contempt – no doubt well earned in some cases – for the clerics. Yet, Huxley was a man of transparent goodness and deeply felt morality, and was happy to recruit such Old Testament prophets as Micah to his cause, provided of course that they were shorn of any religious dimension. Even so, it is not clear that living in a metaphysical vacuum brought him much peace, and towards the end of his life it seems that he dimly discerned that the new science that he had helped so ably and energetically to popularize was opening a Pandora's box.

Thomas Henry Huxley was not the last of the Huxley clan to struggle with the implications of evolution, yet arguably the vagueness never dissipated. Most notably his grandson, Julian Huxley, returned to the fray, but he adopted a view of evolution that was akin to what is referred to as process theology. Here in the younger Huxley's view the grand scheme of things was, as he wrote near the beginning of one work,[18] 'unitary; continuous; irreversible; self-transforming; and generating variety in novelty during its transformations',[19] sentiments echoed at the end of the same volume when he hoped that 'I have given you some feeling of the unity and sweep of the process ... some insight into its nature as a self-transforming process, constantly generating new patterns and novel qualities, building its future by transcending its past.'[20] And, as Greene notes, in his stimulating *Huxley to Huxley* (note 15), this view of life 'excludes from the process all these elements – aim, purpose, creative ground – which run counter to the positivistic grain of modern science and which alone could make such a process [i.e. evolution] intelligible.' Greene continues and concludes that 'The result is that Julian Huxley ... propounds the paradox that nature, though devoid of aim and purpose, yet moves towards ever higher levels of order and value.'[21]

As Julian's grandfather had dimly perceived, the naturalistic programme would open the door to the manipulation of a world from which meaning had fled. Thus the march 'towards ever higher levels of order and value' remained a chimaera. In reality, as Greene observes, 'The multiplication of devices and techniques for controlling the natural environment and influencing human behavior seems only to aggravate man's lack of control over the general course of events and to intensify the moral problems connected with human freedom ... the possibilities for evil multiply as rapidly as the possibilities for good, and history affords little assurance that men or women, either individually or collectively, will choose the general welfare of mankind in preference to immediate personal or national advantage.'[22]

HERESY! HERESY!!

Let me continue this brief review with two other episodes that in their different ways are almost comic, but in at least the first case have a much darker side. Otherwise each is very different. The first concerns Ernst Haeckel, the keeper of the Darwinian evangeliarium in Germany, and the second the farce of the Dayton 'Monkey Trial' in 1925. In the mind of Haeckel, Darwin was the greatest of heroes, and Darwinism the new beacon to lead the world from its benighted obscurantism. There is an amusing, at least to the English, story of Haeckel crossing the Channel and visiting Down House: a metaphorical, if not literal, genuflection between Teutonic adorer and English gentleman scientist. There was, however, no meeting of the minds. The German was effusive, gushing, bombastic, enthusiastic, and Darwin? well ...

Haeckel had seized upon the Darwinian explanation and vigorously and tirelessly promoted it in Germany. Not only was this done with books, but also with lavishly illustrated lectures. Concerning the latter, Daniel Gasman[23] refers to a poster for one such lecture in Berlin, and remarks how this example of Darwinismus-Kunst provided 'a sinister environment for a Darwinian Passion Play.'[24] It is, moreover, Gasman who has done much to reveal the rottenness at the heart of Haeckel's project.[25] As I remarked elsewhere[26] when drawing upon his work:

> The reality is very different and ... much darker. Haeckel's pursuit of Darwinism went far beyond any scientific formulation, even further than Herbert Spencer's uneasy rhetoric usually referred to as Social Darwinism. Haeckel's role as a spokesman for these malign influences was encapsulated in the so-called Monist league, of which he was the effective founder. This mish-mash enshrined a set of pantheistic beliefs of supposedly cosmic importance. Haeckel and his fellow Monists were dedicated believers in organic progress, the end-point of which, unsurprisingly, was the Aryan ideal. Nowhere was Haeckel's influence greater or more charismatic than in his book *The Riddle of the Universe*. Vastly popular, endlessly reprinted and translated, it nevertheless 'appealed to a pseudo-educated mind ... without much sophistication who had sought an authoritative yet simple account of modern science and a comprehensible view of the world'.[27] Behind the bearded sage and devotee of the little town of Jena, was an intolerant mind wedded to racism and antisemitism. After initial reservation, Haeckel's admiration of Bismarck and his autocracies grew. Predictably his support for the German conquest of Europe was fervid. Unrepentant of German aggression he died, still revered, in 1919. His farrago of ideas, however, not only lived on but found a warm reception with the Nazis. Just how much Hitler knew of Haeckel's actual work is not clear, but the influence of his philosophy is obvious. Eavesdropping at Hitler's table talk, with its hypnotic mixture of rant, bluster and threats all set in a half-baked philosophy, constantly echoes the monistical Haeckel.[28]

In the Scopes (or Monkey) Trial in Dayton, Tennessee[29] the issue was as much the long-standing loathing between the principals for the defence and the prosecution, respectively William Jennings Bryan and Clarence Darrow. Darrow was at that time the most famous lawyer in America, already notorious for his risky and ingenious defence of Nathan Leopold and Richard Loeb, two Chicago teenagers who abducted, bludgeoned to death, and then mutilated with acid a schoolboy and neighbour, a luckless fourteen-year-old known as Robert Franks. This murderous pair were bunglers of a high order, but had it seems undertaken the atrocity in the belief that they would escape detection. There was, ostensibly, no other motive. At the trial, Darrow undercut the prosecution by an unexpected change to a plea of 'Guilty', followed by a rhetorical harangue to the effect that Leopold and Loeb retained their innocence because of the environment in which they were nurtured, one of intellectual stimulation, but emotional starvation, and in the case of Leopold, the corruption of a young mind that inevitably results from being made to read Nietzsche. This libertarian argument, delivered as a thrilling speech that concluded with a quotation from Omar Khayyam, has plenty of resonances today. As one writer on the Scopes Trial, Kevin Tierney, notes, at the end of the speech, 'many in the audience were crying. Darrow had taken the case far beyond the bounds of reason and logic ... It was his most masterly oration, rousing his audience to display emotion openly beyond what the conventions allowed ... The spellbinder has cast his spell once again.'[30]

Thus Darrow embraced a moral perspective, as exemplified by the Leopold and Loeb trial, which led to Dayton and his battle against Bryan. Here was a very different man. No fool: as a Democrat, he had been quite close to winning the American presidency, and at the time of Dayton was America's greatest defender of fundamentalist Christianity. Like Darrow he, too, was a brilliant orator. And it was this engagement, between Bryan and Darrow, rather than either the teaching of evolution or its scientific truths, which lay at the heart of this trial. To be sure, the action began as a defence of the protection of civil liberties and the necessary separation of Church and State, and this was initiated as a test case by the American Civil Liberties Union (ACLU). To their dismay, Darrow effectively imposed himself, offering to waive fees in his determination to ridicule and thereby crush the forces of religious obscurantism and, quite incidentally, maintain his public profile. Before long, the ACLU effectively lost control of events as Darrow hijacked the circus for his own good purposes. Opposed to him, with Bryan as its senior spokesman, was a largely rural constituency that possessed a heartfelt, if largely inarticulate, belief in a fundamentalist Christianity that seemed imperilled on all sides by the secular agents. Kevin Tierney spells out the real nature of this trial:

> There was a further, more sinister animosity between Bryan and the defense. Bryan, failing though he was, was still probably the greatest public speaker then living in the United States. The defense team (including Darrow) wanted to pitch themselves against him so as to seize the crown. Darrow recognized Bryan's power over his audience. He was a worthy competitor, to be confronted in a final battle in which Darrow's superiority would be displayed. Thus the entire Scopes trial was colored by a personal rivalry as well as a difference of opinion.[31]

There were (and are) indeed serious objections to fundamentalism, rural or otherwise. But for that matter, so there were to Darrow's childish conception of theology, which was in its way was as dated and credulous as Bryan's.

The trial itself has further ironies, which while quite well known are seldom spelt out, especially by those who regard it as a test case between the shining uplands of science and the snarling religious reactionaries. This is not to defend the idiocies of legislation designed (then or now) to prevent the teaching of evolution or any other science, however uncomfortable the findings might appear to be. Societies that ignore what we discover do so at their peril, but if they imagine that on occasion the discoveries of evolution are neutral in their implications, again societies delude themselves. Even so, the technical defendant, John Thomas Scopes, was a physics teacher at the school in Dayton. It is not even clear if he taught anything on evolution, and his own knowledge of this area was vestigial. So far as Darrow was concerned, Scopes was a pawn, if not a stooge, and once the trial started, he was ignored. The trial was a simple test case. Scopes had received no warning from the School Board not to teach evolution in defiance of the State Legislature; the Board could not care less. The entire dynamic came first from the ACLU and thereafter from a ruthless Darrow. Again to quote Tierney: 'As Darrow had gained in years and confidence, he became not only merely an unbeliever, but a militant agnostic.'[32] Ultimately his animus against religion became almost obsessional, as in later years Darrow 'continued to sift through the religious outpourings of the day, intent upon uncovering literalism, absurdity, and contradiction with the same energy that a devotee of the faith might seek out heresy.'[33] Underpinning the beliefs of what Tierney has aptly described as 'the last village atheist'[34] was a fundamentalist Darwinism. Thus 'The theory of evolution was close to Darrow's philosophy about life in general, not merely to his anti-religious bent. He believed life to be cruel, and mankind to be unregenerate . . . Darwinism also had appealing deterministic aspects. No man could step aside from the march of history, which was inexorable and inevitable.'[35] Yet, in the long term, Darrow's rhetorical brilliance and intelligence proved a poisoned chalice. As the years slipped by he became 'more and more a performer, a mouther of scripts, and less a thinker . . . as he fed his appetite for glory while travelling from city to city for one-night stands.'[36]

The trial, which resulted in the jury being absent for less than ten minutes and Scopes receiving a $100 fine, achieved little, other than to increase the rancour and suspicion between the religious fundamentalists and the scientists. In America the skirmishing continues unabated, with one side unable or unwilling to comprehend the methods of science, and the opposing party all too often exhibiting a lofty arrogance, mingled with contemptuous disdain, which presupposes that any religious instinct is a mental aberration. The former know in their hearts that something is out of kilter, but their cause is hopeless. Not only are they overly simplistic and credulous, but they are not averse to selective quotation divorced from its context, and even more seriously outright twisting and distortion of the evidence. So, too, they seem unaware that theology involves rather more than scriptural inerrancy, especially with respect to the Creation myth. Such a polarization is not only regrettable; it is intellectually poisonous. For those wedded to sociological relativism the solution is to allot each their sphere of influence, but such apparent generosity merely conceals a strategy for sidelining religion and a road to philosophical incoherence. Despite the antagonisms, however, there are also attempts to find common ground between sciences and religions, most notably in the field of cosmology. In part because of the apparent absence of law-like properties, biology and evolution have remained largely excluded from this search for common ground, a stance that has been reinforced by the myth of contingent forces driving the evolution of life. Yet despite this, strange to say, biology and especially genetics have their own fundamentalisms. These, in turn, reveal a disquieting agenda that has curious echoes of the very systems they purport to despise.

GENETIC FUNDAMENTALISM

That biology can be co-opted for agendas, if not ideologies, that promise an ever-more-perfect future, albeit across piles of corpses, is evident from the lunacies adopted by totalitarian states. Such madness is, of course, a thing of the past – or is it? Now new distortions beckon, not least those to be allowed by assigning a protean malleability to life as engendered by genuflection to the primacy of the gene. Now the gene is all-powerful. Susan Oyama[37] describes genes as 'molecular agencies that are immortal, omnipotent, omniscient, and even immaterial'.[38] In a related vein, Peter Koslowski's[39] critique of this view, as it has been offered by Dawkins, defines an approach that 'concedes a faculty for aspiration, intentionality and consciousness to the genes. In doing so, he [Dawkins] falls into a genetic animism, which apportions perception and decision to the genes and oversubscribes by far to the efficiency and the speed of Darwinian selection mechanisms.'[40]

Such critiques differ greatly from the popular notion of the gene, which does indeed seem to be able to act as a universal agency. The spectacular examples of genetically controlled defects, not to mention the generation of ectopic monsters with hideously sprouting organs, reveal the potency of the gene. So, too, in examples of caste structure, especially in eusocial animals (p. 142), the paradox of sterile workers ceaselessly toiling for the benefit of the hive or nest is explicable by a proportion of their genes surviving[41] even though they themselves either cannot or will not reproduce.

A closer examination, however, reveals that these particular examples fall far short of allowing the gene *per se* a universal application. In her essay 'The gene is dead – long live the gene!' Eva Neumann-Held[42] argues that regarding the genes generally as particulate objects of heredity is hopelessly simplistic. The genes make sense only in a known context, but in reality to know in sufficient detail a context that will provide the sought-after predictability may be very difficult, perhaps impossible. That is why it is so misleading, if not dangerous, to speak of genes for, say, schizophrenia or aggression. Given the right set of prompts (who knows, perhaps plenty of junk food?) then the risk could increase – or decrease. And what else are we meant to expect?

Outside its cellular milieu the DNA is biologically inert, if not use-less. Genes may provide a switchboard for life, but the complexity of life will depend on something else: how the same genes may be re-cruited to make different products, how the developmental networks change and evolve, and how apparently trivial events such as gene duplication and protein isoforms open immense new territories for biological exploration. Life may be impossible without genes, but to ascribe to them powers of intentionality misses the mark.

Despite these qualifications, however, the hard-core view that claims primacy of the gene holds sway. In some hands, perhaps most notably those of the sociobiologist E. O. Wilson, it is a vehicle of unbounded faith both in its power and in terms of its implications, not least for the human prospect. This is spelt out, for example, in *Consilience*,[43] which is an extended belief-statement in an overarch-ing system where all will be explained – society, art, religion – by the gene. Wilson expresses himself with fervour and conviction, but a more dispassionate reading of *Consilience* leaves me more impressed by Wilson's faith in the argument, accompanied by leaps in logic, un-warranted assumptions, and over-simplification.[44] And the world pic-ture of genetic primacy has a well-known parallel, that is, the mental equivalents of genes. They too are all-purpose, if elusive, little things that are known as memes. Perhaps that irritating little tune that con-tinues to bounce around your head is a meme? One cannot but no-tice how trivial many of the examples presented are, unless it is to portray the sheer wickedness of religious beliefs inculcated into the brains of morally helpless humans. Happily, the alternative religions of consumerism and shopping are arrived at by the exercise of human dignity, unpersuaded by anything remotely like a meme. So, too, in a way reminiscent of the notion of 'junk' DNA, one cannot help but notice that the discussion of memes is often pejoratively associated with some notion of 'mind-parasites'. But memes are trivial, to be ban-ished by simple mental exercises. In any wider context they are hope-lessly, if not hilariously, simplistic. To conjure up memes not only reveals a strange imprecision of thought, but, as Anthony O'Hear[45] has remarked, if memes really existed they would ultimately deny the reality of reflective thought.

These views on genes and memes matter very much, because granting such molecular (or memal) hegemony puts us back firmly on the path towards the Abolition of Man.[46] And yet the rot started at an earlier stage. As John Greene opines:[47] 'To the very end, he [Darwin] failed to appreciate the morally ambiguous character of human progress. He failed because, like many social scientists today, he had no adequate conception of Man.'[48] As Greene also remarks, humans are very peculiar creatures indeed; clearly a product of evolution, yet a species that has, or has been allowed, to know mental states that transcend (so far as we know) any other sentience on the Earth. Again to quote Greene, 'science becomes pointless and even destructive unless it takes on significance and direction from a religious affirmation concerning the meaning and value of human existence.'[49]

Despite the sleights of hand, special pleading, and sanctimoniousness as the ultra-Darwinists attempt to smuggle back the moral principle through the agency of the gene, only the most hardened cases would suppose that a map of the genome will provide the blueprint of this millennium's equivalent to the Code of Hamurabi. And yet these myths of genetic determinism, set in a dreary world of reductionism, are being used to drive new agendas, most notably in eugenics. At present it is the natural world, which according to some, should be treated as a sort of genetic play-dough. Now vanished is the notion that the world we have been given might have its own integrity and values. Rather the prevailing view of scientism is that the biosphere is infinitely malleable. Again the moral high ground is hijacked on the assumption that all this is for our perceived good, although in reality the benefits are far more likely to fill the coffers of the corporations and erode the diversity of crop species, to be followed by who knows what? There is no doubting that in at least some cases this manipulation is possible. Two research, workers Temple Smith and Harold Morowitz,[50] remark 'As a consequence of reflective thought, we have today within our grasp the ability to assemble genetic combinations that have nearly zero probabilities of ever being sequentially assembled by nature.' Yet, these writers seem unenthused as they remark of this genomic programme, 'now a gambler's game [is being] played by those who may not be fully aware of the stakes'.[51]

What follows from the genetic meddling in maize and soon pigs, will, it may be safely assumed, be applied in due course to humans. It is interesting to recall that T. H. Huxley, in contrast to the ever-confident Galton, whom we met in Chapter 7 doing arithmetic by smell, drew back in horror from this eugenic prospect, arguing simply that no man could possibly know enough to decide. Let us recall, in the words of C. S. Lewis,[52] the prophetic voice of the head of the National Institute of Co-ordinated Experiments, better known as N.I.C.E. Listen to the recently ennobled, and hideous, Lord Feverstone:

'Man has got to take charge of man. That means, remember, that some men have got to take charge of the rest ... we'll get on to biochemical conditioning in the end and direct manipulation of the brain ... A new type of man.' Regrettably there are still a few dunderheads, 'old women of both sexes up in arms and yapping about humanity', but happily all this can be dealt with. Feverstone knows what's going on when he addresses the naive Mark Studdock: 'You are what we need; a trained sociologist with a radically realistic outlook ... We want you to write it *down* – to camouflage it ... It's nothing to do with journalism. Your readers in the first instance would be committees of the House of Commons.' Of course the dunderheads need to be manipulated also, but fortunately there is 'Jules ... a distinguished novelist and scientific populariser whose name always appeared before the public ... He's all right for selling [N.I.C.E.] to the great British public in the Sunday papers and he draws a whacking salary. He's no use for work.'[53]

A PATH TO RECOVERY?

The corrosive view that all in this world is to be bent to our pleasure or whim is hedged in reality with expediencies and half-truths, and in the view of many represents the royal road to catastrophe. So how might we begin to think about, let alone achieve, a Recovery? First, we need to recall the limits to science. It is no bad thing to remind ourselves of our finitude, and of those things we might never know. Practically, we should not be afraid to acknowledge that there are areas that Roger Shattuck calls *Forbidden knowledge*,[54] too dangerous in our present state of understanding to explore. At its simplest it is a precautionary principle, and more significantly a belated acknowledgement that the architecture of the Universe need not be simply physical. We should also recall, as if we needed reminding, that we are mortal and limited, and thus should remember that the old myths of unrestricted curiosity and the corruption of power are not necessarily fables.[55]

Second, for all its objectivity science, by definition, is a human construct, and offers no promise of final answers. We should, however, remind ourselves that we live in a Universe that seems strangely well suited for us. In earlier chapters I dwelt, all too briefly, on the paradoxes of the origin of life (Chapter 4) and the many peculiarities of the Solar System (Chapter 5) which seem to be prerequisites for our existence. On a cosmic scale it is now widely appreciated that even trivial differences in the starting conditions would lead to an unrecognizable and uninhabitable universe. The idea of a universe suitable for us is, of course, encapsulated in the various anthropic principles. These come in several flavours, but they all remind us that the physical world has many properties necessary for the emergence of life. Of these probably the best-known are those connected to the synthesis of carbon in the interior of stars, and the many strange properties of water (and ice).

Less widely appreciated, but of equal moment, is Howard Van Till's[56] insistence that 'It is not simply the *numerical values of certain parameters* that must be "just right" in order for life to develop. No, it's the entire formational economy of the universe that must be "just right". The full menu of the universe's formational capabilities must be sufficiently robust to make possible the actualization of carbon-based life ... I would argue that the *formational capabilities* of the universe are more fundamental than the *numerical values* of certain physical parameters.'[57] Not only is the Universe strangely fit to purpose, but so, too, as I have argued throughout this book, is life's ability to navigate to its solutions.

As is well known the anthropic principle, in whatever guise, has largely attracted the interest (or scepticism) of cosmologists and physicists. Biologists, on the other hand, have generally been content to take such features as carbon or water as givens, with life as an emergent inevitability on any suitable planet. But there are connections, because at the heart of the study of evolution are two things. One, emphasized throughout this book, is the uncanny ability of evolution to navigate to the appropriate solution through immense 'hyperspaces' of biological possibility. The other, equally germane and even more mysterious, is the attempt to explain the origins of sentience, such that the product of ultimately inanimate processes can come to understand both itself, its world, and, as I have already noted, its (and thus our) strange sense of purpose. We need also to remember that scientific explanations need not be all-embracing, and indeed it would be surprising if they were. As Michael Polanyi, a philosopher of science who took religion seriously, noted, other descriptions have their own power. In *Personal knowledge*[58] he writes:

> The book of Genesis and its great pictorial illustrations, like the frescoes of Michelangelo, remain a far more intelligent account of the nature and origin of the universe than the representation of the world as a chance collocation of atoms. For the biblical cosmology continues to express – however inadequately – the significance of the fact that the world exists and that man has emerged from it, while the scientific picture denies any meaning to the world, and indeed ignores all our most vital experience of this world. The assumption that the world has some meaning which is linked to our own calling as the only morally responsible beings in the world, is an important example of the supernatural aspect of experience which Christian interpretations of the universe explore and develop.[59]

So, at some point and somehow, given that evolution has produced sentient species with a sense of purpose, it is reasonable to take the claims of theology seriously. In recent years there has been a resurgence of interest in the connections that might serve to reunify the scientific world-view with the religious instinct. Much of the discussion is tentative, and the difficulties in finding an accommodation remain daunting, but it is more than worth the effort. In my opinion it will be our lifeline.

CONVERGING ON CONVERGENCE

The principal aim of this book has been to show that the constraints of evolution and the ubiquity of convergence make the emergence of something like ourselves a near-inevitability. Contrary to received wisdom and the prevailing ethos of despair, the contingencies of biological history will make no long-term difference to the outcome. Yet the existence of life itself on the Earth appears to be surrounded with improbabilities. To reiterate: life may be a universal principle, but we can still be alone. Whether or not this is literally true may never be established, and, as many of us have argued, it is far more prudent to assume that we are unique, and to act accordingly.[60]

Yet now we are faced with a special dilemma. The very scientific method that allows us to study the natural world, be it interstellar organic molecules or memory in dolphins, also gives us tools that treat the world as endlessly malleable, ostensibly for the common good but as often as not for the enrichment of the few and the impoverishment of the many. Such attitudes fly in the face of traditional wisdoms, and in part explain the existing antagonisms between scientific practices and religious sensibilities. Mutual misunderstandings, fuelled by naivety and ignorance, can only lead to warfare. Although science may emerge triumphant, it will be a pyrrhic victory; the conquered kingdom will lie in ruins, strewn across a plain of infinite melancholy. Constructive approaches are more difficult, and are usually viewed with contempt, but I believe promise far more. In essence, we can ask ourselves what salient facts of evolution are congruent with a Creation. In my judgement, they are as follows:

(1)   its underlying simplicity, relying on a handful of building blocks;

(2)   the existence of an immense universe of possibilities, but a way of navigating to that minutest of fractions which actually work;

(3)   the sensitivity of the process and the product, whereby nearly all alternatives are disastrously maladaptive;

(4)   the inherency of life whereby complexity emerges as much by the rearrangement and co-option of pre-existing building blocks as against relying on novelties *per se*;

(5)   the exuberance of biological diversity, but the ubiquity of evolutionary convergence;

(6)   the inevitability of the emergence of sentience, and the likelihood that among animals[61] it is far more prevalent than we are willing to admit.

Having already quoted G. K. Chesterton once in this chapter, let me return to this wisdom by way of a conclusion. As he writes,[62]

> Turning a beggar from the door may be right enough, but
> pretending to know all the stories the beggar might have narrated
> is pure nonsense; and this is practically the claim of the egoism
> which thinks that self-assertion can obtain knowledge. A beetle
> may or may not be inferior to a man – the matter awaits
> demonstration; but if he were inferior by ten thousand fathoms,
> the fact remains that there is probably a beetle view of things of
> which a man is entirely ignorant. If he wishes to conceive that
> point of view he will scarcely reach it by persistently revelling in

the fact that he is not a beetle. The most brilliant exponent of the egoistic school, Nietzsche, with deadly and honourable logic, admitted that the philosophy of self-satisfaction led to looking down upon the weak, the cowardly, and the ignorant.[63]

So the beetle, no longer the butt of Haldane's jibe,[64] is an example of the richness of a Creation. Whether we shall always remain 'entirely ignorant' of what a beetle (or a bat) thinks is open to discussion. Even if we do, the complexity and beauty of 'Life's Solution' can never cease to astound. None of it presupposes, let alone proves, the existence of God, but all is congruent. For some it will remain as the pointless activity of the Blind Watchmaker, but others may prefer to remove their dark glasses. The choice, of course, is yours.

12 Last word

The vehicle landed at 15.47 GMT, just over a mile from Kimmeridge, on the south coast of England. 'Just in time for tea?' murmured my companion, as we climbed through the long grass, insects rising in the summer air. There, already sitting on the ground, were the three extraterrestrials. As we joined them, I asked, 'Would you like some water?' 'Or perhaps something stronger?' suggested my companion. 'Thank you', came the grave reply. 'We ourselves are thirsty, on such a warm day. And maybe something for our plants?' The chlorophyll of the alien species blended well with the surrounding vegetation, its flowers a deep purple. When our visitor picked up one of the pots, it slipped and in catching it he grazed his finger. Red blood oozed to the surface. 'Haemoglobin, I suppose?' They nodded. Our hands clasped, both warm to the touch. It seemed superfluous to ask, but the beating of a vein hinted at the inevitable dual circulation system and arteries with their elastic proteins. One of them sniffed the air appreciatively; the world smelt beautiful as their and our nasal glomeruli registered the olfactory signals. As the swifts screamed overhead, the minute hairs in our ears and the auditory equivalents of the extraterrestrials acted in the same way, transducing the sound into a register of inner music. 'Observe the pointed wings of those flying animals – swifts, did you say? – clearly migrants, just as at home.' We strolled slowly back down the hill, towards the sea. Despite the steepness of the slope, the aliens were confident, their balancing mechanisms the same, their walking movements controlled by the same neural network. We stopped on the cliff edge; in the bay three dolphins moved westwards. 'Fast and effective swimmers, I see; what other shape could they possibly have?' All of us looked at them with common delight, through camera-eyes, lenses full of crystallins and retinas with opsin molecules making photons into sight. By now dusk was falling, Venus already bright in the sky. It was time to go home. 'Before you leave, may I ask where you are from? Was it a very long journey?' But we already had guessed. 'A long journey? Well, only in some ways. Where is our home? Why, the planet we call Earth, of course. Surely you already knew in your hearts that there is only one Earth?' We were, of course, looking at ourselves.

Notes

A note about the notes. The principal aim is to give citations, chapter and verse, to the many facts, interpretations, and sources I give. Many more could have been given, but I hope they form a useful introduction. Where a reference is repeated the original note is also cited. There are also a few cross-chapter citations.

PREFACE THE CAMBRIDGE SANDWICH

1. J. Maynard Smith's review 'Taking a chance on evolution' was published in the 14 May 1992 issue on pp. 34–6.
2. Quotation is on p. 34, column 4.
3. Quotation is on p. 35, column 1.
4. Quotation is on p. 35, column 4.
5. R. C. Lewontin's review ('Fallen angels') was published in the *New York Review of Books*, 14 June 1990, pp. 3–4, 6–7; quotation is on p. 7, column 1.
6. J. Maynard Smith (1992; citation is in note 1), p. 36, column 1.
7. R. C. Lewontin (1990; citation is in note 5), p. 7, column 1.
8. As with Maynard Smith, Lewontin's remarks are very thoughtful, and it is only fair to put the two quotations in their full context. Thus, in conclusion he writes: 'So evolution may be contingent only in a superficial and uninteresting way. The exact forms that have left descendants visible in fossil remains may indeed be accidental variants of a historically accidental process. But they may all be distinctions without a difference, superficial orthographic variants of a deep structure whose rules we have yet to uncover. A description of all the organisms that have ever been cannot decide the issue ... We cannot know the answer unless we have a theory of biological form that is deduced from some general principles of biological organization, rather than inferred from the collection of objects. Or it may be that no such principles exist, and that in this broadest sense, life has no meaning' (citation is in note 5, p. 7, column 1).
9. Thus the well-known false 'thumb', i.e. the pre-pollex (or radial sesamoid) of the giant panda is, as pointed out by F. Wood-Jones in *Proceedings of the Zoological Society of London*, vol. 109B, pp. 113–129, 1939, equivalent to the sixth digit of the mole hand. In this case the convergence of bone structure does not extend to function; in the moles it appears that the sixth digit serves to widen the hand and so assist in digging, whereas in pandas the 'thumb' is opposable and helps to strip bamboo leaves.
10. See Peter Ward and Donald Brownlee in *Rare earth: Why complex life is uncommon in the Universe* (Springer [Copernicus], New York, 2000).
11. See Stuart Ross Taylor in *Destiny or chance: Our Solar System and its place in the cosmos* (Cambridge University Press, Cambridge, 1998).

1. LOOKING FOR EASTER ISLAND

1. These are a group known as the Myxobacteria, and include the arborescent fruiting body of *Chondromyces crocatus*; see H. Reichenbach in the book *Myxobacteria II* (M. Dworkin and D. Kaiser, eds), pp. 13–62 Washington, DC, American Society for Microbiology (1993).
2. For the extreme accuracy of owl hearing see E. I. Knudsen and M. Konishi in *Journal of Comparative Physiology*, vol. 133, pp. 13–21, 1979; see also notes 265 and 266 in Chapter 7. While the owls have phenomenally acute hearing, so do some insects; see for example the

paper by P. Müller and D. Robert in *Journal of Experimental Biology*, vol. 204, pp. 1039–1052, 2001; see also note 305 in Chapter 7.

3. For this and other remarkable examples of long-distance navigation see F. Papi and P. Luschi in *Journal of Experimental Biology*, vol. 199, pp. 65–71, 1996.

4. See Simon Winchester's book *The river at the centre of the world: A journey up the Yangtze, and back in Chinese time* (Viking [Penguin], London, 1997), pp. 96–97.

5. But not necessarily; A. Eyre-Walker in *Genetics* (vol. 152, pp. 675–683, 1999) provides evidence of selection in the evolution of junk DNA of junk DNA, while W. Makalowski (in *Gene*, vol. 354, pp. 61–67, 2000) argues that it is more like a genomic scrapyard, to be pillaged as and when necessary. Others, such as R. N. Mantegna *et al.* in *Physical Review Letters*, vol. 73, pp. 3169–3172, 1994 suggest that junk DNA encodes information analogous to the structure of language (specifically an equivalent to Zipf's law, where the total numbers of occurrences of words in a text fall on a linear distribution when plotted on a log–log scale, and also find evidence for redundancy whereby meaning is preserved even with deletions). These views are questioned and replied to in a subsequent issue (vol. 76, pp. 1976–1981).

6. See L. Z. Holland and N. D. Holland in *American Zoologist*, vol. 38, pp. 647–658, 1998; W. R. Jackman *et al.* in *Developmental Biology*, vol. 220, pp. 16–26, 2000; H. Toresson *et al.* in *Development, Genes, and Evolution*, vol. 208, pp. 431–439, 1998; T. V. Venkatesh *et al.* in *Development, Genes, and Evolution*, vol. 209, 254–259, 1999; H. Wada *et al.* in *Developmental Biology*, vol. 213, pp. 131–141, 1999; and N. C. Williams and P. W. H. Holland in *Molecular Biology and Evolution*, vol. 15, pp. 600–607, 1998.

7. See T. C. Lacalli *et al.* in *Philosophical Transactions of the Royal Society of London B*, vol. 344, pp. 165–185, 1994, and vol. 351, pp. 243–263, 1996, as well as *Acta Zoologica (Stockholm)*, vol. 80, pp. 113–124, 1999 and *Proceedings of the Royal Society of London B*, vol. 266, pp. 1461–1470, 1999.

8. See P. W. H. Holland in *Development* (Supplement 1994), pp. 125–133, 1994; *American Zoologist*, vol. 38, pp. 829–842, 1998; and *Seminars in Cell & Developmental Biology*, vol. 10, pp. 541–547, 1999.

9. Such examples are known from the Lower Cambrian (Chengjiang fossil assemblages): see D. Shu *et al.* in *Nature*, vol. 384, pp. 157–158, 1996; and from the Middle Cambrian (Burgess Shale): see papers by A. M. Simonetta *et al.* in *Bollettino di Zoologia*, vol. 62, pp. 243–252, 1995 and *Italian Journal of Zoology*, vol. 66, pp. 99–119, 1999.

10. This remarkable article, by Temple Smith and Harold Morowitz, entitled 'Between physics and history', was published in *Journal of Molecular Evolution*, vol. 18, pp. 265–282, 1982.

11. See especially W. M. Elsasser's insights into what he calls very large and immense numbers, in *Reflections on a theory of organisms: holism in biology* (Johns Hopkins University Press, Baltimore, 1998). This equally strange and stimulating book is, as its subtitle indicates, definitely anti-reductionist. Also very relevant to this question is *Information and the origin of life* by B-O Küppers (MIT Press, Cambridge, MA, 1990); see especially Chapters 6 and 7.

12. They are probably most familiar in the form of the digestive enzymes such as trypsin and amylases, but they are essential in all parts of the cellular economy. Other enzymes are essential for dealing with a stiff gin and tonic; here you need alcohol dehydrogenase.

13. T. Smith and H. Morowitz (1982; citation is in note 10), p. 268.

14. T. Smith and H. Morowitz (1982; citation is in note 10); both quotations are on p. 268.

15. See the short article on 'Laws of forms revisited', by M. Denton and C. Marshall in *Nature*, vol. 410, p. 417, 2001. See also M. J. Denton *et al.* in *Journal of Theoretical Biology*, vol. 219, pp. 325–342, 2002.

16. M. Denton and C. Marshall (2001; citation is in note 15), p. 417.

17. See, for example, the paper by Y-H. Lee *et al.* in *Journal of Molecular Evolution*, vol. 45, pp. 278–284, 1997, where they show how pathways through RNA sequence space may be equally parsimonious, but not equally likely.

18. Stephen Freeland reminds me, however, that while this axiom seems sensible, 'the only explicit rationale for this view is Fisher's geometric theorem (in his *The genetical theory of natural selection* [Oxford, Clarendon Press, 1930; see pp. 38–41]); a simple abstract model of evolution that predicts an inversely proportional relationship between the magnitude of effect of a random mutation and the probability that it will represent an adaptive improvement'; see Freeland's article in *Journal of Genetic Programming and Evolvable Machines*, vol. 3, pp. 113–127, 2002.

19. *Symbiosis* means 'living together', and, although the term is often thought to be synonymous with mutual benefit, biologists take a symbiosis to be a neutral term that can, if sufficient information is available, be resolved as benefit (+), loss (−), or neither (0). Parasitism usually has a winner (the parasite) and a loser (the host), thus (+)(−), unless each parasitizes the other, i.e. (−)(−).

20. T. Smith and H. Morowitz (1982; citation is in note 10), p. 280.

21. This conceit of 'Different routes to similar ends' is specifically addressed by P. H. Harvey and L. Partridge in *Nature*, vol. 392, pp. 552–553, 1998, as well as by the first author writing in *Current Biology*, vol. 10, p. R271, 2000.

22. That may be the natural order, but biotechnological manipulation makes it possible to introduce novel amino acids, thus artificially expanding the genetic code. See, for example, L. Wang *et al.* in *Science*, vol. 292, pp. 498–500, 2001, and the following paper (pp. 501–504) by V. Döring *et al.*, as well as the commentary by A. Böch on pp. 453–454; see also M. Ibba *et al.* in *Current Biology*, vol. 11, pp. R563–R565, 2001.

23. See the chapter (pp. 21–27) by J. M. Diamond in *Principles of animal design: the optimization and symmorphosis debate*, edited by E. R. Weibel *et al.* (Cambridge University Press, Cambridge, 1998). The graphic predicament of the lift plunging to its doom is based on remarks made by Diamond in a lecture he gave in Cambridge.

24. See Henry Petroski's *Design paradigms: case histories of error and judgement in engineering* (Cambridge University Press, Cambridge, 1994). The book is an absorbing discussion of the trials, tribulations, and daring of bridge builders; Petroski's remarks on the sociology of engineering knowledge and tradition are also of great, and sometimes alarming, interest.

25. See, for example, G. M. Taylor *et al.* in *Biological Journal of the Linnean Society*, vol. 70, pp. 37–62, 2000. They address safety factors in the claws of crabs, and also give a useful introduction to the literature of this area.

26. Carl Gans 'Momentarily excessive construction as the basis for protoadaptation' is the title of his paper in *Evolution*, vol. 33, pp. 227–233, 1979.

27. See *Kea, bird of paradox: the evolution and behavior of a New Zealand parrot*, by J. Diamond and A. B. Bond (University of California Press, Berkeley, 1999).

28. See John Currey's paper in *Journal of Experimental Biology*, vol. 202, pp. 3285–3294, 1999.

29. See J. D. Currey *et al.* in *Proceedings of the Royal Society of London B*, vol. 268, pp. 107–111, 2001, and also V. de Buffrénil and A. Casinos in *Annales des Sciences naturelles, Zoologie, Paris*, vol. 16, pp. 21–32, 1995.

30. J. D. Currey (1999, citation is in note 28), p. 3289. In his paper Currey also draws attention to the convergence between the teeth of the sea urchins (on which see R. Z. Wang *et al.* in *Philosophical Transactions of the Royal Society of London B*, vol. 352, pp. 469–480, 1997), which rasp the sea floor for edibles, and the gnawing incisors of the rodents.

31. The key papers are by Steve Freeland and Laurence Hurst in *Journal of Molecular Evolution*, vol. 47, pp. 238–248, 1998 and *Molecular Biology and Evolution*, vol. 17, pp. 511–518, 2000a; see also *Trends in Biochemical Sciences*, vol. 25, pp. 44–45, 2000b.

32. Concerning the optimality of two base pairs see E. Szathmary in *Proceedings of the Royal Society of London B*, vol. 245, pp. 91–99, 1991, and *Proceedings of the National Academy of Sciences, USA*, vol. 89, pp. 2614–2618, 1992. See also D. A. Mac Donaill in *Chemical Communications*, vol. 18, pp. 2062–2063, 2002.

33. A. L. Weber and S. L. Miller, in *Journal of Molecular Evolution*, vol. 17, pp. 273–284, 1981, draw on various lines of evidence (including prebiotic synthesis, stability, and function in proteins) to argue that the 20 terrestrial amino acids are the norm.

34. A. L. Weber and S. L. Miller (1981, citation is in note 33), p. 273.

35. A. Jiménez-Sánchez, however, suggests (in *Journal of Molecular Evolution*, vol. 41, pp. 712–716, 1995) that the original arrangement in an 'RNA world' (see note 1, Chapter 4) was still triplets (and one base pair (AU)), coding for seven amino acids (and a stop codon).

36. See J. T. Wong in *Microbiological Science*, vol. 5, pp. 174–181, 1988. R. Amirnovin in *Journal of Molecular Evolution*, vol. 44, pp. 473–476, 1997, concludes that much depends on which amino acids are assumed to have a biosynthetic linkage. M. Di Giulio and M. Medugno, however, present a statistically based critique of Amirnovin's idea in a subsequent issue (vol. 50, pp. 258–263, 2000), reasserting their belief in 'the intimate relationship between the biosynthetic pathways of amino acids and the organization of the genetic code' (p. 263). On the other hand, T. A. Ronneberg *et al.* (in *Proceedings of the National Academy of Sciences, USA*, vol. 97, pp. 13690–13695, 2000) speak strongly against the importance of biosynthetic pathways, although they also emphasize that code expansion almost certainly occurred.

37. S. Freeland and L. Hurst (1998; citation is in note 31), p. 244.

38. S. Freeland *et al.* (2000b; citation is in note 31), p. 45.

39. It should be pointed out that the conclusions reached by Freeland and Hurst are not universally accepted. See M. Di Giulio and M. Medugno in *Journal of Molecular Evolution*, vol. 52, pp. 378–382, 2001, who argue that the effectiveness of the genetic code lies in a co-evolution between the code and the biosynthetic pathways leading to the various amino acids.

40. For example, the codons UAA and UAG, which nearly always mean STOP, have been reassigned in several unrelated groups to code for glycine. The evolvability of the genetic code is reviewed by R. D. Knight *et al.* in *Nature Reviews, Genetics*, vol. 2, pp. 49–58, 2001, who remark (p. 49) 'Curiously, many of the same codons are reassigned in independent lineages, frequently between the same two meanings ... indicating that there may be an underlying predisposition towards certain reassignments. At least one of these changes seems to confer a direct selective advantage.'

41. See, for example, M. Guilio and M. Medugno in *Journal of Molecular Evolution*, vol. 49, pp. 1–10, 1999.

42. See Geoffrey Irwin's book *The prehistoric exploration and colonisation of the Pacific* (Cambridge University Press, Cambridge, 1992).

2. CAN WE BREAK THE GREAT CODE?

1. See, for example, the apt comments (pp. 531–532) by Christian de Duve in *Proceedings of the American Philosophical Society* (vol. 142, pp. 525–532, 1998).

2. See the poignant chapter 19, 'The self marooned in the cosmos' in Walker Percy's *Lost in the Cosmos: The last self-help book* (Arena, London, 1994).

3. Such is argued by P. D. Ward and D. Brownlee in *Rare Earth: Why complex life is uncommon in the Universe* (Copernicus, New York, 2000), which comes to a number of similar conclusions as to the scarcity of habitable planets. These workers, however, suggest that the origination of life is relatively commonplace but that complex life forms, such as pen-wielding humanoids, are very rare.

4. See, for example, Christian de Duve's *Vital dust: life as a cosmic imperative* (HarperCollins [BasicBooks]], New York, 1995).

5. C. de Duve (1998; citation is in note 1), p. 527.

6. Threga IX? My imaginary planet, strangely Earth-like and home to our visitors, whom we finally meet in Chapter 12.

7. See *Blueprints: solving the mystery of evolution* (Oxford University Press, Oxford, 1990) by M. A. Edey and D. C. Johanson.

8. M. A. Edey and D. C. Johanson (1990), p. 295.

9. See Iris Fry's *The emergence of life on Earth: A historical and scientific overview* (Free Association, London, 1999)

10. George Wald, for example, provides a cogent discussion of the merits and otherwise of carbon and silicon in his chapter (pp. 127–142) in *Horizons in biochemistry. Albert Szent-Györgyi dedicatory volume*, edited by M. Kasha and B. Pullman (Academic Press, New York, 1962).

11. See, for example, Robert Shapiro and Gerald Feinberg in their chapter (pp. 248–255) in *Physical cosmology and philosophy*, edited by J. Leslie (Macmillan, New York, 1990).

12. See I. Fry (1999; citation is in note 9), p. 239.

13. See I. Fry (1999; citation is in note 9), p. 241.

14. See F. H. Westheimer in *Science*, vol. 235, pp. 1173–1178, 1987.

15. Francis Crick and Leslie Orgel's paper can be found in *Icarus*, vol. 19, pp. 341–346, 1973.

16. P. Parsons, for example, in *Nature*, vol. 383, pp. 221–222, 1996, reviews some current thinking. He points out that so far as a source for panspermia is concerned, a planet orbiting a red giant might be particularly suitable, given that such a star is relatively cool, emits less ultraviolet radiation (which is extremely damaging, at least to terrestrial biochemistry), and very long-lived, thus increasing the net chance of life evolving on any associated planet.

17. See, for example, F. Egami in *Journal of Molecular Evolution*, vol. 4, pp. 113–120, 1974. In contrast, the related element chromium is very sparse in sea water, and correspondingly finds no use in living organisms. The element selenium, however, is vital, but is available only in trace amounts. See also W. R. Chappell *et al.* and T. H. Jukes in *Icarus* (vol. 21, pp. 513–517, 1974) and A. Banin and J. Navrot in *Science* (vol. 189, pp. 550–551, 1975).

18. For a masterly overview of this general area consult R. J. P. Williams and J. J. R. Fraústo da Silva on *The natural selection of the chemical elements* (Clarendon Press, Oxford, 1996).

19. Christopher Switzer and colleagues' article can be found in *Biochemistry*, vol. 32, pp. 10489–10496, 1993.

20. C. Switzer *et al.* (1993; citation is in note 19), p. 10489.

21. C. Switzer *et al.* (1993; citation is in note 19), p. 10489.

22. C. Switzer *et al.* (1993; citation is in note 19), p. 10489.

23. For example, Switzer *et al.* (1993; citation is in note 19) also note that the type of heterocycle ring that occurs in the oligonucleotides leads to ambiguities in the hydrogen bonding.

24. See C. Mao *et al.* in *Nature*, vol. 386, pp. 137–138, 1997.

25. See E. Winfree *et al.* in *Nature*, vol. 394, pp. 539–544, 1998.

26. See Y. Zhang and N. C. Seeman in *Journal of the American Chemical Society*, vol. 116, pp. 1661–1669, 1994.

27. See A. P. Alivisatos *et al.* in *Nature*, vol. 382, pp. 609–611, 1996; and the preceding paper, on pp. 607–609, by C. A. Mirkin *et al.*

28. See E. Braun *et al.* in *Nature* 391, 775–778, 1998.

29. C. Switzer *et al.* (1993; citation is in note 19), p. 10495.

30. See, for example, B. Bhat *et al.* in *Journal of the American Chemical Society*, vol. 118, pp. 3065–3066, 1996.

31. This paper, entitled 'Enzymatic incorporation of a new base pair into DNA and RNA extends the genetic alphabet', may be found in *Nature*, vol. 343, pp. 33–37, 1990.

32. Joseph Piccirilli *et al.* (1990), p. 33.

33. For another such example, see F. Seela and A. Melenewski in *European Journal of Organic Chemistry* for 1999, pp. 485–496, 1999.

34. Piccirilli *et al.* (1990; citation is in note 32), p. 34.

35. See C. Roberts *et al.* in *Journal of the American Chemical Society*, vol. 119, pp. 4640–4649, 1997.

36. See H. Hashimoto and C. Switzer in *Journal of the American Chemical Society*, vol. 114, pp. 6255–6256, 1992.

37. See J. P. Dougherty in *Journal of the American Chemical Society*, vol. 114, pp. 6254–6255, 1992.

38. See T. L. Sheppard and R. Breslow in *Journal of the American Chemical Society*, vol. 118, pp. 9810–9811, 1996.

39. The key review papers may be found by Albert Eschenmoser and colleagues in *Science* (vol. 284, pp. 2118–2124, 1999) and *Origins of Life and Evolution of the Biosphere* (vol. 24, pp. 389–423, 1994 and vol. 27, pp. 535–553, 1997).

40. See K. D. James and A. D. Ellington in *Origins of Life and Evolution of the Biosphere*, vol. 25, pp. 515–530, 1995; quotation is on p. 520.

41. Thus K. D. James and A. D. Ellington (1995; citation is in note 40) remark on work by the Eschenmoser team that they 'have concluded that very few [alternatives] can potentially form complementary duplexes that would in theory be capable of self-replication ... These conclusions can be extended to nucleic acid bases as well as sugars: experiments with alternate base pairing schemes have suggested that the current set of purines and pyrimidines is in many ways optimal ... the unnatural nucleic acid analogues that have been examined experimentally have proven to be largely incapable of self-replication,' p. 520.

42. 'Why pentose – and not hexose – nucleic acids?' is the main title of the paper by S. Pitsch *et al.* in *Helvetica Chimica Acta*, vol. 76, pp. 2161–2183, 1993, and is answered with respect to 'a whole series of intrinsic steric handicaps' (p. 2163). In his overview Eschenmoser (1999; citation is in note 39) concluded that 'The outcome of these studies has led to the conclusion that for functional reasons, the three hexopyranosyl-(4′-6′) oligonucleotide systems investigated ... could not have acted as viable competitors of RNA in the emergence of nature's genetic system' (p. 2121).

43. Concerning alternatives to ribose, as five-carbon (pentose) sugars, see M. Beier *et al.* in *Science*, vol. 283, pp. 699–703, 1999.

44. See the paper by K.-U. Schöning *et al.* in *Science*, vol. 290, pp. 1347–1351, 2000, as well as a commentary by L. Orgel on pp. 1306–1307.

45. A. Eschemoser (1999; citation is in note 39), p. 2122.

3. UNIVERSAL GOO: LIFE AS A COSMIC PRINCIPLE?

1. C. de Duve (1998; citation is in Chapter 2, note 1).

2. C. de Duve (1998; citation is in Chapter 2, note 1), p. 526.

3. This quotation is given in Robert Shapiro's *Origins: A skeptic's guide to the creation of life on Earth* (Bantam, New York, 1987), p. 187.

4. See my article in *Astronomical Society of the Pacific Conference Series*, vol. 213, pp. 410–419, 2000, p. 417.

5. An overview is given by p. Ehrenfreund and S. B. Charnley in *Annual Review of Astronomy and Astrophysics*, vol. 38, pp. 427–483, 2000.

6. This conceit may be found in Robert Shapiro (1987; citation is in note 3); see p. 230.

7. See S. B. Charnley *et al.* in *Spectrochimica Acta A*, vol. 57, pp. 685–704, 2001. P. Ehrenfreund *et al.* (in *Astrophysical Journal*, vol. 550, pp. L95–L99, 2001) also point out that unless shielded amino acids in deep space are very liable to destruction by ultraviolet radiation.

8. See, for example, L. J. Allamandola's chapter (pp. 81–102) in *The cosmic dust connection*, edited by J. M. Greenberg (Kluwer, Dordrecht, 1996), and also *Astrophysical Journal*, vol. 511, L115–L119, 1999.

9. See L. Becker *et al.* in *Geochimica et Cosmochimica Acta*, vol. 61, pp. 475–81, 1997 and *Earth and Planetary Science Letters*, vol. 167, pp. 71–79, 1999; and M. Zolotov and E. Shock in *Journal of Geophysical Research*, vol. 104 (E6), pp. 14033–14049, 1999 and *Meteoritics & Planetary Science*, vol. 35, pp. 629–638, 2000.

10. See J. P. Bradley *et al.* in *Geochimica et Cosmochimica Acta*, vol. 60, pp. 5149–5155, 1996 and *Meteoritics & Planetary Science*, vol. 33, pp. 765–773, 1998; K. L. Thomas-Keprta *et al.* in *Geochimica et Cosmochimica Acta*, vol. 64, pp. 4049–4081, 2000; P. R. Buseck *et al.* and

D. J. Barber and E. R. D. Scott in *Proceedings of the National Academy of Sciences, USA*, vol. 98, pp. 13490–13495, 2001 and vol. 99, pp. 6556–6561, 2002 respectively.

11. For example, amino acids (see J. L. Bada *et al.* in *Science*, vol. 279, pp. 362–365, 1998) and sulphur isotopes (see J. P. Greenwood *et al.* in *Geochimica et Cosmochimica Acta*, vol. 61, pp. 4449–4453, 1997), not to mention contaminants (e.g. A. Steele *et al.* in *Meteoritics & Planetary Science*, vol. 35, 237–241, 2000).

12. See L. E. Borg *et al.* in *Science*, vol. 286, p. 90–94, 1999, E. R. D. Scott in *Journal of Geophysical Research*, vol. 104 (E2), pp. 3803–3813, 1999; D. C. Golden *et al.* in *Meteoritics & Planetary Science*, vol. 35, pp. 457–465, 2000; and J. M. Eiler *et al.* in *Geochimica et Cosmochimica Acta*, vol. 66, pp. 1285–1303, 2002.

13. See, for example, H. Naraoka *et al.* in *Earth and Planetary Science Letters*, vol. 184, pp. 1–7, 2000.

14. Thus W. F. Hume, in *The Cairo Scientific Journal*, vol. 5, pp. 212–215, 1911, reported from a newspaper account to the effect that one fragment 'fell on a dog at Denshal, leaving it like ashes in a moment', p. 212.

15. See Kevin Yau and his colleagues' article in *Meteoritics*, vol. 29, pp. 864–871, 1994.

16. K. Yau *et al.* (1994), Table 1 (p. 867).

17. See J. S. Lewis's interesting *Rain of iron and ice: the very real threat of comet and asteroid bombardment* (Addison-Wesley [Helix], Reading, MA, 1996), pp. 176–182.

18. See S. Veski *et al.* in *Meteoritics & Planetary Science*, vol. 36, pp. 1367–1375, 2001.

19. See P. G. Brown *et al.* in *Science*, vol. 290, pp. 320–325, 2000, with commentary by J. N. Grossman on pp. 283–284.

20. See a special section edited by P. G. Brown *et al.*, citation is in note 19, of *Meteoritics & Planetary Science*, vol. 37(5), 2002; see also P. Ehrenfreud *et al.* in *Proceedings of the National Academy of Sciences, USA*, vol. 98, pp. 2138–2141, 2001.

21. See G. Kminek *et al.* on pp. 697–701 of P. G. Brown *et al.*, citation is in note 19. See also D. P. Glavin in *Proceedings of the National Academy of Sciences, USA*, vol. 96, pp. 8835–8838, 1999 concerning contamination in the Nakhla meteorite.

22. A leading figure in the study of extraterrestrial amino acids is J. R. Cronin, and work by him and his colleagues can be found in *Journal of Molecular Evolution*, vol. 17, pp. 265–272, 1981, two papers in *Geochimica et Cosmochimica Acta* (vol. 49, pp. 2259–2265, 1985 and vol. 50, pp. 2419–2427, 1986 respectively), as well as *Advances in Space Research* (vol. 15 (3), pp. 91–97, 1995) and *Science* (vol. 275, pp. 951–955, 1997).

23. See also J. F. Kerridge (*Advances in Space Research*, vol. 15, pp. 107–111, 1995), who remarks that the style of amino acid formation is consistent with their synthesis being random and undirected, governed by principles of thermodynamic stability and very different from any organic activity.

24. See G. Cooper *et al.* in *Nature*, vol. 414, pp. 879–883, 2001, and commentary by M. A. Sephton on pp. 857–859.

25. There is an interesting twist here, however, because these amino acids have been found immediately adjacent to the 'fireball layer' marking the huge impact that both terminated (by definition) the Cretaceous and its chief glory, the dinosaurs; see M. Zhao and J. L. Bada in *Nature*, vol. 339, pp. 463–465, 1989). Subsequently K. Zahnle and D. Grinspoon in *Nature* (vol. 348, pp. 157–160, 1990) argued that the amino acids floated to Earth in association with comet dust that extended from the nucleus that actually collided with the Earth. But was it a comet? The latest evidence suggests it was a meteorite: see F. K. Kyte in *Nature*, vol. 396, pp. 237–239, 1998. And perhaps the overwhelming evidence for extraterrestrial mayhem has slanted our view of the source of these strange amino acids. If organic matter was gasified by the intense shock of impact then perhaps these amino acids are a direct result of the impact; see E. S. Olson in *Nature*, vol. 357, p. 202, 1992.

26. See various articles in *Nature* by G. Yuen *et al.* in vol. 307, pp. 252–254, 1984; S. Epstein *et al.* in vol. 326, pp. 477–479, 1987; and M. H. Engel *et al.* in vol. 348, pp. 47–49, 1990.

27. Reviews can be found in a chapter (pp. 819–857) by J. R. Cronin *et al.* in *Meteorites and the early Solar System*, edited by J. F. Kerridge and M. S. Matthews (University of Arizona Press, Tucson, 1988), and *The molecular origins of life: assembling pieces of the puzzle*, edited by A. Brack (Cambridge University Press, Cambridge, 1998); see in particular pp. 119–146.

28. See, for example, P. G. Stoks and A. W. Schwartz in *Geochimica et Cosmochimica Acta*, vol. 45, pp. 563–569, 1981.

29. See D. W. Deamer and R. M. Pashley in *Origins of life and evolution of the biosphere*, vol. 19, pp. 21–38, 1989; see also J. P. Dworkin *et al.* in *Proceedings of the National Academy of Sciences, USA*, vol. 98, pp. 815–819, 2001.

30. Right-handed, or dextral (D-), amino acids do arise in various metabolic processes; see *D-amino acids in sequences of secreted peptides of multicellular organisms*, edited by D. Jollès (Birkhauser, Basel, 1998).

31. J. L. Bada *et al.* (*Nature*, vol. 301, pp. 494–496, 1983) gave a careful critique of M. H. Engel and B. Nagy in *Nature* (vol. 296, pp. 837–840, 1982) and received in return a robust response.

32. See J. R. Cronin and S. Pizzarello in *Science* (vol. 275, pp. 951–955, 1997) and commentary on pp. 942–943 by J. L. Bada; see also S. Pizzarello and J. R. Cronin in *Geochimica et Cosmochimica Acta*, vol. 64, pp. 329–338, 2000 and M. H. Engel & S. A. Macko in *Precambrian Research*, vol. 106, pp. 35–45, 2001. Subsequently, S. Pizzarello and G. W. Cooper (in *Meteoritics & Planetary Science*, vol. 36, pp. 897–909, 2001) reported that alanine was originally racemic, but an L-enantiomeric excess in glutamic acid was probably terrestrial contamination.

33. See, for example, J. Bailey *et al.* in *Science*, vol. 281, pp. 672–674, 1998, and accompanying commentary by R. Irion on pp. 626–627. Their observation is based on infrared radiation, but it is assumed that the UV part of the spectrum, which is energetic enough to break chemical bonds, is also polarized. This is not to say that all such amino acids will be non-racemic. Laboratory studies of the analogue of ultraviolet radiation on interstellar ices produced a variety of amino acids, but these are racemic. See M. P. Bernstein *et al.* and G. M. Muñoz Caro *et al.* in *Nature*, vol. 416, pp. 401–403 and 403–406 respectively, 2002, and commentary by E. L. Shock on pp. 380–381.

34. See the paper in *Nature* (vol. 405, pp. 932–935, 2000) by G. L. J. A. Rikken and E. Raupach, as well as the commentary by L. D. Barron on pp. 895–896.

35. See K. Soai *et al.* in *Nature*, vol. 378, pp. 767–768, 1995. Again the excess is small (about 2%), and the chemistry seems to have rather little relevance to the origin of life.

36. See R. M. Hazen *et al.* in *Proceedings of the National Academy of Sciences, USA*, vol. 98, pp. 5487–5490, 2001. See also the paper by C. A. Orme *et al.* (*Nature*, vol. 411, pp. 775–779, 2001, with commentary by L. Addadi and S. Weiner on pp. 753, 755).

37. For an overview see *Comets and the origin and evolution of life*, edited by P. J. Thomas *et al.* (Springer, New York, 1997), especially chapter 6 (pp. 147–173) by C. F. Chyba and C. Sagan.

38. Produced by ultraviolet irradiation in simulated interstellar conditions of PAH napthalene; see M. P. Bernstein *et al.* in *Meteoritics & Planetary Science*, vol. 36, pp. 351–358, 2001. Quinones are an important building block of chlorophyll, which in Chapter 6 is argued to be the molecule of choice wherever in the Universe photosynthesis occurs.

39. See E. Pierazzo and C. F. Chyba in *Meteoritics & Planetary Science*, vol. 34, pp. 909–918, 1999.

4. THE ORIGIN OF LIFE: STRAINING THE SOUP OR OUR CREDULITY?

1. See, for example, I. Fry's *The emergence of life on Earth: A historical and scientific overview* (Free Association Books, London, 1999).

2. The short article by A. Lazcano and S. L. Miller in *Cell*, vol. 85, pp. 793–798, 1996 provides a useful and crisp overview.

3. See, for example, *The RNA world, second edition: The nature of modern RNA suggests a prebiotic RNA*, edited by R. F. Gesteland *et al.* (Cold Spring Harbor Laboratory Press, Cold Spring Harbor, NY, 1998).

4. See David Bartel and Peter Unrau's article in *Trends in Biochemical Sciences*, vol. 24, pp. M9–M13, 1999 (Millennium issue).

5. Bartel and Unrau (1999; citation is in note 4), p. M9. See also notes 44 and 45.
6. See L. E. Orgel in *Trends in Biochemical Sciences*, vol. 23, pp. 491–495, 1998.
7. See L. E. Orgel in *Trends in Biochemical Sciences*, vol. 23, p. 491, 1998.
8. See, for example, A. Bar-Nun and A. Shaviv in *Icarus*, vol. 24, pp. 197–210, 1975; W. L. Chameides and J. C. G. Walker in *Origins of Life and Evolution of the Biosphere*, vol. 11, pp. 291–302, 1981; and J. P. Ferris and W. J. Hagan in *Tetrahedron*, vol. 40, pp. 1093–1120, 1984.
9. See Manfred Eigen's influential article, entitled 'Selforganization of matter and the evolution of biological macromolecules', in *Die Naturwissenschaften*, vol. 58, pp. 465–523, 1971; and *Steps towards life: A perspective on evolution* (Oxford University Press, Oxford, 1992).
10. See John Maynard-Smith in *Nature*, vol. 280, pp. 445–446, 1979.
11. See D. H. Lee *et al.* in *Current Opinion in Chemical Biology*, vol. 1, pp. 491–496, 1997.
12. See D. H. Lee *et al.* in *Nature*, vol. 390, pp. 591–594, 1997.
13. See Klaus Dose's essay 'The origin of life: More questions than answers' in *Interdisciplinary Science Reviews*, vol. 13, pp. 348–356, 1988. Although this article is more than ten years old, little has changed to suggest we should take a more optimistic view.
14. Dose (1988; citation is in note 13), quotations are on pp. 348 and 349.
15. See John Horgan's stimulating article in *Scientific American*, vol. 264(2), pp. 100–109, 1991.
16. In terms of general scepticism, the key text is Robert Shapiro's book *Origins*, with the telling subtitle *A skeptic's guide to the creation of life on Earth* (Bantam, New York, 1987); I have frequent recourse to Shapiro's ideas later in this chapter. A generally sceptical view can also be found in Chapter 16 of the book *Blueprints: Solving the mystery of evolution* (Oxford University Press, Oxford, 1990) by M. A. Edey and D. C. Johanson.
17. J. Horgan (1991, citation is in note 15), p. 101.
18. See Anthony Keefe and Stanley Miller's paper on 'Potentially prebiotic syntheses of condensed phosphates' in *Origins of Life and Evolution of the Biosphere*, vol. 26, pp. 15–25, 1996.
19. A. Keefe and S. Miller (1996; citation is in note 18), p. 15.
20. Concerning its photochemical production in a prebiotic atmosphere, see J. P. Pinto *et al.* in *Science*, vol. 210, pp. 183–185, 1980.
21. See R. Shapiro's 'Prebiotic ribose synthesis: a critical analysis' in *Origins of Life and Evolution of the Biosphere*, vol. 18, pp. 71–85, 1988; see also note 16.
22. L. E. Shapiro (1988; citation is in note 21), p. 83.
23. See T. Mizuno and A. H. Weiss in *Advances in Carbohydrate Chemistry and Biochemistry*, vol. 29, pp. 173–227, 1974.
24. See R. Larralde *et al.* in *Proceedings of the National Academy of Sciences, USA*, vol. 92, pp. 8158–8160, 1995.
25. See D. Müller *et al.* in *Helvetica Chimica Acta*, vol. 73, pp. 1410–1468, 1990.
26. See G. Zubay's paper in *Origins of Life and Evolution of the Biosphere*, vol. 28, pp. 13–26, 1998.
27. G. Zubay (1998; citation is in note 26), p. 19. This work is returned to by G. Zubay and T. Mui in a subsequent issue of *Origins of Life and Evolution of the Biosphere* (vol. 31, pp. 87–102, 2001), as part of a discussion of nucleotide synthesis. As usual, the general tone is upbeat, but problems remain: 'Limited success has been achieved in the synthesis of inosine ... The situation for adenosine is much worse ... For several years we have tried in vain to improve this system', p. 100.
28. See K-U. Schöning *et al.* in *Science*, vol. 290, pp. 1347–1351, 2000; see also note 43, Chapter 2.
29. See, for example, the papers by J. P. Ferris *et al.* in *Nature*, vol. 381, pp. 59–61, 1996 (with commentary on pp. 20–21 by G. von Kiedrowski) and A. R. Hill *et al.* and R. Liu and L. Orgel in *Origins of Life and Evolution of the Biosphere*, vol. 28, pp. 235–243 and 245–257, 1998 respectively.
30. See Ferris *et al.* (1996), and S. J. Sowerby *et al.* (in *Proceedings of the National Academy of Sciences, USA*, vol. 98, pp. 820–822, 2001), which explored the differential adsorption of the various nucleic acid bases on tiny particles of natural graphite, which the authors duly note

'is not considered a dominant prebiotic material', followed by a prompt appeal to 'zeolites, feldspars and silicas', p. 821.

31. See, for example, A. W. Schwartz and L. E. Orgel on 'Template-directed synthesis of novel, nucleic acid-like structures' in *Science*, vol. 228, pp. 585–587, 1985.

32. R. Liu and L. Orgel (1998; citation is in note 29), pp. 256–257.

33. See Graham Cairns-Smith's *Genetic takeover and the mineral origins of life* (Cambridge University Press, Cambridge, 1982). Nevertheless, so far as I am aware no experimental work has provided a reality check on this hypothesis, and in more recent discussions of the origin of life these ideas typically receive routine genuflection rather than any serious engagement.

34. See F. G. Mosqueira *et al.* in *Origins of Life and Evolution of the Biosphere*, vol. 26, pp. 75–94, 1996.

35. F. G. Mosqueira *et al.* (1996; citation is in note 34), p. 92.

36. See J. D. Bernal in *The physical basis of life* (Routledge & Kegan Paul, London, 1951); Bernal's emphasis was very much on clay minerals, as well as quartz (pp. 33–37).

37. See, for example, M. Levy and S. L. Miller in *Proceedings of the National Academy of Sciences, USA*, vol. 95, 7933–7938, 1998, with specific reference to the thermal stability of the nucleotides.

38. See Robert Shapiro in *Planetary dreams: The quest to discover life beyond Earth* (Wiley, New York, 1999). This book echoes his interests and scepticisms about present attempts to discover how life originated, but it is also both a passionate plea for the exploration of outer space combined with a deep, almost religious, conviction that life is a universal principle. Shapiro writes with a flair and accessibility that is difficult to match.

39. R. Shapiro (1999; citation is in note 38), p. 137.

40. G. Wächtershäuser has promoted his ideas widely and at some length; see, for example, his papers in *Progress in Biophysics and Molecular Biology*, vol. 58, pp. 85–201, 1988; *Microbiological Reviews*, vol. 52, pp. 452–484, 1988; and his chapter (on pp. 206–218) in *The molecular origins of life: assembling pieces of the puzzle*, edited by A. Brack (Cambridge University Press, Cambridge, 1998). An excellent overview of Wächtershäuser's programme is given by Fry (1999; citation is in note 1), pp. 162–172.

41. See, for example, E. Blöchl *et al.* in *Proceedings of the National Academy of Sciences, USA*, vol. 89, pp. 8117–8120, 1992; D. Hafenbradl *et al.* in *Tetrahedron Letters*, vol. 36, pp. 5179–5184, 1995; the two papers by C. Huber and G. Wächtershäuser in *Science*, respectively vol. 276, pp. 245–247, 1997 and vol. 281, pp. 670–672, 1998 (with accompanying commentary by G. Vogel on p. 627 and 629); W. Heinen and A. M. Lauwers in *Origins of Life and Evolution of the Biosphere*, vol. 26, pp. 131–150, 1996; and G. D. Cody *et al.* in *Science*, vol. 289, pp. 1337–1340, 2000 (with commentary by G. Wächtershäuser on pp. 1307–1308, who remarks of these experiments 'It remains to be established whether such conditions [200 MPa, 250 °C] are geophysically possible').

42. Experiments, promoted by those sceptical of the Wächtershäuser school, have looked at prebiotic amino acid and nucleotide synthesis in a FeS/H_2S system and produced negative results; see A. D. Keefe *et al.* in *Proceedings of the National Academy of Sciences, USA*, vol. 92, pp. 11904–11906, 1995; see also the somewhat sceptical tenor of M. A. A. Schoonen *et al.* in *Origins of Life and Evolution of the Biosphere*, vol. 29, pp. 5–32, 1999.

43. The paper by M. J. Russell and A. J. Hall in *Journal of the Geological Society, London*, vol. 154, pp. 377–402, 1997 is a case in point. Despite its encouraging title 'The emergence of life from iron monosulphide bubbles at a submarine hydrothermal redox and pH front', this particular paper is strong on theory and expectations, but which pathways will ultimately prove experimentally tractable is less obvious.

44. See Andrew Ellington's brief overview in *Biological Bulletin*, vol. 196, pp. 315–319, 1999.

45. A. Ellington (1999; citation is in note 44), p. 317.

46. Literature on hydrothermal locales as the site for the shift from abiogenetic chemical processes to life is fast expanding. Relevant material includes J. P. Amend and E. L. Shock in

Science, vol. 281, pp. 1659–1662, 1998; J. P. Ferris in *Origins of life and evolution of the biosphere*, vol. 22, pp. 109–134, 1992; N. G. Holm and E. M. Andersson in *Planetary and Space Science*, vol. 43, pp. 153–159, 1995; E.-i. Imai *et al.* in *Science*, vol. 283, pp. 831–833, 1999; E. L. Shock and M. D. Schulte in *Journal of Geophysical Research*, vol. 103E, pp. 28, 513–28, 527, 1998; and N. G. Holm and J. L. Charlou in *Earth and Planetary Science Letters*, vol. 191, pp. 1–8, 2001. See also note 72.

47. See, for example, N. R. Pace in *Cell*, vol. 65, pp. 531–533, 1991.
48. See, for example, V. Moulton *et al.* in *Journal of Molecular Evolution*, vol. 51, pp. 416–421, 2000, who argue that the necessary folding of RNA is prejudiced at elevated temperatures, although they are careful to note that this does not rule out other stages in the trek from abiogenesis occuring in hot environments.
49. See M. Gogarten-Boekels *et al.* in *Origins of Life and Evolution of the Biosphere*, vol. 25, pp. 251–264, 1995.
50. See, for example, P. Forterre *et al.* in *Origins of Life and Evolution of the Biosphere*, vol. 25, pp. 235–249, 1995. Interestingly, in the context of Chapter 10, they argue that the key enzyme used to stabilize DNA in hyperthermophiles, known as reverse gyrase, is actually a composite and 'originated from the "late" fusion of a bona fide DNA helicase and DNA topoisomerases genes. Indeed many helicases and topoisomerases probably evolved independently from different RNA-metabolizing enzymes of the RNA world ... This is suggested by the existence of several independent superfamilies of these enzymes' (p. 241). See also note 21, Chapter 5.
51. See S. L. Miller and A. Lazcano in *Journal of Molecular Evolution*, vol. 41, pp. 689–692, 1995.
52. From the experimental procedures given by Huber and Wächtershäuser (1998, citation is in note 41), note 2 on p. 672 and Heinen and Lauwers (1996, citation is in note 41), p. 132.
53. See Leslie Orgel's critique in *Proceedings of the National Academy of Sciences, USA*, vol. 97, pp. 12503–12507, 2000.
54. L. E. Orgel (2000; citation is in note 53), p. 12506.
55. A. W. Schwartz and L. E. Orgel (1985, citation is in note 31), p. 586.
56. See S. J. Sowerby and W. M. Heckl in *Origins of Life and Evolution of the Biosphere*, vol. 28, pp. 283–310, 1998.
57. S. J. Sowerby and W. M. Heckl (1998; citation is in note 56), p. 291.
58. S. J. Sowerby and W. M. Heckl (1998; citation is in note 56), p. 292.
59. S. J. Sowerby and W. M. Heckl (1998; citation is in note 56), p. 297.
60. See the paper by P. A. Bachmann *et al.* in *Nature*, vol. 357, pp. 57–59, 1992.
61. See C. Böhler *et al.* in *Origins of Life and Evolution of the Biosphere*, vol. 26, pp. 1–5, 1996, who report on the oligomerization of amino acids by micelles of CTAB, and also remark that this process is achieved at 'relatively concentrated aqueous solutions' (i.e. 0.05 molar), but with only a tenfold dilution 'few long oligomers can be detected', p. 1.
62. See L. E. Orgel in *Nature*, vol. 358, pp. 203–209, 1992.
63. L. E. Orgel (1992; citation is in note 62), p. 204.
64. L. E. Orgel (1992; citation is in note 62), p. 207
65. The original set of experiments by Stanley Miller and Harold Urey was published in *Science*, vol. 117, pp. 528–529, 1953, and *Journal of the American Chemical Society*, vol. 77, pp. 2351–2361, 1955.
66. Overviews by S. L. Miller *et al.* may be found in *Journal of Molecular Evolution*, vol. 9, pp. 59–72, 1976; *Cold Spring Harbor Symposia on Quantitative Biology*, vol. 52, pp. 17–27, 1987; and on pp. 59–85 of *The molecular origins of life: assembling pieces of the puzzle*, edited by A. Brack (Cambridge University Press, Cambridge, 1998).
67. See, for example, R. Stribling and S. L. Miller in *Origins of Life and Evolution of the Biosphere*, vol. 17, pp. 261–273, 1987.
68. S. L. Miller *et al.* (1976; citation is in note 66), p. 64.
69. Toxic and corrosive? Within our biochemical citadels we take oxygen for granted, but rust and raging forest fires are a more immediate reminder of the potency of oxygen.

70. S. L. Miller (1987; citation is in note 66), p. 19.
71. S. L. Miller (1987; citation is in note 66), p. 19.
72. See W. L. Marshall in *Geochimica et Cosmochimica Acta*, vol. 58, pp. 2099–2106, 1994.
73. See J. P. Amend and E. L. Shock in *Science*, vol. 281, pp. 1659–1662, 1998, who calculate the energetics of amino acid synthesis in hydrothermal settings, but note that 'in all the calculations presented here [there] is the assumption that the appropriate enzymes for each step in the amino acid synthesis pathways are present and active', p. 1660.
74. See J. P. Dworkin *et al.* in *Proceedings of the National Academy of Sciences, USA*, vol. 98, pp. 815–819, 2001.
75. R. Shapiro (1987; citation is in note 16).
76. See M. Levy and S. L. Miller (1998; citation is in note 37). Concerning the many other difficulties in the prebiotic synthesis of adenine, see the article by R. Shapiro in *Origins of Life and Evolution of the Biosphere*, vol. 25, pp. 83–98, 1995.
77. These quotations are from pp. 182–184 of R. Shapiro (1987; citation is in note 16).
78. In R. F. Gesteland *et al.* (1998; citation is in note 3).
79. See G. F. Joyce *et al.* in *Nature*, vol. 310, pp. 602–604, 1984.
80. In R. F. Gesteland *et al.* (1998; citation is in note 3), p. 68.
81. In R. F. Gesteland *et al.* (1998; citation is in note 3), p. 72.
82. See, for example, C. M. Dobson *et al.* in *Proceedings of the National Academy of Sciences, USA*, vol. 97, pp. 11864–11868, 2000, as well as V. R. Oberbeck *et al.* in *Journal of Molecular Evolution*, vol. 32, pp. 296–303, 1991.
83. M. Eigen (1971; citation is in note 9), p. 519 (his emphasis).
84. See Iris Fry's paper in *Biology and Philosophy*, vol. 10, pp. 389–417, 1995 (see also note 89).
85. I. Fry (1995; citation is in note 84), p. 405.
86. K. D. James and A. D. Ellington (1995; citation is in note 40, Chapter 2), p. 528.
87. See p. 188 of P. Shapiro (1987; citation is in note 16).
88. See F. Crick's *Life itself: its origin and nature* (Macdonald, London, 1982).
89. F. Crick (1982; citation is in note 88), p. 38.
90. F. Crick (1982; citation is in note 88), p. 88.
91. See George Wald's article in *Scientific American*, vol. 191(2), pp. 44–53, 1954.
92. G. Wald (1954; citation is in note 91), p. 46.

5. UNIQUELY LUCKY? THE STRANGENESS OF EARTH

1. This chapter is an expansion of some brief remarks I made in Chapter 9 of *The Crucible of Creation: The Burgess Shale and the rise of animals* (Oxford University Press, Oxford, 1998). It also has many obvious parallels to *Rare Earth: Why complex life is uncommon in the Universe* by P. D. Ward and D. Brownlee (Springer-Verlag, [Copernicus], New York, 2000). Concerning the thesis of a rare Earth, this chapter is in broad agreement, although as far as possible I have emphasized the areas that Ward and Brownlee dealt with more cursorily, thus aiming to complement rather than duplicate their account.
2. Not surprisingly, given its relative proximity and especially the *Apollo* lunar missions, there is an extensive literature on the Moon and its geological history. Two of the many useful guides are: *Lunar sourcebook: a user's guide to the Moon*, edited by G. Heiken *et al.* (Cambridge University Press, Cambridge, 1991) and the chapter on the Moon by S. R. Taylor on pp. 247–275 of *Encyclopedia of the Solar System*, edited by P. R. Weissman *et al.* (Academic Press, San Diego, 1999).
3. The original observations are available (in Latin) on p. 276 of volume 1 of *The historical works of Gervase of Canterbury*, edited by W. Stubbs (Her Majesty's Stationery Office, London, 1879). A more accessible account may be found on pp. 50–51 of J. S. Lewis's *Rain of iron and ice: the very real threat of comet and asteroid bombardment* (Addison-Wesley [Helix Books],

Reading, MA, 1996). See also J. Hartung in *Meteoritics*, vol. 11, pp. 187–194, 1975, and *Lunar Science*, vol. VII, pp. 348–350, 1976. This interpretation was, however, queried by H. H. Nininger and G. I. Huss in *Meteoritics*, vol. 12, pp. 21–25, 1977. They suggested that it might more plausibly be interpreted as a meteorite that had entered the Earth's atmosphere, but happened to pass in front of the Moon's disc, so being mistaken for an impact. There is a further problem because such a major collision would lead to the ejection of huge quantities of rubble, which would easily escape from the weak gravitational field of Moon and subsequently be captured by the Earth. As P. Withers pointed out (in *Meteoritics & Planetary Sciences*, vol. 36, pp. 525–529, 2001), this would have produced a spectacular week-long meteor 'storm', for which there appears to be no historical evidence. Unless conditions were very overcast in England during June 1178, the date of the inferred impact, it is unlikely that Gervase would not have made some comment. This is because although Gervase's chronicles are mostly to do with matters of Church and State, he evidently had an eye for natural phenomena. These included eclipses and, more remarkably, an earthquake in 1158 that was felt across England. This is immediately followed by a report that in London the Thames dried out so that people could cross it dry-shod, but this apparently is an unrelated observation.

4 See W. K. Hartman *et al.* on pp. 493–512 of the book *Origin of the Earth and the Moon*, edited by R. M. Canup and K. Righter (University of Arizona Press, Tucson, 2000). In their judicious review they stress that the standard account of lunar bombardment, especially the intense major episode at *c*. 3.8–4.2 Ga cannot be accepted uncritically, and various alternatives need to be kept in mind. Evidence consistent with the late heavy bombardment episode, based on the ages of meteorites that have come from the Moon, is given by B. A. Cohen *et al.* in *Science*, vol. 290, pp. 1754–1756, 2000, while D. A. Kring and B. A. Cohen, in *Journal of Geophysical Research*, vol. 107 (E2), pp. 4-1-4-6, 2002, review the evidence for this cataclysmic episode throughout the inner Solar System.

5. See K. J. Zahnle and N. H. Sleep in their chapter (pp. 175–208) of *Comets and the origin of evolution of life*, edited by P. J. Thomas *et al.* (Springer, New York, 1997).

6. A. Morbidelli *et al.* in *Meteoritics & Planetary Science*, vol. 36, pp. 371–380, 2001, suggest that important components of the impactors were actually relatively small bolides, rather than the monsters usually envisaged, whose high-inclination orbits relative to the planets of the inner Solar System imparted high impact velocities.

7. See S. Chang on pp. 10–23 of *Early Life on Earth*, Nobel Symposium No. 84, edited by S. Bengtson (Columbia University Press, New York, 1994).

8. These details are taken from the paper by N. H. Sleep and K. Zahnle in *Journal of Geophysical Research*, vol. 103, pp. 28529–28544, 1998.

9. See, for example, A. P. Nutman *et al.* in *Geochimica et Cosmochimica Acta*, vol. 61, pp. 2475–2484, 1997; *Chemical Geology*, vol. 141, pp. 271–287, 1997; and *Precambrian Research*, vol. 78, pp. 1–39, 1996, as well as C. M. Fedo and M. J. Whitehouse in *Science*, vol. 296, pp. 1448–1452, 2002.

10. Claims for actual fossils, such as yeast-like organisms described by H. D. Pflug in *Naturwissenschaften*, vol. 65, pp. 611–615, 1978, have won almost no support.

11. Claims for an organic origin of the carbon, such as by S. J. Mojzsis *et al.* in *Nature*, vol. 384, pp. 56–59, 1996 (with a cautious commentary by J. M. Hayes on pp. 21–22), have been criticized; see J. M. Eiler *et al.* in vol. 386, p. 665, 1997. See also Fedo and Whitehouse (2002, citation is in note 9) and M. A. van Zuilen *et al.* in *Nature*, vol. 418, pp. 627–630, 2002. R. Schoenberg *et al.*, in *Science*, vol. 418, pp. 403–405, 2002, suggest some of the carbon may have a meteoric origin; see also A. D. Anbar *et al.* in *Journal of Geophysical Research*, vol. 106E, pp. 3219–3236, 2001.

12. The isotopic studies of $\delta^{13}C$ (i.e. the ratio of ^{12}C to ^{13}C) in the Isua Supergroup were pioneered by M. Schidlowski (see *Nature*, vol. 333, pp. 313–318, 1988), and are consistent with more recent work by M. T. Rosing in *Science*, vol. 283, pp. 674–676, 1999.

13. Metamorphic gneisses from north-west Canada are dated at 3960 Ma by S. A. Bowring *et al.* in *Geology*, vol. 17, pp. 971–975, 1989; mineral grains of zircons date back to *c.* 4400 Ma; see S. A. Wilde *et al.* in *Nature*, vol. 409, pp. 175–178, 2001. Although these grains have been reworked into younger rocks they still provide some indication of what the very early Earth was like, and notably indicate the existence of some sort of ocean; see also S. J. Mojzsis *et al.* in *Nature*, vol. 409, pp. 178–181, 2001.

14. See K. A. Maher and D. J. Stevenson in *Nature*, vol. 331, pp. 612–614, 1988; see also V. R. Overbeck and G. Fogleman in vol. 339, p. 434, 1989, and N. H. Sleep *et al.* in vol. 342, pp. 139–142, 1989. A more recent overview is given by K. J. Zahnle and N. H. Sleep (1997; citation is in note 5).

15. Respective estimates are given by B. A. Cohen *et al.* (2000) and D. A. Kring and B. A. Cohen (2002; citations in note 4).

16. See, for example, M. Gogarten-Boekels *et al.* in *Origins of Life and Evolution of the Biosphere*, vol. 25, pp. 251–264, 1995; see also N. H. Sleep *et al.*, in *Proceedings of the National Academy of Sciences, USA*, vol. 98, pp. 3666–3672, 2001.

17. See V. R. Oberbeck and G. Fogleman in *Origins of Life and Evolution of the Biosphere*, vol. 19, pp. 549–560, 1989; L. E. Orgel in a later issue (vol. 28, pp. 91–96, 1998) provides a characteristically level-headed assessment of this topic.

18. See A. Lazcano and S. L. Miller in *Journal of Molecular Evolution*, vol. 39, pp. 546–554, 1994.

19. See D. M. Raup and J. W. Valentine in *Proceedings of the National Academy of Sciences, USA*, vol. 80, pp. 2981–2984, 1983.

20. See K. O. Stetter on pp. 1–18 (including the discussion) of a Ciba Foundation Symposium (no. 202) *Evolution of hydrothermal ecosystems on Earth (and Mars?)*, edited by G. R. Bock and J. A. Goode (Wiley, Chichester, 1996).

21. See, for example, N. Galtier *et al.* in *Science* (vol. 283, pp. 220–221, 1999), as well as earlier remarks by P. Forterre, for example, in *Cell* (vol. 85, pp. 789–792, 1996) and *Current Opinion in Genetics & Development* (vol. 7, pp. 764–770, 1997); see also note 49, Chapter 4.

22. See N. H. Sleep and K. Zahnle (1998; citation is in note 8).

23. See J. W. Head *et al.* in *Science* (vol. 286, pp. 2134–2137, 1999). Other evidence for a much more equable early climate on Mars is given by J. B. Pollack *et al.* in *Icarus*, vol. 71, pp. 203–224, 1987; F. Forget and R. T. Pierrehumbert in *Science*, vol. 278, pp. 1273–1276, 1997; R. M. Huberle in *Journal of Geophysical Research*, vol. 103E, pp. 28467–28479, 1998; and M. A. Mischna in *Icarus*, vol. 145, pp. 546–554, 2000.

24. One such recent announcement is of a discovery from the Oman desert, announced by E. Gnos *et al.* in *Meteoritics & Planetary Sciences*, vol. 37, pp. 835–854, 2002.

25. Citations are in notes 9–12, Chapter 3.

26. These data and the following information are derived from B. Gladman's paper in *Icarus*, vol. 130, pp. 228–246, 1997.

27. See P. Davies on pp. 304–317 (including discussion) of G. R. Bock and J. A. Goode (1996; citation is in note 20).

28. See R. M. E. Mastrapa *et al.* in *Earth and Planetary Science Letters*, vol. 189, pp. 1–8, 2001.

29. B. Gladman (1997; citation is in note 26) comments that the thickness of the Earth's atmosphere and the relatively small target size of Mars make both dispatch and destination respectively more problematic. See also C. Mileikowsky *et al.* in *Icarus*, vol. 145, pp. 391–427, 2000.

30. The exact quotation is 'So I tell my students: learn your biochemistry here and you will be able to pass examinations on Arcturus', p. 16 of *Life beyond Earth and the mind of Man*, edited by R. Berendzen, NASA Scientific & Technical Information Office, 1973.

31. The title of Norman Pace's paper in *Proceedings of the National Academy of Sciences, USA*, vol. 98, pp. 805–808, 2001.

32. N. Pace (2001; citation is in note 31), p. 806.

33. This is a very fast-moving field, with new results constantly being posted on websites: www.exoplanets.org and www.obspm.fr/planets for example. Much of the information on these extra-solar planets given below is derived from these sources. Published reviews include those by J. T. Lunine in *Proceedings of the National Academy of Sciences, USA*, vol. 96, pp. 5353–5356, 1999 and G. W. Marcy in his chapter (pp. 1285–1311) in *Protostars and planets*, edited by V. Mannings *et al.* (University of Arizona Press, Tucson, 2000). Relevant books include the fairly technical (S. Clark's *Extrasolar planets: The search for new worlds* (John Wiley [Praxis], Chichester, 1998); and, more accessible, D. Goldsmith's *Worlds unnumbered: the search for extrasolar planets* (University Science Books, Sausalito, CA, 1997).

34. See A. Vidal-Madjar *et al.* in *Planetary and Space Science*, vol. 46, pp. 629–648, 1998, who suggests that gravitational perturbations and an occultation of the star hint at the presence of giant planets within the disc.

35. Direct detection of a planet orbiting the star HD 209458 using photometry is reported by D. Charbonneau *et al.* in *Astrophysical Journal*, vol. 529, pp. L45–L48, 2000. This planet had already been detected by the radial velocity technique, but transit observations of the light flux from the star provide other data. For example, it is significantly less dense than Saturn, the least dense of the Solar System planets.

36. See R. P. Butler *et al.* in *Astrophysical Journal*, vol. 526, pp. 916–927, 1997, and D. Goldsmith's report in *Science*, vol. 296, p. 1951, 2002.

37. So far as I am aware a coherent consideration of life in a supergravity environment is not available. The well-known essay *On being the right size* by J. B. S. Haldane (published in his *Possible worlds and other essays* (Chatto and Windus, London, 1927)), as are various sections in C. J. Pennycuick's *Newton rules biology: A physical approach to biological problems* (Oxford University Press, Oxford, 1992), see especially p. 53. The role of hypergravity (and its reverse, weightlessness) has received extensive study with respect to rocket-borne ascent and exploration of the Solar System; see, for example, *Proceedings of the Annual Meetings of the IUPS Commission on Gravitational Physiology*, published as supplements to *The Physiologist* (including vol. 23 (Suppl. 6), 1980; vol. 24 (Suppl. 6), 1981; vol. 26 (Suppl. 6), 1983; vol. 28 (Suppl. 6), 1985; vol. 35 (Suppl. 1), 1992; vol. 36 (Suppl. 1), 1993). G. A. Cavagna *et al.* in *Journal of Physiology*, vol. 528, pp. 657–668, 2000 (see also the commentary by A. E. Minetti in *Nature*, vol. 409, pp. 467–468, 2001) assess the role of gravity on human walking, both at lower values (equivalent to Mars) and at one-and-a-half times the Earth's gravitational field.

38. See A. C. Cameron in *Astronomy & Geophysics*, vol. 43, pp. 4.21–4.25, 2002.

39. See, for example, J. E. Chambers and G. W. Wetherill in *Icarus*, vol. 136, pp. 304–327, 1998.

40. 'Making more terrestrial planets' is the title of J. E. Chambers's paper in *Icarus*, vol. 152, pp. 205–224, 2001.

41. J. E. Chambers (2001; citation is in note 40), p. 212.

42. J. E. Chambers (2001; citation is in note 40), p. 223.

43. See D. N. Lin *et al.* in *Nature*, vol. 380, pp. 606–607, 1996; see also J. C. B. Paploizou and J. D. Larwood in *Monthly Notices of the Royal Astronomical Society*, vol. 315, pp. 823–833, 2000. In at least one case there is also compelling evidence that this process was both early and relatively rapid; see J. I. Lunine in *Proceedings of the National Academy of Sciences, USA*, vol. 98, pp. 809–814, 2001.

44. See G. Israelian *et al.* in *Nature*, vol. 411, pp. 163–166, 2001. The evidence is based on the presence of a lithium isotope (^{6}Li), which is expected to occur in the early atmospheres of giant planets, but is later destroyed as the star evolves.

45. See F. A. Rasio and E. B. Ford in *Science*, vol. 274, 954–956, 1996; S. J. Weidenschilling and F. Marzari in *Nature*, vol. 384, pp. 619–621, 1996; N. Murray *et al.* in *Science*, vol. 279, pp. 69–72, 1998; and F. Marzari and S. J. Weidenschilling in *Icarus*, vol. 156, pp. 570–579, 2002. In this last paper a 'jumping Jupiter' model is proposed to explain the 'hot Jupiter' close

to its sun and to infer the existence of a more distant planet forming a stable system arising from a chaotic beginning.

46. S. J. Weidenschilling and F. Marzari (1996; citation is in note 45), p. 620.

47. For an interesting discussion of the possibility of habitable moons orbiting extra-solar giant planets see D. M. Williams *et al.* in *Nature*, vol. 385, pp. 234–236, 1997 (and the accompanying commentary by C. F. Chyba on p. 201, which in part is also a paean for Carl Sagan).

48. G. Gonzalez *et al.* in *Icarus*, vol. 152, pp. 185–200, 2001.

49. See, for example, papers by K. K. Khurana *et al.* in *Nature*, vol. 395, pp. 777–780, 1998 (and accompanying commentary on pp. 749, 751 by F. Neubauer); R. T. Pappalardo *et al.* in *Journal of Geophysical Research*, vol. 104E, pp. 24015–24055, 1999; J. S. Kargel *et al.* in *Icarus*, vol. 148, pp. 226–265, 2000; and M. G. Kivelson *et al.* in *Science*, vol. 289, pp. 1340–1343, 2000; and X-P. Wu *et al.* in *Geophysical Research Letters*, vol. 28, pp. 2245–2248, 2001. There is also a possibility that some sort of sub-surface ocean exists on another Galilean satellite, Callisto; see J. Ruiz in *Nature*, vol. 412, pp. 409–411, 2001, and the commentary by K. A. Bennett on pp. 395–396.

50. See C. K. Gessmann and D. C. Rubie in *Earth and Planetary Science Letters*, vol. 184, pp. 95–107, 2000, and A. G. W. Cameron in *Meteoritics & Planetary Science*, vol. 36, pp. 9–22, 2001; the latter account provides a helpful introduction to the earlier literature.

51. See C. Alibert *et al.* in *Geochimica et Cosmochimica Acta*, vol. 58, pp. 2921–2926, 1994.

52. See the papers by M. T. McCulloch in *Earth and Planetary Science Letters*, vol. 126, pp. 1–13, 1994; C. J. Allègre and *et al.* in *Geochimica et Cosmochimica Acta*, vol. 59, pp. 1445–1456, 1995; M. Ozima and F. A. Podosek in *Meteoritics & Planetary Sciences*, vol. 33 (4, Suppl), p. A120, 1998; and F. A. Podosek in *Science*, vol. 283, pp. 1863–1864, 1999.

53. See, for example, D-C. Lee *et al.* in *Earth and Planetary Science Letters*, vol. 198, pp. 267–274, 2002.

54. An argument based on tungsten isotopes; see A. N. Halliday and D-C. Lee in *Geochimica et Cosmochimica Acta*, vol. 63, pp. 4157–4179, 1997.

55. See R. M. Canup and E. Asphaug in *Nature*, vol. 412, pp. 708–712, 2001, as well as the accompanying commentary by J. Melosh on pp. 694–695.

56. See A. G. W. Cameron and W. Benz in *Icarus*, vol. 92, pp. 204–216, 1991.

57. This paper, by A. G. W. Cameron, the fifth in a series exploring the so-called Giant Impact Hypothesis, is in *Icarus*, vol. 126, pp. 126–137, 1997. For an update and overview see A. G. W. Cameron's paper in *Meteoritics & Planetary Science*, vol. 36, pp. 9–22, 2001.

58. See R. M. Canup and L. W. Esposito in *Icarus*, vol. 119, pp. 427–426, 1996.

59. R. M. Canup and L. W. Esposito (1996; citation is in note 58), p. 429.

60. See S. Ida *et al.* in *Nature*, vol. 389, pp. 353–357, 1997.

61. This is the title of a commentary on the paper by S. Ida *et al.* (1997; citation is in note 60), published in the same issue of *Nature*, on pp. 327–328.

62. A. G. W. Cameron and W. Benz (1991; citation is in note 56).

63. A. G. W. Cameron and W. Benz (1991; citation is in note 56), p. 215.

64. See P. D. Ward and D. Brownlee (2000; citation is in note 1), pp. 222–234.

65. See J. Laskar and R. Robutel in *Nature*, vol. 361, pp. 615–617, 1993.

66. The principal exponent of these ideas is G. Williams, who has published extensively on this topic, including a recent paper in *Sedimentary Geology*, vol. 120, pp. 55–74, 1998; see also *Nature*, vol. 396, pp. 453–455, 1998.

67. See J. Laskar and P. Robutel, in *Nature*, vol. 361, pp. 608–612, 1993.

68. See Neil Comins's intriguing book *What if the Moon didn't exist: voyages to Earths that might have been* (HarperCollins, New York, 1993), which in addition to the title's consideration looks at a series of other questions, such as what the Earth would be like if its obliquity were much greater (as for Uranus), what might happen if a black hole met the Earth, and if the Earth were smaller; this last topic is returned to briefly below.

69. See J. P. Vanyo and S. M. Awramik in *Precambrian Research*, vol. 29, pp. 121–142, 1985.

70. See W. Benz *et al.* in *Icarus*, vol. 74, pp. 516–528, 1988.

71. See, in particular, D. J. Stevenson in *Annual Review of Earth and Planetary Sciences*, vol. 15, pp. 271–315, 1987.

72. Mars has two moons (Deimos and Phobos), but these are very small, respectively with maximum lengths of 16 and 26 km. See P. C. Thomas on pp. 309–314 of *Encyclopedia of the Solar System*, edited by P. R. Weissman (Academic Press, San Diego, 1999). These moons are usually interpreted as captured asteroids, but, as Thomas notes, this is very difficult to explain given their equatorial orbits. In addition, the long-term future of Phobos is uncertain, and calculations suggest that its orbit will decay by tidal friction, with the moon crashing onto the Martian surface within 100 Ma.

73. His stimulating article *How common are habitable planets?* may be found on pp. C11–C14 in a special issue of *Nature* (vol. 402, 1999). N. F. Comins (see note 68) also explores in some detail the consequences of living on a planet smaller than the Earth.

74. G. Gonzalez in *Astronomy and Geophysics*, vol. 40, pp. 318–320, 1999 explores some of these points, remarking on both the roundness of the Moon (and the Sun) and the exactness of the apparent sizes of each body (a tiny 3.7 arc minutes) allowing the Moon to block the bright photosphere but not the dramatic chromosphere. Gonzalez also reminds us of the various happy coincidences – the nature of the Sun, the Earth's axial stability, the origin of the Moon by impact, and the size of the Earth – that are probably prerequisites for astronomers to evolve.

75. See pp. 210–213 of M. E. Bakich's gazetteer to the Solar System, *The Cambridge Planetary Handbook* (Cambridge University Press, Cambridge, 2000). See also D. H. Levy's *Impact Jupiter: The crash of comet Shoemaker–Levy 9* (Plenum, New York, 1995).

76. George Wetherill's paper 'Possible consequences of absence of "Jupiters" in planetary systems' was published in *Astrophysics and Space Science*, vol. 212, pp. 23–32, 1994; see also his commentary in *Nature*, vol. 373, p. 470, 1995. The role of Jupiter as 'goal-keeper' is also stressed by Ward and Brownlee (2000; citation in note 1), see pp. 235–242.

77. See P. R. Weissman in *Nature*, vol. 344, pp. 825–830, 1990, as well as his earlier paper (pp. 272–282) in *Dynamics of the Solar System*, edited by R. L. Duncombe (Reidel, Dordrecht, 1979).

78. G. Wetherill (1994; citation is in note 76), p. 23.

79. Two accessible introductions are W. Alvarez's *T. rex and the crater of doom* (Princeton University Press, Princeton, 1997) and C. Frankel's *The end of the dinosaurs: Chicxulub crater and mass extinctions* (Cambridge University Press, Cambridge, 1999).

80. See J. W. Kirchner and A. Weil in *Nature*, vol. 404, pp. 177–180, 2000, and commentary by D. Erwin on pp. 129–130.

81. See the paper by J. K. Schubert and D. J. Bottjer in *Geology*, vol. 20, pp. 883–886, 1992, and the summary by D. J. Bottjer *et al.* in *Geological Society of London Special Publication*, vol. 83, pp. 7–26, 1995.

82. See P. B. Wignall and R. J. Twitchett in *Sedimentology*, vol. 46, pp. 303–316, 1999. Other parallels to sedimentary features typical of the earlier Precambrian environment might include large-scale precipitation of sea-floor carbonate cements; see A. D. Woods *et al.* in *Geology*, vol. 27, pp. 645–648, 1999.

83. See S. D'Hondt *et al.* in *Science*, vol. 282, pp. 276–279, 1998.

84. See the reports by G. A. Logan *et al.* in *Nature*, vol. 376, pp. 53–56, 1995, with an accompanying commentary on pp. 16–17 by M. Walter; and in *Geochimica et Cosmochimica Acta*, vol. 61, pp. 5391–5409, 1997.

85. See J. S. Lewis (1996; citation is in note 3), Chapter 14.

86. Concerning displacement of bodies from the asteroid belt see, for example, the papers by P. Farinella *et al.* in *Nature*, vol. 371, pp. 314–317, 1994 and B. J. Gladman *et al.* in *Science*, vol. 277, pp. 197–201, 1997.

87. See, for example, the interesting studies by J. J. Sepkoski on pp. 211–255 of *Evolutionary paleobiology: in honor of James W. Valentine*, edited by D. Jablonski *et al.* (University of Chicago Press, Chicago; 1996), and with others in *Paleobiology*, vol. 26, pp. 7–18, 2000.

88. Concerning the final destiny of the Earth and the Solar System see the discussion by K. P. Rybicki and C. Denis in *Icarus*, vol. 151, pp. 130–137, 2001. See also notes 105–107.

89. Evidence that the bulk of the Earth's water derives from the outer asteroid belt is addressed by A. Morbidelli *et al.* in *Meteoritics & Planetary Science*, vol. 35, pp. 1309–1320, 2000.

90. See J. I. Lunine (2001; citation is in note 43).

91. See C. F. Chyba and C. Sagan on pp. 147–173 of *Comets and the origin and evolution of life*, edited by P. J. Thomas *et al.* (Springer, New York; 1997), and A. H. Delsemme in *Advances in Space Research*, vol. 15(3), pp. 49–57, 1995.

92. C. F. Chyba in *Advances in Space Research*, 15 (3), pp. 45–48, 1995.

93. See papers by Z. Sekanina in *Icarus*, vol. 27, pp. 123–133, 1976, and more especially T. A. McGlynn and R. D. Chapman in *The Astrophysical Journal*, vol. 346, pp. L105–L108, 1989.

94. Evidence for this is discussed by A. Vidal-Madjar *et al.* in one of their series of papers on β Pictoris, specifically in *Astronomy and Astrophysics* (vol. 290, pp. 245–258, 1994) (see also in the same journal A. M. Lagrange-Henri *et al.*, vol. 190, pp. 275–282, 1988, and H. Beust *et al.*, vol. 236, pp. 202–216, 1989).

95. See S. A. Stern *et al.* in *Nature*, vol. 345, pp. 305–308, 1990.

96. See S. A. Stern *et al.* in *Icarus*, vol. 91, pp. 65–75, 1991; this search entailed 17 nearby stars and the authors emphasize the problems of detectability with present technology.

97. See G. J. Melnick *et al.* in *Nature*, vol. 412, pp. 160–163, 2001.

98. Although Chyba (1995; citation is in note 92) reminds us that it is possible to envisage a solar system whose inner planets have received volatile delivery, but lacks an equivalent of the Oort Cloud.

99. Three such episodes have been identified in the Precambrian, one about 2200 Ma ago (see, for example, D. A. Evans *et al.* in *Nature*, vol. 386, pp. 262–266, 1997; D. McB. Martin in *Geological Society of America Bulletin*, vol. 111, pp. 189–203, 1999; and J. L. Kirschvink *et al.* in *Proceedings of the National Academy of Sciences, USA*, vol. 97, pp. 1400–1405, 2000), and two at *c.* 700 Myr and 600 Myr ago (see P. Hoffman *et al.* in *Science*, vol. 281, pp. 1342–1346, 1998, and supporting blasts in *Nature*, vol. 397, p. 384, 1999 and vol. 400, p. 708, 1999; but see also G. S. Jenkins and C. R. Scotese in *Science*, vol. 282, pp. 1644–1645, 1998 (response by Hoffman *et al.* follows in pp. 1645–1646), G. S. Jenkins and L. A. Frakes in *Geophysical Research Letters*, vol. 25, pp. 3525–3528, 1998; W. T. Hyde *et al.* in *Nature*, vol. 405, pp. 425–429, 2000; and R. A. Kerr in *Science*, vol. 287, pp. 1734–1736, 2000; and M. J. Kennedy *et al.* in *Geology*, vol. 29, pp. 443–446, 2001).

100. See H. C. Jenkyns and P. A. Wilson in *American Journal of Science*, vol. 299, pp. 341–392, 1999.

101. See, for example, James Lovelock's *Gaia: A new look at life on Earth* (Oxford University Press, Oxford, 1979) and *The ages of Gaia: A biography of our living Earth* (Oxford University Press, Oxford, 1988).

102. See, for example, Michael Hart's paper in *Icarus*, vol. 37, pp. 351–357, 1979.

103. A topic explored by J. Laskar in *Nature*, vol. 338, pp. 237–238, 1989.

104. See J. F. Kasting *et al.* in *Icarus*, vol. 101, pp. 108–138, 1993 and J. F. Kasting in *Origins of Life and Evolution of the Biosphere*, vol. 27, pp. 291–307, 1997.

105. See S. Franck *et al.* in *Naturwissenschaften*, vol. 88, pp. 416–426, 2001; see also the papers by S. Franck *et al.* in *Journal of Geophysical Research*, vol. 105E, pp. 1651–1658, 2000 and *Planetary and Space Science*, vol. 48, pp. 1099–1105, 2000 (see also note 88).

106. See the papers by S. Franck *et al.* in *Chemical Geology*, vol. 159, pp. 305–317, 1999, and *Tellus*, vol. 52B, pp. 94–107, 2000.

107. Franck *et al.* (2001; citation is in note 105), p. 420. One should note that they qualify this by remarking that chemolithotrophic hyperthermophiles might persist for longer but 'all higher

forms of life would certainly be eliminated'. Not all authors are quite so pessimistic; T. M. Lenton and W. von Bloh (in *Geophysical Research Letters*, vol. 28, pp. 1715–1718, 2001) give an extra 300 Ma (and perhaps a little more).

108. See C. H. Lineweaver in *Icarus*, vol. 151, pp. 307–313, 2001.

109. See for example, G. Gonzalez *et al.* in the *Astrophysical Journal*, vol. 511, pp. L111–L114, 1999. Although this correlation of metallicity and the presence of planets is strong, it is not inevitable. In *Astronomy & Geophysics* (vol. 40, pp. 5.25–5.29, 1999) G. Gonzalez has also suggested that in a number of respects, including its metallicity, our Sun may be anomalous. However, I. N. Reid (in *Publications of the Astronomical Society of the Pacific*, vol. 114, pp. 306–329, 2002) suggests that while there is a strong correlation between stellar metallicity and the presence of planets, our Sun falls in the mid-range of such metallicities.

110. However, S. J. Vogt *et al.* report (in *Astrophysical Journal*, vol. 536, pp. 902–914, 2000) six new planetary systems, of which two are metal poor.

111. See, for example, R. Jimenez *et al.* in *Astrophysical Journal*, vol. 561, pp. 171–177, 2001.

112. See the article 'How old is ET?' by R. P. Norris on pp. 103–105 of *When SETI succeeds: The impact of high-information contact*, edited by A. Tough (Foundation for the Future, Bellevue WA, 2000).

113. R. P. Norris (2000; citation is in note 112), p. 105.

114. See James Annis's article in *Journal of the British Interplanetary Society*, vol. 52, pp. 19–22, 1999.

115. J. Annis (1999; citation is in note 114), p. 22.

116. See the paper by G. Gonzalez *et al.* in *Icarus*, vol. 152, pp. 185–200; 2001a, as well as a more popular overview in *Scientific American*, vol. 285 (4), pp. 52–59, 2001b. In the context of a special position of the Solar System with respect to our galaxy, see also the earlier paper by L. S. Marochnik in *Astrophysics and Space Science*, vol. 89, pp. 61–75, 1983.

117. G. Gonzalez *et al.* (2001b; citation is in note 116), p. 59, left-hand column.

118. G. Gonzalez *et al.* (2001b; citation is in note 116), p. 59, right-hand column.

6. CONVERGING ON THE EXTREME

1. This conceit is taken from my article in *Proceedings of the Royal Institution of Great Britain*, vol. 70, pp. 21–62, 1999, see p. 24.

2. See *The crucible of creation: The Burgess Shale and the rise of animals* (Oxford University Press, Oxford, 1998).

3. This has not prevented, of course, speculation of possible alternatives. In terms of biology there is D. Dixon's entertaining *After Man: A zoology of the future* (London, Granada; 1981). Not surprisingly most emphasis, however, has been on historical counterfactuals, such as *If it had happened otherwise: lapses into imaginary history* (Longmans, Green, London, 1931), *Virtual history: alternatives and counterfactuals* (Macmillan, London, 1998) and *What if, military historians imagine what might have been* (London, Pan, 2001) edited respectively by J. C. Squire, N. Ferguson, and R. Cowley. The realm of alternative histories has also attracted novelists, such as Kingsley Amis in his witty *The alteration* (Penguin, London, 1988), and the chilling *The sound of his horn* (Tartarus, Horam, 1999) by Sarban (John William Wall). Others have considered how past events might be manipulated; of these, Orson Scott Card's poignant *Pastwatch: the redemption of Christopher Columbus* (Tom Doherty, New York, 1996) is a haunting example of a world that might have been.

4. After a rather dull period, matters have become quite polarized between enthusiasts for the Search for Extra-Terrestrial Intelligence (SETI) engaged in debate with the sceptics, which include not only myself but such individuals as I. Crawford; see part 4, pp. 24–26, and part 6, p. 19 in *Astronomy & Geophysics*, vol. 38, 1997, as well as *Scientific American*, vol. 283 (1), pp. 28–33, 2000.

5. A. Y. Mulkijanian and W. Junge, in *Photosynthesis Research*, vol. 51, pp. 27–42, 1997, suggest that the origins of photosynthesis lie in protective measures of pigmented proteins against

ultraviolet radiation. Given that this UV flux will be the norm on primitive planets, and the fact that such building blocks of chlorophyll as quinones are synthesized in interstellar space (see Chapter 3), so chlorophyll may be a universal molecule (note 13).

6. The general consensus seems to be that this endosymbiotic capture of the cyanobacteria/chloroplast (and the comparable case of α-proteobacteria captured to yield in due course the mitochondria) were unique events (see, for example, A. J. Roger and C. F. Delwiche respectively in *American Naturalist*, vol. 154, pp. S146–S163 and S164–S177, 1999; but see J. W. Stiller *et al.* in *Journal of Phycology*, vol. 39, pp. 95–105, 2003). Even so, although rare, there are a number of other cases of what is called secondary endosymbiosis. This occurs when two separate eukaryotes (or at least the plastid component) combine to form a single organism. See, for example, papers by S. P. Gibbs in *Annals of the New York Academy of Sciences*, vol. 361, pp. 193–207, 1981, S. E. Douglas *et al.* in *Nature*, vol. 350, pp. 148–151, 1991; S. Köhler *et al.* in *Science*, vol. 275, pp. 1485–1489, 1997, Z-d. Zhang *et al.* in *Nature*, vol. 400, pp. 155–159, 1999; K. Takishita and A. Uchida in *Phycological Research*, vol. 47, pp. 207–216, 1999, and *Journal of Molecular Evolution*, vol. 51, pp. 26–40, 2000 respectively; and J. F. Saldarriaga *et al.* in *Journal of Molecular Evolution*, vol. 53, pp. 204–213, 2001 (who also present evidence for multiple loss of plastids). Helpful overviews of this area are provided by G. McFadden and P. Gilson in *Trends in Ecology & Evolution*, vol. 10, pp. 12–17, 1995 and T. Cavalier-Smith in *Trends in Plant Sciences*, vol. 5, pp. 174–182, 2000. Finally, and to reinforce the likelihood that symbiotic unions between species are, in the grand order of things, overwhelmingly probable, see the case in the insects where the symbiotic bacteria have themselves symbiotic bacteria; see C. D. von Dohlen *et al.* in *Nature*, vol. 412, pp. 433–436, 2001 (see also note 48).

7. Interestingly, the first major glaciations appear to date from the same interval (see, for example, J. L. Kirschvink *et al.* in *Proceedings of the National Academy of Sciences, USA*, vol. 97, pp. 1400–1405, 2000). One possibility is that as levels of oxygen increased so the amount of reactive methane, a powerful greenhouse gas, declined. This in turn led to planetary cooling and ultimately severe, perhaps catastrophic, glaciation (see A. A. Pavlov *et al.* in *Journal of Geophysical Research*, vol. 105E, pp. 11981–11990, 2000).

8. See, however, Chapter 5 (pp. 72–73) for a more extended discussion of the significance of the oldest sedimentary carbon.

9. The third carbon isotope, carbon 14 (^{14}C), is radioactive and is produced principally by cosmic rays striking atoms of nitrogen. The relatively short half-life of ^{14}C means that it plays no effective part in the biochemistry of photosynthesis and the geological burial of carbon.

10. As ever, this statement requires scientific qualification. Fractionation of carbon isotopes, denoted by δ^{13}C (the ratio of ^{12}C to ^{13}C against an agreed standard that by definition has a δ^{13}C of zero), occurs in other biochemical systems, such as methane generation by some bacteria (methanogens). To the first approximation different metabolic processes imprint separable isotopic signatures, with photosynthetic fractionation averaging *c.* 27‰ δ^{13}C.

11. For helpful reviews of RuBisCO see the articles by F. C. Hartman and M. R. Harpel in *Annual Review of Biochemistry*, vol. 63, pp.197–234, 1994 and G. Schneider *et al.* in *Annual Review of Biophysics and Biomolecular Structure*, vol. 21, pp. 119–143, 1992. Some interesting aspects of its molecular structure and possibly severe constraints on its function are addressed by E. A. Kellogg and N. D. Juliano in *American Journal of Botany*, vol. 84, pp. 413–428, 1997.

12. The prokaryotic alga *Acaryochloris* lives within the tissues of tunicates and contains significant proportions of chlorophyll *d*; see H. Miyashita *et al.* in *Nature*, vol. 383, p. 402, 1996, H. Schiller *et al.* in *FEBS Letters*, vol. 410, pp. 433–436, 1997. Concerning a similar feature in the green alga *Ostreobium*, see P. Halldal in *Biological Bulletin*, vol. 134, pp. 411–424, 1968, and B. Koehne *et al.* in *Biochimica et Biophysica Acta*, vol. 1412, pp. 94–107, 1999. In part, at least, the red-shift is due to the shaded habitats of these photosynthesizers.

13. This quotation (pp. 13–14) is taken from George Wald's stimulating and thoughtful paper 'Fitness in the Universe: Choices and necessities' in *Origins of Life and Evolution of the Biosphere*, vol. 5, pp. 7–27, 1974.

14. R. D. Wolstencroft and J. A. Raven, in *Icarus*, vol. 157, pp. 535–548, 2002, have an interesting discussion of how likely it is that photosynthesis has evolved on remote planets, and the degree of similarity there might be to Earth. They conclude that the development of an oxygenated atmosphere is very likely, although the energetics of photon capture and specifically the number required to reduce a molecule of CO_2 might differ.

15. D. J. DeRosier gives a helpful introduction to the flagellar motors of bacteria in *Cell*, vol. 93, pp. 17–20, 1998, as do R. M. Berry and J. P. Armitage in *Advances in Microbial Physiology*, vol. 41, pp. 291–337, 1999. See also in a thematic series 'The molecular physics of biological movement', in *Philosophical Transactions of the Royal Society of London B*, vol. 355, 2000 the papers by H. C. Berg (pp. 491–501) and R. M. Berry (pp. 503–509). Rotary activity is also reported in eukaryotic cells, albeit involving the microtubules within the flagellum; see C. K. Omoto and G. B. Witman in *Nature*, vol. 290, pp. 708–710, 1981.

16. More typical speeds are of the order of 200 rev/second, but Y. Magariyama *et al.* report speeds of up to 1700 rev/second in the bacterium *Vibrio alginolyticus*; see *Nature*, vol. 371, p. 752, 1994.

17. See, for example, the account of how the filament is constructed given by K. Yonekura *et al.* in *Science*, vol. 290, pp. 2148–2152, 2000, with an accompanying commentary by R. M. Macnab on pp. 2086–2087, where he remarks that the assembly of this structure by the bacterium 'is a much more sophisticated process than any of us could have envisaged', p. 2087.

18. See, in particular, M. J. Behe's *Darwin's black box: The biochemical challenge to evolution* (The Free Press, New York, 1996). Unsurprisingly, the views of Behe have come under fierce and justified criticism. As might be expected, orthodox Darwinists are willing to wade in, but so are those who are keen to reconcile science and religion: see, in particular, K. R. Miller's thoughtful *Finding Darwin's God: A scientist's search for common ground between God and evolution* (HarperCollins [Cliff Street Books], New York, 1999), especially Chapter 5.

19. This is F_1-ATPase, a component of the enzyme ATP synthase, which is located in the mitochondria. See K. Kinosita *et al.* in *Cell*, vol. 93, pp. 21–24, 1998, two papers in *Philosophical Transactions of the Royal Society of London B*, vol. 355, 2000, by A. G. W. Leslie and J. E. Walker (pp. 465–472) and K. Kinosita *et al.* (pp. 473–489) respectively; and M. L. Hutcheon *et al.* in *Proceedings of the National Academy of Sciences, USA*, vol. 98, pp. 8519–8524, 2001.

20. See the paper *Locomotion like a wheel?* by R. Full *et al.* in *Nature*, vol. 365, p. 495, 1993. An analogy of a kind can also be found in the aquatic larvae of mosquitoes, which execute a series of somersaults and still travel in a straight line through the water; see J. Brackenbury in *Journal of Experimental Biology*, vol. 202, pp. 2521–2529, 1999 for a revealing description of this machine-like behaviour.

21. One should also note the convergences between the pangolin (an Old World group) and the New World xenarthrans, which include the anteaters and armadillos; see F. Delsuc *et al.* in *Proceedings of the Royal Society of London B*, vol. 268, pp. 1605–1615, 2001, where their phylogenetic analysis 'confirms that morphological similarities between xenarthrans and pangolins are a spectacular example of convergence in relation to myrmecophagy [ant-eating]' (p. 1612). The extent of the similarities between pangolins and xenarthrans is emphasized by K. Z. Reiss in *American Zoologist*, vol. 41, pp. 507–525, 2001, where she has an interesting discussion of the nature of convergences in both a phylogenetic and adaptational framework. Importantly she comments that 'there is no evidence that phylogenetic constraints have facilitated or prevented the evolution of cranial muscle specializations for myrmecophagy' (p. 520). These conclusions on convergence echo earlier work by K. D. Rose and R. J. Emry on pp. 81–102 of *Mammal phylogeny: Placentals*, edited by F. S. Szalay *et al.* (Springer, New York, 1993). Such critical assessments are of considerable importance in phylogenetics,

because prior analysis executed by hard-line cladists had strongly supported a link between the pangolins and xenarthrans. As Rose and Emry comment, a credulous oversimplification of 'character states to facilitate algorithm-driven analyses often leads to unwarranted confidence in some phylogenetic conclusions – for instance, the conclusion of synapomorphy where homoplasy is more probable' (p. 82). In the context of convergence among the anteaters it is also worth remarking that a number of fossil forms, variously assigned to ant-eating groups, may well be mistakenly allied because of unappreciated convergences in form and function. Thus, *Ernanodon* from the Paleocene of China is probably convergent on the xenarthrans (Rose and Emry 1993, p. 96; see also T. J. Gaudin in *Fieldiana: Geology*, vol. 41 (new series), pp. 1–38, 1999), while the much-discussed *Eurotamandua* from the famous Messel oil shale (and also the Geiseltal lignite) of Eocene age in Germany poses serious problems concerning the biogeographic dispersal of the xenarthran anteaters, problems that as Delsuc *et al.* (2001) remark can be resolved if 'the striking morphological resemblances between this taxon [*Eurotamandua*] and Myrmecophagidae [anteaters] might be the result of adaptive convergence towards fossoriality and ant feeding. This example once again underlines how morphological adaptation to similar ecological niches could be positive misleading in terms of phylogenetic signal' (p. 1613).

22. A useful escape reaction, not least because the aboriginal tribe regards the pangolin as a great delicacy. Details (of the escape tactics, not the recipe) are given by R. R. Tenaza in *Journal of Mammalogy*, vol. 56, p. 257, 1975.

23. These examples are mentioned by R. L. Caldwell in *Nature*, vol. 282, pp. 71–73, 1979, a paper whose principal purpose is to draw attention to the backward somersaulting abilities of the stomatopods (see also note 20).

24. G. A. Gill and A. G. Coates, in *Lethaia*, vol. 10, pp. 119–134, 1977 remark, it is less clear whether rolling of these more or less spherical coral colonies is achieved only when nudged by another animal, say a fish, or even whether the colony remains *in situ* and grows by periodic 'press-ups'; see also J. B. Lewis's discussion of spherical *Siderastrea* (*S. radians*) in *Coral Reefs* 7, pp. 161–167, 1989.

25. LaBarbera's paper is entitled 'Why the wheels won't go', published in *American Naturalist*, vol. 121, pp. 395–408, 1983. Also of interest are the remarks in 'Why legs and not wheels' by I. Walker in *Acta Biotheoretica*, vol. 39, pp. 151–155, 1991, although her emphasis is more on the contrast between biological and man-made systems.

26. Philip Pullman, in the last book of his neo-secular fantasy, *The amber spyglass* (Scholastic, London, 2000), provides an ingenious portrayal of the wheeled malefa. Their wheels, however, are derived from seed-pods, and the axle is effectively formed by a powerful claw with an extending spur that locks into the pod. The reliance on the trees and their pods leads to an ecological interdependence that provides Pullman with some graphic and imaginative writing.

27. See R. A. Fortey's book *Life: An unauthorized biography* (HarperCollins, London, 1997).

28. R. A. Fortey (1997; citation is in note 27), p. 325.

29. R. A. Fortey was not the first to speculate on the existence of giant floating organic bladders. Carl Sagan, for example, in *Cosmos* (Macdonald, London, 1981, see pp. 40–43) conjures up what he dubs 'floaters' living high in the turbulent atmosphere of Jupiter-like planets. They would, he proposed, either consume organic material synthesized abiotically in the reducing atmosphere (in a Miller–Urey like fashion, see Chapter 4, pp. 60–62) or make their own food in a way analogous to photosynthetic plants. These organisms, unlike the terrestrial Fortean bladders, had, of course, a ready supply of hydrogen. Jupiter, too, is conceivably an abode of life, but as R. D. MacElroy stresses in his chapter (pp. 69–93) in *Chemical evolution of the giant planets*, edited by C. Ponnamperuma (Academic Press, New York, 1976) there are very considerable difficulties. So far as Fortean bladders are concerned, the extreme atmospheric turbulence would seem to make their survival problematic. As MacElroy writes, 'We surmise that the origin of life on Earth required surfaces and some degree of stability. Stability appears not to be a general characteristic of Jupiter, except for the Great Red Spot. Postulating

surfaces with stability in a turbulent atmosphere requires imagination and faith, but there is historical precedence for the use and overuse of these qualities. Fanciful description of life in places not visited has been one of the most wonderful and revealing aspects of man's history. Wonderful, because the descriptions show a basic belief that anything imaginable is possible; revealing, because of the fundamental acceptance that the natural order of the Universe is to be populated with live creatures. Inevitably, it is harshly revealed that the historical descriptions of such magnificent beasts as unicorns and mermaids have been the result of naïveté and the absence of facts. A biologist considering the possibilities of life on other planets and their satellites is the most naïve of scientists' (p. 83).

30. See also my article 'Palaeodiversifications: mass extinctions, 'clocks', and other worlds' in *GeoBios*, vol. 32, pp. 165–174, 1999.

31. Useful accounts of this organ are given by R. McN. Alexander in *Biological Reviews*, vol. 41, pp. 141–176, 1966, and J. B. Steen on pp. 413–443 in vol. 4 of *Fish Physiology*, edited by W. S. Hoar and D. J. Randall (Academic Press, New York, 1970).

32. G. N. Lapennas and K. Schmidt-Nielsen provide a helpful review in *Journal of Experimental Biology*, vol. 67, pp. 175–196, 1977. The crystals of guanine are very thin, and when the swim-bladder contracts they can strongly fold.

33. See his review on pp. 70–72 of *Marine biology: its accomplishment and future prospect*, edited by J. Mauchline and T. Nemoto (Hokusen-Sha, Tokyo, 1991). Guanine, which is one of the four bases of DNA, also acts as a reflective agent on the surface of fish, as well as in the eyes of both sharks and scallops; see E. J. Denton in *Philosophical Transactions of the Royal Society of London B*, vol. 258, pp. 286–313, 1970.

34. Counter-current systems are widespread and convergent in animals, especially those involving heat exchange and thermoregulation. A well-known example is in long-legged birds, whereby the feet, perhaps perched on the ice of a frozen lake, are markedly colder than the rest of the body because the heat carried in the blood descending the legs is transferred by the counter-current system to the ascending stream.

35. This interesting octopus is described by A. Packard and M. Wurtz in *Philosophical Transactions of the Royal Society of London B*, vol. 344, pp. 261–275; 1994. More speculatively A. Seilacher and M. LaBarbera have proposed (in *Palaios*, vol. 10, pp. 493–506, 1995) that the buoyancy of the ammonites, whose closest living relatives are squids, depended upon a bladder-like arrangement with an associated gas gland. Seilacher and LaBarbera suggest that the last-formed chamber retained a flexible wall, only later to calcify as a subsequent chamber was delimited. In association with a putative gas gland this would have allowed it to act as an analogue of the swim-bladder.

36. Concerning the siphonophores see the review by G. O. Mackie *et al.* in *Advances in Marine Biology*, vol. 24, pp. 97–262, 1987.

37. Details of the gas gland are provided by D. E. Copeland in *Biological Bulletin*, vol. 135, pp. 486–500, 1968. He suggests that carbon monoxide is produced because it is less markedly soluble than carbon dioxide and so 'more readily retained by the highly hydrated float tissue' (p. 497). Other siphonophores also secrete carbon monoxide (see, for example, G. V. Pickwell *et al.* in *Science*, vol. 144, pp. 860-862, 1964), as does the bladder kelp, an enormous brown alga; see G. B. Rigg and B. S. Henry in *American Journal of Botany*, vol. 22, pp. 362–365, 1935. In this seaweed, *Nereocystis*, the distal end bears a float with a capacity of four litres; see N. L. Nicholson in *Journal of Phycology*, vol. 6, pp. 177–182, 1970. There is another interesting convergence because as Rigg and Henry (1935) also remark 'The inner surface of this float is covered with a cobweb-like layer of tubular cells which bear such a striking resemblance to the sieve tubes of higher plants in both structure and function that they [too] are commonly referred to as sieve tubes' (p. 363). Various other convergences between these giant brown algae and other plants are addressed on pp. 19–20 of K. J. Niklas's *Plant biomechanics: An engineering approach to plant form and function* (University of Chicago Press, Chicago, 1992).

38. A detailed description of this siphonophore is given by P. R. Pugh in *Transactions of the Royal Society of London B*, vol. 301, pp. 165–300, 1983. *Thermopalia* is very delicate, but spends most of its time attached to the basaltic sea floor like a tethered balloon, anchored by its float. Pugh provides a helpful review of siphonophore floats, and notes that the mechanisms whereby the respiratory processes in the adjacent tissues are not poisoned by the carbon monoxide are not really understood, although he suggests that the chitinous lining provides a barrier to diffusion.

39. See the review by A. E. Walsby in *Microbiological Reviews*, vol. 58, pp. 91–144, 1994.

40. The gas helium, which is much preferable to hydrogen for reasons of safety, is much less likely to be employed as it is difficult to imagine a way of concentrating the trace amounts leaking out of the interior of Earth; see K. A. Farley and E. Neroda in *Annual Review of Earth and Planetary Sciences*, vol. 26, pp. 189–218, 1998.

41. See the paper by R. Conrad in *Advances in Microbial Ecology*, vol. 10, pp. 231–283, 1988.

42. The unexpectedly high production of hydrogen by microbial mats, and the possible consequences for changes in the composition of the atmosphere of the early Earth (see also note 7), is described by T. M. Hoehler *et al.* in *Nature*, vol. 412, pp. 324–327, 2001, with the accompanying commentary by B. B. Jorgensen on pp. 286–287, 289.

43. Organelles are tiny structures in the eukaryotic cell, and are probably most familiar in the form of chloroplasts (containing chlorophyll), which are derived from the endosymbiosis of cyanobacteria-like cells (see note 6) and mitochondria, also of endosymbiotic origin.

44. Hydrogenosomes are probably derived from mitochondria, and are employed in anaerobic environments by such single-celled organisms as ciliates, various flagellate protistans, and chytridiomycete fungi; see M. Müller in *Journal of General Microbiology*, vol. 139, pp. 2879–2889, 1993; T. M. Embley *et al.* in *Proceedings of the Royal Society of London B*, vol. 262, pp. 87–93, 1995 and *Trends in Ecology & Evolution*, vol. 12, pp. 437–441, 1997; J. H. P. Hackstein *et al.* in *Trends in Microbiology*, vol. 7, pp. 441–447, 1999; and M. Benchimol in *Tissue & Cell*, vol. 32, pp. 518–526, 2000. Because hydrogenosomes generally lack a genome, establishing their origins has not been straightforward, but in one ciliate the evidence for a mitochondrial origin is now very strong; see A. H. A. M. van Hoek *et al.* in *Molecular Biology and Evolution*, vol. 17, pp. 202–206, 2000. Hydrogenosomes must have evolved on numerous occasions; J. D. Palmer in *Science*, vol. 275, pp. 790–791, 1997, writes, 'Hydrogenosomes are a spectacular example of the repeated evolution of biochemically similar organelles,' p. 790.

45. Concerning hydrogenase, see the review by L. Casalot and M. Rousset in *Trends in Microbiology*, vol. 9, pp. 228–237, 2001. So, too, in an article on the origin and evolution of hydrogenase, D. S. Horner *et al.* (in *Molecular Biology and Evolution*, vol. 17, pp. 1695–1709, 2000) remark, 'The extraordinary capacity of eukaryotes to repeatedly evolve organelles capable of producing hydrogen (hydrogenosomes) is a fascinating biological puzzle,' p. 1706.

46. Thus hydrogenase has been found in the green algae *Chlamydomonas*; see the papers in *European Journal of Biochemistry* by T. Happe and J. D. Naber, and T. Happe *et al.* in vol. 214, pp. 475–481, 1993 and vol. 222, pp. 769–774, 1994, respectively.

47. A helpful summary on the vestimentiferans and their remarkable metabolism is given by C. L. van Dover in Chapter 7 of *The ecology of deep-sea hydrothermal vents* (Princeton University Press, Princeton, 2000); see also C. Arndt *et al.* in *Journal of Experimental Biology*, vol. 204, pp. 741–750, 2001.

48. Concerning the associations between insects and symbiotic bacteria see, for example, E. J. Houk and G. W. Griffiths in *Annual Review of Entomology*, vol. 25, pp.161–187, 1980, and N. A. Moran and A. Telang in *BioScience*, vol. 48, pp. 295–304, 1998. Moran and Telang emphasize that such insect–bacterial associations have evolved a number of times independently.

49. Sure enough, the collected issue of the *Fortean Times* 42–46, edited by P. Sieveking, is entitled *If pigs could fly* (J. Brown, London, 1994). The title is based on a brief report in volume 45, p. 36, and involves the unexpected discovery of a giant turd on the first-floor

balcony of a dwelling in the Huntington Beach area, California. The report continues, 'a vet analysed a sample and concluded it was the work of a huge pig', which presumably had detached itself from the passing flotilla in order to relieve itself.

50. The problem of envisaging a biological equivalent to an airship is addressed in his *Cat's paws and catapults: Mechanical worlds of nature and people* (Norton, New York, 1998); see especially pp. 151–152. S. Vogel's entertaining and informative books on biomechanics are very well worth reading. So, too, is his succinct chapter 'Convergence as an analytical tool in evolutionary design' (pp. 13–20) in *Principles of animal design: The optimization and symmorphosis debate*, edited by E. R. Weibel *et al.* (Cambridge University Press, Cambridge, 1998).

51. See, for example, the paper by R. A. J. Taylor and D. Reling in *Environmental Entomology*, vol. 15, pp. 431–435, 1986. To be sure, there are local concentrations of an 'aerial plankton', especially in connection with atmospheric convergences, but as V. A. Drake and R. A. Farrow (in *Trends in Ecology & Evolution*, vol, 4, pp. 381–385, 1989) emphasize, these are temporary, and are usually associated with migration. Locusts are perhaps the most familiar example.

52. An excellent introduction to the spiders, with a helpful description of the complexity of the silk glands and associated apparatus, is given in R. F. Foelix's *Biology of spiders* (Oxford University Press, Oxford, 1996 [Second edition]).

53. See the article by J. W. Schultz in *Biological Reviews*, vol. 62, pp. 89–113, 1987.

54. For a helpful overview of the mechanical design of spider silks, see J. M. Gosline *et al.* in *Journal of Experimental Biology*, vol. 202, pp. 3295–3303, 1999.

55. To a first approximation silk proteins are arranged either as α-helices or parallel β-pleated sheets, the latter more familiar in the form of spider silk and commercial silk. A third variety is the cross β-pleated sheet; see Craig (1997, note 61) for a helpful overview.

56. The gene family that encodes the fibroins is reported by P. A. Guerette *et al.* in *Science*, vol. 272, pp. 112–115, 1996.

57. Conservation and convergence of spider silk fibroin sequences is reported by J. Gatesy *et al.* in *Science*, vol. 291, pp. 2603–2605, 2001.

58. See C. L. Craig *et al.* in *Evolution*, vol. 48, pp. 287–296, 1994; quotation is from p. 293.

59. So far as I am aware the extrusion of liquid proteins to form silks is found only in the arthropods. But there are other parallels, notably in the formation of the dogfish egg case. This is composed of the protein collagen, which is extruded in a liquid crystalline state, and like silk hardens with the loss of water. Helpful accounts of the egg-case production are given by D. Feng *et al.* and D. P. Knight *et al.* in *Philosophical Transactions of the Royal Society of London B*, vol. 343, pp. 285–302, 1994 and vol. 351, pp. 1205–1222, 1996 respectively, as well as by D. P. Knight *et al.* in *Biological Reviews*, vol. 71, pp. 81–111, 1996. In addition, it is worth noting that while silk proteins in arthropods are typically thought of in the context of fibroins, 'a collagen-like silk is produced by sawflies in the family Perigidae (Hymenoptera)' (Craig, 1997; citation is in note 61, p. 244), while this family and the related Argidae also employ polyglycine, an apparent precursor of collagen (see also the earlier review of arthropod silks by K. M. Rudall and W. Kenchington in *Annual Review of Entomology*, vol. 16, p. 73–96, 1971, especially pp. 81–84).

60. This is *Bombyx mori*, which as E. B. Ford explains in *Moths* (Collins, London, 1955) produces a silken cocoon of three layers of which the middle can be unwound to form 'a single unbroken thread about 1500 feet in length' (p. 101).

61. The evolution of silks in arthropods is reviewed by C. L. Craig in *Annual Review of Entomology*, vol. 42, pp. 231–267, 1997.

62. See J. Brackenbury in *Physiological Entomology*, vol. 21, pp. 7–14, 1996, where he describes how certain caterpillars can descend from danger on a silk lifeline.

63. For an account of the New Zealand glow-worm, see J. E. Lloyd's chapter on insect bioluminesence (pp. 241–272) in *Bioluminescence in action*, edited by P. J. Herring (Academic Press, London, 1978).

64. See J. F. Jackson's interesting account of web-building in other dipterans, and their connection to *Arachnocampa*, published in *American Midland Naturalist*, vol. 92, pp. 240–245, 1974. This paper is of additional interest because it addresses a macro-evolutionary fantasy put forward earlier by Richard Goldschmidt, which he subsequently abandoned, to explain how web-building, bioluminescence, and carnivory could have all evolved simultaneously.

65. See B. M. Walshe's account of chironomid larval feeding in *Proceedings of the Zoological Society of London*, vol. 121, pp. 63–79, 1951. The web is spun at frequent intervals, and then consumed with its trapped contents.

66. A review of filter-feeding in aquatic insects is given by J. B. Wallace and R. W. Merritt in *Annual Review of Entomology*, vol. 25, pp. 103–132, 1980.

67. J. M. Edington, in *Journal of Animal Ecology*, vol. 37, pp. 675–692, 1968, reports nets of *Plectrocnemia conspersa* as being built of a series of irregular strands, up to *c.* 7.5 cm across.

68. See A. G. Hildrew and C. R. Townsend in *Journal of Animal Ecology*, vol. 45, pp. 41–57, 1976.

69. This information on weaver ants is from *The ants* by B. Hölldobler and E. O. Wilson (Springer, Berlin, 1990), see pp. 620–622. Silk production in ants is typically associated with the larvae, but as B. L. Fisher and H. G. Robertson demonstrate (in *Insectes Sociaux*, vol. 46, pp. 78–83, 1999), adults of *Melissotarsus* also secrete silk threads.

70. Ants, incidentally, are by no means the only arthropods to make tents by the expedient of folding together leaves. Various caterpillars of butterflies construct tents and other shelters by bringing leaves, including those of grass, together with silk threads (see E. B. Ford's *Butterflies* (Collins, London, 1945, pp. 89–90); so, too, does the orthopteran *Camptonotus* (see Craig 1997, note 61, p. 251).

71. Concerning this behaviour see, for example, the papers by A. P. Brooke in *Journal of Zoology, London*, vol. 221, pp. 11–19, 1990 and T. H. Kunz and G. F. McCracken in *Journal of Tropical Ecology*, vol. 12, pp. 121–137, 1996. E. Cholewa *et al.* explain, in *Biological Journal of the Linnean Society*, vol. 72, pp. 179–191, 2001, how damage to the leaves by the bats, necessary for folding and achieved by chewing, is contrived so that water supply is maintained and the leaf remains fresh.

72. Not that this prevents other animals using silk for their own purposes, such as birds employing spider silk in their nest construction; a trait evidently arrived at independently several times. See *Bird nests and construction behaviour* by M. Hansell (Cambridge University Press, Cambridge, 2000).

73. See George Ledyard Stebbins's paper *Natural selection and the differentiation of angiosperm families* in *Evolution*, vol. 5, pp. 299–324, 1951.

74. Their work is published on pp. 283–294 in *Constructional morphology and evolution*, edited by N. Schmidt-Kittler and K. Vogel (Springer, Berlin, 1991), and as a parallel paper in *Evolution*, vol. 47, pp. 341–360, 1993. Thomas *et al.* (in *Science*, vol. 288, pp. 1239–1242, 2000) extended this work to an analysis of how quickly the 'skeleton space' was occupied in the Cambrian 'explosion'. By reference to the Burgess Shale fauna, which shows an exceptionally complete cross-section of Cambrian life, they conclude that occupation was very rapid.

75. Such explorations can be demonstrated, for example, in the case of the aperture of the protistan foraminiferans (see V. Mikhalevich and J-P. Debenay in *Journal of Micropalaeontology*, vol. 20, pp. 13–28, 2001). This seemingly minor example is important because there have been relatively few investigations of such evolutionary 'explorations'; it also illustrates the general principle.

76. See the paper 'Stingray jaws strut their stuff' by A.P. Summers *et al.* in *Nature*, vol. 395, pp. 450–451, 1998.

77. The bone-like structure in *Ibla* is described by H. A. Lowenstam and S. Weiner in *Proceedings of the National Academy of Sciences, USA*, vol. 89, pp. 10573–10577, 1992. These authors emphasize the convergence with lamellar bone, but are also careful to note various differences. Barnacles themselves also show interesting convergences, one of which is a striking trend in the reduction in the number of parietal plates in the acorn (or balanomorph)

barnacles, a point noted by Darwin in his famous monograph on the cirripedes. A. R. Palmer, in *Paleobiology*, vol. 8, pp. 31–44, 1982, argues that this reduction is a direct result of predation pressure exerted by drilling gastropods, whose preferred mode of attack is at plate margins: fewer plates, fewer margins and greater safety.

78. See D. A. Kelly in *Integrative and Comparative Biology*, vol. 42, pp. 216–221, 2002. Kelly has some interesting comments on the nature of convergence, and also remarks 'If, as hypothesized, mammals and turtles independently evolved inflatable penises from ancestors who lacked intromittent organs, the similarity of the anatomical designs ... to produce penile stiffness is astonishing', p. 219. Interestingly, many mammals also possess a penis bone, known as the baculum. They have their uses: a zoological club, to which I was invited to give a lecture, keeps order by having to hand the baculum of a large sea-mammal, a walrus as I recall.

79. The basic cycle of burrowing shown by many groups of invertebrates, entailing alternate formation of the penetration and terminal anchors in the sediment, is strongly convergent; see E. R. Trueman's *The locomotion of soft-bodied animals* (Edward Arnold, London, 1975).

80. This group is relatively diverse today, but has a very poor fossil record; see S. E. Evans and D. Sigogneau-Russell in *Palaeontology*, vol. 44, pp. 259–273, 2001.

81. The paper by J. C. O'Reilly *et al.* on hydrostatic locomotion in the caecilians is in *Nature*, vol. 386, pp. 269–272, 1997.

82. That, in essence, is the thesis of *Pavane* by Keith Roberts (reprinted by VGSF, London, 1995) a counterfactual science fiction novel: twentieth-century England is still firmly Catholic, and the much-delayed scientific revolution is emerging in a Europe where there is still a chance of avoiding the disasters that have arisen in the real world, where it is assumed that rather than our being the custodians of God's great creation, nature is infinitely malleable to our whims.

83. See A. Donovan's *Antoine Lavoisier: Science, administration and revolution* (Cambridge University Press, Cambridge, 1993). Lavoisier's laboratory in Paris was one of the most advanced of its type, equipped with superbly constructed instruments and a team of enthusiastic and dedicated acolytes. Asked to advise on the atrocious state of French gunpowder, he transformed both its quality and methods of production. A footnote to history? Possibly not. It can be argued that the American War of Independence, which depended critically on French supplies to the rebels, would have failed. An interesting thought, and not for the usual reasons of injured British pride. It has been argued that an important aim of the British government was to restrict the westward expansion of the colonists (see R. Harvey's *A few bloody noses: The American War of Independence* (John Murray, London, 2001)), one result of which was the genocide of indigenous tribes and nations, with the destruction of a cultural richness that is our common loss.

84. See Ferguson (1998; citation is in note 3) for interesting discussions of how the British might have avoided the disaster of the trenches. John Adamson addresses Charles the First's crucial mistake in not attacking the Scots' army on the Tweed in 1639. Andrew Roberts and Niall Ferguson explore the consequences of England falling in 1940 by Nazi invasion.

85. Key papers are those by the groups led by R. E. Lenski and M. Travisano. The initial paper in this series, by R. E. Lenski *et al.*, was published in *American Naturalist*, vol. 138, pp. 1315–1341, 1991. An overview by R. E. Lenski and M. Travisano is provided in *Proceedings of the National Academy of Sciences, USA*, vol. 91, pp. 6808–6814, 1994, and more recently by M. Travisano in *Current Biology*, vol. 11, pp. R440–R442, 2001.

86. The paper, entitled *Experimental tests of the roles of adaptation, chance, and history in evolution*, is published in *Science*, vol. 267, pp. 87–90, 1995.

87. See the paper by F. Vasi *et al.* in *American Naturalist*, vol. 144, pp. 432–456, 1994, where they looked at the effects of fluctuating resources: feast and famine. Taking twelve independently evolving lines they asked whether the 'derived genotypes exhibit the same suite of adaptive changes in their life histories, or did they find alternate solutions to the challenges imposed by the seasonal environment?' They remark that superficially the evidence points to

'substantial divergence of the derived genotypes in the life-history bases of their improved fitness', but they continue, 'it is equally important to emphasize that *direction* of evolutionary change was highly parallel for most of the fitness components and life-history traits' (their emphasis). They concluded 'there was near-perfect uniformity in the directional responses exhibited by the replicate populations' (quotations on p. 449).

88. A. Joshi (in *Current Science*, vol. 72, pp. 944–949, 1997) looked at how *E. coli* responded to changes in temperature, again weighing up the relative roles of chance, history, and adaptive convergence. He concluded, 'In the case of adaptation to novel temperature, too, the contribution of adaptation to fitness at 20 °C [the bacteria having been 'trained' to 'enjoy' substantially higher temperatures of between 32 °C and 42 °C, either constant or fluctuating] was significantly greater than the combined contribution of history and chance, despite a significant increase over 1000 generations in the latter', p. 946.

89. See, for example, the discussion by M. Travisano *et al.* in *Evolution*, vol. 49, pp. 189–200, 1996. This concerns the response of the bacteria to sugar substrates, where there are a number of important qualifications when the bacteria encounter novel, i.e. non-glucose, environments. Here the variation in evolved fitness was much greater, possibly owing to heterogeneous pleiotropy.

90. The general topic of evolutionary 'dead ends' and how they are avoided is perhaps somewhat neglected. Especially interesting is the problem of specialization, seemingly reducing the range of evolutionary possibilities, and thereby increasing the risk of extinction. A good example concerns the chrysomeline leaf beetles, as described by A. Termonia *et al.* in *Proceedings of the National Academy of Sciences, USA*, vol. 98, pp. 3909–3914, 2001. As larvae these animals synthesize chemical compounds to assist in defence. Convergently, however, at least two groups of these beetles have learnt how to derive chemicals from a specific group of host plants, known as the salicaceans, that confer defence. The advantage is that metabolic energy does not need to be wasted by the insect in making defensive chemicals, and as a bonus these larvae also derive a source of sugar. The problem, however, is that they seemingly irretrievably commit themselves to the salicacean plants, so entering an evolutionary cul-de-sac. This, however, did not happen because one group was able to shift its metabolic pathways in a novel direction and so open up a whole new set of host–plant associations. Another example of how specialization does not necessarily lead to an evolutionary dead end is given by C. D'Haese in *Cladistics*, vol. 16, pp. 255–273, 2000. It concerns another group of insects, the primitive and wingless collembolans. Here, too, convergence is rife, and involves multiple occupation of habitats by generalized descendants of specialized ancestors.

91. Just how dynamic the genome is during evolution is addressed by D. Papadopoulos *et al.* in *Proceedings of the National Academy of Sciences, USA*, vol. 96, pp. 3807–3812, 1999. They studied genome evolution in *E. coli* for 10 000 generations, which occurs over 1500 days. A number of interesting results emerged, including the decoupling of genomic and phenotypic (e.g. cell size, fitness) evolution. Thus adaptive changes slowed down markedly after the 2000th generation, but genomic evolution continued unabated.

92. Moreover, in experiments concerned with adaptation to temperature and a wider variety of metabolic substrates (see M. Travisano and R. E. Lenski in *Genetics*, vol. 143, pp. 15–26, 1996) the results are more complex and reflect various sorts of divergence.

93. See number 6 in the series, published by M. Travisano in *Genetics*, vol. 146, pp. 471–479, 1997.

94. Specifically those selected for glucose and now transferred to maltose showed substantial genetic variation but no improvement in fitness, whereas those adapted to maltose and transferred to glucose showed effectively the opposite, i.e. equal fitness improvement irrespective of whether it was glucose or maltose, but little genetic variation.

95. M. Travisano (1997; citation is in note 93), p. 476.

96. A. Joshi (1997; citation is in note 88), p. 947.

97. The experiments are described on pp. 947–948 of Joshi (1997). He notes 'the fact that for fitness (or a trait strongly correlated with fitness), the effect of current ongoing selection tends to obliterate the effects of historical and random divergence among lineages', p. 948.

98. Joshi (1997), p. 949. The quote continues, 'With a change in selection pressure, the genetic structure of a population is rapidly shuffled and reorganized, much as what happens to a particular sequence of sand dunes upon a sudden change in the direction of the wind'. The wind bloweth where it will, but sand dunes still have a structure and so does the evolution of life.

99. See H. Teotónio and M. R. Rose in *Nature*, vol. 408, pp. 463–466, 2000 (with commentary by J. J. Bull on pp. 416–417).

100. H. Teotónio and M. R. Rose (2000; citation is in note 99), p. 465.

101. D. Schluter's paper on 'Tests for similarity and convergence of finch communities' is published in *Ecology*, vol. 67, pp. 1073–1085, 1986.

102. The question as to whether entire communities of organisms can converge in any exact sense is generally greeted with negatives (see, for example, J. Blondel *et al.* in *Evolutionary Biology*, vol. 18, pp. 141–213, 1984). Even so, the United States/International Biological Program (US/IBP) Synthesis Series 3 and 5, respectively *Convergent evolution in warm deserts: An examination of strategies and patterns in deserts of Argentina and the United States*, edited by G. H. Orians and O. T. Solbrig, and *Convergent evolution in Chile and California: Mediterranan climate ecosystems*, edited by H. A. Mooney (Dowden, Hutchinson & Ross, Stroudsburg, PA, 1977) have not only many examples of pair-wise comparisons within particular groups, but also some interesting remarks on the overall degree of convergence between the North and South American communities. An overview of community convergence is given by G. H. Orians and R. T. Paine in their chapter (pp. 431–458) in *Coevolution*, edited by D. J. Futuyma and M. Slatkin (Sinauer, Sunderland, MA; 1983). For an interesting example of community parallelisms in the fossil record, specifically between Devonian brachiopod communities, see P. Wallace in *Lethaia*, vol. 11, pp. 259–272, 1978.

103. See his interesting discussion in *Systematic Biology*, vol. 41, pp. 403–420, 1992. Losos also notes, perhaps surprisingly, that in two of the islands (Puerto Rico and Jamaica) 'the assembly of anole faunas … followed a highly similar evolutionary sequence' (p. 415), suggesting in turn that the evolutionary trajectories are channelled and thereby directive. Losos is careful to point out the origins of such a 'decision' might lie in the existence of a particular ecomorph, such as an arboreal generalist, and that another starting point, e.g. a grass-bush ecomorph, might 'exhibit a quite different trajectory of community assembly' (p. 416). He also has an interesting discussion of certain ecomorphs that 'ought' to exist on some islands but do not, such as the absence of a grass-bush ecomorph in Jamaica. Subsequently Losos *et al.* (in *Science*, vol. 279, pp. 2115–2118, 1998, with a commentary by G. Vogel on p. 2043) extended this analysis of anolid lizards and concluded that: 'adaptive radiation in similar environments can overcome historical contingencies to produce strikingly similar evolutionary outcomes' (p. 2115). On the other hand, among the aquatic anolid lizards, a habit that has evolved several times, there is no evidence for convergence; see M. Leal *et al.* in *Evolution*, vol. 56, pp. 785–791, 2002. These workers suggest that the lack of convergence may be because there are several distinct habitats, or alternatively 'more than one way may exist to adapt to a single habitat' (p. 787).

104. A. Ben-Moshe *et al.*, in *American Naturalist*, vol. 158, pp. 484–495, 2001, have compared the community organization (and morphology in terms of teeth and skull patterns) in Old and New World rodents, and identified patterns of ecomorphological convergence, that 'strongly indicates general rules governing ecological communities or guilds', p. 484.

105. See, for example, the paper by R. Korona in *Genetics*, vol. 143, pp. 637–644, 1996, looking at fitness convergence in a soil bacterium. Here, too, genetic variability was maintained even though the replicates converged towards similar adaptive end points. See also R. C. McLean and G. Bell in *American Naturalist*, vol. 160, pp. 569–581, 2002.

106. Amelie Karlstrom and colleagues' paper is published in *Proceedings of the National Academy of Sciences, USA*, vol. 97, pp. 3878–3883, 2000.

107. Both crustaceans and trilobites are arthropods, but trilobites disappeared 250 Ma ago, rather insignificant victims of the Permian mass extinctions. The Jurassic trilobite-mimic is described by A. Radwanski in *Acta Geologica Polonica*, vol. 45, pp. 9–25, 1995. Trilobites show their own convergences. N. C. Hughes *et al.* (in *Evolution & Development*, vol. 1, pp. 24–35, 1999) give a particularly interesting example where a proetid trilobite (*Aulacopleura konincki*) converges in form on the older olenimorph trilobites. It appears that the proetid is reoccupying a specific habitat, low in oxygen. This paper is also important because Hughes *et al.* provide cogent arguments in support of adaptational explanations in the context of a trend to a more efficient mechanism, specifically protective enrolment.

108. The appropriately named *Archimedes* has a characteristic spiral column from which lace-like fans arise. It is a widespread form, best known in Carboniferous strata. P. D. Taylor and F. K. McKinney describe a mimic (in *Journal of Paleontology*, vol. 70, pp. 218–229, 1996), from the Eocene of North Carolina. The second author also describes in the same journal (vol. 72, pp. 819–826, 1998) another interesting case of convergence in the bryozoans, specifically in the independent invention of structures that functionally mimic the specialized, non-feeding zooids known as the avicularia.

109. Palaeontologists usually refer to taxa that bow out of the fossil record and then at some later point in the stratigraphic column reappear as either Lazarus or Elvis taxa, terms coined respectively by D. Jablonski in his chapter (pp. 183–229) in *Dynamics of extinction*, edited by D. K. Elliott (Wiley [Wiley-Interscience], New York, 1986) and D. H. Erwin and M. L. Droser in *Palaios*, vol. 8, pp. 623–624, 1993. The former are taxa that survive an extinction event, but remain undetected, that is 'dead' to the fossil record (see the article by E. Fara in *Geological Journal*, vol. 36, pp. 291–303, 2001 for an overview), only to re-emerge at some later date. Elvis taxa, on the other hand, are those that which look like the pre-catastrophe forms, but are mere imitations.

110. See the paper by D. H. Erwin and H-Z. Pan on pp. 223–229 of *Biotic recovery from mass extinction events* (edited by M. B. Hart), *Geological Society of London Special Publication* 102, 1996.

111. Despite the similarities that define any evolutionary convergence, there are also subtle and interesting differences, a point of very considerable importance for those interested in explaining the realities of evolution: destinations are similar, seldom identical. Even so, on occasion the degree of similarity is quite astonishing. K. C. Emberton (see *Evolution*, vol. 49, pp. 469–475, 1995), for example, notes a convergence between two land snails so precise that only a dissection allows them to be distinguished

112. This comparison is discussed by H. Ulrich in *Natur und Museum*, vol. 95, pp. 499–508, 1965.

113. A helpful review of mantispid biology is given by K. E. Redborg in *Annual Review of Entomology*, vol. 43, pp. 175–194, 1998.

114. See the paper by U. Aspöck and M. W. Mansell in *Systematic Entomology*, vol. 19, pp. 181–206, 1994. Writing of the Rhachiberothidae they comment, 'We assume that the raptorial forelegs have evolved independently of those of the Mantispidae. Parallel evolution in related taxa is certainly more frequent than generally assumed, and quite natural due to a similar gene pool' (p. 204).

115. The mechanics of the predatory strike are described by P. T. A. Gray and P. J. Mill in *Journal of Experimental Biology*, vol. 107, pp. 245–275, 1983.

116. See S. N. Gorb and R. G. Beutel in *Naturwissenschaften*, vol. 88, pp. 530–534, 2001, as well as a more extended treatment by R. G. Beutel and S. N. Gorb in *Journal of Zoological Systematics and Evolutionary Research*, vol. 39, pp. 177–207, 2001. In addition, similar attachment structures have evolved independently in other arthropods; and these authors comment on the interesting analogies with attachment devices in the vertebrates. It is also worth mentioning the evidence for 'remarkable convergence' (p. 1509) in digital adhesive pad

design in some lizards; see E. E. Williams and J. A. Peterson in *Science*, vol. 215, pp. 1509–1511, 1982, who draw attention to similarities with the beetles: as ever, different starting point, same destination.

117. See the paper by D. Tshudy and U. Sorhannus in *Journal of Paleontology*, vol. 74, pp. 474–486, 2000, who note the independent evolution of pectinate claws in at least four lineages of crustaceans. This example is quite telling because the claws are often found isolated as fossil specimens and 'Convergence in this claw form developed to the extent that isolated fossil claws ... commonly have been misidentified at high taxonomic levels', p. 474.

118. See C. L. Morrison *et al.* in *Proceedings of the Royal Society of London B*, vol. 269, pp. 345–350, 2002. These authors suggest that the trend towards 'crab-ness' is irreversible, and they have an interesting discussion as to whether this form represents some sort of key innovation or whether it is facilitated by developmental pathways.

119. Even so, C. L. Morrison *et al.* (2002; citation is in note 118) remark that in at least some cases the key step in the transition to a crab morphology appears to have occurred in shallow water.

120. There is an extensive literature on the sabre-tooth cats. Papers by W. A. Akersten in *Los Angeles County Museum Contributions to Science*, vol. 356, pp. 1–22, 1985; M. Antón *et al.* in *Zoological Journal of the Linnean Society*, vol. 124, pp. 369–386, 1998; and M. Antón and A. Galobart in *Journal of Vertebrate Paleontology*, vol. 19, pp. 771–784, 1999, provide helpful introductions.

121. Concerning the thylacosmilids, see the papers by W. D. Turnbull on pp. 399–414 in *Development, function and evolution of teeth*, edited by P. M. Butler and K. A. Joysey (Academic Press, London, 1978), and C. S. Churcher in *Australian Mammalogy*, vol. 8, pp. 201–220, 1985.

122. A very useful overview of the sabre-tooth cats is given by A. Turner and M. Antón in their book *The big cats and their fossil relatives: An illustrated guide to their evolution and natural history* (Columbia University Press, New York, 1997). They point out that sabres have evolved at least three times in the placentals, twice in the more advanced nimravids (e.g. *Barbourofelis*) and felids (e.g. *Homotherium* and *Smilodon*), as well as in the more primitive creodont oxyaenids (e.g. *Apataelurus*).

123. See Bob Bakker's paper published in *Gaia* (Lisbon), vol. 15, pp. 145–158, 1998. One obvious difference between the mammalian sabre-tooth cats and the allosaurids is that the latter do not develop the giant canines, but Bakker suggests 'that the entire upper jaw of allosaurs functioned as one huge, saw-edged Samoan war-club, with each small, individual tooth acting as a mega-serration. Polynesian warriors glued shark teeth to clubs, and the individual shark tooth functioned as a mega-serration that was, on a much smaller scale, also serrated. Such a club inflicts a long-jagged wound with concomitant trauma and blood loss, especially if the mega-serrated blade is pulled backwards as it strikes its target', pp. 152–153.

124. See, for example, J. A. W. Kirsch's thoughtful paper on the adaptedness of marsupials in *American Scientist*, vol. 65, pp. 276–288, 1977. Interestingly he suggests that the diagnostic marsupium, that is, the pouch in which the young develop, actually arose several times independently within the marsupials, an argument that presupposes that primitively this group of mammals lacked the pouch.

125. This judgement is based in part on the near extirpation of the South American marsupial faunas, including the thylacosmilids, following the so-called Great American Interchange when island bridges (at *c.* 6 Myr) and then land bridges (at *c.* 2 Ma) were established that linked this continent to North America. It now seems, however, that this is too simplistic, and human activities may have also played a significant part in these extinctions. See, for example, the paper assessing extinction patterns in the late Pleistocene mammals of South America by E. P. Lessa and R. A. Farina in *Palaeontology*, vol. 39, pp. 651–662, 1996.

126. See, for example, W. V. Koenigswald and F. Goin's remarks on enamel structure in *Palaeontographica, Abteilung* A, vol. 255, pp. 129–168, 2000. In passing it is also worth

drawing attention to a striking convergence between the tooth enameloid of two groups of actinopterygian fish (characidids and sphyraenidids) and modern sharks, involving the independent development of highly ordered, bending resistant structure; see the paper by W-E. Reif in *Scanning Electron Microscopy*, 1979, pp. 547–554, 1979.

127. Specifically H. N. Bryant and C. S. Churcher write 'Diverse dental anatomies among sabre-tooths suggest that, despite necessary functional similarities, precise adaptations varied. The hunting adaptations of these animals were surely diverse'; see *Nature*, vol. 325, p. 488, 1987.

128. See his chapter (pp. 157–278) in *The biology of marsupials*, edited by D. Hunsaker (Academic Press, New York, 1977). The examples given concern (a) the pronounced lamination of cells and fibres found in the lateral geniculate nucleus of various marsupials and also primates and carnivores, which as Johnson remarks, 'raise many hypotheses about arboreality and visual evolution' (p. 262); (b) the occurrence of small cells in distinct subgroups in the phalanger *Trichosaurus* that 'resemble in many ways the "barrels" first described in sensory cortex of mice and related to projections of vibrissae [whiskers] and other regions of high receptor density in rats' (pp. 221–222); and (c) 'The formation of gyri and sulci at predictable loci in the sensory representation in neocortex' (p. 262) of both marsupials and placentals. Johnson concludes by remarking, 'Marsupials represent the great alternative case of mammalian adaptive radiation, and when the same result happens in two such separate phylogenetic lines, we can begin to identify the determining factors in brain evolution' (p. 262).

L. Krubitzer, in *Trends in Neurosciences*, vol. 18, pp. 408–417, 1995, also has an interesting discussion of some convergences in brain structure between placental and marsupial mammals. It is also worth mentioning convergences of brain structure *within* the placentals, of which one of the best known is the independent evolution of cytochrome oxidase-rich 'blobs' in the primary visual cortex of primates and cats; see T. M. Preuss in *Brain, Behavior and Evolution*, vol. 55, pp. 287–299, 2000.

129. See, for example, W. R. Scott in *Journal of Morphology*, vol. 5, pp. 301–406, 1891.

130. See the series of comparisons drawn by G. Dubost in *Revue écologie (Terre Vie)*, vol. 22, pp. 3–28, 1968.

131. These parallels among placental mammals are addressed by O. Madsen *et al.* in *Nature*, vol. 409, pp. 610–614, 2001. The following paper (pp. 614–618) by W. J. Murphy *et al.* is also directly relevant.

132. See R. T. Bakker's chapter (pp. 350–382) in *Coevolution*, edited by D. J. Futuyma and M. Slatkin (Sinauer, Sunderland, MA, 1983).

133. R. T. Bakker (1983; citation is in note 132) p. 354.

134. See Kirk Winemiller's paper in *Ecological Monographs*, vol. 61, pp. 343–365, 1991; see also N. Lamouroux *et al.* in *Ecology*, vol. 83, pp. 1792–1807, 2002, who document community convergence amongst freshwater fish on the basis of stream hydraulics, especially in terms of the dimensionless Froude number.

135. The paper by R. Hanel and C. Sturmbauer is in *Journal of Molecular Evolution*, vol. 50, pp. 276–283, 2000; see also J. J. Day in *Biological Journal of the Linnean Society*, vol. 76, pp. 269–301, 2002.

136. The description of replicate evolution in Lake Tanganyika cichlids is given by L. Rüber *et al.* in *Proceedings of the National Academy of Sciences, USA*, vol. 96, pp. 10230–10235, 1999; see also L. R. Rüber and D. C. Adams in *Journal of Evolutionary Biology*, vol. 14, pp. 325–332, 2001. This paper makes a specific attempt to remove the 'burden of history', whereby a particular group has a limited evolutionary choice because of phylogenetic 'decisions' by its ancestors. See also the general overview of the Rift Lake cichlids by M. L. J. Stiassny and A. Meyer in *Scientific American*, vol. 280 (2), pp. 44–49, 1999, and the figure on p. 48 with the remark that 'distantly related cichlids...evolved to become uncannily alike.'

137. See R. Felger and J. Henrickson in *Haseltonia*, 1997 (part 5), pp. 77–85, 1997.

138. For a helpful review of the topic of plant adaptation to desert environments, see the chapter (pp. 67–106) in Orians and Solbrig (1977; citation is in note 102). R. K. Peet, in *American*

Naturalist, vol. 112, pp. 441–444, 1978, offers a short but incisive discussion on various aspects of convergence, including the adaptations shown by various desert plants. His point that there may be more than one adaptive solution to a problem (such as extreme aridity) is important, and a useful counterbalance to over-simplistic thinking. Thus he points out that in Australia native succulents are rare, but have flourished when introduced, e.g. the prickly pear. In the end, what matters is how many adaptive solutions exist, and the extent to which prior conditions govern final outcomes.

139. See the papers by R. M. Cowling *et al.*, for example in *Vegetatio*, vol. 43, pp. 191–197, 1980, *Journal of Biogeography*, vol. 21, pp. 651–664, 1994, and *Australian Journal of Ecology*, vol. 19, pp. 220–232, 1994, as well as H. A. Mooney and E. L. Dunn in *Evolution*, vol. 24, pp. 292–303, 1970.

140. This is known as myrmecochory. An interesting example of convergence between Australia and South Africa species is discussed by A. V. Milewski and W. J. Bond on pp. 89–98 of *Ant-plant interactions in Australia*, edited by R. C. Buckley (Junk, The Hague, 1982).

141. This overview of convergence across the globe is provided by P. B. Reich *et al.* in *Proceedings of the National Academy of Sciences, USA*, vol. 94, pp. 13730–13734, 1997. As they remark, 'What selection pressures would lead to such common "solutions" among phylogenetically different groups? We hypothesize that there are interrelated constraints set by biophysics and natural selection' (p. 13733). Thus, in this case certain types of leaf simply cannot exist, while others are effectively inevitable.

142. Many other examples of convergence in plants are given in K. J. Niklas's authoritative *The evolutionary biology of plants* (University of Chicago Press, Chicago, 1997); see in particular his Chapter 7. See also Vogel (1998, citation is in note 50) for a series of examples including twist-to-bend ratios, vascular fluid flow, and leaf shape. More specifically, T. J. Givnish *et al.* (in *Molecular Phylogenetics and Evolution*, vol. 12, pp. 360–385, 1999) give an extensive list of convergences in the commelinoid flowering plants, ranging from type of pollination, flower structure, fruits, silica bodies, and xeric adaptations.

143. See the paper by K. J. Niklas in *Evolution*, vol. 45, pp. 734–750, 1991, which looks at the biomechanical convergence in the petioles of flowering plants (angiosperms) and ferns.

144. The tree habit has evolved multiple times; for a specific example involving tree-ferns see J. Galtier and F. M. Hueber in *Proceedings of the Royal Society of London B*, vol. 268, pp. 1955–1957, 2001. Such comparisons also span geological time; H. W. Pfefferkorn *et al.* (in *Historical Biology*, vol. 15, pp. 235–250, 2001), for example, draw attention to convergences between trees growing in areas subject to flooding and high sedimentation rates in South America today and the Carboniferous deltas of Laurasia. Many other types of plant habit have also converged. Consider the Mediterranean-like evergreen sclerophyll shrubs, where Mooney and Dunn (1970; citation is in note 139) explore the convergences imposed by drought, fire, and mineral shortages.

145. Pollen and ovule each have half the complement of chromosomes, and hence are haploid, in the same manner as egg and sperm. Successful fertilization combines the chromosomes, to give the diploid seed that germinates and grows into the plant in the garden (known as the sporophyte) where ultimately the sex cells are produced, female ovule to await male pollen, and so the cycle continues . . .

146. Together with his famous metaphor of life's diversity as a 'tangled bank', this phrase must be one of the most quoted of Darwin's remarks; first published in *More letters of Charles Darwin. A record of his work in a series of hitherto unpublished letters*, edited by F. Darwin and A. C. Seward (John Murray, London, 1903), see vol. II, pp. 20–21 for an extract from a letter to J. D. Hooker written on 22 July 1879.

147. Represented by three genera: *Ephedra*, *Gnetum*, and *Welwitschia*.

148. Helpful overviews of the male and female reproductive structures in Gnetales are given respectively by L. Hufford and P. K. Endress in *International Journal of Plant Sciences*, vol. 157 (Supplement 6), pp. S95–S112 and S113–S125, 1996. To a first approximation the gnetalacean flowers bear separate sexes, but various degrees of bisexuality are known.

149. The identification and significance of the gnetalacean double fertilization is addressed by J. S. Carmichael and W. E. Friedman in *American Journal of Botany*, vol. 83, pp. 767–780, 1996.

150. See, for example, the review paper by P. R. Crane *et al.* in *Nature*, vol. 374, pp. 27–33, 1995.

151. In the context of the water vascular system it is also worth noting that it, too, appears to be convergent, having evolved independently in the bryophytes; see R. Ligrane *et al.* in *New Phytologist*, vol. 156, pp. 491–508, 2002. Significantly, this convergence is not simple, but entails considerable biochemical complexity.

152. See, for example, A. Hansen *et al.* in *Molecular Biology and Evolution*, vol. 16, pp. 1006–1009, 1999; T. Kh. Samigullin in *Journal of Molecular Evolution*, vol. 49, pp. 310–315, 1999; and the three papers in *Proceedings of the National Academy of Sciences, USA* by K.-U. Winter *et al.* (vol. 96, pp. 7342–7347, 1999); S-M. Chaw *et al.* (vol. 97, pp. 4086–4091, 2000); L. M. Bowe *et al.* (vol. 97, pp. 4092–4097, 2000); Y-L. Qiu *et al.* in *Nature*, vol. 402, pp. 404–407, 1999 (as well as the preceding paper by P. S. Soltis on pp. 402–404; Y.-L. Qiu *et al.* in *Molecular Biology and Evolution*, vol. 18, pp. 1745–1753, 2001; S. Shindo *et al.* in *Evolution & Development*, vol. 1, pp. 180–190, 1999, and *International Journal of Plant Sciences*, vol. 162, pp. 1199–1209, 2001; E. Gugerli *et al.* in *Molecular Phylogenetics and Evolution*, vol. 21, pp. 167–175, 2001; S. Magallon and M. J. Sanderson in *American Journal of Botany*, vol. 89, pp. 1991–2006, 2002; and D. E. Soltis *et al.* in *American Journal of Botany*, vol. 89, pp. 1670–1681, 2002.

153. Interestingly, it is now known that at least some basal angiosperms have a diploid, rather than triploid, endosperm; see J. H. Williams and W. E. Friedman in *Nature*, vol. 415, pp. 522–526, 2002. It is also possible that the triploid condition evolved twice.

154. In support of the angiosperm–Gnetales connection the following characters in common are often listed: lignin biochemistry (marked by the distinctive Mäule reaction (see R. D. Gibbs' chapter (pp. 269–312) in *The physiology of forest trees*, edited by K. V. Thimann (Ronald, New York, 1958), although he emphasizes that the corresponding lignin biochemistry has probably evolved several times (p. 307)); a double integument; tunica formation in the apical meristem; pollen wall (exine) structure; and lack of archegonia; see, for example, J. A. Doyle and M. J. Donoghue in *Botanical Review*, vol. 52, pp. 321–431, 1986, p. 356, and P. Crane *et al.* (1995; citation is in note 150), Box 1 on p. 29.

155. Thus S. Carlquist in *International Journal of Plant Sciences*, vol. 157 (Supplement 6), pp. S58–S76, 1996 provides evidence to support the idea that xylem vessels in the Gnetales are evolutionarily independent of those developed in the angiosperms.

156. The idea of a common ancestry to double fertilization is effectively the argument presented by W. E. Friedman and S. K. Floyd in *Evolution*, vol. 55, pp. 217–231, 2001, who reasonably remind us that key information in a number of plant groups is still wanting, especially with respect to the origin of the endosperm.

157. There are, however, some dissenting voices. For example, M. Schmidt *et al.* (in *Journal of Molecular Evolution*, vol. 54, pp. 715–724, 2002) argue on the bases of phytochrome genes that Gnetales are basal to gymnosperms. This does not make Gnetales and angiosperms a sister-group, but in principle it could mean that their common ancestor shared reproductive organs preadapted for evolution towards flowers. A close relationship between Gnetales and conifers was also questioned by C. Rydin *et al.* (in *International Journal of Plant Sciences*, vol. 163, pp. 197–214, 2002).

158. See P. K. Endress (1996; citation is in note 148).

159. The question, of course, arises that if the triploid endosperm of angiosperms has a clear nutritive value for the adjacent developing embryo, what then is the function of the diploid supernumerary zygote in the gnetalaceans? The standard explanation (see, for example, W. E. Friedman in *Proceedings of the National Academy of Sciences, USA*, vol. 92, pp. 3913–3917, 1995) looks to the evolutionary phenomenon known as inclusive fitness, whereby the embryo (and thereby its genes) receive the benefit of a nutritional supplement by the 'sacrifice' of a genetically similar (or identical) partner.

160. Thus W. E. Friedman and S. K. Floyd (2001; citation is in note 156) remark, 'Thus, it would not be surprising if double fertilization, per se, represents an "apomorphic tendency"

[effectively the repeated likelihood that a particular character will appear; for further discussion see P. D. Cantino in *Systematic Botany*, vol. 10, pp. 119–122; 1985] ... of seed plants that evolved several times, predicated upon the fact that all seed plants form pairs of both male and female nuclei that may participate in separate fertilization events ... If the evolution of double fertilization is homoplasious among seed plant lineages, this phenomenon would stand out as a premier example of an apomorphic tendency predicated upon what has been referred to as an "underlying synapomorphy"... in essence, a case of parallelism that results from an underlying developmental constraint or bias' (pp. 225–226). This, of course, opens an indefinite regression: are seeds, land plants, and sex all evolutionary inevitable? If so, then sooner or later flowers.

161. S. C. Tucker, in *International Journal of Plant Sciences*, vol. 158 (Supplement 6), pp. S143–S161, 1997, has an interesting discussion of convergences in flower structure. See also L. Hufford's analysis of floral homoplasies in the same issue, pp. S65–S80. More specific examples of convergence in flowering plants are discussed by various authors, including P. K. Endress *et al.* in *Nordic Journal of Botany*, vol. 3, pp. 293–300, 1983 (addressing the appearances of apocarpy) and R. Dahlgren in *Botaniska Notiser*, vol. 123, pp. 551–568, 1970 (concerning South African leguminoseans).

162. See, for example, M. Proctor *et al.*: *The natural history of pollination* (HarperCollins, London, 1996), as well as more recent literature such as P. Bernhardt's discussion of convergence in beetle-pollinated flowers in *Plant Systematics and Evolution*, vol. 222, pp. 239–320, 2000, and E. L. Borba and J. Semir's discussion of convergence in fly pollination of orchids in *Annals of Botany*, vol. 88, pp. 75–88, 2001.

163. See, for example, papers by V. Bretagnolle in *The American Naturalist* in vol. 142, pp. 141–173, 1993, and P.-A. Crochet *et al.* in *Journal of Evolutionary Biology* (vol. 13, pp. 47–57, 2000).

164. This topic is discussed by M. Mönkkönnen in *Evolutionary Ecology* (vol. 9, pp. 520–528, 1995), who points out that not only has long-distance bird migration evolved repeatedly, but there is a strong tendency in the main feathers (the primaries) for the distal set to elongate and the proximal ones to shorten, thereby bringing the wing-tips closer to the leading edge. This topic was returned to in considerable detail by R. Lockwood *et al.* in *Journal of Avian Biology*, vol. 29, pp. 273–292, 1998.

165. The osteology of migrant birds is addressed by R. G. Calmaestra and E. Moreno in *Journal of Zoology, London*, vol. 252, pp. 495–501, 2000. The convergences are not in themselves particularly surprising because all are related to the muscle insertions of the major flight muscles.

166. The paper, specifically exploring convergences between carduelid finches and turdid chats, is by A. Landmann and N. Winding in *Oikos*, vol. 73, pp. 237–250, 1995.

167. See the thoughtful paper by A. Barbosa and E. Moreno in *Netherlands Journal of Zoology*, vol. 45, pp. 291–304, 1995, who analyse the bills, wings, and legs of four families of aerially foraging birds. Their analysis is important because they weigh the relative influences of adaptation for this demanding way of life against the constraints to change imposed by the phylogenetic 'burden' inherited from the ancestral form that did not pursue this way of life. Also relevant in the context of aerial insectivory is the paper by A. Keast *et al.* in *The Auk*, vol. 112, pp. 310–325, 1995 on convergence, of a sort, between the flycatchers (tyrannids) and the American Redstart.

168. This concerns the proposed convergence between the neotropical Honey Creepers, specifically the nectar-adapted warblers (Parulidae) and tanagers (Thraupidae), a topic addressed by W. J. Beecher in *The Wilson Bulletin*, vol. 63, pp. 247–287, 1951.

169. This unexpected convergence emerged as part of a wider study of bird phylogeny by M. Van Tuinen *et al.* in *Proceedings of the Royal Society B*, vol. 268, pp. 1345–1350, 2001. The considerable morphological similarities of grebes and loons had led them, by the irrefutable certainties of cladistic analysis, to be regarded as closely allied. The molecular data, on the other hand, strongly indicates that the similarities are the result of convergence. As these

authors trenchantly remark, 'We propose that this unusual alliance of birds has been overlooked because of the exceptional adaptations to their respective aquatic niches have obscured evolutionary history,' p. 1349.

170. See the chapter (pp. 149–188) on adaptive radiations in birds by R. W. Storer in *Avian biology*, vol 1, edited by D. S. Farner *et al.* (Academic Press, New York, 1971). Among the many examples he gives are those of graviportal locomotion, climbing, webbing, diving (see also citation in note 169), carnivory including multiple evolution of shrike-like forms, the carrion eaters (e.g. vultures), and nectar feeders. Evidence for repeated convergence of feeding types is also stressed by A. H. Bledsoe in *The Auk*, vol. 105, pp. 504–515, 1988 on the basis of molecular data that cut radically across various morphological assumptions.

171. See, for example, the remarks by G. A. Bartholomew in his chapter (pp. 11–37) in *New directions in ecological physiology*, edited by M. E. Feder *et al.* (Cambridge University Press, Cambridge, 1987). In addition to noting that 'Sphinx moths and hummingbirds are aerodynamically similar and resemble each other when in flight despite totally different morphological organization,' he also shows that 'The energy costs of hovering flight in these two structurally and physiologically different types of organisms are remarkably similar' (p. 21). He continues, 'One can also translate this similarity into a general statement of evolutionary interest. The aerodynamic constraints imposed by the combination of forward flight, while searching for flowers, and sustained hovering, while feeding on nectar, are associated with the evolution of similar body size, body shape, and wing shape in sphingids and hummingbirds despite fundamentally different patterns of morphological organization. The energy expenditures during hovering flight in these convergent insects and birds are virtually identical and extremely high ... sphingids and hummingbirds are approaching the biological limit of aerodynamic performance for animals of their size' (p. 22). In another context R. K. Colwell, in *American Naturalist*, vol. 156, pp. 495–510, 2001, illustrates an example of convergence (specifically Rensch's rule $\male > \female$ in larger species, $\female > \male$ in smaller species) between hummingbirds and flower mites, both of which compete for the same resource and are subject to similar constraints in terms of sexual selection and courtship.

172. The discovery of the toxin homobatrachotoxin in the skin and feathers of the bird *Pitohui* is reported by J. P. Dumbacher *et al.* in *Science*, vol. 258, pp. 799–801, 1992; see also the informative commentary by J. M. Diamond in *Nature*, vol. 360, pp. 19–20, 1992, where he remarks that this 'discovery provides an astonishing example of convergent evolution', p. 19.

173. The bird in question, *Ifrita*, is probably not closely related to *Pitohui*, but its systematic position is still enigmatic. It is unknown, therefore, whether its ability to make this alkaloid toxin is an example of convergence; see J. P. Dumbacher *et al.* in *Proceedings of the National Academy of Sciences, USA*, vol. 97, pp. 12970–12975, 2000.

174. The convergence in chemical ecology between heteropterans and the bird *Aethia cristatellais* is reported by H. D. Douglas *et al.* in *Naturwissenschaften*, vol. 88, pp. 330–332, 2001. They point out that the ultimate source of the toxin is not known, but it may act to repel ectoparasites. Nor is this the only example of such a chemical convergence. An ability to sequester pyrrolizidine alkaloids (PA) and cardiac glycosides (CG) from plants, to be employed for defence or sometimes as pheromones, is well developed in arctiidid and ctenuchid moths but has also evolved independently in nymphalid butterflies; see M. Wink and E. von Nickirsch-Rosenegk in *Journal of Chemical Ecology*, vol. 23, pp. 1549–1568, 1997. In addition, these authors remark on an extension of this example of convergence with the independent evolution of PA sequestration in such beetles as *Oreina*. This example is important because 'The biochemical mechanisms that enable insects to sequester PAs or CGs are certainly intricate' (p. 1552), thus emphasizing that convergences are far from simplistic.

175. See the summary (pp. 213–214) of D. J. Futuyama's contribution (pp. 207–231) in *Coevolution*, edited by D. J. Futuyama and M. Slatkin (Sinauer, Sunderland, MA, 1983).

176. See W. Cooper in *Journal of Zoology, London*, vol. 257, pp. 53–66, 2002.

177. See the remarks on p. 60 of G. G. Simpson's paper in *Bulletin of the Museum of Comparative Zoology*, vol. 139 (1), pp. 1–86, 1970, as well as an earlier paper, specifically on *Necrolestos* by B. Patterson, in *Breviora*, vol. 94, pp. 1–14, 1958.

178. A key worker in this area is Eviatar Nevo. Particularly helpful is his *Mosaic evolution of subterranean mammals: Regression, progression and global convergence* (Oxford University Press, Oxford, 1999). This is an outstanding source book, not only for a detailed discussion of convergences in the burrowing mammals, but also because of Nevo's emphatic and repeated emphases on the reality of adaptation and its central role in evolutionary biology.

179. Concerning powerful claws see also the paper by M. A. Coombs in *Transactions of the American Philosophical Society*, vol. 73 (7), pp. 1–96, 1983, where she reviews the occurrences of large mammalian herbivores with prominent claws. Three principal categories of activity are identified (digger/tearer, bipedal browser, and climber), and each is identified by a complex of associated characters.

180. 'How to eat a carrot?' is the intriguing main title of the paper by H. Burda *et al.* in *Naturwissenschaften*, vol. 86, pp. 325–327, 1999. Concerning the use of smell by subterranean rodents to detect and discriminate against plant food, see G. Heth *et al.* in *Behavioral Ecology and Sociobiology*, vol. 52, pp. 53–58, 2002, who emphasize its convergent nature.

181. For overviews, see M. J. Mason and P. M. Narins in *American Zoologist*, vol. 41, pp. 1171–1184, 2001, and *Journal of Comparative Psychology*, vol. 116, pp. 158–163, 2002. In Chapter 8, we shall see how the burrowing termites and ants also employ seismic communication (Chapter 8, notes 78 and 79).

182. There are, however, gradients of fierceness, and in desert environments behaviour in the mole rat *Spalax* is unusually pacific, probably as an adaptation 'to minimize overheating, water and energy expenditure'; see E. Nevo *et al.* in *Behaviour*, vol. 123, pp. 70–76, 1992. The physiological basis underlying this tendency towards pacifism is rather interesting, and may involve the steroid corticosterone. In threatening situations, which may lead to overt aggression, this chemical is generated to release metabolic energy. In doing so, however, it also leads to elevated rates of filtration by the kidneys and thereby increases water consumption, a problem for mole-rats living in arid environments who also need to avoid overheating; see G. Ganem and E. Nevo in *Behavioural Ecology and Sociobiology*, vol. 38, pp. 245–252, 1996. Ganem also suggests, in a subsequent issue (vol. 42, pp. 365–367, 1998), that another group of mole rats (the bathyergids) may have followed a similar pathway to pacifism (see also ensuing discussion by H. Burda on pp. 369–370). This is very much the exception, and as the writers also note the 'animals were easily handled by us and caressed freely like pets, an inconceivable act with other known *Spalax* species' (p. 72).

183. See the paper by K. D. Rose and R. J. Emry in *Journal of Morphology*, vol. 175, pp. 33–56, 1983, who note the phylogenetic confusion that these animals have sown whereby convergence is mistaken for phyletic affinities to such groups as the xenarthrans (citation is in note 21).

184. See *The biology of the naked mole-rat*, edited by P. W. Sherman *et al.* (Princeton University Press, Princeton, 1991), p. viii.

185. The independent evolution of eusociality in bathyergid mole-rats is described by J. U. M. Jarvis and N. C. Bennett in *Behavioural Ecology and Sociobiology*, vol. 33, pp. 253–260, 1993, with specific reference to the Damaraland mole-rat (see also J. U. M. Jarvis *et al.* in *Trends in Ecology & Evolution*, vol. 9, pp. 47–51, 1994, and C. G. Faulkes *et al.* in *Proceedings of the Royal Society of London B*, vol. 264, pp. 1619–1627, 1997). The review by H. Burda *et al.*, in *Behavioural Ecology and Sociobiology*, vol. 47, pp. 293–303, 2000, has an interesting discussion as to why eusociality evolves in these mammals.

186. See the brief report by N. G. Solomon in *Trends in Ecology & Evolution*, vol. 9, p. 264, 1994.

187. See B. J. Crespi (in *Nature*, vol. 359, pp. 724–726, 1992); B. D. Kranz *et al.* (in *Insectes Sociaux*, vol. 48, pp. 315–323, 2001); and T. W. Wills *et al.* (in *Naturwissenschaften*, vol. 88, pp. 526–529, 2001), as well as a general review by Crespi *et al.* in *Annual Review of Entomology*, vol. 42, pp. 51–71, 1997.

188. Eusociality in an ambrosia beetle (*Australoplatypus incompertus*) is documented by D. S. Kent and J. A. Simpson in *Naturwissenschaften*, vol. 79, pp. 86–87, 1992. This group is important in another context, that of the independent evolution of fungal agriculture (see Chapter 8).

189. The physical dimorphism which distinguishes the reproductives and helpers in a mole-rat colony is described by M. J. O'Riain *et al.* in *Proceedings of the National Academy of Sciences, USA*, vol. 97, pp. 13194–13197, 2000, and is marked by a lengthening of the vertebrae and corresponding enlargement of the abdominal region, which in turns allows more space for both reproductive and digestive tissue in the breeding female. These authors conclude by remarking that 'Termite, ant, and naked mole-rat "queens" with their enlarged abdomens and long life-spans provide a striking example of convergent evolution in response to such selective pressure' (p. 13197), which they argue will be ecological and extrinsic.

190. See J. E. Duffy in *Nature*, vol. 381, pp. 512–514, 1996, with a commentary by J. Seger and N. A. Moran on pp. 473–474.

191. This was demonstrated by J. E. Duffy *et al.* in *Evolution*, vol. 54, pp. 503–516, 2000, who emphasized the possible link between eusociality in these shrimps and the role of competition and enemy pressure; see also J. E. Duffy *et al.* in *Behavioral Ecology and Sociobiology*, vol. 51, pp. 488–495, 2002.

192. See A. Perle *et al.* in *Nature*, vol. 362, pp. 623–626, 1993, and associated commentary by A. C. Milner on p. 589. A more complete description of *Mononykus* is given by A. Perle *et al.* in *American Museum Novitates*, no. 3105, pp. 1–29, 1994.

193. See, for example, the overview by L. M. Chiappe in *Nature*, vol. 378, pp. 349–355, 1995. Concerning *Mononykus*, he is upbeat about its avian affinities.

194. Thus Perle *et al.* (1993) write, 'The highly modified forelimb of *Mononychus* [a name subsequently changed to *Mononykus*; see Perle *et al.* 1994, p. 3; citation is in note 192] is similar to that of digging animals. [whereas] ... The short forelimb and long, gracile hindlimb are, paradoxically, incongruous with a burrowing habitus' (p. 625), a view reiterated by Perle *et al.* (1994, p. 26; citation is in note 192).

195. See his article in *The Auk*, vol. 112, pp. 958–963, 1995. A reply to Zhou is offered by L. Chiappe *et al.* in a subsequent issue, vol. 114, pp. 300–302, 1997. Even if *Mononykus* is a bird, a number of its features must be convergent with the modern birds; see Perle *et al.* (1994, pp. 20–24; citation is in note 192).

196. See G. M. Taylor *et al.* in *Brain, Behavior and Evolution*, vol. 45, pp. 96–109, 1995, who remark of the fossorial frog brain, 'A reduced optic tectum and enlarged main olfactory bulb and torus semicircularis would appear to be in response to the unique selection pressures of living underground' (p. 105).

197. See the paper by M. S. Y. Lee in *Biological Journal of the Linnean Society*, vol. 65, pp. 369–453, 1998, as well as the references therein. The author offers several interesting remarks on the role of convergence, emphasizing the functional constraints that lead to adaptive complexes associated with reptilian burrowing ('small body size, a robust skull, a long slender body, and reduced appendages', p. 419), as well as the dangers of treating such characters independently in a phylogenetic analysis.

198. The expert is E. N. Arnold, and useful introductions to his careful and imaginative work may be found in *Philosophical Transactions of the Royal Society of London B*, vol. 344, pp. 277–290, 1994; *Journal of Zoology, London*, vol. 235, pp. 351–388, 1995; and *Bulletin of the Natural History Museum of London (Zoology)*, vol. 65, pp. 165–171, 1999. This work is also of particular importance because Arnold casts his analysis of evolutionary trajectories and convergence in a statistical context that makes it possible to assess the order in which character traits are acquired in different lineages: evolution is real (hence the order varies), but the destination is often the same.

199. Arnold (1999), p. 165. With specific respect to iguanian lizards he notes extensive ear reduction has occurred independently five times, and reiterates the important point 'that the sequence of ear modification has been different in some groups even though the end results

have substantial similarity' (p. 170). While ear reduction in these lizards is understandable as an adaptation to a burrowing mode of life (protection, and perhaps also better transmission of sound), Arnold is also careful to point out that such reduction occurs in some tree-dwelling forms.

200. This is the famous lizard, *Basiliscus*, which lives in areas of central and South America, and in many regions is referred to by the local inhabitants as Largarto Jesus Cristo, the Jesus Christ lizard. For an account of its aquatic bipedalism see J. Laerm in *American Midland Naturalist*, vol. 89, pp. 314–333, 1973, and *Ecology*, vol. 55, pp. 404–411, 1974.

201. C. Luke, in *Biological Journal of the Linnean Society*, vol. 27, pp. 1–16, 1986, provides evidence that lizard toe fringes have evolved independently at least 26 times. While emphasizing convergence in an adaptive context, such as the 12 (or 13) independent originations of triangular toe-fringes among lizard inhabiting wind-blown sand, Luke is careful to explore what may be non-adaptive aspects. Thus, two other fringe types are also found in the habitat of wind-blown sand.

202. See J. J. Wiens and J. L. Slingluff in *Evolution*, vol. 55, pp. 2303–2318, 2001, with specific respect to anguid lizards. Interestingly, they suggest that this trend to a snake-like form is gradual and there is no need to invoke any macroevolutionary jumps.

203. A helpful overview of convergences among desert mammals is given by M. A. Mares *et al.* on pp. 107–163 in Orians and Solbrig (1977; citation is in note 102), in a chapter that also has interesting comparisons of various desert dwelling groups, including the amphibians, lizards and ants. A parallel investigation with emphasis on macroecological similarities (or otherwise) is given by M. L. Cody *et al.* on pp. 144–192 in Mooney *et al.* (1977; citation is in note 102).

204. Key papers by M. A. Mares on convergence in the desert rodents may be found in *Proceedings of the National Academy of Sciences, USA*, vol. 72, pp. 1702–1706, 1975; *Paleobiology*, vol. 2, pp. 39–63, 1976; *Bulletin of the Carnegie Museum of Natural History*, vol. 16, pp. 1–51, 1980; and *BioScience*, vol. 43, pp. 372–379, 1993.

205. Specifically, Mares (1980) writes, 'it is likely that future evolution within these deserts will be that of refinement within a guild, rather than the development of a new, exploited adaptive zone. Only in the Monte and in Australia do there seem to be vacant zones awaiting exploitation by rodents. Given the evenness with which such zones have been filled in other desert areas, I feel that the evolution of such species in these deserts is highly probable,' p. 42.

206. The relationships of these somewhat enigmatic marsupials are discussed by M. R. Sanchez-Villagra in *Zoological Journal of the Linnean Society*, vol. 131, pp. 481–496, 2000, and *Palaeontology*, vol. 43, pp. 287–301, 2001.

207. G. G. Simpson (1970; citation is in note 177), p. 49. Other aspects of this convergence include the common possession in the argyrolagids and kangaroo rats of spaces within the palate, for which no functional explanation appears to exist (it may be adaptive or possibly coincidental); Simpson remarks that it is 'baffling' (see p. 50). Another shared feature is a very globular bulla, which may be linked to hearing in arid environments.

208. See M. A. Mares (1975; citation is in note 204).

7. SEEING CONVERGENCE

1. So far as humans are concerned probably not, and in Chapter 9 I review how each step leading towards humans has its parallels elsewhere. Concerning the woodpecker, J. M. Diamond has stressed (e.g. in Chapter 12 of *The rise and fall of the third chimpanzee* (Radius, London, 1991)) how only this one group of birds has developed the complex of adaptations, such as specialized beak, sawdust filters, skull structure, and musculature that together make percussive attack on trees possible. He uses this example to argue that, in an analogous way, humans (at least in a radio-building capacity) are also unique.

2. Vangids are a group of birds endemic to the island of Madagascar. They show a spectacular adaptive radiation, one that rivals the Galapagos finches or Hawaiian honeycreepers.

S. Yamagishi and K. Eguchi note in *Ibis*, vol. 138, pp. 283–290, 1996 (see also S. Yamagishi *et al.* in *Journal of Molecular Evolution*, vol. 53, pp. 39–56, 2001) that 'The bill shapes in this family vary so widely that it is difficult to believe these vangids belong to a single family' (p. 284). Their paper is an exploration of the wide range of foraging ecologies. In particular, analysis of the Sickle-billed Vanga (*Falculea palliata*) 'suggests that this species is a true substitute for woodpeckers'. Yamagishi and Eguchi are careful to add that 'Even with this species, however, the niche of woodpeckers, with special adaptations for boring through the concealing wood and for probing the exposed tunnel to catch the retreating insect, is not completely occupied. Thus we may conclude that several vangid species each partially fill the role of woodpeckers on this island' (p. 288). M. Cartmill, in his chapter (pp. 655–670) in *Prosimian biology*, edited by R. D. Martin *et al.* (Duckworth, London, 1974), also presents a list of other islands, uncolonized by true woodpeckers, that in one way or another possess an avifauna that has 'developed adaptations analogous to the picid [woodpecker] specialisations for tunnelling and probing in wood' (p. 657).

3. Specifically these are the strongly convergent aye-aye (*Daubentonia*), certain marsupial possums (*Dactylopsila*) (see Cartmill 1974), and an extinct group known as the apatemyids (see W. v. Koenigswald and H.-P. Schierning in *Nature*, vol. 326, pp. 595–597, 1987). Of *Daubentonia* Cartmill wrote that it 'forages for wood-boring grubs in the rain forest . . . of Madagascar. In feeding, it begins by tapping the surface of a branch . . . turning its large ears towards the spot being investigated. Detecting an insect, it probes the crevices of the infested plant with a grotesquely attentuated third finger, exposing deep-burrowing larvae by biting and tearing at the wood with its rodent-like incisor teeth' (p. 657). Cartmill continues by addressing the similarities to *Dactylopsila*, remarking, 'it must be assumed that all the peculiar cranial traits [of *Daubentonia* and *Dactylopsila*] are . . . functionally related to their other adaptations for preying on wood-boring insects. This assumption is further supported by the fact that comparable peculiarities . . . are characteristic of woodpeckers' (p. 662). In the opinion of T. Iwano and C. Iwakawa (*Folia Primatologica*, vol. 50, pp. 136–142, 1988) the aye-aye is better regarded as being convergent on squirrels, a group that is also absent from Madagascar. So woodpecker or squirrel mimic? In one sense both, because N. Garbutt points out in *Mammals of Madagascar* (Pica Press, Sussex, 1999) that the aye-aye is adept at opening nuts and scooping out the contents with its extraordinary middle finger, and is also known to raid coconut plantations. This animal also spends considerable time searching for insect grubs, as described above. In any event, J. M. Diamond is less than impressed with these convergences on woodpeckers (see his chapter (pp. 157–164) in *Extraterrestrials: Where are they?*, edited by B. Zuckerman and M. H. Hart (Cambridge University Press, Cambridge, 1995 [Second edition]) He has a point, but to infer an absence of extraterrestrial intelligences on the basis of woodpecker monophyly, and against the mass of other evidence for convergences at all levels, suggests that it is perhaps a belief system that is being called into question.

4. This is the celebrated Bombardier Beetle. A helpful introduction is given by W. Agosta in his book *Bombardier beetles and fever trees: A close-up look at chemical warfare and signals in animals and plants* (Addison- Wesley, Reading, MA, 1996), see pp. 38–41. The mechanism of bombardment depends on combining a mixture of hydroquinones plus hydrogen peroxide, and driving the chemical reaction with enzymes. It produces a scaldingly hot ($100\,°C$) and toxic defensive spray; see D. J. Aneshansley *et al.* in *Science* (vol. 165, pp. 61–63, 1969). In one group of Bombardiers, the paussine beetles, the accuracy of the discharge is dependent on the so-called Coanda Effect. This is familiar from the irritating habit of certain brands of teapot to dribble the tea down the spout rather than into the cup. In the case of the Bombardier, however, small flanges, that hitherto seemed to be without an adaptive function, 'serve as launching guides' for the corrosive fluid when the animal chooses to shoot forwards; see T. Eisner and D. J. Aneshansley in *Science*, vol. 215, pp. 83–85, 1982, as well as a subsequent article by T. Eisner *et al.* in *Psyche*, vol. 96, pp. 153–160, 1989. Whether or not the bombardier habit evolved independently in a variety of beetles is still debated; see for example T. Eisner

et al. in *Chemoecology*, vol. 2, pp. 29–34, 1991. Squirting hot and toxic irritants in the faces of your foes is not the only way to deal with trouble. Chapter 15 of *The biology of the Coleoptera* (Academic Press, London, 1981) by R. A. Crowson gives many other fascinating examples of how beetles defend themselves. Explosive discharges of protective chemicals are not confined to the insects; some millipedes spray quinone-containing secretions, while the polydesmid millipedes produce a protective cloud of hydrogen cyanide; see the description by T. Eisner *et al.* of the defensive secretions of millipedes on pp. 41–72 of *Arthropod venoms*, edited by S. Bettini (Springer, Berlin, 1978). Another fascinating example of defence is the release of fine hairs that entangle a would-be predator, such as occurs in the larvae of the dermestid beetles (more familiar as the 'woolly bear' and carpet beetle; my brother, unfamiliar with these beasts, recently sent me a specimen for identification drowned in gin) (see M. Ma *et al.* in *Annals of the Entomological Society of America*, vol. 71, pp. 718–723, 1978). An interesting convergent arrangement has been arrived at by the polyxenid millipedes; see T. Eisner *et al.* in *Proceedings of the National Academy of Sciences, USA*, vol. 93, pp. 10848–10051, 1996.

5. A group known as the tunicates (or sea-squirts), which are quite closely related to the vertebrates, have a thick outer tunic which contains a proportion of the biopolymer cellulose; see D. D. P. Delmer in *Annual Review of Plant Physiology and Plant Molecular Biology*, vol. 50, pp. 245–276, 1999. Cellulose is otherwise unknown in animals but is typical of plants (as well as algae and some bacteria; see Y. van Daele *et al.* in *Bulletin de Société Royale Sciences de Liège*, vol. 59, pp. 329–417, 1990, and *Biology of the Cell*, vol. 76, pp. 87–96, 1992; and S. Kimura *et al.* in *Protoplasma*, vol. 194, pp. 151–163, 1996, vol. 204, pp. 94–102, 1998, and vol. 216, pp. 71–74, 2001). It is not clear whether cellulose has evolved more than once. Given, however, that cellulose is just one example of a carbohydrate polymer (others being chitin, which is typical of insect exoskeletons, and hyaluronan, which is important in skin and cartilage), and that there is some evidence that hyaluronan synthase has probably evolved twice (see A. P. Spicer and J. A. McDonald in *Journal of Biological Chemistry*, vol. 273, pp. 1923–1932, 1998), then it would not be particularly surprising if the same held true for cellulose (see p. 1930).

6. As discussed below the acuity of the compound eye is substantially less than that of the camera-eye, and it is perhaps unlikely that the eyes of a mosquito could specifically resolve my image. However, wasps have a visual acuity that permits identification of other individuals; see E. A. Tibbetts in *Proceedings of the Royal Society of London B*, vol. 269, pp. 1423–1428, 2002.

7. The details of convergence of vision, hearing, and olfaction are returned to below. The similarities of motor control of locomotion in vertebrates and insects are reviewed by K. G. Pearson in *Annual Review of Neuroscience*, vol. 16, pp. 265–297, 1993, and *Current Opinion in Neurobiology*, vol. 5, pp. 786–791, 1995. In the former article, for example, he notes that 'Many of the observations on the afferent control of the stance to swing transition in the cat are closely paralleled by findings in the walking systems of crustacea and insects' (p. 279). (Citation is in note 28.)

8. See D. Sandeman's absorbing paper 'Homology and convergence in vertebrate and invertebrate nervous systems', published in *Naturwissenschaften*, vol. 86, pp. 378–387, 1999. This article is a rich

source of insights on convergences, some of which are returned to below. As Sandeman remarks, the topic remains of perennial interest because 'Convergences provide us with a glimpse of what may be a "best way" to achieve an end', p. 386.

9. D. Sandeman (1999; citation is in note 8), p. 383.

10. The phylogenetic position of the strepsipterans has long been problematic, but molecular data published by A. Rokas *et al.* in *Insect Molecular Biology*, vol. 8, pp. 527–530, 1999, suggests this group is not close to dipterans, leaving little doubt that the evolution of halteres, as they remark, is 'a remarkable case of convergent evolution' (p. 529). (See also note 11.)

11. More details of these rather remarkable insects may be found in the review by J. Kathirithamby in *Systematic Entomology*, vol. 14, pp. 41–92, 1989. Additional papers by Kathirithamby on the strepsipterans include those on the elusive females (in *Zoological Journal of the Linnean Society*, vol. 128, pp. 269–287, 2000), and a special issue (number 1) of vol. 27 of *International Journal of Insect Morphology and Embryology*, 1998. The last paper (pp. 53–60) in this issue, by M. F. Whiting, considers the phylogenetic position of the strepsipterans, and contrary to Rokas *et al.* (1999) concludes that they are close to the dipterans (but see the discussion by M. F. Whiting and J. P. Huelsenbeck in *Systematic Biology*, vol. 47, pp. 134–138 and 519–537 respectively, 1998; see also W. C. Wheeler *et al.* in *Cladistics*, vol. 17, pp. 113–169, 2001).

12. J. Kathirithamby (1989; citation is in note 11) notes, 'Structurally and functionally the mesothoracic wings of Strepsiptera are analogous to the dipteran halteres' (p. 52), and this comparison is pursued in greater detail by W. Pix *et al.* in *Naturwissenschaften*, vol. 80, pp. 371–374, 1993.

13. Elytra are modified hind-wings characteristic of beetles, and although sometimes involved in flight their principal function seems to be to provide a protective wing-case. However, in the Uganda lymexylid beetle (*Atractocercus brevicornis*), which unusually for this group is a fast flyer, the elytra have been modified into haltere-like organs. If they are removed, stable flight is no longer possible. Interestingly, despite the use of these natural gyroscopes, the sensory equipment of this African beetle is simpler than that seen in the dipterans. Details of this remarkable beetle may be found in P. L. Miller's paper in *The Entomologist*, vol. 104, pp. 105–110, 1971.

14. See A. D. Imms's *A general textbook of entomology including the anatomy, physiology, development and classification of insects* (9th edition, revised by O. W. Richards and R. G. Davies) (Methuen, London, 1957).

15. A. D. Imms (1957; citation is in note 14), p. 452.

16. An overview of this type of gravity perception, and especially the minerals employed, may be found on pp. 190–196 in H. A. Lowenstam and S. Weiner's *On biomineralization* (Oxford University Press, New York, 1989); see also M. D. Ross in *Advances in Space Research*, vol. 4 (12), pp. 305–314, 1984.

17. These statocysts occur in the so-called Müller vesicles, found in a group of ciliates known as the loxodids; see T. Fenchel and B. J. Finlay in *Journal of Protozoology*, vol. 33, pp. 69–76, 1986.

18. For a brief resumé see D. Sandeman (1999; citation is in note 8). More detailed accounts are available in two consecutive papers by D. Sandeman and C. Janse, published in *Journal of Comparative Physiology*, vol. 130, pp. 95–111, 1979, and D. Sandeman in *Fortschritte der Zoologie*, vol. 28, pp. 213–229, 1983.

19. D. Sandeman (1983; citation is in note 8), p. 228.

20. Reference to the literature on eyes is given below, including a number of general reviews. In this latter category should also be added vol. 2 (*Evolution of the eye and visual system*), edited by J. R. Cronly-Dillon and R. L. Gregory, of *Vision and visual dysfunction* (Macmillan, London, 1991), *Animal eyes* by M. F. Land and D-E. Nilsson (Oxford University Press, Oxford, 2002), the paper in *Annual Review of Neuroscience*, vol. 15, pp. 1–29, 1992 by M. F. Land and R. D. Fernald simply entitled *The evolution of eyes*, and the review with the same title by M. F. Land in *Berlin-Brandenburgische Akademie der Wissenschaften Berichte und Abhandlungen*, vol. 8, pp. 311–334, 2000.

21. I should emphasize that there are many other types of eye (see references and citations note 20). These demonstrate a whole series of remarkable adaptations, not least the amazing triple-lens system of the copepod *Pontella*. Although I touch on a few such examples my emphasis here is on convergences and the limits of possibility.

22. A helpful overview of cephalopod eyes (and the other sense-organs) is given by B. U. Budelmann in *Marine and Freshwater Behaviour and Physiology*, vol. 25, pp. 13–33, 1994.

23. Not all cephalopods have lenses; for an interesting discussion of the primitive pinhole eye of the cephalopod *Nautilus* see J. B. Messenger on p. 374 of Cronly-Dillon and Gregory (1991; citation is in note 20). Land (2000; citation is in note 20) expresses surprise that *Nautilus* has failed to evolve a lens: 'The problem, from an evolutionary point of view is that almost any blob of jelly, placed in the region behind the pupil, would improve both image quality and light gathering power ... Why did this not happen in *Nautilus*? I still find this a bigger mystery than the origin of really good eyes, which so concerned Darwin' (p. 321). This is a fair point, but may in part be answered by reference to the olfactory capabilities of *Nautilus*; see J. A. Basil *et al.* in *Journal of Experimental Biology*, vol. 203, pp. 1409–1414, 2000, as well as its crepuscular habit and low metabolic rate (concerning the latter see R. G. Boutilier *et al.* in *Nature*, vol. 382, pp. 534–536, 1996). When *Nautilus* 'needs' a lens, so it will be 'provided'.
24. See the paper by B. Willekens *et al.* in *Tissue & Cell*, vol. 16, pp. 941–950, 1984, where they document the lens structure of *Sepiola*.
25. For more information on the cephalopod lens see the review by J. M. Arnold on pp. 265–311 of *Experimental embryology of marine and fresh-water invertebrates*, edited by G. Reverberi (North-Holland, Amsterdam, 1971). Other relevant papers by J. M. Arnold are in *Journal of Ultrastructure Research*, vol. 14, pp. 534–539, 1966 and vol. 17, pp. 527–543, 1967 respectively, as well as T. J. C. Jacob and G. Duncan in *Nature*, vol. 290, pp. 704–706, 1981. Other helpful reviews are in chapters 17 (by H. R. Saibil, pp. 371–397) and 18 (by I. A. Meinertshagen, pp. 399–419) in *Squid as experimental animals*, edited by D. L. Gilbert *et al.* (Plenum, New York, 1990).
26. The idea that focusing depends on the movement of the cephalopod lens is raised by J. G. Sivak in *Journal of Comparative Physiology*, vol. 147A, pp. 323–327, 1982, although this idea goes back many years; see J. S. Alexandrowicz in *Archives de zoologie expérimentale et générale*, vol. 66, pp. 71–134, 1927.
27. The focal properties of the cephalopod lens are addressed by J. G. Sivak in *Canadian Journal of Zoology*, vol. 69, pp. 2501–2506, 1991.
28. So, too, flying insects have a sophisticated and integrated control system that not only links specific sensory inputs to motor control, but has fail-safe mechanisms whereby input of faulty data does not lead to catastrophic error. See, for example, H. Reichert and C. H. F. Rowell in *Trends in Neurosciences*, vol. 9, pp. 281–283, 1986, who note, 'It would not be surprising if similar principles of neuronal organization were found in all animals that need to integrate a rhythmic motor output with non-phase-locked sensory input', p. 283.
29. See the papers by B. U. Budelmann and J. Z. Young in *Philosophical Transactions of the Royal Society of London, B*, vol. 306, pp. 159–189, 1984, the overview by B. U. Budelmann 1994 (citation is in note 22) as well as the detailed tabular comparison between cephalopod and vertebrate vestibulo-oculomotor reflexes by B. U. Budelmann and Y. Tu in *Vie et Milieu*, vol. 47, pp. 95–99, 1997. Details of the statocyst organization in octopus are addressed by B. U. Budelmann *et al.* in *Philosophical Transactions of the Royal Society of London B*, vol. 315, pp. 305–343, 1987. These authors emphasize the complexity of the system, and while noting the differences from the vertebrate system they also stress 'striking analogies', p. 340.
30. See H. Collewijn's paper comparing eye movement of cuttlefish and rabbit, in *Journal of Experimental Biology*, vol. 52, pp. 369–384, 1970. In addition A. McVean in *Comparative Biochemistry and Physiology*, vol. 78A, pp. 711–718, 1984, emphasizes that in comparison with the mammals, the extra-ocular muscles in the octopus are inferior in such matters as speed of contraction and muscle viscosity. Interestingly, the arrangement of the musculature that moves the head of the fly is analogous to the optic muscles so that the entire head capsule, with its paired compound eyes, is equivalent to the vertebrate eye; see N. J. Strausfeld *et al.* and J. J. Milde *et al.* in *Journal of Comparative Physiology*, vol. 160A, pp. 205–224 and 225–238 respectively, 1987.
31. Earthworms are oligochaetes, and although once thought to be the most primitive annelids, the almost entirely marine polychaetes are usually cast in this role. The leeches, famous for their blood-sucking abilities, are thought to have arisen from the oligochaetes.

32. Key papers are those by C. O. Hermans and R. M. Eakin in *Zeitschrift für Morphologie der Tiere*, vol. 79, pp. 245–267, 1974, and G. Wald and S. Raypart in *Science*, vol. 196, pp. 1434–1439, 1977. Wald and Raypart are enthusiastic in their identification of the convergences (with words such as 'remarkable' and 'extraordinary'), whereas although Hermans and Eakin (1974) accept convergence, at the same time they struggle to find an underlying homology between the alciopid camera-eye and that of the cephalopods.

33. Annelids and molluscs, as well as various other groups such as the brachiopods, nemerteans, and platyhelminthes, are placed in a group known as the lophotrochozoans. Such similarities as there are in terms of the camera-eyes are due to similar modes of embryological development, and not because those of cephalopod and alciopid polychaete evolved from a common ancestor that possessed such an eye.

34. G. Wald and S. Raypart (1977; citation is in note 32) remark, 'The presence of accessory retinas in alciopid eyes offers a prime instance of the phenomenon of evolutionary convergence', p. 1437. They suggest that they may function to detect light of long wavelength, which penetrates more deeply into the sea, and could thus be used to judge depth.

35. A general overview of the camera-like eyes in both the heteropods (such as *Pterotrachea*) and also *Littorina* and *Strombus* (see notes 37 and 38 respectively), is given by G. H. Charles on pp. 455–521 of vol. II of *Physiology of Mollusca*, edited by K. M. Wilbur and C. M. Yonge (Academic Press, London, 1966); see also the superb review by M. F. Land on pp. 471–592 of *Handbook of sensory physiology*, vol. VII/6B: *Invertebrate visual centers and behavior I*, edited by H. Autrum (Springer, Berlin, 1981), and also the chapter (pp. 364–397) by Messenger in Cronly-Dillon and Gregory (1991; citation is in note 20).

36. Although reviewed in more recent literature, the key reference on heteropods was published more than a century ago by R. Hesse in *Zeitschrift für Wissenschaftliche Zoologie*, vol. 68, pp. 379–477, 1900. The complex structure of the photoreceptors in *Pterotrachea*, which have some resemblances to those of vertebrates, is addressed by P. N. Dilly in *Zeitschrift für Zellforschung und mikroskopische Anatomie*, vol. 99, pp. 420–429, 1969. A curiosity of the heteropods is that although the eye is camera-like, the retina consists of a narrow ribbon. In life there are rapid scanning movements in which the eye is rapidly flicked downwards and then slowly returns to the horizontal position. M. F. Land, in *Journal of Experimental Biology*, vol. 96, pp. 427–430, 1982, suggests that this scanning makes possible the detection of stationary objects. Interestingly, in the heteropod *Atlanta* the movement of the eyes is synchronous and their overlap suggests the possession of binocular vision (see notes 86 and 89).

37. Concerning the well-known winkle *Littorina littorea*, see the paper by G. F. Newell in *Proceedings of the Zoological Society of London*, vol. 144, pp. 75–86, 1965. The eye structure of *L. irrorata* is discussed by P. V. Hamilton *et al.* in *Journal of Comparative Physiology*, vol. 152A, pp. 435–445, 1983.

38. Aspects of the sophistication of the eye of *Strombus* are addressed by H. L. Gillary in *Journal of Experimental Biology*, vol. 60, pp. 383–396, 1974; vol. 66, pp. 159–171, 1977; and vol. 107, pp. 243–310, 1983. Further details of the ultrastructure of these eyes are given by H. L. Gillary and E. W. Gillary in *Journal of Morphology*, vol. 159, pp. 89–116, 1979; the light-sensitive pigments (rhodopsin and retinochrome) are discussed by K. Ozaki *et al.* in *Vision Research*, vol. 26, pp. 691–705, 1986.

39. For accounts of the structure of the cubozoan eye, see V. J. Martin in *Canadian Journal of Zoology*, vol. 80, pp. 1703–1722, 2002, G. Laska and M. Hündgen in *Zoologische Jahrbücher Abteilung für Anatomie und Ontogenie der Tiere*, vol. 108, pp. 107–123, 1982, and J. S. Pearse and V. B. Pearse in *Science*, vol. 199, p. 458, 1978. See also J. Piatigorsky *et al.* in *Journal of Comparative Physiology*, vol. 164A, pp. 577–587, 1989.

40. The suggestion that the lens might be adjustable was made by F. W. Berger in *Memoirs from the Biological Laboratory of the Johns Hopkins University*, vol. 4(4), pp. vi + 84, 1900; see pp. 44 and 58–60, see also Piatigorsky *et al.* (1989).

41. See, for example, the work by M. J. F. Blumer *et al.* in *Zoomorphology*, vol. 115, pp. 221–227, 1995. Most probably the light-sensitive protein employed will be rhodopsin, the importance of which is returned to below. C. Musio *et al.* report (in *Journal of Comparative Physiology*, vol. 187A, pp. 79–81, 2001) a rhodopsin-like molecule in *Hydra*, but this animal lacks eye-spots, and, curiously, expression of the protein is widely distributed across the animal.

42. Evidence for the visual abilities of cubozoans is given by W. M. Hamner *et al.* in *Marine Freshwater Research*, vol. 46, pp. 985–990, 1995, and they cite some intriguing observations by an earlier investigator, J. H. Barnes. R. F. Hartwick (in *Hydrobiologia*, vol. 216/217, pp. 171–179, 1991) also has some interesting comments on the possible role of vision in the cubozoans, especially with respect to search for mates. G. I. Matsumoto (in *Marine and Freshwater Behavior and Physiology*, vol. 26, pp. 139–148, 1995) also reviews the optical acuity of the cubozoans, and leaves little room for doubt that these animals genuinely see objects. See also S. E. Stewart in *Marine and Freshwater Behavior and Physiology*, vol. 27, pp. 175–188, 1996.

43. This rather remarkable behaviour, for a cnidarian, is discussed by B. Werner in *Marine Biology*, vol. 18, pp. 212–217, 1973, and entails transfer of the spermatophores to the female in a series of specific steps. See also Hartwick (1991).

44. Comparable remarks apply also to the other cnidarians, such as *Hydra*, with their nerve nets. For example, G. O. Mackie (in *American Zoologist*, vol. 30, pp. 907–920, 1990) points to various under-appreciated complexities, including pacemakers and giant axons, as well as quite sophisticated behaviours, remarking, 'most modern workers would have to agree with Parker [a distinguished biologist] in dismissing as inadequate both von Uexkull's description of sea anemones as "a bundle of reflexes" and, at the other extreme, Gosse's ["hero" of Edmund Gosse's *Father and son*] picture of them as creatures endowed by consciousness and will', p. 917.

45. Unless one labels a pacemaker as a brain, and this is still remote from the bilaterian brain. Nick Strausfeld has suggested to me that the earliest function of the brain was 'to assess asymmetries in the sensory surround and to compensate for these by appropriate motor efferent reply. From this basal condition brains might have evolved specialized connections to exploit asymmetries of the sensory world by mediating appropriate downstream asymmetries of motion, which would provide for goal-directed behavior' (personal communication; 10/02/2002).

46. See Matsumoto (1995, citation is in note 42), p. 146. He also hints that the unique lens proteins (the crystallins; see Piatigorsky *et al.* (1989, citation is in note 39; see also note 114) developed by the cubozoans might also be a significant factor. While the origin of these cubozoan crystallins is, unsurprisingly, separate from other animals (see J. Piatigorsky *et al.* in *Journal of Biological Chemistry*, vol. 268, pp. 11894–11901, 1993; see also J. Piatigorsky (in *Proceedings of the National Academy of Sciences, USA*, vol. 98, pp. 12362–12367, 2001), these novel lens proteins actually provide an excellent example of molecular convergence (citation is in note 114 and Chapter 10).

47. The 'unexpected complexity' refers specifically to the enteric nervous system of echinoderms as documented by J. E. García-Arrarás *et al.* in *Journal of Experimental Biology*, vol. 204, pp. 865–873, 2001.

48. Richard Satterlie and Thomas Nolan's paper is in *Journal of Experimental Biology*, vol. 204, pp. 1413–1419, 2001.

49. R. Satterlie and T. Nolen (2001), p. 1418.

50. However speculative intelligence based on a nerve net might be, so far as brains are concerned we shall see in the following chapters that there are indeed several but convergent routes to 'orthodox' intelligence.

51. An account of the predatory behaviour of a Panamanian species of *Dinopis* is given by M. H. Robinson and B. Robinson in *The American Midland Naturalist*, vol. 85, pp. 85–96, 1971; the behaviour of an Australian species is addressed by A. D. Austin and A. D. Blest in *Journal of*

Zoology, London, vol. 189, pp.145–156, 1979. The sensitivity of the eyes of *Dinopis* to its night-time prey is discussed below, but this spider may have a further trick up its 'sleeve'. As Austin and Blest note, 'There is a suggestion that *Dinopis* may enhance prey/background contrast by an ingenious device ... *Dinopis* always defecates on completing snare-construction ... the white splash of faeces makes a conspicuous spot on the ground roughly in the middle of the field which the spider surveys, and over which the prey must run', p. 156.

52. This account of the *Dinopis* eye is taken from the paper by A. D. Blest and M. F. Land in the *Proceedings of the Royal Society of London B*, vol. 196, pp. 197–222, 1977. Further insights into the rather remarkable sensitivity of this eye are given by S. Laughlin *et al.* in *Journal of Comparative Physiology*, vol. 141A, pp. 53–65, 1980.

53. See the two consecutive papers by M. F. Land in *Journal of Experimental Biology*, vol. 51, pp. 443–470 and 471–493, 1969, as well as the review on pp. 53–78 of *Neurology of arachnids*, edited by F. G. Barth (Springer, Berlin, 1985).

54. See D. L. Clark and G. W. Uetz in *Animal Behaviour*, vol. 40, pp. 884–890, 1990.

55. See D. S. Williams and P. McIntyre in *Nature*, vol. 288, pp. 578–580, 1980. In addition to describing the retinal structure that makes a telephoto component possible, they also note a similarity to certain birds, remarking, 'Both a group of vertebrates and invertebrates have therefore adopted the same strategy to improve the visual acuity despite a restricted cephalic space', p. 580. Further, these workers demonstrate that the lenses of the jumping spiders are able to correct for spherical aberration in the same way as the net-casting spiders.

56. Concerning the spectral sensitivities of these jumping spiders, see A. D. Blest *et al.* in *Journal of Comparative Physiology*, vol. 145A, pp. 227–239, 1981; S. Yamashita on pp. 103–117 of Barth (1985; citation is in note 53); and A. G. Peaslee and G. Wilson in *Journal of Comparative Physiology*, vol. 164A, pp. 359–363, 1989.

57. M. F. Land (1981; citation is in note 35), p. 515. Ten years later, Messenger, in Cronly-Dillon and Gregory (1991; citation is in note 20), could write regarding *Strombus*, 'Indeed we have no clues to the function of these remarkable eyes', p. 376.

58. Evidence of biotic interaction and displacement of the strombids against the aporrhaids is reviewed by K. Roy in *Paleobiology*, vol. 22, pp. 436–452, 1996.

59. K. Roy (1996; citation is in note 58), p. 441. Further information on strombid behaviour and ecology, especially its remarkable activity, is given by C. J. Berg in *Behaviour*, vol. 51, pp. 274–322, 1974, and *Bulletin of Marine Science*, vol. 25, pp. 307–317, 1975.

60. See F. Evans, in *Proceedings of the Zoological Society of London*, vol. 137, pp. 393–402, 1961, who describes the various behaviours of some Ghanaian winkles.

61. See P. V. Hamilton in *Marine Behaviour and Physiology*, vol. 5, pp. 255–271, 1978.

62. One should note there are other ways of dealing with the problem of seeing on featureless tidal flats. In some shore-dwelling crabs, such as the fiddler crabs, the compound eyes are mounted on long stalks like periscopes and have a high resolving power conferred by a specific arrangement of the components (ommatidia) that provides a so-called acute zone; see J. Zeil *et al.* in *Journal of Comparative Physiology*, vol. 159A, pp. 801–811, 1986. These workers also remark that 'visual systems adapted to the geometry of vision in a flat world' (p. 810) have evolved independently in the insects, among the hemipterans and dipterans, and an analogy can also be found between the acute zone of arthropods and the visual streak found in the eyes of various vertebrates that live in open habitats.

63. Further discussion in the optical acuity of winkles can be found in the paper by P. V. Hamilton and M. A. Winter in *Animal Behaviour*, vol. 30, pp. 752–760, 1982. Subsequently these authors extended their study in vol. 32, pp. 51–57, 1984 to a variety of other snails, emphasizing a range of visual abilities. *Littorina irrorata* was significantly better at shape recognition than even a related snail, living in similar habitats, known as *Tectarius muricatus*.

64. A. de Queiroz has an interesting discussion as to whether image-forming eyes promote evolutionary diversifications; see his article in *Evolution*, vol. 53, pp. 1654–1664, 1999. He concluded that no such relationships could be established in terms of net speciation, but did acknowledge that a connection existed between image-forming capability and activity.

65. See the article by R. G. Northcutt and J. H. Kaas in *Trends in Neurosciences*, vol. 18, pp. 373–379, 1995, where they emphasize the independent evolution of visual areas in cats and monkeys.

66. In the context of convergence in fibre optics, it is worth mentioning that in the light organs of some squid the diffuse area of luminosity 'is the result of a hemispherical cushion of light guide fibres overlying the photogenic tissue' (p. 213 of Herring 1978; citation is in note 117).

67. See the paper by E. M. Kampa in *Vision Research*, vol. 5, pp. 475–481, 1965, where he notes that the eyes of the deep-sea crustaceans are 'strikingly similar to each other' (p. 476), not only in terms of ommatidial structure but also in their developmental pattern.

68. T. H. Oakley and C. W. Cunningham in *Proceedings of the National Academy of Sciences, USA*, vol. 99, pp. 1426–1430, 2002 present a phylogenetic analysis based on molecular data that indicates how the compound eyes of the mydocopids, the only group of ostracods to possess such eyes, must have arisen independently of the other arthropods. The likelihood of this convergence was also posited by G. Fryer in *Biological Journal of the Linnean Society*, vol. 58, pp. 1–55, 1996; see pp. 37–40.

69. F. B. Krasne and P. A. Lawrence describe the compound eye of the sabellid *Branchiomma* in *Journal of Cell Science*, vol. 1, pp. 239–248, 1966. Further information on their microstructure is given by A. Kerneis in *Journal of Ultrastructure Research*, vol. 53, pp. 164–179, 1975.

70. A detailed description of the compound eye in the arc-shell (*Arca*) is given by W. Patten in *Mitteilungen aus der Zoologischen Station zu Neapel*, vol. 6, pp. 542–756, 1886, a topic that is returned to by T. R. Waller in *Smithsonian Contributions to Zoology*, vol. 313, pp. iii + 58, 1980. An overview of bivalve eyes and their evolution is given by B. Morton in *Oceanography and Marine Biology: an Annual Review*, vol. 39, pp. 165–205, 2001. Interestingly, he remarks, 'It thus seems that there is a general picture emerging of increasing sophistication with regard to the bivalve pallial eye ... There also seems to be a general picture of increasing sophistication with time' (p. 193), although he also stresses that not all simple eyes will inevitably evolve towards greater complexity.

71. Their occurrence in the genera *Bispira* and *Megalomma* is noted by K. Fitzhugh (in *Zoological Journal of the Linnean Society*, vol. 102, pp. 305–322, 1991), as part of a cladistic analysis of the sabellids. In an earlier paper (*Bulletin of the American Museum of Natural History*, no. 192, 1–104, 1989) Fitzhugh also noted that the evolution of the compound eyes in these sabellid worms is convergent; see p. 54.

72. See D-E. Nilsson's description of the sabellid and bivalve eyes and their functional significance in *Philosophical Transactions of the Royal Society of London B*, vol. 346, pp. 195–212, 1994.

73. In passing one should note that not all arthropod eyes are compound, and some are simpler eye-spots, including the curious example of those butterflies that have eyes located on their genitalia (where no doubt they have their uses). Don't believe me? See K. Arikawa and K. Aoki in *Nature*, vol. 288, pp. 700–702, 1980, and *Journal of Comparative Physiology*, vol. 148A, pp. 483–489, 1982. See also Y. Miyako *et al.* in *Journal of Comparative Neurology*, vol. 327, pp. 458–468, 1993.

74. Discussion of the nature and degree of convergence between the eyes of the strepsipterans (and to a lesser extent some other living arthropods) and trilobites is given by G. Horvath *et al.* in *Historical Biology*, vol. 12, pp. 229–263, 1997, as well as by E. Buschbeck *et al.* in *Science*, vol. 286, pp. 1178–1180, 1999. The similarities are certainly striking, but they do not encompass the unique calcitic mineralogy of the trilobite eye. Concerning the optics of strepsipteran eyes and the likely differences between trilobite and strepsipteran eyes, see W. Pix *et al.* in *Journal of Experimental Biology*, vol. 203, pp. 3397–3409, 2000.

75. See the paper by J. Aizenberg *et al.* in *Nature*, vol. 412, pp. 819–822, 2001 (and commentary by R. Sambles on p. 783). Nor is this the first report of the possible role of the calcitic skeleton in brittle-stars for optical sensitivity. For example, S. Johnsen in *Journal of Experimental Biology*, vol. 195, pp. 281–291, 1994 suggests that parts of the skeleton are sensitive to polarized light, and he subsequently reported the presence of probable rhodopsin at the arm-tips (see *Biological Bulletin*, vol. 193, pp. 97–105, 1997).

76. A standard laboratory demonstration is to place a large crystal of clear calcite, known as iceland spar, above a dot. Viewed through the crystal the dot becomes double, and as the crystal rotates so one dot remains stationary, while the other dot rotates around it. If the crystal is viewed along the optic axis, then the two dots coalesce into one.

77. For further information on the sensitivity of some brittle-stars, especially spectacular colour changes, see G. Hendler in *Marine Ecology*, vol. 5, pp. 379–401, 1984.

78. Schizochroal eyes are particularly characteristic of the phacopid trilobites, and consist of large biconvex lenses set in cup-like sclera and surrounded by calcitic skeleton.

79. Concerning the investigation of the doublet structure of schizochroal eyes in trilobites the key figure is E. N. K. Clarkson, whose papers in *Palaeontology*, vol. 22, pp. 1–22, 1979, and with J. Miller in *Philosophical Transactions of the Royal Society of London B*, vol. 288, pp. 461–480, 1980, provide the necessary details. A further development in this area is the suggestion, by J. Gál *et al.* in *Vision Research*, vol. 40, pp. 843–853, 2000, that this type of trilobite lens could also act as a bifocal.

80. K. M. Towe, in *Geological Society of America Abstracts with Program*, vol. 11, p. 529, 1979, questions whether the supposed doublet structure is not an artefact of diagenesis, specifically recrystallization and migration of ions.

81. This famous experiment was undertaken by K. M. Towe and reported in *Science*, vol. 179, pp. 1007–1009, 1973, and involved capturing images of 'happy faces'; see his Fig. 1j.

82. See the paper by D-E. Nilsson and R. F. Modlin in *Journal of Experimental Biology*, vol. 189, pp. 213–236, 1994.

83. D-E. Nilsson and R. F. Modlin (1994; citation is in note 82), p. 216.

84. D-E. Nilsson and R. F. Modlin (1994; citation is in note 82), p. 227.

85. See K. Kirschfeld's chapter on pp. 354–370 of *Neural principles in vision*, edited by F. Zettler and R. Weiler (Springer, Berlin, 1976), as well as the additional remarks in Land (1981; see note 35 for citation) on pp. 551–553. More details on what arthropod eyes are, and are not, capable of is given by E. J. Warrant and P. D. McIntyre in *Progress in Neurobiology*, vol. 40, pp. 413–461, 1993. The relative ineffectiveness of the compound eye has, moreover, been long appreciated; see the paper by A. Mallock in *Proceedings of the Royal Society of London*, vol. 55, pp. 85–90, 1894.

86. See, for example, the remarks by J. D. Pettigrew on the evolution of binocular vision (pp. 271–283) in Cronly-Dillon and Gregory (1991; citation is in note 20). He also remarks that such vision may extend at least to fish and amphibians. For an interesting exception to avian binocular vision see note 231.

87. See M. F. Land's paper in *Journal of Insect Physiology*, vol. 38, pp. 939–951, 1992.

88. M. F. Land (1992; citation is in note 87), p. 947.

89. See S. Rossel in *Nature*, vol. 302, pp. 821–822, 1983, and Y. Toh and J.-Y. Okamura in *Journal of Experimental Biology*, vol. 204, pp. 615–625, 2001.

90. See, for example, the remarks by N. J. Strausfeld and J.-K. Lee in *Visual Neuroscience*, vol. 7, pp. 13–33, 1991, which not only document some of the extraordinary complexity of the visual centres in the insect brain, but also touch on the canonical properties of such nervous systems as are shared by insect and primate.

91. See the fascinating paper by J. D. Pettigrew *et al.* in *Current Biology*, vol. 9, pp. 421–424, 1999, as well as the commentary by M. V. Srinavasan in *Nature*, vol. 399, pp. 305–306, 1999. Concerning 'This remarkable convergence' (p. 424) Pettigrew *et al.* present a list (their Table 1) of the convergent features between the optics and behaviours of the chameleons and

sand-lances. Thirteen are listed, of which five are not otherwise known in any other lizard or fish. Srinivasan remarks 'in these two animals, eye design cannot be predicted by evolutionary origin – rather, it has been crafted almost exclusively by environmental constraints', p. 305.

92. Rapid strikes at prey by using a tongue are also familiar in the frogs, where as K. C. Nishikawa (in *BioScience*, vol. 47, pp. 341–354, 1997) notes that the more derived and sophisticated method known as 'inertial elongation of the tongue [has] evolved as many as eight times independently', and that 'Surprisingly, novel sensory fibers (i.e. afferents) in the tongue have evolved independently four to five times in frogs that use inertial elongation to protract their tongues', p. 344.

93. Luitfried v. Salvini-Plawen and Ernst Mayr's much-cited paper 'On the evolution of photoreceptors and eyes' was published in *Evolutionary Biology*, vol. 10, pp. 207–263, 1977.

94. L. v. Salvini-Plawen and E. Mayr (1977; citation is in note 93), p. 249.

95. See the helpful overview by M. F. Land in *Scientific American*, vol. 239 (part 6), pp. 88–99, 1978, as well as M. F. Land (1981), pp. 536–538, and Messenger (1991; both citations are in note 35).

96. See his review on pp. 118–135 of J. R. Cronly-Dillon and R. L. Gregory (1991; citation is in note 20).

97. M. F. Land (1991; citation is in note 20), p. 124 (his emphasis).

98. Guanine, however, is not a universal reflector, and in at least some cephalopods the material employed appears to be chitin; see the paper by E. J. Denton and M. F. Land in *Proceedings of the Royal Society of London B*, vol. 178, pp. 43–61, 1971.

99. In his 1981 paper M. F. Land (citation is in note 35) remarks on the surprising fact that the deep-sea squids lack an equivalent tapetum, despite the manifest advantages of some such light-gathering mechanism in the inky depths of this environment. Land, however, goes on to note that one group of invertebrates, the lycosid spiders, has managed this feat.

100. This is particularly pronounced in the hatchet fish; see the paper by E. J. Denton on pp. 59–86 of *Marine biology: its accomplishment and future prospect*, edited by J. Mauchline and T. Nemoto (Hokusen-Sha, Tokyo, 1991), as well as the paper by E. J. Denton *et al.* in *Proceedings of the Royal Society of London B*, vol. 225, pp. 63–97, 1985.

101. There is a wide literature on protistan eye-spots. Useful overviews are given by K. W. Foster and R. D. Smyth in *Microbiological Reviews*, vol. 44, pp. 572–630, 1980 and J. D. Dodge on pp. 323–340 of Cronly-Dillon and Gregory (1991; citation is in note 20). More specific papers include those by K. W. Foster *et al.* in *Nature*, vol. 311, pp. 756–759, 1984; G. Kreimer *et al.* in *FEBS Letters*, vol. 293, pp. 49–52, 1991; W. Deininger *et al.* in *EMBO Journal*, vol. 14, pp. 5849–5858, 1995; and E. Ebnet *et al.* in *The Plant Cell*, vol. 11, pp. 1473–1484, 1999. Information on the genes connected to eye-spot formation, which has to occur every time the cell divides, is given by M. R. Lamb *et al.* and D. G. W. Roberts *et al.* in *Genetics*, respectively in vol. 153, pp. 721–729, 1999, and vol. 158, pp. 1037–1049, 2001.

102. What P. A. Kivic and P. L. Walne describe in the title of their paper as 'multiple parallel evolutions' of protistan eye-spots is discussed in *BioSystems*, vol. 16, pp. 31–38, 1983. They also emphasize that nearly always the photosensory apparatus arises as a modification of a pre-existing system. Even bacteria have eye-spots; see P. Albertano *et al.* in *Micron*, vol. 31, pp. 27–34, 2000.

103. See K. Schaller and R. Uhl in *Biophysical Journal*, vol. 73, pp. 1573–1578, 1997, where they note that the reflectance is concentrated in the yellow part of the spectrum. The preceding and companion paper by Schaller *et al.* (pp. 1562–1572) shows the remarkable phototactic response of *Chlamydomonas*.

104. These are more familiar, and more unwelcome than is sometimes realized because dinoflagellate 'blooms' are responsible for so-called red tides and in some instances for the release of toxins whose transfer to shellfish can lead to dire consequences for the gourmet. Information on these toxins is given by Y. Shimizu on pp. 282–315 of *The biology of dinoflagellates*, edited by F. J. R. Taylor (Blackwell, Oxford, 1987).

105. For a succinct review of dinoflagellate eye-spots see Dodge (1991; citation is in note 20).
106. Dinoflagellate 'eyes' are typical of the warnowiids, and interesting descriptions are given by C. Greuet in *Protistologica*, vol. 4, pp. 209–230, 1968, and vol. 13, pp. 127–143, 1977; and *Cytobiologie*, vol. 17, pp. 114–136, 1978. An overview is provided by the same author on pp. 119–142 of Taylor (1987; citation is in note 104).
107. The dioptric properties of the crystalline lens are discussed by D. Francis in *Journal of Experimental Biology*, vol. 47, pp. 495–501, 1967.
108. In his review of dinoflagellate evolution F. J. R. Taylor in *BioSystems*, vol. 13, pp. 65–108, 1980, remarks, 'The ocelli exhibit an uncanny parallelism to the structure of metazoan eyes, with all the major functional components except for conduction of signals, constructed entirely at the *subcellular* level', p. 76 (his emphasis).
109. See the chapter (pp. 207–227) by J. D. Dodge in *The chromophyte algae: Problems and perspectives*, edited by J. C. Greene *et al.* (Oxford University Press, Oxford, 1989). It also appears that in the dinoflagellates the chloroplasts were acquired by secondary endosymbiosis, which is again a convergent feature (see Chapter 6, note 6).
110. Ester Piccinni and Pietro Omodeo's paper is in *Bollettino di Zoologia*, vol. 42, pp. 57–79, 1975.
111. E. Piccinni and P. Omodeo (1975; citation is in note 110), p. 72. This remark is echoed by Taylor (1980; citation is in note 108) when he writes of the dinoflagellate eye-spot 'What can they do with a focussed image without a nervous system?', p. 76.
112. Entanglement of prey is achieved by thread-like cnidocysts that are stored in specialized organelles; see the papers by L. Mornin and D. Francis in *Journal de Microscopie (Paris)*, vol. 6, pp. 759–772, 1967 and C. Greuet in *Protistologica*, vol. 7, pp. 345–355, 1971; see also the paper by C. Greuet and R. Hovasse in *Protistologica*, vol. 13, pp. 145–149, 1977, and Greuet (1987; citation is in note 106).
113. The cnidocysts are convergent on the nematocysts of the cnidarians, which are painfully familiar from jellyfish stings. What is probably another independent invention of the cnidocyst is found in the myxozoans. Long thought to be protistans, it is now recognized they represent parasitic metazoans. The cnidocyst-like structures, known as polar capsules, seemed consistent with their being degenerate cnidarians (see, for example, M. E. Siddall *et al.* in *Journal of Parasitology*, vol. 81, pp. 961–967, 1995 and M. Schlegel *et al.* in *Archiv für Protistenkunde*, vol. 147, pp. 1–9, 1996), but it is now clear they are some sort of bilaterian, as argued by J. F. Smothers *et al.* in *Science*, vol. 265, pp. 1719–1721, 1994, and by B. Okamura *et al.* in *Parasitology*, vol. 124, pp. 215–223, 2002 (see also A. S. Monteiro *et al.* in *Molecular Biology and Evolution*, vol. 19, pp. 968–971, 2002). Okamura *et al.* remark, 'The resemblance in morphogenesis and final structure of cnidarian nematocysts and myxozoan polar capsules remains to be explained. We suggest the possibility that this may have arisen by independent incorporation of eukaryotic symbionts into Cnidaria and Myxozoa which then evolved as nematoblasts and capsulogenic cells respectively', p. 222. These cnidocysts are only one of a variety of so-called extrasomes, dischargeable thread-like structures (e.g. trichocysts, toxicysts, etc.), found in various protistans; see K. Hausmann in *International Review of Cytology*, vol. 52, pp. 197–276, 1978. His comments on the parallels are interesting: 'Surprisingly, some enigmatic similarity exists between extrasomes and organelles of systematically widely differing organisms such as ... toxicysts of ciliates and the nematocysts of cnidarians', pp 267–268.
114. There is a rich and absorbing literature on the crystallins. Useful introductions can be found in N. J. Clout *et al.* (*Nature Structural Biology*, vol. 4, p. 685, 1997); H. Janssens and W. J. Gehring (*Developmental Biology*, vol. 207, pp. 204–214, 1999); J. Piatigorsky *et al.* (*Journal of Comparative Physiology*, vol. 164A, pp. 577–587, 1989; *Journal of Biological Chemistry*, vol. 267, pp. 4277–4280, 1992; *Science*, vol. 252, pp. 1078–1079, 1991; and especially their wide-ranging review in *European Journal of Biochemistry*, vol. 235, pp. 449–465, 1996); P. J. L. Werten *et al.* (*Proceedings of the National Academy of Sciences, USA*, vol. 97, pp. 3282–3287, 2000); G. Wistow *et al.* (*Proceedings of the National Academy of Sciences,*

USA, vol. 87, pp. 6277–6280, 1990; and *Trends in Biochemical Sciences*, vol. 18, pp. 301–306, 1993); and H. Chang *et al.* (*Biophysical Journal*, vol. 78, pp. 2070–2080, 2000).

115. The explanation for lens transparency was arrived at only relatively recently: see the paper in *Nature* (vol. 302, pp. 415–417, 1983) by M. Delaye and A. Tardieu, as well as the accompanying commentary by G. Benedek on pp. 383–384.

116. See the paper by J. Piatigorsky *et al.* in *Journal of Biological Chemistry*, vol. 275, pp. 41064–41073, 2000. There appears to be only one crystallin in this bivalve mollusc, and it is homologous to one of those found in the cephalopod eye.

117. A helpful review of these photophores is given by P. J. Herring in his chapter (pp. 199–240) in *Bioluminescence in action*, edited by P. J. Herring (Academic Press, London, 1978). A remarkable range of animals, not only in the sea but on land (e.g. fireflies) can produce light, and Herring emphasizes how 'luminescence has been developed on numerous separate occasions' (p. 239). Light production is also well known from dinoflagellates, sometimes giving a beautiful oceanic phosphorescence, while glowing woods are illuminated by bioluminescent fungi.

118. See M. K. Montgomery and M. J. McFall-Ngai in *Journal of Biological Chemistry*, vol. 267, pp. 20999–21003, 1992. Nor is this the only such example of transformation of musculature into a bioluminescent organ because I. A. Johnston and P. J. Herring describe such an example (in *Proceedings of the Royal Society of London B*, vol. 225, pp. 213–218, 1985) in the scopelarchid fish *Benthalbella*. It does not appear to be known whether this latter transformation involves production of crystallins. The evolutionary versatility of the modification of musculature extends yet further. This includes the development of electrical organs, which will be considered below. In addition, rather remarkably an eye muscle of some fish is modified as a heater organ to keep the eye and adjacent parts of the brain warm. See B. A. Block and G. de Metrio *et al.* in *Journal of Morphology*, vol. 190, pp. 169–189, 1986 and vol. 234, pp. 89–96, 1997 respectively.

119. For example, Piatigorsky *et al.* (1989; see note 114) remarks 'The heterogeneity of proteins that have been used as lens crystallins in different vertebrate and invertebrate species is astounding. It is not known whether this is a consequence of evolutionary pragmatism with many proteins being able to fulfil the requirements of lens transparency or whether each species has differences which need lens proteins with special properties. In any case, the diversity of lens crystallins throughout the animal kingdom is consistent with the occurrence of convergent evolution' (p. 584). A more specific example of convergence comes from the crystallins found in the eye-lenses of frogs and lizards (geckos). These have been independently recruited from the same protein superfamily, specifically a group of stress-related enzymes known as the aldo-keto reductases; see M. A. M. van Boekel *et al.* in *Journal of Molecular Evolution*, vol. 52, pp. 239–248, 2001. The geckos also show their own example of ocular convergence. This concerns the shift from nocturnal vision to a diurnal existence. This happened independently at least three times, and is established on the basis of a study of the crystallins; see B. Röll in *Naturwissenschaften*, vol. 88, pp. 293–296, 2001.

120. See, for example, J. Nathans in *Biochemistry*, vol. 31, pp. 4923–4931, 1992.

121. Not to mention the story (or is it a myth?) of the pilots of the Luftwaffe being encouraged to eat carrots (rich in vitamin A) before embarking on another night of flying in an attempt to destroy the Royal Air Force.

122. See the papers by G. H. Jacobs *et al.* in *Nature*, vol. 382, pp. 156–158, 1996, and P. M. Kainz *et al.* in *Vision Research*, vol. 38, pp. 3315–3320, 1998.

123. Evidence for trichromatic vision in Coquerel's sifaka and the red ruffed lemur is presented by Y. Tan and W.-H. Li in *Nature*, vol. 402, p. 36, 1999. These authors suggest trichromacy is more primitive than has been thought, but the likelihood is that at least some features of trichromatic vision in the primates are convergent; see G. K. Jacobs in *Vision Research*, vol. 38, pp. 3307–3313, 1998, who emphasizes the case of the howler monkey (See also note 122).

124. Thus C. A. Arrese *et al.* (in *Current Biology*, vol. 12, pp. 657–660, 2002) suggest 'the potential for trichomacy in marsupials may have a different evolutionary origin from that in primates', p. 659.

125. A popular suggestion for the origin of trichromatic vision in these primates was to facilitate the recognition of coloured fruits against the green background of the ancestral jungle habitat; see the papers by B. C. Regan *et al.* in *Vision Research*, vol. 38, pp. 3321–3327, 1998, and N. G. Caine and N. I. Mundy in *Proceedings of the Royal Society of London B*, vol. 267, pp. 439–444, 2000.

126. See N. J. Dominy and P. W. Lucas in *Nature*, vol. 410, pp. 363–366, 2001, who emphasize the maintenance of trichromatic vision (and its independent evolution in the New World howler monkeys) as being linked to selection for young leaves, which in the tropics are coloured red, rich in protein, and tender, and can provide an alternative in times of scarcity of fruit.

127. Vision in many of the New World monkeys, including the squirrel monkey and marmoset, is therefore polymorphic (see J. D. Mollon *et al.* in *Proceedings of the Royal Society of London B*, vol. 222, pp. 373–399, 1984) making them suitable for colour discrimination experiments, such as those conducted by Caine and Mundy (2000; citation is in note 125). Despite their classic status as dichromatic/trichromatic a number of details concerning the genetic regulation are only now being worked out; see S. Kawamura *et al.* in *Gene*, vol. 269, pp. 45–51, 2001. For an overview of the evolutionary genetics of primate colour vision, see W-H. Li *et al.* in *Evolutionary Biology*, vol. 32, pp. 151–178, 2000. A similar polymorphism leading to both dichromatic and trichromatic vision is also inferred in some prosimians; see G. H. Jacobs *et al.* in *Vision Research*, vol. 42, pp. 11–18, 2002. It is uncertain whether this condition is convergent with that found in the platyrrhine monkeys; see also note 123.

128. See the paper by S.-K. Shyue *et al.* in *Science*, vol. 269, pp. 1265–1267, 1995, where they discuss the evidence for polymorphic trichromacy evolving independently in the marmosets and squirrel monkeys.

129. See the discussion by M. J. Tovée *et al.* in *Vision Research*, vol. 32, pp. 867–878, 1992, see especially p. 877.

130. See the anonymous remarks in a wartime issue of *Nature*, vol. 146, p. 266, 1940. The short article is more to do with the implications of camouflage for the colour-blind, but this issue of *Nature* also has a lively correspondence on the competence (or otherwise) of the military in this area.

131. See E. B. Ford's *Moths* (Collins, London, 1955) where he remarks that on visiting moorland in search of the larvae of the Emperor moth their cryptic coloration makes their discovery quite difficult even when fully exposed, yet the colour-blind can spot them 'when 20 or 30 yards distant' (p. 95). Ford goes on to point out the potentially adaptive advantage of this sort of colour blindness.

132. A useful overview of colour vision in mammals is given by G. H. Jacobs in *Biological Reviews*, vol. 68, pp. 413–471, 1993.

133. Among the vertebrates ultraviolet (UV) vision is widespread, and has been documented in fish, amphibians, reptiles, birds, and mammals; see G. H. Jacobs in *American Zoologist*, vol. 32, pp. 544–554, 1992. See also note 144.

134. Male red–green colour blindness is because of a linkage to the X sex chromosome; see the classic paper by J. Nathans *et al.* in *Science*, vol. 232, pp. 193–202, 1986, as well as a recent update by S. Yokoyama and F. B. Radlwimmer in *Genetics*, vol. 153, pp. 919–932, 1999.

135. The spectral tuning of the protein necessary for red–green colour vision is discussed by various authors, including J. Neitz *et al.* in *Science*, vol. 252, pp. 971–974, 1991.

136. Evidence for spectral tuning at the key sites being sometimes dependent on significant changes in the molecular size of the amino acid is given by S-K. Shyue *et al.* in *Journal of Molecular Evolution*, vol. 46, pp. 697–702, 1998.

137. See G. G. Kochendoerfer *et al.* in *Trends in Biochemical Sciences*, vol. 24, pp. 300–305, 1999. In effect the absorption depends on very localized reconfigurations within the protein molecule that alter either the electrostatic or dipolar electric charges.

138. See R. Yokoyama and S. Yokoyama in *Proceedings of the National Academy of Sciences, USA*, vol. 87, pp. 9315–9318, 1990. Curiously the fish in question is a blind cave-dweller, but it has evidently migrated to this habitat recently and the rhodopsin is still functional. It is also worth noting that red–green vision is retained in the blind mole-rat (see p. 142, Chapter 6). Here it is suggested that it has been co-opted for the maintenance of circadian rhythms. Details of this changed functionality are given by Z. K. David-Gray *et al.* in *Nature Neuroscience*, vol. 1, pp. 655–656, 1998, who also show that the sensitivity is strongly red-shifted, perhaps because the reduced eye is covered with skin and red blood vessels.

139. See Yokoyama and Radlwimmer (1999; citation is in note 134), where they review the 'five-site rule' for the mammals, reptiles (chameleon), birds (pigeon), and goldfish, and a further update by these authors in *Genetics*, vol. 158, pp. 1697–1710, 2001.

140. For the various subtleties of substitution leading to red–green vision of mammals see the paper by S. Yokoyama and F. B. Radlwimmer in *Molecular Biology and Evolution*, vol. 15, p. 560–567, 1998.

141. Sensitivity to blue light in the Conger eel is discussed by S. Archer and J. Hirano in *Proceedings of the Royal Society of London B*, vol. 263, pp. 761–767, 1996, who also review parallel site substitutions in humans, birds, and other fish. See also B. S. W. Chang *et al.* in *Molecular Phylogenetics and Evolution*, vol. 4, pp. 31–43, 1995, who emphasize convergences at sites in the rhodopsin associated with a blue shift.

142. 'For whales and seals the ocean is not blue' is the main title of the article by L. Peichl *et al.* in *European Journal of Neurosciences*, vol. 13, pp. 1520–1528, 2001. Not only is the loss of blue sensitive s-cones convergent (and may have also occurred twice within the whales), but it probably confers an adaptive advantage.

143. Six sites are identified by A. J. Hope *et al.* (in *Proceedings of the Royal Society of London B*, vol. 264, pp. 155–163, 1997), of which two are of key importance.

144. In the case of birds the sensitivity to UV radiation can be achieved by a single amino acid substitution, but the mechanisms adopted by other groups evidently differ, although they depend on a limited number of site changes; see Y. Shi in *FEBS Letters*, vol. 486, pp. 167–172, 2000; and S. Yokoyama *et al.* and Y. Shi *et al.* in *Proceedings of the National Academy of Sciences, USA*, vol. 97, pp. 7366–7371, 2000, and vol. 98, pp. 11731–11736, 2001 respectively.

145. An overview of visual pigments in the invertebrates is, however, given by W. Gärtner and P. Towner in *Photochemistry and Photobiology*, vol. 62, pp. 1–16, 1995.

146. Details of squid rhodopsin are addressed by A. Davies *et al.* in *Journal of Molecular Biology*, vol. 314, pp. 455–463, 2001. There are some interesting differences from the mammals, not least the high degree of ordering of the rhodopsin in the photoreceptor membranes. This arrangement is linked to the ability both to detect polarized light and to optimize light capture.

147. This example of convergence between squid and mammal rhodopsin is documented by A. Morris *et al.* in *Proceedings of the Royal Society of London B*, vol. 254, pp. 233–240, 1993.

148. Concerning spectral sensitivities in the hymenopterans see D. Peitsch *et al.* in *Journal of Comparative Physiology* 170A, pp. 23–40, 1992. See also B. S. W. Chang *et al.* in *Gene*, vol. 173, pp. 215–219, 1996, where they report the detailed structure of honeybee rhodopsin.

149. See the paper by A. D. Briscoe in *Molecular Biology and Evolution*, vol. 18, pp. 2270–2279, 2001. Briscoe suggests that in both these cases the molecular substitution may well be under positive Darwinian selection, a topic returned to in its molecular context in Chapter 10. In addition, Briscoe draws attention to parallels with crayfish (see also note 150), and subsequently (*Molecular Biology and Evolution*, vol. 19, pp. 983–986, 2002) addressed the parallel substitutions for red–green sensitivity in butterflies and bees.

150. The ability to absorb specific wavelengths of light in various species of freshwater crayfish is discussed by K. A. Crandall and T. W. Cronin in *Journal of Molecular Evolution*, vol. 45, pp. 524–534, 1997. One should also note the adaptations for vision in deep-water shrimps, where an otherwise unexpected near-ultraviolet sensitivity is linked to bioluminesence; see

T. W. Cronin and T. M. Frank in *Proceedings of the Royal Society of London B*, vol. 263, pp. 861–865, 1996.

151. See the paper by J. P. Carulli *et al.* in *Journal of Molecular Evolution*, vol. 38, pp. 250–262, 1994.

152. See, for example, the use of oil droplets to confer ultraviolet vision in hummingbirds, described by T. H. Goldsmith in *Science*, vol. 207, pp. 786–788, 1980.

153. See J. Marshall *et al.* in *Current Biology*, vol. 9, pp. 755–758, 1999.

154. For example, T. W. Cronin and N. J. Marshall, in *Nature*, vol. 339, pp. 137–140, 1989, report ten or more types of photoreceptor sensitive to different wavelengths of light.

155. See the paper by D.-E. Nilsson and E. J. Warrant, in *Current Biology*, vol. 9, pp. R535–R537, 1999, which is a commentary on Marshall *et al.* (1999, note 153).

156. See R. D. Fernald in *Current Opinion in Neurobiology*, vol. 10, pp. 444–450, 2000.

157. R. D. Fernald (2000; citation is in note 156), p. 446.

158. There is a large literature on bacteriorhodopsin; for a helpful introduction see the article by U. Haupts *et al.* in *Annual Review of Biophysics and Biomolecular Structure*, vol. 28, pp. 367–399, 1999.

159. This is a slight simplification, inasmuch as there are also varieties that pump chloride (Cl) ions (halorhodopsin) and are phototactic (sensory rhodopsin); the latter is returned to below (citation is in note 171) in the context of colour discrimination.

160. See, for example, the paper by H. Luecke *et al.* in *Journal of Molecular Biology*, vol. 291, pp. 899–911, 1999; and K. Palczewski *et al.* in *Science*, vol. 289, pp. 739–745, 2000 (and commentary by H. R. Bourne and E. C. Meng on pp. 733–734).

161. See the papers by O. Béjà *et al.* in *Science*, vol. 289, pp. 1902–1906, 2000 (and accompanying commentary by E. Pennisi on p. 1869), and *Nature*, vol. 411, pp. 786–789, 2001.

162. Recent papers on rhodopsin in the green algae include a review by J. L. Spudich *et al.* (*Israel Journal of Chemistry*, vol. 35, pp. 495–513, 1995), as well as by M. Beckmann and P. Hegemann (*Biochemistry*, vol. 30, pp. 3692–3697, 1991), W. Deininger *et al.* (1995), E. Ebnet *et al.* (1999; both citations are in note 101), and L. Barsanti *et al.* in *FEBS Letters*, vol. 482, pp. 247–251, 2000.

163. See W. Deininger *et al.* 1995 (citation is in note 101). The non-involvement of chlamyrhodopsin in photoreception is addressed by M. Fuhrmann *et al.* in *Journal of Cell Science*, vol. 114, pp. 3857–3863, 2001.

164. See G. Nagel *et al.* in *Science*, vol. 296, pp. 2395–2398, 2002 and O. A. Sineshchekov *et al.* in *Proceedings of the National Academy of Sciences, USA*, vol. 99, pp. 8689–8694, 2002, as well as the overview by K. D. Ridge in *Current Biology*, vol. 12, pp. R588–R590, 2002.

165. The paper describing fungal phototaxis is by J. Saranak and K. W. Foster in *Nature*, vol. 387, pp. 465–466, 1997. Rhodopsin has also been identified in other fungi, notably *Neurospora*, but here its function in terms of light regulation is less obvious, see J. Bieszke *et al.* in *Proceedings of the National Academy of Sciences, USA*, vol. 96, pp. 8034–8039, 1999a. See also the same group's report of rhodopsin from a yeast in *Biochemistry*, vol. 38, pp. 14138–14145, 1999b.

166. See the papers by R. Henderson and G. F. X. Schertler in *Philosophical Transactions of the Royal Society of London B*, vol. 326, pp. 379–389, 1990, and J. Soppa in *FEBS Letters*, vol. 342, pp. 7–11, 1994. Soppa does not rule out descent from a common ancestor, but emphasizes the lack of sequence similarities.

167. Most notably the pocket within the bacteriorhodopsin that binds the retinal has no convincing counterpart in animal rhodopsin; see, for example, R. Henderson *et al.* in *Journal of Molecular Biology*, vol. 213, 899–929, 1990.

168. See J. L. Spudich *et al.* (in *Annual Review of Cell and Developmental Biology*, vol. 16, pp. 365–392, 2000, see pp. 379–380).

169. J. L. Spudich *et al.* (2000; citation is in note 168), p. 385. Despite the likelihood of convergence between the rhodopsins it is only fair to point out that others regard them as derived from a common protein; see, for example, Bieszke *et al.* (1999a), who tentatively

suggest that their discovery of rhodopsin in *Neurospora* might provide a link, although in their 1999b paper (both citations in note 165) they prefer to emphasize the differences.

170. Reviewed by H. G. Kohorana in *Proceedings of the National Academy of Sciences, USA*, vol. 90, pp. 1166–1171, 1993; see also W. Zhang *et al.* in a subsequent issue, vol. 93, pp. 8230–8235, 1996.

171. Evidence for this colour discrimination is given by J. L. Spudich and R. A. Bogomolni in *Nature*, vol. 312, pp. 509–513, 1984, with a commentary by L. Stryer on pp. 498–499. A mutation analysis of this sensory rhodopsin is given by K-H. Jung and J. L. Spudich in *Journal of Bacteriology*, vol. 180, pp. 2033–2042, 1998. Spectral sensitivity is also reported in the proteorhodopsin; see Béjà *et al.* (2001; citation is in note 161).

172. The molecular basis of circadian rhythms is addressed by J. C. Dunlap in *Cell*, vol. 96, pp. 271–290, 1999.

173. For an overview, albeit emphasizing their occurrence in mice, see A. Sancar in *Annual Review of Biochemistry*, vol. 69, pp. 31–67, 2000.

174. See, for example, papers by Y. Miyamoto and A. Sancar in *Proceedings of the National Academy of Sciences, USA*, vol. 95, pp. 6097–6102, 1998, and G. T. J. van der Horst *et al.* in *Nature*, vol. 398, pp. 627–630, 1999. Similar functions in *Drosophila* are addressed by P. Emery *et al.* in *Cell*, vol. 95, pp. 669–679, 1998.

175. See, for example, S. Folkard in *Philosophical Transactions of the Royal Society of London B*, vol. 327, pp. 543–553, 1990. As Folkard notes (using the 24-hour clock), night is the realm of accidents. Three Mile Island? 04.00; Chernobyl? 01:23; Bhopal? 'just after midnight'. Single-vehicle accidents are three times more likely to occur between 21.00 and 09.00 than they are during the rest of the day, notwithstanding lower traffic density. If this is taken into account, then the increase in vehicle accidents rises to twelve times. So far as the avoidance of major catastrophes is concerned, Folkard suggests that 'more extreme measures may be required. The best solution here would appear to be create a nocturnal sub-society that not only works at night, but also remains on a nocturnal routine on rest days', p. 103. I assume this suggestion is tongue-in-cheek, although I must say I am so glad I am an alpha ...

176. The possibility is reviewed by A. Sancar (2000; citation is in note 173), see p. 62.

177. D. B. Small *et al.* in *Plant Molecular Biology*, vol. 28, pp. 443–454, 1995 discuss cryptochromes in the alga *Chlamydomonas*.

178. See, for example, H-Q. Yang *et al.* in *Cell*, vol. 103, pp. 815–827, 2000.

179. This evidence for molecular convergence is addressed by A. R. Cashmore *et al.* in *Science*, vol. 284, pp. 760–765, 1999.

180. See A. Sancar in *Biochemistry*, vol. 33, pp. 2–9, 1994.

181. The evidence against photolyases in humans is addressed by Y. F. Li *et al.* in *Proceedings of the National Academy of Sciences, USA*, vol. 90, pp. 4389–4393, 1993.

182. So how much time do we need? The famous article by D-E. Nilsson and S. Pelger, entitled 'A pessimistic estimate of the time required for an eye to evolve' (in *Proceedings of the Royal Society of London B*, vol. 256, pp. 53–58, 1994) suggests that the transformation from simple eye-spot to fully functioning camera-eye can be achieved in substantially less than a million years.

183. See his *Hereditary Genius: An enquiry into its laws and consequences*: first published in 1869, and its second edition in 1892. The latter is available as a reprint, with an introduction by C. D. Darlington, and published by Fontana in 1962.

184. Chesterton's book is entitled *Eugenics and other evils* (Cassell, London, 1922).

185. Francis Galton's book entitled *Memories of my life* (Methuen, London, 1908); unfortunately it has been long out of print.

186. Evident also from his panache and chutzpah when travelling through remote regions, including his journey in Southern Africa, described in *Narrative of an explorer in tropical South Africa, being an account of a visit to Damaraland in 1851* (Ward Lock, London, 1889).

Even better is his *The art of travel; or shifts and contrivances available in wild countries* (John Murray, London, 1860 [3rd edition]).

187. Colin Beavan's book is *Fingerprints: Murder and the race to uncover the science of identity* (Fourth Estate, London, 2002).

188. C. Beaven (2002; citation is in note 187), p. 98.

189. This article was published in *The Psychological Review*, vol. 1, pp. 61–62, 1894.

190. F. Galton (1908; citation is in note 185), p. 284.

191. For an interesting discussion of some similarities between the ways olfactory information is processed and language perceived, see T. S. Lorig in *Neuroscience and Biobehavioral Reviews*, vol. 23, pp. 391–398, 1999, who suggests that there is at least some cortical overlap in the brain.

192. Details of how the burrows are modified to allow habitation of such unpromising substrates is given by G. C. Hickman in *Canadian Journal of Zoology*, vol. 61, pp. 1688–1692, 1983, adaptations that he regarded as admirable. The exceptional diving abilities of the star-nosed mole, and its respiratory implications, are addressed by I. W. McIntyre *et al.* in *Journal of Experimental Biology*, vol. 205, vol. 45–54, 2002.

193. For an engaging account, with excellent illustrations, see the article by T. L. Yates in *Natural History*, vol. 92 (11), pp. 54–61, 1983. Yates suggests that the divergence of the star-nosed moles from other talpids is quite ancient, perhaps 30 million years ago, but that migration to America may have occurred no earlier than the later Pliocene. Elsewhere, however, the star-nosed mole is extinct.

194. The principal publications by K. C. Catania can be found in *Nature*, vol. 375, pp. 453–454, 1995 and *Journal of Comparative Physiology*, vol. 185A, pp. 367–372, 1999; see also the paper by K. C. Catania and J. H. Kaas in *BioScience*, vol. 46, pp. 578–586, 1996.

195. For a review of Eimer's organ in the mole group see K. C. Catania in *Brain, Behavior and Evolution*, vol. 56, pp. 146–174, 2000. He emphasizes that although these remarkable structures are effectively unique to the talpids, not only can their epidermal origin be inferred with some confidence, but convergently derived structures do occur in other mammals, notably the push-rod mechanoreceptors of the duck-billed platypus and echidna. Concerning these latter structures see, for example, P. R. Manger and J. D. Pettigrew in *Brain, Behavior and Evolution*, vol. 48, pp. 27–54, 1996, who also remark on the similarity with the Eimer's organ.

196. The way in which this fovea-like region gets a head-start in capturing a significant portion of the nervous supply is described by K. C. Catania in *Nature Neuroscience*, vol. 4, pp. 353–354, 2001. He also notes that similar circumstances are found in the development of the fovea of primate eyes; see P. Azzopardi and A. Cowey in *Nature*, vol. 361, pp. 719–721, 1993. See also R. N. S. Sachdev and K. C. Catania in *Journal of Neurophysiology*, vol. 87, pp. 2602–2611, 2002.

197. K. C. Catania (1999; citation is in note 194), p. 367. The sensory representations of the star-nosed mole in the overall context of mammalian cortical maps is addressed by J. H. Kaas and K. C. Catania in *BioEssays*, vol. 24, pp. 334–343, 2002.

198. See the paper by K. C. Catania and M. S. Remple in *Proceedings of the National Academy of Sciences, USA*, vol. 99, pp. 5692–5697, 2002.

199. P. W. Lucas and P. F. Roche, in *Monthly Notices of the Royal Astronomical Society*, vol. 314, pp. 858–864, 2000, present such evidence for free-floating planets (and also brown dwarfs) in Orion; see also M. R. Z. Osorio *et al.* in *Science*, vol. 290, pp. 103–107, 2000, with commentary by R. Irion on p. 26.

200. Concerning the navigational abilities of the blind mole-rats and the possible role of a magnetic compass, see the papers by T. Kimchi and J. Terkel in *Animal Behaviour*, vol. 61, pp. 171–180, 2001 and *The Journal of Experimental Biology*, vol. 204, pp. 751–758, 2001.

201. Just how widespread the construction of such cognitive maps might be in the animal kingdom may yet yield some surprises. Desert ants (*Cataglyphis*), for example, are famous for

returning to their nests across hundreds of metres in a straight line. Evidently the ants know how to sum up their outward journey and perform a path integration that allows them, in ways not fully understood, to expedite their return; see S. Wohlgemuth *et al.* in *Nature*, vol. 411, pp. 795–798, 2001 (and commentary by M. V. Srinivasan on pp. 752–755); see also S. Åkesson and R. Wehner in *Journal of Experimental Biology*, vol. 205, pp. 1971–1978, 2002.

202. 'Why do snakes have eyes?' ask X. Bonnet *et al.* in *Behavioral Ecology and Sociobiology*, vol. 46, pp. 267–272, 1999, where they document the ecological flexibility of these blinded snakes. And how did the tiger snakes get to Carnac Island? Evidently they were released there by a travelling showman, after his wife was bitten by one and subsequently died.

203. Among the most famous of these is the infrared sensitivity, such as in the pit vipers. Rather remarkably some insects have independently evolved a heat-sensitive organ, which has some striking similarities to that of the snakes; see H. Schmitz *et al.* in *Naturwissenschaften*, vol. 89, pp. 226–229, 2002.

204. See J. S. Edwards *et al.* in *Journal of Neurobiology*, vol. 20, pp. 101–114, 1989.

205. See K. C. Catania (2000, citation is in note 195).

206. See the paper by K. C. Catania *et al.* in *Journal of Experimental Biology*, vol. 202, pp. 2719–2726, 1999 (and also Catania, 2000; citation is in note 195).

207. Changes in the masticatory apparatus of the star-nosed mole are discussed by T. E. Grand *et al.* in *Journal of Mammalogy*, vol. 79, pp. 492–501, 1998. Concerning convergence amongst the xenarthran and pangolin anteaters, see note 21, Chapter 6.

208. The glomeruli are dense, more-or-less spherical, aggregations of nervous tissue that consist of first-order synaptic neuropil. It is clear that specific parts of the glomerulus are dedicated to specific types of olfaction, such as the legendary sensitivity of some insects to sex pheromones. There is an extensive literature on this topic. An introduction can be found in such papers as those by B. S. Hansson *et al.* in *Science*, vol. 256, pp. 1313–1315, 1992; L. A. Oland and L. P. Tolbert in *Journal of Neurobiology*, vol. 30, pp. 92–109, 1996; N. J. Vickers *et al.* and P. P. Laissue *et al.* in *Journal of Comparative Neurology*, vol. 400, pp. 35–56, 1998, and vol. 405, pp. 543–552, 1999 respectively; and R. Ignell *et al.* in *Brain, Behavior and Evolution*, vol. 57, pp. 1–17, 2001.

209. For an overview see L. Dryer in *Bioessays*, vol. 22, pp. 803–810, 2000. In the fruit-fly some 59 genes have been identified as coding for the seven trans-membrane domain proteins used in olfaction. See, for example, the paper by A. Gao and A. Chess in *Genomics*, vol. 60, pp. 31–39, 1999, where they note that the total number of olfactory genes involved could climb substantially. While emphasizing the differences with vertebrates, Gao and Chess also note some similarities, including 'zoning' of receptors. Vertebrates have about ten times as many known olfactory genes, but they also code for transmembrane domain proteins, albeit with little structural similarity to those of the insects; for an overview see P. Mombaerts in *Annual Review of Neuroscience*, vol. 22, pp. 487–509, 1999 and *Science*, vol. 286, pp. 707–711, 1999.

210. See, for example, the paper entitled 'Mechanisms of olfactory discrimination: converging evidence for common principles across phyla' in *Annual Review of Neuroscience*, vol. 20, pp. 595–631, 1997 by J. G. Hildebrand and G. M. Shepherd, as well as the paper entitled 'Olfactory systems: common design, uncommon origins?' by N. J. Strausfeld and J. G. Hildebrand in *Current Opinion in Neurobiology*, vol. 9, pp. 634–639, 1999.

211. In contrast to the insects, and most famously the moths, the action of sex pheromones in vertebrates is less explored, but P. W. Sorenson *et al.* (in *Current Opinion in Neurobiology*, vol. 8, pp. 458–467, 1998) emphasize the likely analogies in the use of these chemical languages.

212. N. Strausfeld and J. G. Hildebrand (1999; citation is in note 210), p. 634.

213. N. Strausfeld and J. G. Hildebrand (1999; citation is in note 210), p. 635 (all quotations).

214. The realization that insects probably emerged from the aquatic crustaceans has only become apparent in recent years, largely thanks to molecular biology; see, for example, the papers by

J. L. Boore *et al.* in *Nature*, vol. 392, pp. 667–668, 1998; E. García-Machado *et al.* in *Journal of Molecular Evolution*, vol. 49, pp. 142–149, 1999; and K. Wilson *et al.* in *Molecular Biology and Evolution*, vol. 17, pp. 863–874, 2000. The ramifications of this proposal, not least in terms of the origin of the insect wings from some sort of aquatic appendage (see M. Averof and S. M. Cohen in *Nature*, vol. 285, pp. 627–630, 1997), and the hitherto unappreciated similarities between the brains of crustaceans and insects (see, for example, M. Utting *et al.* in *Journal of Comparative Neurology*, vol. 416, pp. 245–261, 2000) has led to some fertile research.

215. This possibility is raised by G. Laurent and H. Davidowitz in their paper on insect olfaction and oscillatory neural assemblies in *Science*, vol. 265, pp. 1872–1875, 1994 (see also the accompanying commentary by D. W. Tank *et al.* on pp. 1819–1820). An update and overview, with continued emphasis on the similarities between not only the vertebrates and insects, but also the molluscs (in the form of the slug), are given by A. Gelperin in *Journal of Experimental Biology*, vol. 202, pp. 1855–1864, 1999.

216. See, for example, the paper by V. V. Gurevich in *Journal of Biological Chemistry*, vol. 273, pp. 15501–15506, 1998.

217. The shared presence of seven trans-membrane domain proteins is important because these olfactory proteins belong to the same class (G protein-coupled receptors; for an overview see H. G. Dohlman *et al.* in *Annual Review of Biochemistry*, vol. 60, pp. 653–688, 1991) as rhodopsin (used in vision), and also function in taste (see M. A. Hoon *et al.* in *Cell*, vol. 96, pp. 541–551, 1999). If proteins themselves are universal, so perhaps there is a strong preference to use specific protein architectures to see, smell, and taste in much the same way.

218. See C. E. Merrill *et al.* in *Proceedings of the National Academy of Sciences, USA*, vol. 99, pp. 1633–1638, 2002, and the commentary by A. Nighorn and J. G. Hildebrand on pp. 1113–1114.

219. See J. Riesgo-Escovar *et al.* in *Proceedings of the National Academy of Sciences, USA*, vol. 92, pp. 2864–2868, 1995, who document a shared function in *Drosophila* vision and olfaction for a phospholipase C.

220. Helpful introductions to this fascinating topic can be found in papers by M. B. Fenton in *Quarterly Review of Biology*, vol. 59, pp. 33–53, 1984; U. M. Norberg and J. M. V. Rayner in *Philosophical Transactions of the Royal Society of London B*, vol. 316, pp. 335–427, 1987; and E. C. Teeling *et al.* in *Nature*, vol. 403, pp. 188–192, 2000. Concerning the phylogenetic context of bat echolocation, see M. S. Springer *et al.* in *Proceedings of the National Academy of Sciences, USA*, vol. 98, pp. 6241–6246, 2001. Bat echolocation has its examples of convergence, notably the independent evolution of nasal-emitting sounds in rhinolophids and phyllostomids; see S. C. Pedersen's chapter (pp. 174–213) in *Ontogeny, functional ecology and evolution of bats*, edited by R. A. Adams and S. C. Pedersen (Cambridge University Press, Cambridge, 2000); see also E. C. Teeling *et al.* in *Proceedings of the National Academy of Sciences, USA*, vol. 99, pp. 1431–1436, 2002. So, too, two genera in these groups (*Rhinolophus* and *Pteronotus parnelli*) have convergently arrived at Doppler-sensitive sonar; see M. Vater on p. 145 of the same volume.

221. M. Ruedi and F. Mayer, in *Molecular Phylogenetics and Evolution*, vol. 21, pp. 436–448, 2001, discuss evidence for such convergence, and emphasize that 'Independent adaptive radiations among species of the genus *Myotis* therefore produced strikingly similar evolutionary solutions in different parts of the world' (p. 436). Various other examples of convergence are presented, with the important corollary that 'morphological similarity amongst these bats is a poor predictor of their genetic similarities' (p. 444). Ruedi and Mayer are unlikely to please those biologists who appeal to contingent factors in evolution, given that the title of their paper includes reference to 'deterministic ecomorphological convergences'.

222. This ability of the blind to navigate by reflected sound is very variable, but the exceptional few who surge through city streets, past crowds, obstacles, and traffic, are – at least to the sighted – a matter of some astonishment. In Chapter 12 his book *Listening in the dark: the*

acoustic orientation of bats and men (Yale University Press, New Haven, 1958) D. R. Griffin has a very interesting discussion of this topic. In the 1974 reprint (Dover, New York) Griffin draws attention to more recent work, including two papers in *Science* on the use of sonar by blind humans, respectively by W. N. Kellogg, vol. 137, pp. 399–404, 1962 and C. E. Rice in vol. 155, pp. 656–664, 1967.

223. J. P. Rauschecker and U. Kniepert (in *European Journal of Neuroscience*, vol. 6, pp. 149–160; 1994) discuss how in 'visually deprived cats' the auditory skills are markedly enhanced.

224. For further details see the paper by J. Simmons *et al.* in *Biological Bulletin*, vol. 191, pp. 109–121, 1996.

225. Concerning the perception in moths of the ultrasonic sound frequencies that the bats generate see the papers by K. D. Roeder in *Journal of Experimental Zoology*, vol. 134, pp. 127–157, 1957 (with A. E. Treat) and *Animal Behaviour*, vol. 10, pp. 300–304, 1962. On the topic of evasion by the hunted moths see M. B. Fenton and J. H. Fullard in *American Scientist*, vol. 69, pp. 266–275, 1981 and T. K. Werner in *Canadian Journal of Zoology*, vol. 59, pp. 525–529, 1981. What may well be a convergent example of such sensitivity is the evidence for ultrasound detection by a clupeid fish (the American shad), which would allow evasive action on hearing the ultrasonic noises projected by the echolocatory and predatory cetaceans; see D. A. Mann *et al.* in *Nature*, vol. 389, p. 341, 1997 and *Journal of the Acoustical Society of America*, vol. 104, pp. 562–568, 1998.

226. This topic is reviewed by W. E. Connor in *Journal of Experimental Biology*, vol. 202, pp. 1711–1723, 1999. He points out that while jamming of the bat's echolocation system is a widely accepted explanation, alternatives include startling the bat or advertising the moth's distastefulness (see, for example, D. C. Dunning and M. Krüger in *Biotropica*, vol. 27, pp. 227–231, 1995). A helpful overview of the tactics and counter-tactics of insects and bats, prey and predator, associated with sound and hearing is given by L. A. Miller and A-M. Surlykke in *BioScience*, vol. 51, pp. 570–581, 2001. (See also note 256).

227. For a succinct account of this topic see chapters 5–7 of H. C. Hughes' enjoyable book *Sensory exotica: A world beyond human experience* (MIT Press, Cambridge; 1999).

228. D. R. Griffin (1958; citation is in note 222).

229. D. R. Griffin (1958; citation is in note 222), p. 287.

230. M. Konishi and E. I. Knudsen discuss how oilbirds hear and echolocate in *Science*, vol. 204, pp. 425–427, 1979.

231. Concerning the visual system of oilbirds see the article by J. D. Pettigrew and M. Konishi in *National Geographic Society Research Report*, vol. 16, pp. 439–450, 1984, as well as comments by Pettigrew (1991; citation is in note 86). In the former paper they remark that vision is more important than echolocation, and also draw parallels to the optics of other nocturnal birds.

232. Details of these experiments are given by D. R. Griffin and R. A. Suthers in *Biological Bulletin*, vol. 139, pp. 495–501, 1970. Somewhat similar conclusions were arrived at by M. B. Fenton in *Biotropica*, vol. 7, pp. 1–7, 1975, where he noted that even obstacles only 1.5 mm thick could be avoided. More recent work on echolocation in the swiftlets includes papers by D. R. Griffin and D. Thompson in *Behavioral Ecology and Sociobiology*, vol. 10, pp. 119–123, 1982, and D. M. Smyth and J. R. Roberts in *Ibis*, vol. 125, pp. 339–345, 1983. The actual mechanism of production of the echolocating clicks is discussed by R. A. Suthers and D. H. Hector in *Journal of Comparative Physiology*, vol. 148A, pp. 457–470, 1982.

233. Thus Smyth and Roberts (1983) remark, 'It is obvious that sensitivity of echolocation [in the swiftlets] does not match the spectacular feats demonstrated by insectivorous bats in detecting, ranging and tracking minute objects and in distinguishing between a variety of shapes and textures' (p. 345), although more precise comparisons can be made with some of the fruit-bats, such as *Rousettus*. This animal belongs to the so-called mega-bats that as a group evidently lost their ability to echolocate, so suggesting that its presence in the fruit-bat

Rousettus is a secondary acquisition; see Springer *et al.* (2001; citation in note 220). One might also note that among the fruit-bats nectarivory has evolved at least twice, and probably several times; see J. A. W. Kirsch *et al.* in *Australian Journal of Zoology*, vol. 43, pp. 395–428, 1995 and L. J. Hollar and M. S. Springer in *Proceedings of the National Academy of Sciences, USA*, vol. 94, pp. 5716–5721, 1997.

234. Electrosensitivity in the star-nosed mole has been posited by Grand *et al.* (1998; see note 207 for citation; see also *Journal of Mammalogy* vol. 74, pp. 108–116, 1993), but this proposal receives short shrift from Catania (2000; see note 195 for citation).

235. Thus the duck-billed platypus is electroreceptive, using glands in its bill; see the two papers in *Nature* by respectively H. Scheich *et al.* (vol. 319, pp. 401–402, 1986) and J. E. Gregory *et al.* (vol. 326, pp. 386–387, 1987), as well as the article by U. Proske *et al.* in *Journal of Comparative Physiology*, vol. 173A, pp. 708–710, 1993. So also is the related echidna; see, for example, J. E. Gregory *et al.* in *Journal of Physiology*, vol. 414, pp. 521–538, 1989; K. H. Andres *et al.* in *Anatomy and Embryology*, vol. 184, pp. 371–393, 1991; and P. R. Manger *et al.* in *Proceedings of the Royal Society of London B*, vol. 264, pp. 165–172, 1997. This electrical sensitivity arose independently of other vertebrates, notably the mormyrid and gymnotid fish that are discussed below. Details of the electroreceptor structure are given by Manger and Pettigrew (1996; citation is in note 195), and they also have a helpful discussion of the similarities and differences in electroreception in the fish and monotremes. They list three significant differences, but emphasize the more numerous similarities and stress 'that particular constraints are acting upon the evolution of electroreception' (p. 49), where the convergence emerges by the combination of physical constraints and adaptation to the environment.

236. Concerning the basic similarities of the strongly convergent electric organs of the fish see the review by H. H. Zakon and G. A. Unquez in *Journal of Experimental Biology*, vol. 202, pp. 1427–1434, 1999, as well as *Electric fishes: history and behavior* (Chapman and Hall, London, 1995), a book written largely by P. Moller but with various contributions by a number of his colleagues.

237. In the context of the early recognition of electric fish, especially with respect to medicine, see P. Kellaway's essay in *Bulletin of the History of Medicine*, vol. 20, pp. 112–137, 1946, and also D'Arcy Thompson's discussion of the torpedo ray in *A glossary of Greek fishes* (Oxford University Press, London, 1947), pp. 169–171. A more recent overview of the historical importance of electric fish, including the debate between Galvani and Volta (hence galvanic and volts), is given in the first chapter of Moller (1995).

238. The only surviving work by Scribonius Largus is his *Compositiones medicae*; see the edition prepared by S. Sconocchia (BSB B. G. Teubner, Leipzig, 1983). Reference to the electrical uses of the torpedo fish may be found in sections 11 and 162, translations of which are given by Kellaway (1946, p. 130).

239. See, for example, J. M. Riddle's *Dioscorides on pharmacy and medicine* (University of Texas Press, Austin, 1985). Rather little is known of Dioscorides, other than that he was born in Asia Minor, in Anazarbus, and he studied botany and pharmacology in Tarsus, 'no mean city', as St Paul observed (Acts 22.39). Riddle casts doubt on the popular legend that Dioscorides marched with the Roman legions.

240. See the edition of *The Greek herbal of Dioscorides. Illustrated by a Byzantine A.D. 512. Englished by J. Goodyer* (Oxford University Press, Oxford, 1933), p. 97; 'The sea torpedo being applyed in griefs in long continuance about ye head, doth assuage the fiercenesse of the grief: the same too applyed doth stay up the seate, being either overturned, or else fallen downe', p. 97.

241. Not that the treatment fell out of fashion for many centuries. Thus, in his *Panzoologicomineralogia. Or a compleat history of animals and minerals* (Godwin, Oxford, 1661) Robert Lovell writes of the aptly named Cramp-fish, i.e. torpedo, with reference to Dioscorides that 'Applied to the head, they help old paines thereof, and restraine the falling out of the fundament', p. 191.

242. See Alison Winter's *Mesmerized: Powers of mind in Victorian Britain* (University of Chicago Press, Chicago, 1998).

243. For an account of these experiments see vol. III (May 26, 1836 – Nov. 9, 1839) of *Faraday's diary, being the various philosophical notes of experimental investigation made by Michael Faraday, etc.*, edited by T. Martin (G. Bell and Sons, London, 1933).

244. P. Kellaway (1946; citation is in note 237), also gives some graphic descriptions of treatment of various ailments in South America using the electric eel, which he refers to as 'heroic remedies' (p. 136). Moreover, a report in B. Kramer (1996; citation is in note 246) says that 'according to natives, the *puraqué* (local name for electric eel) is said to harvest palm fruit which it has been observed to eat, by electroshocking the base of a tree', p. 18.

245. The phylogenetic relationships of the gymnotids and mormyrids are addressed by J. A. Alves-Gomes in *Journal of Experimental Biology*, vol. 202, pp. 1167–1183, 1999, and a detailed assessment of the gymnotids is provided by J. S. Albert in *Miscellaneous Publications Museum of Zoology, University of Michigan*, no. 190, vi + 127 pp., 2001. As Alves-Gomes points out 'these two lineages of fish separated more than 140×10^6 years ago, and current evidence suggests that their common ancestor was neither electroreceptive nor electrogenic. It is quite remarkable, therefore, that after this prolonged period of independent evolution, these two fish clades have evolved a number of very elaborate similarities associated with their sensory and motor biology, completely independently of one another. The electrogenic and electrosensory systems ... are used for electrolocation and communication, and the design, physiology and *modus operandi* of the two systems are extraordinarily similar, if not identical, in several ways', p. 1180.

246. The topic of electric fish and mormyrids specifically has attracted many outstanding scientists, but specific mention should be made of Hans Lissman, a refugee from Nazi Germany and a distinguished Cambridge physiologist, who pioneered this field; see, in particular, his papers in *Nature*, vol. 167, pp. 201–201, 1951, and *Journal of Experimental Biology*, vol. 35, pp. 156–191 (in which he emphasizes the various convergences) and pp. 457–486, 1958, the latter with K. F. Machin. Otherwise, the key reference to this topic is the book *Electroreception*, edited by T. H. Bullock and W. Heiligenberg (Wiley, New York, 1986). Also helpful are special issues of *Journal of Comparative Physiology*, vol. 173A, pp. 657–763, 1993, and *Journal of Experimental Biology*, vol. 202, pp. 1167–1458, 1999; see also B. Kramer's *Electroreception and communication in fishes*, as vol. 42 of *Progress in Zoology* (Fischer, Stuttgart, 1996). The topic of electrical generation in fish is an area of considerable research activity; key groups include those led by C. D. Hopkins (see, for example, papers in *American Zoologist*, vol. 21, pp. 211–222, 1981; *Current Opinion in Neurobiology*, vol. 5, pp. 769–777, 1995); M. Kawasaki (see *Journal of Neuroscience*, vol. 17, pp. 1761–1768, 1997; *American Zoologist*, vol. 33, pp. 86–93, 1993; *Journal of Comparative Physiology*, vols. 173A, pp. 9–22, 1993 and 174A, pp. 133–144, 1994 respectively; *Journal of Neuroscience*, vols. 16, pp. 380–391, 1996 and 18, pp. 7599–7611, 1998 respectively); and B. Kramer (see, for example, *Ethology*, vol. 103, pp. 404–420, 1997, and *Naturwissenschaften*, vol. 84, pp. 119–121, 1997).

247. This is with reference to the active detection of objects near the fish; see G. von der Emde in *Journal of Experimental Biology*, vol. 202, pp. 1205–1215, 1999.

248. This ability to deceive the fish was established by Hans Lissman in 1951 (citation is in note 246), where he fed the fish's own electrical pulses back into the tank, eliciting an attack on the electrodes. Concerning the employment of an artificial dipole, see Chapter 10 by P. Moller and J. Serrier, in P. Moller (1995, citation is in note 236).

249. In one group of gymnotids, known as the apteronotids, the electric organ derives from the nervous system; see S. G. Waxman *et al.* in *Journal of Cell Biology*, vol. 53, pp. 210–224, 1972.

250. A general introduction to the process of electrogeneration in fish can be found in M. V. L. Bennett's chapter (pp. 347–491) in vol. 5 of *Fish physiology*, edited by W. S. Hoar and D. J. Randall (Academic Press, New York, 1971). Also helpful is Zakon and Unquez (1999; citation is in note 236).

251. Comments on the four principal types of electric organ in the mormyrid fish and their evolution are provided by J. Alves-Gomes and C. D. Hopkins in *Brain, Behavior and Evolution*, vol. 49, pp. 324–351, 1997. The question of the phylogenetic relationships within the mormyrids and the evolution of the various electrocytes is pursued further by S. Lavoué *et al.* in *Molecular Phylogenetics and Evolution*, vol. 14, pp. 1–10, 2000, and J. P. Sullivan *et al.* in *Journal of Experimental Biology*, vol. 203, pp. 665–683, 2000.

252. Much of the literature cited in this section of the book is directly relevant to the type of electrical signals generated by these fish, but a thorough overview is given by Moller (1995; citation is in note 236).

253. This possibility is explored by P. A. Aguilera *et al.* in *Journal of Experimental Biology*, vol. 204, pp. 185–198, 2000.

254. The role of predation in gymnotid, and probably mormyrid, diversification is addressed by P. K. Stoddard in *Nature*, vol. 400, pp. 254–256, 1999.

255. The detection by catfish of a mormyrid, the bulldog fish, is discussed by S. Hanika and B. Kramer in *Naturwissenschaften*, vol. 86, pp. 286–288, 1999, and *Behavioral Ecology and Sociology*, vol. 48, pp. 218–228, 2000. In the latter paper, they discuss (pp. 226–227) the inferences drawn by Stoddard (1999) on electric fish diversification but are evidently more sceptical, at least so far as the mormyrids are concerned.

256. P. K. Stoddard (1999; citation is in note 254) points out that in this example of signalling there is a parallel in the ctenuchine moths. These insects use an ultrasonic acoustic signal to advertise themselves to bats, who evidently associate the sound with the disgusting taste of the moths (they are wasp mimics and contain toxic chemicals), but in addition the signalling has been co-opted for sexual attraction. This ultrasonic signalling has in turn evolved independently several times in this group of moths; see papers by W. E. Conner *et al.* in *Journal of Insect Behavior*, vol. 8, pp. 19–31, 1995 and vol. 9, pp. 909–919, 1996 respectively, as well as a helpful overview (W. E. Connor, 1999; citation is in note 226) on acoustic communication in moths.

257. See, for example, the paper by P. Belbenoit in *Journal de Physiologie (Paris)*, vol. 59, pp. 344–345, 1967. The 'electric images' that the fish senses are both depicted and discussed in *Journal of Experimental Biology* by C. Assad *et al.* (in vol. 202, pp. 1185–1193, 1999) and R. Budelli and A. A. Caputi (in vol. 203, pp. 481–492, 2000) respectively, as well as by G. von der Emde and S. Schwarz in *Philosophical Transactions of the Royal Society of London B*, vol. 355, pp. 1143–1146, 2000.

258. In a series of experiments on mormyrid fish P. Cain *et al.*, in *Ethology*, vol. 96, pp. 33–45, 1994, found that once accustomed to a novel environment they no longer depended on active electrolocation, with the interesting implication that familiar environments are stored within the brain as a sort of cognitive map (see also Kawasaki (1997; citation is in note 263) for evidence of 'an internal representation' (p. 474) with respect to the processing of electrical signals in the context of jamming avoidance. Subsequently P. Cain (in *Ethology*, vol. 99, pp. 332–349, 1995) extended this work, and suggested that the relative hydrostatic pressure, and thereby water depth was an important cue in mormyrid navigation.

259. See the chapter (pp. 313–393) by M. V. L. Bennett in the book *Lateral line detectors*, edited by P. H. Cahn (Indiana University Press, Bloomington, 1967). In describing these receptors he writes 'The similarities of tonic [for d.c. reception] and phasic [a.c. reception] electroreceptors in gymnotids and mormyrids constitute an instance of convergent evolution as remarkable as that of the electric organs themselves' (p. 386). His earlier contribution in *Cold Spring Harbor Symposia on Quantitative Biology*, vol. 30, pp. 245–262, 1965 is also helpful.

260. An apt phrase used by C. D. Hopkins in his review of electrical signals in a mormyrid fish community, published in *American Zoologist*, vol. 21, pp. 211–222, 1981, see p. 216.

261. See the paper by P. K. McGregor and G. W. M. Westby in *Animal Behaviour*, vol. 43, 977–986, 1992. There is also evidence, in at least one gymnotid, for plasticity in the electrical signal; produced, most probably as a sexual signal; see C. R. Franchina *et al.* in *Journal of Comparative Physiology*, vol. 187A, pp. 45–52, 2001.

262. C. D. Hopkins (1981; citation is in note 260), p. 216.

263. The jamming avoidance response (JAR) is discussed by W. Heiligenberg *et al.* in *Journal of Comparative Physiology*, vol. 127, pp. 267–286, 1978, and more recently by M. Kawasaki (and colleagues) in papers published in *Biological Bulletin*, vol. 191, pp. 103–108, 1996; *Current Opinion in Neurobiology*, vol. 7, pp. 473–479, 1997; and *Journal of Experimental Biology*, vol. 202, pp. 1377–1386, 1999; and in the same journal by W. Metzer (vol. 202, pp. 1365–1375) and B. Kramer (vol. 202, pp. 1387–1398); see also his earlier paper in *Journal of Experimental Biology*, vol. 130, pp. 39–62, 1987, where the JAR is discussed in the context of sexual dimorphism.

264. Concerning the sensitivity and timing of these signals see, for example, C. D. Hopkins and A. H. Bass in *Science*, vol. 212, pp. 85–87, 1981, and G. Rose and W. Heiligenberg in *Nature*, vol. 318, pp. 178–180, 1985, as well as the overview and update given in M. Kawasaki (1997; citation is in note 263).

265. See M. Konishi (in *Journal of Comparative Physiology*, vol. 173A, pp. 700–702, 1993), who draws attention to the similar neural algorithms possessed by the electric fish, specifically the gymnotid *Eigenmannia*, and owls. In the fish the problem is to resolve different signals that might lead to jamming; in owls to distinguish sounds received by the asymmetrical ears (which have evolved independently at least five times; see R. Å. Norberg in *Philosophical Transactions of the Royal Society of London B*, vol. 280, pp. 375–408, 1977) to enable localization of source, say, a mouse. Konishi concludes that his 'findings show that there are computational rules that transcend different sensory systems and animals', p. 702.

266. It is worth noting that 'sound-processing circuits associated with ear asymmetries in owls [see also note 265] ... may have evolved as many as five to seven times independently', see K. C. Nishikawa in *BioScience*, vol. 47, pp. 341–354, 1997, p. 341, reporting work by S. F. Volman in his chapter (pp. 292–314) on directional hearing in owls in *Perception and motor control in birds: an ecological approach*, edited by M. N. O. Davies and P. R. Green (Springer, Berlin, 1994).

267. Once again this convergence has elicited a sense of surprise. For example, C. C. Bell and T. Szabo remark in their chapter (pp. 375–421) in *Electroreception* (see note 246 for citation) that 'The similarities [between gymnotiform and mormyrid fish] are so striking that it is difficult to believe in their independent evolution. But the taxonomic distance ... is so large ... electroreception must have evolved independently' (p. 407).

268. So, too, does the electrosensitive duck-billed platypus. This animal is a highly effective predator in the streams of Australia, emerging from its burrow to hunt at night, with its eyes closed; see U. Proske *et al.* (1993; citation is in note 235).

269. Details of the retinal structure are discussed by M. R. McEwan in *Acta Zoologica, Stockholm* vol. 19, pp. 427–465, 1938.

270. M. R. McEwan (1938; citation is in note 269), p. 442.

271. A more recent discussion of the visual abilities of mormyrid fish is given by C. Teyssedre and P. Moller in *Zeitschrift für Tierpsychologie*, vol. 60, pp. 306–312, 1982, and an overview of mormyrid vision may be found on pp. 368–370 of Moller (1995, note 236).

272. 'just another example of convergent development' as F. Kirschbaum puts it, as part of the title of his article in *Environmental Biology of Fishes*, vol. 10, pp. 3–14, 1984.

273. Concerning this convergent evolution of ecology, and especially feeding, in the gymnotids and mormyrids, see T. R. Roberts in *Bulletin of the Museum of Comparative Zoology*, vol. 143, pp. 117–147, 1972, and in *Environmental Biology of Fishes* the papers by C. Marrero and K. O. Winemiller (vol. 38, pp. 299–309, 1993) and K. O. Winemiller and A. Adite (vol. 49, pp. 175–186, 1997) respectively.

274. The role of the electric field of the gymnotid *Apteronotus* in the capture of its small prey is discussed by M. E. Nelson and M. A. MacIver in *Journal of Experimental Biology*, vol. 202, pp. 1195–1203, 1999. See also the subsequent paper by M. A. MacIver *et al.* in vol. 204, pp. 543–557, 2001, which dwells on the various behavioural characteristics that enhance successful prey interception.

275. T. R. Roberts (1972; citation is in note 273), p. 141.

276. See C. C. Bell's paper, entitled 'Memory-based expectations in electrosensory systems', in *Current Opinion in Neurobiology*, vol. 11, pp. 481–487, 2001.

277. See, in particular, the discussion of distance perception in *Gnathonemus* by G. von der Emde *et al.* in *Nature*, vol. 395, pp. 890–894, 1998, as well as the commentary by W. Metzner on pp. 838–839.

278. See B. R. Brown in *Journal of Experimental Biology*, vol. 205, pp. 999–1007, 2002.

279. Integration of visual and electroreceptive sensory information in the gymnotid *Apteronotus* is addressed by J. Bastian in *Journal of Comparative Physiology*, vol. 147, pp. 287–297, 1982. An overview of this topic is given on pp. 376–379 of Moller (1995, note 233); see also R. Rojas and P. Moller in *Brain, Behavior and Evolution*, vol. 59, pp. 211–221, 2002.

280. See the paper by R. Budelli and A. A. Caputi in *Journal of Experimental Biology*, vol. 203, pp. 481–492, 2000, where the authors model the generation of electrical images according to the impedance of the perceived objects. They draw on analogies of sight and light, and thereby propose that the system of electroreception allows the discrimination of classes of objects, i.e. recognition is distant invariant, which they term 'electrical colors' (p. 490).

281. See, for example, *Synaesthesia: Classic and contemporary readings*, edited by S. Baron-Cohen and J. E. Harrison (Blackwell, Oxford, 1997). That synaesthesia, despite its relative rarity, might be more than a curiosity is made clear in this book, with implications (perhaps) in the areas of consciousness and artistic appreciation, which can be taken to include even the ephemeral but strangely satisfying experience of fireworks (see pp. 29–30). As our understanding of animal sentience grows, as well as the various sensory modalities that are accommodated in their different brains, so we may find some deeper clues as to the presently intangible nature of qualia and other sense perceptions.

282. See, for example, R. Yaka *et al.* in *European Journal of Neuroscience*, vol. 11, pp. 1301–1312, 1999.

283. See M. J. Weissburg *et al.* in *Journal of Comparative Neurology*, vol. 440, pp. 311–320, 2001.

284. Concerning these possible parallels see E. Mugaini and L. Males *et al.* in *Journal of Comparative Physiology*, vol. 173A, pp. 683–685, 1993. The preceding paper (pp. 682–683) by S. Coombs *et al.* on possible neurological similarities of the lateral line, electrosensory systems, and audition is also directly relevant. While emphasizing both possible differences (and lack of information) Coombs *et al.* also comment that 'Regardless of how these parallels [between the three sensory modalities] may have arisen, they suggest that there may be some fundamental principles of organization and cellular interactions that apply to all three systems' (p. 682). In a related review J. C. Montgomery *et al.* (in *Auditory Neuroscience*, vol. 1, pp. 207–231, 1995) and C. Bell *et al.* (in *Brain, Behavior and Evolution*, vol. 50 (Supplement 1), pp. 17–31, 1997) draw attention to the remarkable similarities in the operation of cerebellum-like structures in four groups of fish (including the gymnotids and mormyrids), as well as the dorsal cochlear nucleus of the mammalian auditory system and the optic tectum of teleost fish.

285. See the papers by G. Diesselhorst and E. Stipetic in *Zeitschrift für Vergleichende Physiologie*, vol. 25, pp. 748–783, 1938 and vol. 26, pp. 740–752, 1939 respectively.

286. See H. Y. Yan and W. S. Curtsinger in *Journal of Comparative Physiology*, vol. 186A, pp. 595–602, 2000 and L. B. Fletcher and J. D. Crawford in *Journal of Experimental Biology*, vol. 204, pp. 175–183, 2001.

287. Sound production in this mormyrid is discussed by J. D. Crawford and X-f. Huang in *Journal of Experimental Biology*, vol. 202, pp. 1417–1426, 1999. It is also interesting to note that the sound frequencies generated by this mormyrid imply exceptionally fast contraction of the drumming muscles, perhaps without rival among the vertebrates.

288. This is one of the triumphs of palaeontological investigation, revolving around key work on the Upper Devonian *Acanthostega* (and related forms) by J. A. Clack; see her *Gaining ground: The origin and evolution of tetrapods* (Indiana University Press, Bloomington, 2002).

289. Concerning aspects of the potentially radical transformations between aquatic sound pressure perception and subaerial hearing from a neurological perspective, see B. Fritsch in *Brain, Behavior and Evolution*, vol. 50, pp. 38–49, 1997.

290. See the overview by J. A. Clack in *Brain, Behavior and Evolution*, vol. 50, pp. 198–212, 1997, as well as J. A. Clack (2002; citation is in note 288).

291. See, for example, the article by R. Nobili *et al.* entitled 'How well do we understand the cochlea?' in *Trends in Neurosciences*, vol. 21, pp. 159–167, 1998.

292. Concerning these sensory cells, G. A. Manley *et al.*, in *Journal of Comparative Physiology*, vol. 164A, pp. 289–296, 1989, wrote 'Considering the independent evolutionary origin of these two vertebrate classes from different groups of reptile ancestors, it is remarkable to find indications of a clear functional parallel between avian and mammalian hair-cell populations.' They then continue with the important observation that 'The parallel evolution of such hair-cell populations suggests that there are fundamental properties of hair cells which lend themselves to a particular kind of division of labour.' An update and overview is given by G. A. Manley and C. Köppl in *Current Opinion in Neurobiology*, vol. 8, pp. 468–474, 1998.

293. The question of the adaptive optimal sound for vocal communication in the tunnels of the blind mole-rat is addressed by G. Heth *et al.* in *Experientia*, vol. 42, pp. 1287–1289, 1986.

294. Seismic communication is reviewed by E. Nevo in *Mosaic evolution of subterranean mammals: Regression, progression, and global convergence* (Oxford University Press, Oxford, 1999), see especially pp. 105–107; see also M. J. Mason and P. M. Narins, in *American Zoologist*, vol. 41, pp. 1171–1184, 2001.

295. The similarities in the method of sound conduction is addressed by H. Burda *et al.* in *Journal of Morphology*, vol. 214, pp. 49–61, 1992. A notable exception to this example of convergence is the stapes ear-bone, which shows remarkable shape variation, even within a genus.

296. See the paper by T. Lindenlaub *et al.* in *Journal of Morphology*, vol. 224, pp. 303–311, 1995. See also note 200.

297. The astonishing acoustic sensitivity of the mosquito, especially in the male to whom the wing-beats of the female spell a special magic, is reported by M. C. Göpfert and D. Robert in *Proceedings of the Royal Society of London B*, vol. 267, pp. 453–457, 2000. Göpfert and Robert also point out that while the female's hearing is less acute, it still far outstrips that of most other insects.

298. See their paper, 'Active auditory mechanics in mosquitoes', in *Proceedings of the Royal Society of London B*, vol. 268, pp. 333–339, 2000.

299. M. C. Göpfert and D. Robert (2000; citation is in note 298), pp. 337–338.

300. See the paper by J. de Wilde in *Archives Neerlandaises de Physiologie de l'homme et des animaux*, vol. 25, pp. 381–400, 1941.

301. For an overview see R. R. Hoy and D. Robert in *Annual Review of Entomology*, vol. 41, pp. 433–450, 1996. Hoy and Robert also touch on the question of convergence, a conclusion well supported by the remarkably diverse positions of the tympanal organ across the body surface of the insect.

302. Interestingly, the chordotonal organ (see L. H. Field and T. Matheson in *Advances in Insect Physiology*, vol. 27, pp. 1–228, 1998) appears to have been co-opted from its primary function, which is to act as a stretch receptor for movement of various parts of the body, e.g. segments. Concerning this derivation see G. S. Boyan in *Journal of Insect Physiology*, vol. 39, pp. 187–200, 1993, and M. J. van Staaden and H. Römer in *Nature*, vol. 394, pp. 773–776, 1998.

303. See, for example, the discussion of tympanal (and atympanal) hearing in hawkmoths, by M. C. Göpfert *et al.* in *Proceedings of the Royal Society of London B*, vol. 269, pp. 89–95, 2002. This example is relevant to the discussion of convergence because these ears 'evolved independently on the basis of homologous structures', p. 93.

304. Although this is not to say that crickets cannot defend themselves from digger wasps that are attempting to parasitize them by one of several mechanisms, including kicking; see W. Gnatzy and R. Heusslein in *Naturewissenschaften*, vol. 73, pp. 212–215, 1986.

305. See the paper by D. Robert *et al.* in *Science*, vol. 258, pp. 1135–1137; 1992. As they aptly remark 'This example of convergent evolution in a hearing organ demonstrates the constraints on morphological design that are imposed by behavioral function as well as by principles of physical acoustics' (p. 1135). An analysis of this ear's morphological origins from pre-existing structures is provided by R. S. Edgecomb *et al.* in *Cell & Tissue Research*, vol. 282, pp. 251–268, 1995, and in a subsequent issue (vol. 301, pp. 447–457) D. Robert and U. Willi discuss the histological architecture of the ormiinid ear, and also echo the earlier remarks when they comment 'It is quite remarkable that in insects the sense of hearing has probably evolved many times independently in at least seven orders, and that, in each instance, however diverse the auditory organs are, they are of the chordotonal type endowed with scolopidial mechanoreceptors', pp. 454–455.

306. D. Robert *et al.* (1992; citation is in note 305), p. 1137.

307. A suggestion made by D. Robert *et al.* in *Cell & Tissue Research*, vol. 284, pp. 435–448, 1996.

308. See the paper by A. C. Mason *et al.* in *Nature*, vol. 410, pp. 686–690, 2001, and the accompanying commentary by P. M. Narins on pp. 644–645.

309. Thus D. Robert *et al.* (1996; citation is in note 307) note 'Remarkably, morphological and behavioral evidence points to the presence of prosternal hearing organs in two calyptrate fly species from another big family of parasitoids, the Sarcophagidae ... Hence, it is likely that acoustic parasitism evolved twice independently within calyptrate flies', p. 446.

310. See the paper by R. G. Walker *et al.* in *Science*, vol. 287, pp. 2229–2234, 2000. Their specific example touches on mechanosensory transduction in the fruit-fly and a homologue in worms (i.e. *C. elegans*), but the link to vertebrate hair cells is also addressed.

311. The role of *Math 1* in the generation of the inner hair cells is addressed by N. A. Bermingham *et al.* in *Science*, vol. 284, pp. 1837–1841, 1999.

312. For the activity of *atonal* as a proneural gene, see A. P. Jarman *et al.* in *Development*, vol. 121, pp. 2019–2030, 1995.

313. The role of *atonal* in the glomeruli of the olfactory lobe is addressed by D. Jhaveri and V. Rodriques in *Development*, vol. 129, pp. 1251–1260, 2002.

314. The literature on the *Pax-6* gene is very extensive; it includes the following papers: R. L. Chow *et al.* in *Development*, vol. 126, pp. 4213–4222, 1999; A. Cvekl and J. Piatigorsky in *BioEssays*, vol. 18, pp. 621–630, 1996; C. Desplan in *Cell*, vol. 19, pp. 861–864, 1997; G. Halder *et al.* in *Science*, vol. 267, pp. 1788–1792, 1995; R. Quiring *et al.* in *Science*, vol. 265, pp. 785–789, 1994; S. I. Tomarev *et al.* in *Proceedings of the National Academy of Sciences, USA*, vol. 94, pp. 2421–2426, 1997; and C. Punzo *et al.* in *Genes & Development*, vol. 15, pp. 1716–1723, 2001.

315. For this suggestion, and a consideration of the early evolution of the *Pax* genes see D. J. Miller *et al.* in *Proceedings of the National Academy of Sciences, USA*, vol. 97, pp. 4475–4480, 2000.

316. The role of *Pax-6* in nasal development is addressed by J. C. Grindley *et al.* in *Development*, vol. 121, pp. 1433–1442, 1995; see also M-D. Franco *et al.* in *Journal of Experimental Biology*, vol. 204, pp. 2049–2061, 2001, who among other matters point out that *Pax-6* appears to play different roles in development of olfactory epithelia in frog and mouse (see p. 2059).

317. *Pax-6* expression in the brain is addressed by M. Takahashi and N. Osumi in *Development*, vol. 129, pp. 1327–1338, 2002; P. Callaerts *et al.* in *Journal of Neurobiology*, vol. 46, pp. 73–88, 2001; and several earlier papers in *Development*, see vol. 113, pp. 1435–1450, 1991 (C. Walther and P. Gruss), vol. 126, pp. 5569–5579, 1999 (P. Chapouton *et al.*), vol. 127, pp. 2357–2365 (E. Matsunaga *et al.*), and vol. 127, pp. 5267–5178, 2000 (R. Pratt *et al.*), as well as M. Schwarz *et al.* in *Mechanisms of Development*, vol. 82, pp. 29–39, 1999.

318. See C. Kioussi *et al.* in *Proceedings of National Academy of Sciences, USA*, vol. 96, pp. 14378–14382, 1999.

319. The role of *Pax-6* in the glucagon producing α cells of the pancreas is addressed by L. St-Onge *et al.* in *Nature*, vol. 387, pp. 406–409, 1997.

320. D. Arendt *et al.* in *Development*, vol. 129, pp. 1143–1154, 2002, present some interesting data on polychaete eyes, noting *Pax-6* activity only in the larval eyes, and not those of the adult.

321. See, for example, D. A. Nelson and P. Marler's discussion of categorical perception of birdsong in wild swamp sparrows in *Science*, vol. 244, pp. 976–978, 1989.

322. See R. A. Wyttenbach *et al.* in *Science*, vol. 273, pp. 1542–1544, 1996.

323. R. A. Wyttenbach *et al.* (1996; citation is in note 322), p. 1543.

324. See R. Nieuwenhuys and C. Nicholson's chapter (pp. 107–134) in *Neurobiology of cerebellar evolution and development*, edited by R. Llinás (American Medical Association, Chicago, 1969). The following chapter (pp. 135–169), by the same authors, provides details of the histology of the mormyrid brain. More recent discussions are in the chapter (pp. 375–421) by C. C. Bell and T. Szabo in Bullock and Heiligenberg (1986; see note 246 for citation), and an article by Szabo in *Acta Morphologica Hungarica*, vol. 31, pp. 219–234, 1983.

325. R. Nieuwenhuys and C. Nicholson (1969; citation is in note 324), p. 107.

326. See the paper by G. E. Nilsson in *Journal of Experimental Biology*, vol. 199, pp. 603–607, 1996. He stresses that the extraordinary energetic cost in the mormyrids is not only because of a large brain, but also because the fish are ectothermic. Moreover, given that these fish may find themselves in poorly aerated water, it is not surprising to learn of their ability to scavenge oxygen effectively, although L. J. Chapman and K. G. Hulen (in *Journal of Zoology, London*, vol. 254, pp. 461–472, 2001) question whether even increase in gill size is really sufficient.

327. See J. S. Albert *et al.* on pp. 647–656 of *Proceedings of the 5th Indo-Pacific Fish Conference, Nouméa, 1997*, edited by B. Séret and J-Y. Sire (Société Française d'Ichtyologie, Paris, 1999). Interestingly, among the fish it is the chondrichthyans, notably the galeomorph sharks and batoid rays, which have proportionally the largest of brains.

328. See the chapter (pp. 319–373) by C. E. Carr and L. Maler in T. H. Bullock and W. Heiligenberg (1986; citation is in note 246).

329. See C. Bell *et al.* (1997; citation is in note 284).

8. ALIEN CONVERGENCES?

1. These two genera belong to a larger group of ants, known as the attines (Tribe Attini), all of which are fungus-growers. This section very largely concentrates on the advanced leaf-cutters.

2. Useful introductions are available in K. Dumpert's book *The social biology of ants* (Pitman, Boston, 1981); the masterly work by B. Hölldobler and E. O. Wilson entitled simply *The ants* (Springer, Berlin, 1990); R. C. Buckley's overview on pp. 111–114 (excluding bibliography) of *Ant–plant interactions in Australia*, edited by R. C. Buckley (Junk, The Hague, 1982); N. A. Weber's article forming vol. 92 of *Memoirs of the American Philosophical Society, Philadelphia* (1972); and J. K. Wetterer's chapter (pp. 309–328) in *Nourishment and evolution in insect societies*, edited by J. H. Hunt and C. A. Nalepa (Westview Press, Boulder, CO, 1994).

3. More primitive attine species are more opportunistic and collect a wide variety of other material, including flowers, seeds, faeces, and insect corpses; see I. R. Leal and P. S. Oliveira in *Insectes Sociaux*, vol. 47, pp. 376–382, 2000.

4. See the article on *Acromyrmex* in the Arizona desert by J. K. Wetterer *et al.* in *Sociobiology*, vol. 37, pp. 633–649, 2001.

5. The species in question is found in Paraguay, and details are given by H. G. Fowler and S. W. Robinson in *Ecological Entomology*, vol. 4, pp. 239–247, 1979. Further information is provided by S. P. Hubbell *et al.* in *Biotropica*, vol. 12, pp. 210–213, 1980. These researchers noted that almost never did the tree-top harvesters carry any material back down to the ground, and that the ants lugging the plant material back to the nest were significantly more robust than their companions dropping the pieces from above. F. Roces, in *Biological Bulletin*, vol. 202, pp. 306–313, 2002, has an interesting discussion of the balance between colony-wide activity and individual autonomy.

6. See Hubbell *et al.* (1980). The adaptive benefits of this leaf transfer, in the context of speeds of locomotion, are addressed by C. Anderson and J. L. V. Jadin in *Insectes Sociaux*, vol. 48, pp. 404–405, 2001. An overview of insect 'bucket brigades', which have evidently evolved independently several times, is given by C. Anderson *et al.* in *Insectes Sociaux*, vol. 49, pp. 171–180, 2002.

7. The benefits and drawbacks of leaf caching by the ant *Atta* are addressed by A. G. Hart and F. L. W. Ratnieks in *Animal Behaviour*, vol. 62, pp. 227–234, 2001.

8. Concerning the energetic cost to the colony of clearing and maintaining these trails, see J. J. Howard's analysis in *Behavioral Ecology and Sociobiology*, vol. 49, pp. 348–356, 2001.

9. The highway patrol undertaken by the attine minors is documented by W. O. H. Hughes and D. Goulson in *Behavioral Ecology and Sociobiology*, vol. 49, pp. 503–508, 2001.

10. See T. A. Linksvayer *et al.* in *Biotropica*, vol. 34, pp. 93–100, 2002.

11. Although the cut surfaces do provide a source of nutriment via the plant sap; see, for example, M. Littledyke and J. M. Cherrett in *Bulletin of Entomological Research*, vol. 66, pp. 205–217, 1976.

12. See R. J. Quinlan and J. M. Cherrett in *Ecological Entomology*, vol. 2, pp. 161–170, 1977.

13. Interestingly, but perhaps not surprisingly, the enzymatic activity of the fungi and that in the ants' digestive system are distinct, each tending to attack different substrates. Such a lack of overlap is consistent with this mutualism system being ancient. It is also possible that one of the enzymatic properties the larvae possess has been transferred from the fungus. See P. D'Ettorre *et al.* in *Journal of Comparative Physiology*, vol. 172B, pp. 169–176, 2002.

14. Details of this process are given by C. R. Currie and A. E. Stuart in *Proceedings of the Royal Society of London B*, vol. 268, pp. 1033–1039, 2001. Weeding makes horizontal (or lateral) transmission between nests substantially more difficult, but helps to ensure vertical transmission (from generation to generation) within the nest; see A. N. M. Bot *et al.* in *Evolution*, vol. 55, pp. 1980–1991, 2001.

15. As F. Roces and C. Kleineidam observe in *Insectes Sociaux* (vol. 47, pp. 348–350, 2000) 'Relocation of fungus gardens to promote [fungal] growth, like the repotting of flowering plants by humans, clearly illustrates the skill of leaf-cutting ants as true "gardeners",' p. 350.

16. At least on the basis of experimental work; see M. Bollazzi and F. Roces in *Insectes Sociaux*, vol. 49, pp. 153–157, 2002.

17. See M. Bass and J. M. Cherrett in *Functional Ecology*, vol. 10, pp. 55–61, 1996.

18. Manuring is addressed by M. M. Martin on pp. 163–166 of his chapter reviewing the use of ingested enzymes in insects in *Invertebrate–microbial interactions*, edited by J. M. Anderson *et al.* (Cambridge, Cambridge University Press, 1984).

19. The last of these compounds includes sterols that are essential for hormone synthesis; see P. Maurer *et al.* in *Archives of Insect Biochemistry and Physiology*, vol. 20, pp. 13–21, 1992.

20. The threats posed to the attine ant colonies by this parasitic fungus are graphically brought out by C. R. Currie in *Oecologia*, vol. 128, pp. 99–106, 2001. See also C. R. Currie *et al.* in *Science*, vol. 299, p. 386–388, 2003. Smaller colonies are at particular risk, whereas the larger ones remain clear of infection.

21. In fact the use of antibiotics is widespread in the social insects and termites; see R. B. Rosengaus *et al.* in *Journal of Chemical Ecology*, vol. 26, pp. 21–39, 2000.

22. For more details see the paper by C. R. Currie *et al.* in *Nature*, vol. 398, pp. 701–704, 1999, as well as the overview by C. R. Currie in *Annual Review of Microbiology*, vol. 55, pp. 357–380, 2001.

23. These include secretions from the metapleural glands, which are especially well developed in the attines. See, for example, the papers by D. Ortius-Lechner *et al.* in *Journal of Chemical Ecology*, vol. 26, pp. 1667–1683, 2000; A. N. M. Bot *et al.* in *Insectes Sociaux*, vol. 48, pp. 63–66, 2001; and M. Poulsen *et al.* in *Behavioural Ecology and Sociobiology*, vol. 52, pp. 151–157, 2002. Experiments in which a fungicide is introduced quickly lead to a change in

foraging behaviour, which may in addition lead to this behaviour being transmitted to another colony; see R.D. North *et al.* in *Physiological Entomology*, vol. 24, pp. 127–133, 1999.

24. See M. Poulsen *et al.* in *Insectes Sociaux*, vol. 49, pp. 15–19, 2002, who document these alternative defence mechanisms in two species of *Acromyrmex*.

25. This case of agro-predation by the ant *Megalomyrmex* is described by R. M. M. Adams *et al.* in *Naturwissenschaften*, vol. 87, pp. 549–554, 2000. In this situation these ants have, of course, to manipulate the fungi, and perhaps not surprisingly their 'behavior closely resembles that of attine ants when they add substrate to or reconstruct gardens ... suggesting that it has been convergently derived in *Megalomyrmex sp.* ants' (p. 551). These authors also point out that the continued health of the occupied gardens suggests that '*Megalomyrmex sp.* may have evolved a similar repertoire of maintenance behaviors and chemical secretions' (p. 550) to those of the attine ants.

26. For a graphic account of the defence and eventual overwhelming of an attine colony by the army ant *Nomamyrex* see the article by M. B. Swartz in *Biotropica*, vol. 30, pp. 682–684, 1998.

27. See, in particular, W. H. Gotwald's *Army ants: the biology of social predation* (Cornell University Press [Comstock], Ithaca; 1995). The best-documented convergence of army ant adaptive syndrome is between the New World *Eciton* and Old World taxa such as *Dorylus* and *Aenictus*. Gotwald also draws attention to at least two more instances of army ant-like behaviour, these being in two species of *Pheidologeton* where his comments include the observation that 'their foraging behavior is uncanny in its resemblance to that of army ants' (p. 38), and also in the minuscule leptanillinine ants. A helpful summary of the behaviour of leptanillinine ants and *Pheidologeton* is given on pp. 590–592 and pp. 593–594 of B. Hölldobler and E. O. Wilson (1990; citation is in note 2).

28. The reality versus the folklore of army ants is addressed by Gotwald (1995), see pp. 237–238, 249, 252. In agriculture army ants may be helpful in pest-control, but they can also be destructive of crops. One should note that the mandibles of ants have been used to suture both natural and surgical wounds, pp. 249–250.

29. See N. R. Franks' article 'Army ants: A collective intelligence' in *American Scientist*, vol. 77, pp. 138–145, 1989. So, too, S. C. Pratt *et al.* (in *Behavioral Ecology and Sociobiology*, vol. 52, pp. 117–127, 2002) describe how in the context of colony emigration another ant (*Leptothorax*) society behaves 'as a single information-processing unit', (p. 117). These authors draw attention to similar quorum sensing, not only in other social insects, notably bee (see note 36), but even bacteria and immune response T-cells.

30. Well, not all of them, because the army ants have their own predators, including birds and such anteaters as the pangolins.

31. As N. R. Franks *et al.* (in *Proceedings of the Royal Society of London B*, vol. 266, pp. 1697–1701, 1999) show, the net result in New World *Eciton* and Old World *Dorylus* is achieved by rather different means. In essence *Dorylus* uses more and smaller workers to carry collectively larger loads, but at a slower speed. Further insights into the convergent teamwork in *Eciton* and *Dorylus* are provided by N. R. Franks *et al.* in *Animal Behaviour*, vol. 62, pp. 635–642, 2001. Here they show that the division of labour typically involves a team with an unusually large and unusually small ant. As the authors point out, an analogy for this exists in the form of the penny-farthing bicycle, where each wheel has a separate sub-task, the 'penny' providing load-bearing and drive, the 'farthing' stability. The net result is a synergistic contraption whose effectiveness is more than the sum of the parts.

32. This is only one example of the recurrent property of self-assembly in social insects, a topic reviewed by C. Anderson *et al.* in *Insectes Sociaux*, vol. 49, pp. 99–110, 2002. These workers identify at least 18 different ways in which insect individuals link themselves together to form structures as disparate as bivouacs, bridges, ladders, and rafts. Not only are many of these behaviours convergent, but, as importantly, they reflect new levels of complexity unavailable to simpler societies.

33. This topic is addressed by N. R. Franks in *Physiological Entomology*, vol. 14, pp. 397–404, 1989.

34. See G. P. Boswell *et al.* in *Proceedings of the Royal Society of London B*, vol. 268, pp. 1723–1730, 2001.

35. See, for example G. J. Vermeij's article in *American Naturalist*, vol. 120, pp. 701–720, 1982, especially pp. 710–713.

36. The complexity and sophistication of the bee colony is probably still underestimated, and their degree of integration with selection operating at the group level of the colony (see, for example, the article by T. D. Seeley in *American Naturalist*, vol. 150, pp. 522–541, 1997) makes them, at least potentially, an interesting evolutionary alternative to the emergence of advanced cognizance. Consider, for example, how a bee swarm dispatches scout bees to numerous sites that are then 'assessed' before a collective decision is arrived at; see S. Camazine *et al.* in *Insectes Sociaux*, vol. 46, pp. 348–360, 1999 and T. D. Seeley and S. C. Buhrmann in *Behavioral Ecology and Sociobiology*, vol. 45, pp. 19–31, 1999.

37. For example, T. S. Collett *et al.*, in *Journal of Comparative Physiology*, vol. 181A, pp. 343–353, 1997, present evidence for contextural learning in honey-bees, and M. Giurfa *et al.* (in *Nature*, vol. 410, pp. 930–933, 2001) show that not only can bees place the outside world in a context and memorize it, but engage in higher-order functions, that is they can 'master abstract interrelationships, such as sameness and difference', p. 930.

38. See the papers by W. Kaiser in *Nature*, vol. 301, pp. 707–709, 1983 (with J. Steiner-Kaiser) and *Journal of Comparative Physiology*, vol. 163A, pp. 565–584, 1988. In the latter paper Kaiser remarks how four characteristics – decreased motility, lowered body temperature, reduced muscle tone, and raised reaction threshold – 'strongly resemble [those] characteristic features of sleep in humans, mammals and birds' (p. 565). So do bees dream?

39. See, for example, G. Buzsaki in *Journal of Sleep Research*, vol. 7 (Supplement 1), pp. 17–23, 1998, as well as more recent contributions in *Nature Neuroscience*, vol. 3, pp. 1237–1238 (R. Stickgold *et al.*) and pp. 1335–1339 (S. Gais *et al.*), 2000. A useful overview of the adaptive contexts of sleep is given by J. L. Kavanau in *Animal Behaviour*, vol. 62, pp. 1219–1224, 2001, and in his earlier article in *Brain Research Bulletin*, vol. 42, pp. 245–264, 1997, where the potential importance of endothermy (see p. 223 of this Chapter) is stressed.

40. The honey bees themselves show an interesting convergence with another eusocial species, the common wasp. This entails so-called 'worker policing' in which any eggs laid by the workers, which will be parthenogenetic and male, are quickly destroyed by other members of the squad – egg production of course being the prerogative of the queen. As K. R. Foster and F. L. W. Ratnieks, in *Proceedings of the Royal Society of London B*, vol. 268, pp. 169–174, 2001, remark 'The worker policing by mutual egg-eating in *V. vulgaris* (the common wasp) is strikingly similar to that found in the honeybee *A. mellifera* ... This is not due to common ancestry since the vespine wasps and honeybees belong to lineages that have evolved eusociality independently' (p. 172). There are, however, differences between these insects in terms of their genetic relatedness of the workers: effectively honeybee queens are polyandrous, that is they receive multiple matings on their nuptial flights, so workers are not closely related and there is a premium in destroying any workers' eggs ('not your genes, destroy! destroy!'). In the wasps, however, relatedness of the workers is higher, and so another explanation for the policing has to be found. Here Foster and Ratnieks suggest that the benefit may arise from reducing intra-colony reproductive aggression. They also remark that such policing may well be a feature of leaf-cutter and other ant colonies; see also remarks by P. Villesen *et al.* (in *Proceedings of the Royal Society of London B*, vol. 269, pp. 1541–1548, 2002) on the convergent emergence of multiple matings in leaf-cutter ants and other very advanced social insects.

41. Stingless they may be, but they are highly effective in defending the colony, 'locking their mandibles in catatonic spasms so that before the grip is broken, their heads tear loose from the body. The *Trigona flaveola* group of species in tropical America also eject a burning liquid

from their mandibles which in Brazil has earned them the name *cagafogos*, meaning 'fire defecators', p. 89 of E. O. Wilson's *The insect societies* (Belknap, Cambridge, MA; 1971). For an interesting account of the highly sophisticated methods of communication in the stingless bees, see the article by J. C. Nieh in *American Scientist*, vol. 87, pp. 428–435, 1999, which he notes has probably evolved independently of the honey-bee; see also note 43.

42. On this topic see the paper by M. S. Engel in *Proceedings of the National Academy of Sciences, USA*, vol. 98, pp. 1661–1664, 2001.

43. See the paper by S. A. Cameron and P. Mardulyn in *Systematic Biology*, vol. 50, pp. 194–214, 2001. Their conclusions are highly resonant with the thrust of this book: 'If indeed the highly eusocial behavior of Apini and Meliponini evolved twice independently – the hypothesis we support as being more likely – then the comparative behavior of these bees takes on an added interest from an evolutionary perspective. The apparent convergent similarities in behavior between these two tribes suggest that highly eusocial organization has limited permutations. Each tribe has been channeled along a similar behavioral track, responding in similar fashion to similar contingencies', p. 209. The notion of convergence in advanced eusocial behaviour actually goes back many years, see, for example, M. L. Winston and C. D. Michener in *Proceedings of the National Academy of Sciences, USA*, vol. 74, pp. 1135–1137, 1977. These authors are also struck by the implications of convergence; they write, 'The development of such [highly social] behavior, probably the most elaborate of any invertebrate, is remarkable enough; independent parallel development of two highly eusocial systems would be even more noteworthy' (p. 1135). Engel (2001), however, regards advanced eusociality in bees as monophyletic (as does F. B. Noll in *Cladistics*, vol. 18, pp. 137–153, 2002). He bases his analysis on the fossil record, especially amber preservation. S. A. Cameron and P. Mardulyn (2001) come to the same conclusion on the basis of morphological data, but this is in conflict with molecular data.

44. The general consensus is that advanced eusocial species, with such features as division of labour and colony homeostasis, are effectively 'locked in' to this arrangement. It should be pointed out that there is evidence that in the case of less organized social behaviours this does not preclude a return to a solitary mode of life and that this has occurred independently a number of times. See, for example, the papers by R. Gadagkar (in *Current Science*, vol. 72, pp. 950–956, 1997) and W. T. Wcislo and B. N. Danforth (in *Trends in Ecology & Evolution*, vol. 12, pp. 468–474, 1997), as well as the useful overview by B. J. Crespi, on pp. 253–287 of *Phylogenies and the comparative method in animal behavior*, edited by E. P. Martins (Oxford University Press, New York, 1996).

45. M. S. Engel (2001; citation is in note 42); see p. 1663. This writer is careful to emphasize that competition need not be the sole cause of extinction in the eusocial bees, and that climate cooling might also play some role.

46. Part of the title of G. P. Boswell *et al.* (2001; citation is in note 34).

47. N. R. Franks (1989; citation is in note 29), p. 139.

48. Ulrich Mueller reminds me that pheromones are not the only currency of exchange in ant communication, and that tactile contact and food exchange can also be important.

49. N. R. Franks (1989; citation is in note 29), p. 144.

50. See his chapter (pp. 281–298) in *Insect movement: mechanisms and consequences*, edited by I. P. Woiwod *et al.* (CABI, Wallingford, 2001, p. 286, his emphases). See also the papers by E. Bonabeau entitled 'Social insect colonies as complex adaptive systems' in *Ecosystems*, vol. 1, pp. 437–443, 1998, and B. J. Cole in *Biological Bulletin*, vol. 202, pp. 256–261, 2002.

51. See the stimulating article by A. D. M. Rayner and N. R. Franks on 'Evolutionary and ecological parallels between ants and fungi' in *Trends in Ecology & Evolution*, vol. 2, pp. 127–133, 1987.

52. C. Kleinidam *et al.* (in *Naturwissenschaften*, vol. 88, pp. 301–305, 2001) have confirmed that the ventilation is wind-induced, a not surprising necessity given the large quantities of oxygen required by the inhabitants and their fungi.

53. See, for example, the remarkable photograph (Fig. 20c) in the paper on nest structure by J. C. M. Jonkman in *Zeitschrift für Angewandte Entomologie*, vol. 89, pp. 217–246, 1980.

54. See A. G. Hart and F. L. W. Ratnieks in *Behavioral Ecology and Sociobiology*, vol. 49, pp. 387–392, 2001, and also *Behavioral Ecology*, vol. 13, pp. 224–231, 2002.

55. Waste dumps are frequently infected with the parasitic fungus *Escovopsis* (see note 20) and mortality of the workers can be significantly increased if the waste material is not quickly cleared; see A. N. M. Bot *et al.* in *Ethology, Ecology and Evolution*, vol. 13, pp. 225–237, 2001. Bot *et al.* also remark that 'There are a number of interesting parallels between waste management in agricultural societies of ant and human. Both isolate waste in specific places: ants use refuse chambers or piles where we use landfill sites ... natural selection has probably led to a series of general attitudes of disgust towards different types of waste products ... Apparently, natural selection and evolution have produced converged methods by which to address these problems in both human and insect societies,' p. 235.

56. See the paper by U. G. Mueller *et al.* in *Science*, vol. 281, pp. 2034–2038, 1998, as well as the stimulating commentary by J. Diamond on pp. 1974–1975, where he explores the similarities to and differences from human agriculture. The likely origins of the attine ant-fungus association are addressed by U. G. Mueller *et al.* in *Quarterly Review of Biology*, vol. 76, pp. 169–197, 2001.

57. See *The origins of agriculture: An international perspective*, edited by C. W. Cowan and P. J. Watson (Smithsonian, Washington, DC, 1992). In particular, the chapter (pp. 143–171) by E. M. de Tapia specifically rejects the generally popular notion of population increase in the instance of Mesoamerica.

58. See U. G. Mueller *et al.* (2001; citation is in note 56).

59. N. A. Weber (1972; citation is in note 2), p. 1.

60. J. C. M. Jonkman in *Zeitschrift für Angewandte Entomologie*, vol. 89, pp. 135–143, 1980 provides some interesting figures on both population sizes and biomass accumulated by the colony.

61. This topic is addressed by R. M. M. Adams *et al.* in *Naturwissenschaften*, vol. 87, pp. 491–493, 2000; see also A. M. Green *et al.* in *Molecular Ecology*, vol. 11, pp. 191–195, 2002.

62. Quoted (p. 763) in the article by L. Ariniello in *Bioscience*, vol. 49, pp. 760–763, 1999.

63. See B. D. Farrell *et al.* in *Evolution*, vol. 55, pp. 2011–2027, 2001, and the review by R. A. Beaver on pp. 121–143 in *Insect–fungus interactions*, edited by N. Wilding *et al.* (Academic Press, London, 1989). See also C. C. Labandeira *et al.* in *American Journal of Botany*, vol. 88, pp. 2026–2039, 2001.

64. See A. R. Berryman's chapter (pp. 145–159) in Wilding *et al.* (1989).

65. A. R. Berryman (1989), p. 149.

66. See the overview by T. D. Paine *et al.* in *Annual Review of Entomology*, vol. 42, pp. 176–206, 1997.

67. Evolution of defensive compounds, such as resin and latex in plants, is rampantly convergent. B. D. Farrell *et al.*, in *American Naturalist*, vol. 138, pp. 881–900, 1991, estimate that 'Canal systems containing latex or resin appear to have originated at least 40 times,' p. 889.

68. B. D. Farrell *et al.* (2001; citation is in note 63), p. 2017.

69. See the chapter (pp. 105–130) by J. P. E. C. Darlington in J. H. Hunt and C. A. Nalepa (1994; citation is in note 2). See also D. K. Aanen *et al.* (in *Proceedings of the National Academy of Sciences, USA*, vol. 99, pp. 14887–14892, 2002), who document the evolution of this mutualism, as well as the differences and similarities between termites and attine ants, and J. Korb and D. K. Aanen in *Behavioural Ecology and Sociobiology*, vol. 53, pp. 65–71, 2003).

70. See the description by J. P. E. C. Darlington in *Journal of Zoology, London*, vol. 198, pp. 237–247, 1982, which together with Darlington (1994) provides the information given here; see also G. Schuurman and J. M. Dangerfield in *Sociobiology*, vol. 27, pp. 29–38, 1996.

71. The example of convergent evolution between the *Echidnophora* and Termitoxeniiae is given by R. H. L. Disney in *Systematic Entomology*, vol. 20, pp. 195–206, 1995.

72. The molecular phylogenetics of *Termitomyces* are addressed by C. Rouland-Lefèvre *et al.* in *Molecular Phylogenetics and Evolution*, vol. 22, pp. 423–429, 2002.

73. R. Heim's book *Termites et champignons: Les champignons termitophiles d'Afrique Noire et d'Asie méridionale* (Boubé, Paris, 1977) is principally devoted to *Termitomyces*.

74. R. Heim (1977; citation is in note 73), p. 43.

75. J. P. E. C. Darlington (1994; citation is in note 69 and in note 2), p. 124.

76. In fact such evidence is now available: see, for example, P. Mora and C. Lattaud in *Insect science and its application*, vol. 19, pp. 51–55, 1999, who report the activity of the enzyme laccase (employed in lignin degradation) in both fungus-cultivating termites and *Termitomyces*, as well as the overview by C. Rouland-Lefèvre on pp. 289–306 of *Termites: Evolution, sociality, symbioses, ecology*, edited by T. Abe *et al.* (Kluwer, Dordrecht, 2000).

77. J. P. E. C. Darlington (1994; citation is in note 2), pp. 125–126.

78. W. H. Kirchner *et al.* discuss vibrational alarm communication in the New World termite *Zootermopsis*, in *Physiological Entomology*, vol. 19, pp. 187–190, 1994. This is a relatively primitive termite, inhabiting rotten wood, and is not a fungus cultivator.

79. See H. Markl in *Science*, vol. 149, pp. 1392–1393, 1965 and *Zeitschrift für vergleichende Physiologie*, vol. 57, pp. 299–330, 1967. More recently F. Roces and B. Hölldobler in *Behavioral Ecology and Sociobiology*, vol. 39, pp. 293–299, 1996, have also demonstrated that this stridulation serves to attract workers to the leaf-cutting site.

80. See A. Röhrig *et al.* in *Insectes Sociaux*, vol. 46, pp. 71–77, 1999, who note that such long-distance communication is unique among the insects. Röhrig *et al.* also remark that the intensity of drumming determines the type of reaction, at its most intense leading to a mass retreat, with the intriguing possibility 'that alarm signals help to avoid costly direct interactions between [adjacent termite] colonies, battles that can kill thousands of soldiers from nests', p. 76.

81. See the chapter by S. E. Kleinfeldt on pp. 283–294 of *Insects and the plant surface*, edited by B. Juniper and R. Southwood (Edward Arnold, London, 1986).

82. This area is reviewed by U. G. Mueller *et al.* (2001; citation is in note 56), pp. 176–177.

83. See D. W. Davidson *et al.* in *Ecology*, vol. 69, pp. 801–808, 1988.

84. Importing vertebrate faeces (p. 1146) and the possibility of pathogen control (p. 1151) are mentioned in a general review of ant-garden ecology by D. W. Davidson in *Ecology*, vol. 69, pp. 1138–1152, 1988.

85. A helpful summary is given by W. H. Gotwald (1995; citation is in note 27); an earlier paper on dorylomimine mimics is by D. H. Kistner in *Annals of the Entomological Society of America*, vol. 59, pp. 320–340, 1966.

86. C. H. Seevers, in *Fieldiana: Zoology*, vol. 47, pp. 137–351, 1965, suggests a total of at least seven transformations of staphylinids to army-ant mimics.

87. See the chapter (pp. 339–413) by D. H. Kistner in *Social Insects* (volume I), edited by H. R. Hermann (Academic Press, New York, 1979), where he discusses the colour mimicry between the beetle *Ecitomorpha nevermanni* and army ant *Eciton burchelli*. Subsequently, Kistner and H. R. Jacobson, in *Sociobiology*, vol. 17, pp. 333–465, 1990, point out that colour variation in a given colony is somewhat wider than previously realized.

88. D. H. Kistner (1979; citation is in note 87), p. 357.

89. D. H. Kistner (1979; citation is in note 87), p. 357, see also his Fig. 8.

90. A. P. Mathew in *Journal of the Bombay Natural History Society*, vol. 52, pp. 249–263, 1958.

91. A. P. Mathew (1958; citation is in note 90), p. 250.

92. The observation of the spider mimicking an ant in distress is based on laboratory investigations. Field observations by A. P. Mathew (1958; citation is in note 90) also indicate this spider stalking and pouncing on ants that stray too far.

93. For a review of this insect–spider mimicry, see J. D. McIver and G. Stonedahl in *Annual Review of Entomology*, vol. 38, pp. 351–379, 1993.

94. This is known as transformational mimicry. For example, A. P. Mathew, in *Journal of the Bombay Natural History Society*, vol. 37, pp. 803–813, 1935, explains how the hemipteran bug *Riptortus pedestris*, in a series of moults, successively resembles the ant *Prenolepis longicornis* ('almost indistinguishable', p. 804), then *Plagiolepis* ('mimicry at this stage is almost perfect', p. 804), and then the already encountered *Oecophylla smaragdina* 'which presented a very perfect resemblance to the Red Ant ... even more perfect than the mimicry which I was observing and studying in the Attid ants' (p. 803). J. D. McIver and G. Stonedahl (1993; citation is in note 93) provide an overview on this topic, with further citations.

95. See J. D. McIver and G. Stonedahl (1993), pp. 364–365.

96. A helpful overview is given by L. E. Gilbert in his chapter (pp. 263–281) in *Coevolution*, edited by D. J. Futuyama and M. Slatkin (Sinauer, Sunderland, MA, 1983), who concludes by remarking 'Mimicry is possibly the most compelling evidence that community patterns are more than random noise,' p. 281.

97. But see, for example, the paper by J. Berger and J. Kaster (in *Evolution*, vol. 33, pp. 511–513, 1979). Berger and Kaster argue that a striking resemblance between a caddisfly larva (*Helicopsyche borealis*) and an aquatic snail (*Physa integra*) arises from a common selection pressure imposed by the hydrodynamic regime rather than a mimicry imposed by predation.

98. See the interesting overview by C. Anderson and D. W. McShea in *Insectes Sociaux*, vol. 48, pp. 291–301, 2001. One might draw attention to the ant equivalent of 'cow-sheds' to house honeydew-yielding aphids, and 'pens' to look after caterpillars. In their conclusion the authors remark, 'What is certainly needed is more detail about construction mechanisms. Where such information was available, we observe a good example of convergent evolution: trenches and arcades appear to be similar in both structure and construction behavior in new world army ants, such as *Labidus*, old world African driver ants, such as *Dorylus*, and Asian marauder ants, *Pheidologeton*. Arcade construction [in the last genus] has certainly arisen independently,' p. 299.

99. See K. Wada in *Ethology*, vol. 96, pp. 270–280, 1994 and *Crustaceana*, vol. 68, pp. 524–526, 1995 (with J. K. Park); and J. Kitaura *et al.* in *Molecular Biology and Evolution*, vol. 15, pp. 626–637, 1998.

100. See the paper by G. A. Croll and J. B. McClintock in *Journal of Experimental Marine Biology and Ecology*, vol. 254, pp. 109–121, 2000.

101. As an aside on the topic of convergence, I cannot resist the example of the discharge of ink by various cephalopods to baffle their attackers. And the parallel? The discharge of a reddish cloud by dwarf sperm-whales, albeit from the anus; see M. D. Scott and J. G. Cordaro in *Marine Mammal Science*, vol. 3, pp. 353–354, 1987.

102. A useful review of the brains of octopuses and other advanced cephalopods is given by B. U. Budelmann on pp. 115–138 of *The nervous system of invertebrates: An evolutionary and comparative approach* edited by O. Breidbach and W. Kutsch (Birkhäuser, Basel, 1995); see also the chapter (pp. 399–413) by B. U. Budelmann *et al.* in *Cephalopod neurobiology: Neuroscience studies in squid, octopus, and cuttlefish*, edited by N. J. Abbott *et al.* (Oxford University Press, Oxford, 1995) and J. Z. Young's monograph *The anatomy of the nervous system of* Octopus vulgaris (Oxford, Clarendon Press, 1971). In describing the cerebellum-like structure in the squid brain, J. Z. Young (in *Nature*, vol. 264, pp. 572–574, 1976) remarked how the pattern of nervous control associated with movement of the eye as well as the whole animal 'shows remarkable similarities between cephalopods and vertebrates, in spite of the fact that the gross anatomical entities are totally different', p. 574.

103. See the stimulating paper by A. Packard entitled *Cephalopods and fish: the limits of convergence* in *Biological Reviews* 47, 241–307, 1972. The question of relative brain sizes is addressed on pp. 265–267. Packard's overall conclusion is that the cephalopods are good, but not as good as the fish. Even so, if the fish eventually win out it is going to take time. J. J. Madan and M. J. Wells also point out, in *Nature*, vol. 380, p. 590, 1996, that the deep-sea

squid have a refuge of sorts by inhabiting sea water with low levels of oxygen, a mode of life made possible by having very thin – and delicate – gills.

104. See chapter 27 (pp. 446–457) by M. Bundgaard *et al.* in N. J. Abbott *et al.* (1995; citation is in note 102), as well as two companion papers by N. J. Abbott *et al.* in *Journal of Physiology*, vol. 368, pp. 197–212 and 213–226, 1985.

105. M. Bundgaard *et al.* (1995; citation is in note 104), p. 446.

106. See H. Jaaro *et al.* in *Trends in Neurosciences*, vol. 24, pp. 79–85, 2001. Interestingly, the most advanced sort of brains in the arthropods, found in the insects, apparently lack neurotrophins, at least to judge from a survey of the fruit-fly genome. This is not to say that insect brains are simple; the reverse is very much the case. Jaaro *et al.* suggest, however, that neurotrophins predetermine the 'differences between a "plastic" nervous system with a potential for evolving complexity, versus a more constrained nervous system that cannot evolve beyond certain limitations' (p. 84). It may be unwise, however, to underestimate the plasticity of insect brains; see, for example, the overview by I. A. Meinertzhagen in *Advances in Insect Physiology*, vol. 28, pp. 84–167, 2001.

107. Chromatic expression is controlled by a specific area of the brain and is achieved by pigmented areas known as chromatophores; see, for example, the papers by E. Florey in *American Zoologist*, vol. 9, pp. 429–442, 1969; R. T. Hanlon and J. B. Messenger in *Philosophical Transactions of the Royal Society of London B*, vol. 320, pp. 437–487, 1988, and *Cephalopod behaviour* (Cambridge University Press, Cambridge, 1996); and A. Packard in chapter 21 (pp. 331–367) of Abbott *et al.* (1995; citation is in note 102). J. B. Messenger (in *Biological Reviews*, vol. 76, pp. 473–528, 2001) remarks 'Cephalopods are such complex and behaviourly advanced animals ... that they must be able to transmit messages of a far subtler kind [than straightforward messages] and it seems likely that future work will reveal this' (p. 520). See also A. C. Crook *et al.* (in *Philosophical Transactions of the Royal Society of London B*, vol. 357, pp. 1617–1624, 2002), who note that in contrast to other animal communication systems (see note 140, Chapter 9) the visual signals of cuttlefish do not follow Zipf's law.

108. Relevant literature includes papers by G. Fiorito and P. Scotto in *Science*, vol. 256, pp. 545–547, 1992; J. Z. Young in *Biological Bulletin*, vol. 180, pp. 200–208, 1991a, *Visual Neuroscience*, vol. 7, pp. 1–12, 1991b; and J. G. Boal *et al.* in *Journal of Comparative Psychology*, vol. 114, pp. 246–252, 2000. Fiorito and Scotto show how one octopus can learn by observing the activities of another octopus. J. Z. Young in his 1991a paper demonstrates how there are effectively two memory systems, one chemo-tactile and one visual. He remarks on the complexity of the brain structure, noting that the learning ability 'is distributed with high redundancy in a series of matrices networks, with recurrent circuitry, up to a late stage where funneling to a few cells occurs' (p. 206). Elsewhere (in chapter 26 (pp. 432–443) of N. J. Abbott *et al.* (1995; citation is in note 102)), Young remarks that 'The information already available shows that the octopus memory involves selection during learning among large number of parallel pathways. The capacity to set up and hold the memory is distributed among many distinct matrices. This seems to be the case also in the complex memories of mammals, birds, and insects', p. 442.

109. T. Moriyama and Y-P. Gunji (in *Ethology*, vol. 103, pp. 499–513, 1997), show how the octopus can choose a seemingly inefficient action (e.g. reduced swimming speed) the more easily to achieve its final objective, say a tasty piece of shrimp.

110. On octopus living on coral atolls in French Polynesia, see J. W. Forsythe and R. T. Hanlon in *Journal of Experimental Marine Biology and Ecology*, vol. 209, pp. 15–31, 1997.

111. J. W. Forsythe and R. T. Hanlon (1997; citation is in note 110), p. 26.

112. See the paper by J. A. Mather and R. C. Anderson in *Journal of Comparative Psychology*, vol. 107, pp. 336–340, 1993. Their analysis depends on the difference between octopus individuals, their patent ability to learn in the face of novel environments, and temperamental behaviour, which also define in part human personality. This work is

extended by D. L. Sinn *et al.* in the same journal (vol. 115, pp. 351–364, 2001). They suspect that the range of octopus temperaments may not be fully explored, while Mather and Anderson comment that there may be commonalities across phyla; addressing an octopus by name need not be completely daft.

113. The rubber-like proteins in the octopus aorta are documented by R. E. Shadwick and J. M. Gosline in *Journal of Experimental Biology*, vol. 114, pp. 239–257, 1985. Such proteins, of which the vertebrate elastin, insect resilin, and bivalve mollusc abductin are another three examples, are possibly convergent. These authors note that 'The tremendous variation in amino acid composition among the four known protein rubbers ... suggests to us that each protein arose independently during evolution in response to selection for similar mechanical design', p. 256.

114. R. E. Shadwick notes in *Journal of Experimental Biology*, vol. 202, pp. 3305–3313, 1999, that 'In animals as different as *Nautilus* [a cephalopod] and the rat, there is a striking similarity in the microscopic appearance of the aortic wall', p. 3309. See also his earlier articles in *Marine Behavior and Physiology*, vol. 25, pp. 69–85, 1994, and *Journal of Experimental Biology*, vol. 114, pp. 259–284, 1985 (with J. M. Gosline); and R. Schipp *et al.* in *Zoomorphology*, vol. 118, pp. 79–85, 1998. It is also worth mentioning that the convergence in arterial wall structure also extends to the crustaceans; see R. E. Shadwick *et al.* in *Physiological Zoology*, vol. 63, pp. 90–101, 1990.

115. Helpful reviews of the circulatory system of cephalopods are given by M. J. Wells in pp. 239–290, volume 5 of *The Mollusca*, edited by A. S. M. Saleuddin and K. M. Wilbur (Academic Press, New York, 1983), and in *Experientia* vol. 43, pp. 487–499, 1987 (with P. J. S. Smith), where they remark, 'The basic design of the molluscan circulatory system was ... well suited to a slow-moving animal ... but a disaster for large, high-metabolic rate invertebrates, competing with fish and other vertebrates. Natural selection works on the material available, and the cephalopod circulatory system is a miracle of making the best of a bad lot' (pp. 497–498). They continue, 'Jet propulsion is inescapably expensive ... It costs a squid ... something like twice as much energy, to travel half as fast, as a salmon. The difference would be even more extreme if comparisons were made with mackeral or tuna ... Add to these problems the appallingly low carrying capacity [of oxygen] of cephalopod blood, and it is evident that the circulatory pumps must be achieving something rather remarkable in terms of cardiac output' (p. 498). A similar conclusion is reached by R. E. Shadwick *et al.* (in *Journal of Experimental Biology*, vol. 130, pp. 87–106, 1987) where they also note that amongst the large squids the arterial haemodynamics is likely to be much more similar to that of the active vertebrates. R. Schipp also has a short, but helpful, update in *Vie et Milieu*, vol. 47, pp. 111–116, 1997.

116. The topic of these branchial hearts is addressed in the literature cited in note 115, and also by A. Fielder and R. Schipp in *Experientia*, vol. 43, pp. 544–553, 1987.

117. These are not the only differences. Despite the systemic heart in the cephalopods, many of the other veins are also muscular and contractile. They also lack true capillaries, but this might be because the blood protein consists of aggregates of haemocyanin, whereas in the vertebrates the squeezing of the red blood corpuscles through the narrow capillaries facilitates oxygen release.

118. This perichondrial tissue is described by A. Bairati *et al.* in *Tissue & Cell*, vol. 27, pp. 515–523, 1995. In a general review of invertebrate cartilages, P. Person and D. E. Philpott, in *Biological Reviews*, vol. 44, pp. 1–16, 1969, remark of cephalopod cartilage 'The cell processes penetrate the matrix, and often give rise to extensive systems of intercellular canaliculae, very similar to those seen in vertebrate bone and cartilage', p. 8.

119. A helpful overview is given in *The mechanosensory lateral line: neurobiology and evolution* edited by S. Coombs *et al.* (Springer, New York, 1989).

120. The lateral line analogue in the cephalopods is known as the epidermal lines; see G. Sundermann in *Cell and Tissue Research*, vol. 232, pp. 669–677; 1983. Confirmation of a sensory role was given by B. U. Budelmann and H. Bleckmann in *Journal of Comparative*

Physiology, vol. 164A, pp. 1–5, 1988, where they noted it is 'another fascinating example, beside the eyes [see notes 22–27, Chapter 7] and equilibrium receptor systems [see note 28, Chapter 7], of convergent evolution between a sophisticated cephalopod and vertebrate sensory system' (p. 4). More details were given by H. Bleckmann *et al.* in vol. 168A, pp. 247–257, 1991 of the same journal. More recent contributions are an overview by B. U. Budelmann in *Marine and Freshwater Behaviour and Physiology*, vol. 25, pp. 13–33, 1994; S. Lenz *et al.* in *Zoologisches Anzeiger*, vol. 234, pp. 145–157, 1995; and S. Lenz in *Vie et Milieu*, vol. 47, pp. 143–147, 1997.

121. See S. Coombs *et al.* in *Philosophical Transactions of the Royal Society of London B*, vol. 355, pp. 1111–1114, 2000.

122. See R. L. Reep *et al.* in *Brain, Behavior and Evolution*, vol. 59, pp. 141–152, 2002.

123. See E. J. Denton and J. Gray in *Proceedings of the Royal Society of London B*, vol. 226, pp. 249–261, 1985.

124. See Jared Diamond's chapter (pp. 3–8) in *Ecological restoration of New Zealand islands*, edited by D. R. Towns *et al.* (Conservation Sciences Publication, No. 2, 1990).

125. J. Diamond (1990; citation is in note 124), p. 3.

126. See the review by R. A. Cooper and P. R. Millener in *Trends in Ecology & Evolution*, vol. 8, pp. 429–433, 1993.

127. The quotation, which paraphrases a sentence in the article 'Invertebrate mice' by G. W. Ramsay in *New Zealand Entomologist*, vol. 6, p. 400, 1978, is from a stimulating review paper by C. H. Daugherty *et al.* in *Trends in Ecology & Evolution*, vol. 8, pp. 437–442, 1993, see p. 439.

128. Further information on the amazing wetas is available in a paper by A. M. Richards in *Journal of Zoology*, vol. 169, pp. 195–236, 1973. One weta, *Deinacrida heteracantha*, can grow up to 82 mm and thus is one of the largest insects in the world. See also *The biology of wetas, king crickets and their allies*, edited by L. H. Field (CAB International, Wallingford, 2001), where the mouse-like characters of wetas are remarked upon by M. McIntyre, on p. 226 of her chapter (pp. 225–242). I should emphasize, however, that the editor, Larry Field, is dismissive of wetas being regarded as ecological analogues of mice (personal communication).

129. J. Diamond (1990, citation is in note 124), p. 4. This wren is extinct, eaten by a cat. J. Diamond also remarks on the attempts of short-tailed bats and wetas to 'become' mice.

130. See C. H. Daugherty *et al.* (1993; citation is in note 127). C. H. Daugherty *et al.* also remark on the other parallels, with large flightless birds occupying the niches elsewhere taken by browsing and grazing mammals.

131. A brief account of this flightless parrot, which can however glide, is available in *A field guide to the birds of New Zealand and outlying islands* (Collins, London, 1966) by R. A. Falla *et al.* (see pp. 170–171).

132. An overview of moa species and their ecological impact is given by A. Cooper *et al.* in *Trends in Ecology & Evolution*, vol. 8, pp. 433–437, 1993, and a general account is given by A. Anderson in *Prodigious birds: Moas and moa-hunting in prehistoric New Zealand* (Cambridge University Press, Cambridge, 1989).

133. See A. Cooper *et al.* in *Proceedings of the National Academy of Sciences, USA*, vol. 89, pp. 8741–8744, 1992, and also A. Cooper *et al.* in *Nature*, vol. 409, pp. 704–707, 2001. Cooper *et al.* suggest that the kiwis and moas represent separate colonizations, with kiwis arriving earlier. Giant, flightless birds have evolved repeatedly. Other examples in addition to those in New Zealand include the Madagascan elephant birds and the fearsome South American phorusrhacids. Another independent development, from the Eocene of Egypt, is reported by D. T. Rasmussen *et al.* in *Palaeontology*, vol. 44, pp. 325–337, 2001. This occurrence is important because Africa appeared to be the only significant landmass without an endemic large, ground-dwelling bird (the ostriches migrated to Africa from Asia in the Miocene).

134. Despite earlier scepticism there is now little doubt that the Maoris were an important element in the extinction of the moa, by both direct and profligate hunting, and also burning of forest habitats. See the chapters by M. Trotter and B. McCullock (pp. 708–727) and

A. Anderson (pp. 728–740) in *Quaternary extinctions: a prehistoric revolution*, edited by P. S. Martin and R. G. Klein (University of Arizona Press, Tucson, 1984).

135. See the papers by W. A. Calder in *Scientific American*, vol. 239 (1), pp. 102–110, 1978, and *BioScience*, vol. 29, pp. 461–467, 1979.

136. See R. A. Falla *et al.* (1996; citation is in note 131), pp. 18–20.

137. See the note by S. M. Monro in *Avicultural Magazine*, vol. 80, p. 202, 1974. The budgie's owner noticed that it looked a bit under the weather, and she was even more surprised when the bird apparently gave birth to 'two perfectly formed chicks'. In particular, no sign of any shells was noted. Whatever the budgie thought of it is obscure, not least because shortly after this event the bird died.

138. See the paper entitled 'Why are there no viviparous birds?' by D. G. Blackburn and H. E. Evans in *American Naturalist*, vol. 128, pp. 165–190, 1986. See also R. L. Dunbrack and M. A. Ramsay, in vol. 133, pp. 138–148, 1989 of the same journal, who argue that an absence of effective lactation has been an important factor against ovoviviparity.

139. See D. J. Anderson *et al.* in *American Naturalist*, vol. 130, pp. 941–947, 1987.

140. W. A. Calder (1978, citation is in note 135), p. 110.

141. See Richard Shine's chapter (pp. 361–369) in the book *Egg incubation: its effects on embryonic development in birds and reptiles*, edited by D. C. Deeming and M. W. J. Ferguson (Cambridge University Press, Cambridge, 1991). He has an interesting discussion of the factors that predispose the squamates to this mode of reproduction, and why it has never evolved in the crocodiles, turtles, or birds.

142. Daniel Blackburn's overview of both viviparity and matrotrophy (that is supply of nutrients by the mother) is given in *American Zoologist*, vol. 32, pp. 313–321, 1992; see also M. H. Wake's article in the same issue, on pp. 256–263.

143. See, for example, the extraordinary case of a Namibian toad (*Nectophrynoides occidentalis*) reported by F. Xavier on pp. 545–552 in *Major patterns in vertebrate evolution*, edited by M. K. Hecht *et al.* (Plenum, New York, 1977). Xavier remarks that it is 'quite remarkable to find the same structures and the same hormones in the Nimba toad and in mammals, along with the same functions, i.e. to contain fetuses in the maternal uterus, to feed them and then to expel them' (p. 550). He also emphasizes, however, there are fundamental differences, of which the most obvious is the absence of a placenta. In addition, the hormones are used in a different way from the mammalian system.

144. For useful reviews see J. P. Wourms in *American Zoologist*, vol. 21, pp. 473–515, 1981 and *Israel Journal of Zoology*, vol. 40, pp. 551–568, 1994, and J. P. Wourms and J. Lombardi in *American Zoologist*, vol. 32, pp. 276–293, 1992. Wourms (1994) estimates that fish viviparity has evolved independently at least 42 times, and in this review comments both on the major changes and on the many advantages of viviparity (see p. 554). It is especially frequent among the sharks and rays (see also Dulvy and Reynolds 1997; citation is in note 154), but it also known in the teleost fish, as well as the famous 'living fossil', the coelacanth *Latimeria*. The paper by Wourms and Lombardi (1992) also remarks on various other sorts of convergence seen in the viviparous fish. These include the so-called trophotaenia, projections of the embryonic hind-gut that facilitate nutrient transfer, and seemingly have evolved 'in four distantly related orders of teleosts' (p. 287). Of even greater interest is 'The pericardial amniochorion which occurs in the embryos of poeciliid fishes [see also B. D. Grove and J. P. Wourms in *Journal of Morphology*, vol. 209, pp. 265–284, 1991 and vol. 220, pp. 167–184, 1994 respectively], some goodeids [see also I. Kokkala and J. P. Wourms, and F. Hollenberg and J. P. Wourms in *Journal of Morphology*, vol. 219, pp. 35–46 and pp. 105–129 respectively, 1994] and in a modified form ... in *Anableps* is of particular interest because of its striking similarity to the anterior amnio-chorionic fold of amniotes. This is an apparent anomaly in that the amnion and chorion first appear in the tetrapod lineage as one of the major steps in the evolution of the cleidoic egg ... Besides being a remarkable example of evolutionary convergence in an extraembryonic membrane, it helps one to understand the developmental basis for the evolution of the amnion and chorion in amniotes,' p. 286.

145. An extensive review of reptile placentation is given by Z. Yaron on pp. 527–603 in vol. 15 of *Biology of the Reptilia*, edited by C. Gans and G. Billett (Wiley, New York, 1985). See also D. G. Blackburn in *Journal of Experimental Zoology*, vol. 266, pp. 414–430, 1993, where he remarks that 'The evolution of placentomes in three separate lineages [of squamate] represents an extraordinary manifestation of convergence' (p. 427). It has long been recognized that the development of the reptile placenta is quite independent of the mammals and might also indicate how it arose in the latter group. See also D. N. Reznick *et al.* in *Science*, vol. 298, pp. 1018–1020, 2002 (with commentary by V. Morell on p. 945) who document the independent evolution of a placenta in the fish *Poeciliopsis*, and also make some interesting remarks on the rapid emergence of complex biological structures.

146. Daniel Blackburn and his colleagues' paper is published in *Proceedings of the National Academy of Sciences, USA*, vol. 81, pp. 4860–4863, 1984.

147. D. G. Blackburn *et al.* (1984; citation is in note 146), p. 4862.

148. Concerning the placenta and viviparity of *Mabuya bistriata*, see L. J. Vitt and D. G. Blackburn in *Copeia*, for 1991, pp. 916–927, 1991. A helpful overview of reproduction in *Mabuya* is given by D. G. Blackburn and L. J. Vitt on pp. 150–164 of *Reproductive biology of South American vertebrates*, edited by W. C. Hamlett (Springer, New York, 1992), where they write, 'The independent evolution of these features among mammals and lizards represents one of the most striking cases of evolutionary convergence in reproductive specializations to be documented among terrestrial vertebrates', p. 161.

149. See D. G. Blackburn in *Amphibia–Reptilia*, vol. 3, pp. 185–205, 1982.

150. See D. G. Blackburn (1992, citation is in note 142). Further details of this remarkable placenta are given by D. G. Blackburn in *Journal of Morphology*, vol. 216, pp. 179–195, 1993.

151. J. P. Wourms and J. Lombardi (1992; citation is in note 144), p. 290, as well as earlier remarks by J. P. Wourms and colleagues in the first chapter (pp. 1–134) of *Fish physiology*, vol. XIB, edited by W. S. Hoar and D. J. Randall (Academic Press, San Diego, 1988). In the latter work they remark 'The selachian yolk-sac placenta has probably obtained the pinnacle of its evolutionary development in the spadenose shark ... Here it appears to function with the same degree of efficiency as a mammalian placenta.' (p. 69). See also note 154.

152. D. G. Blackburn (1992; citation is in note 142), p. 320.

153. D. G. Blackburn (1992; citation is in note 142), p. 320.

154. See the paper by N. K. Dulvy and J. D. Reynolds in *Proceedings of the Royal Society of London B*, vol. 264, pp. 1309–1315, 1997. See also M. S. Y. Lee and R. Shine, in *Evolution*, vol. 52, pp. 1441–1450, 1998, who present evidence for reversals in the tendency towards viviparity in the reptiles. While such reversals are not ruled out absolutely, in reality there are only a handful of possible examples. Such entrenchments impose a strong directionality on the history of life.

155. See D. G. Blackburn (1982; citation is in note 149). See also R. M. Andrews (in *Physiological and Biochemical Zoology*, vol. 75, pp. 145–154, 2002), who explores the link between coolness and oxygen availability in determining the likelihood of viviparity, and especially R. Shine (in *American Naturalist*, vol. 160, pp. 582–593, 2002), who presents sound adaptive reasons why cooler thermal regimes favour *in utero* development of reptile eggs.

156. N. K. Dulvy and J. D. Reynolds (1997; citation is in note 154), p. 1314.

157. See B. Heinrich's *The hot-blooded insects. Strategies and mechanisms of thermoregulation* (Springer, Berlin, 1993), and *The thermal warriors: Strategies of insect survival* (Harvard University Press, Cambridge, MA, 1996).

158. See the papers by B. A. Block *et al.* in *Science*, vol. 260, pp. 210–214, 1993 (and commentary by T. Watson on pp. 160–161), and B. A. Block and J. R. Finnerty in *Environmental Biology of Fishes*, vol. 40, pp. 283–302, 1994.

159. See the paper by F. G. Carey *et al.* in *Memoirs, Southern California Academy of Sciences*, vol. 9, pp. 92–108, 1985, where they also note similarities with tuna.

160. Reviewed in B. A. Block and J. R. Finnerty (1994; citation is in note 158).

161. See the paper by B. A. Block and F. G. Carey in *Journal of Comparative Physiology*, 156B, pp. 229–236, 1985.

162. B. A. Block and F. G. Carey (1985; citation is in note 161), p. 234.

163. B. A. Block *et al.* (1993; citation is in note 158), p. 213.

164. The grim topic of malignant hyperthermia in humans and its equivalent in pigs (porcine stress syndrome) is addressed by D. H. MacLennan and M. S. Phillips in *Science*, vol. 256, pp. 789–794, 1992.

165. The similarities between malignant hyperthermia and 'tuna burn' are reviewed by B. A. Block in *Annual Review of Physiology*, vol. 56, pp. 535–577, 1994; see also B. A. Block and J. R. Finnerty (1994; citation is in note 158).

166. C. G. Farmer's paper, entitled 'Parental care: The key to understanding endothermy and other convergent features in birds and mammals' was published in *American Naturalist*, vol. 155, pp. 326–334, 2000.

167. C. G. Farmer (2000; citation is in note 166), p. 330; as given here the quotation has been rewritten by combining parts of a paragraph.

168. Convergences of maternal care also occur *within* the mammals, such as between the placental ungulates and macropod marsupials, e.g. kangaroo. See D. O. Fisher *et al.*, in *Evolution*, vol. 56, pp. 167–176, 2002.

169. C. G. Farmer (2000; citation is in note 166), p. 330.

170. *The descent of Man, and selection in relation to sex* (Murray, London, 1889, second edition), p.181. Or is this an 'urban myth', like the last Cornish monoglot, Dolly Pentreath (d. 1777) and her apocryphal posthumous parrot?

171. See P. M. Gray *et al.* in *Science*, vol. 291, pp. 52–54, 2001.

172. This concerns the male palm cockatoo that selects a fresh stick and shapes it (by removing foliage, chewing the ends off and stripping bark) into a drumstick suitable for drumming on a hollow log; see G. A. Wood in *Corella*, vol. 8, pp. 94–95, 1984.

173. P. M. Gray *et al.* (2001; citation is in note 171), p. 52.

174. See I. M. Pepperberg in *Animal Learning and Behavior*, vol. 27, pp. 15–17, 1999 (citations are in notes 137 and 139 of Chapter 9).

175. Allison Doupe and Patricia Kuhl's paper, entitled 'Birdsong and human speech: Common themes and mechanisms' is in *Annual Review of Neuroscience*, vol. 22, pp. 567–631; 1999. The realization that there are important parallels between birdsong and human speech include, for example, second language acquisition (or bilingualism) (see I. M. Pepperberg and D. M. Neapolitan in *Ethology*, vol. 77, pp. 150–168, 1988), and evidence for cultural transmission (see R. B. Payne *et al.* in *Behaviour*, vol. 77, pp. 199–221, 1981).

176. A. Doupe and P. Kuhl (1999; citation is in note 175), p. 567.

177. A. Doupe and P. Kuhl (1999; citation is in note 175), p. 597.

178. A. Doupe and P. Kuhl (1999; citation is in note 175), p. 594.

179. A. Doupe and P. Kuhl (1999; citation is in note 175), p. 567.

180. See G. F. Striedter in *Journal of Comparative Neurology*, vol. 343, pp. 35–56, 1994; subsequently, E. D. Jarvis *et al.* (in *Nature*, vol. 406, pp. 628–632, 2000) reported striking similarities in the forebrain region of hummingbirds, songbirds, and parrots. This idea is further developed by M. A. Farrier in *Brain, Behavior and Evolution*, vol. 58, pp. 80–100, 2001, who writes, 'nature has conducted what are in effect three independent experiments in the evolution of vocal learning in birds: parrots, hummingbirds, and songbirds. By studying the evolution of similar behavioral capacities from similar starting points, it might be possible to address some important questions in evolutionary biology ... For example, how much freedom does evolution have to find different solutions to the same problem?' p. 96.

181. There are plenty of examples of convergence in such grinding teeth, such as those between the molars of the rhinoceros *Elasmotherium* and horses (see W. B. Scott in *Journal of Morphology*, vol. 5, pp. 301–406, 1891), and within the elephants (see M. O. Thomas *et al.* in *Proceedings of the Royal Society of London B*, vol. 267, pp. 2493–2500, 2000).

182. In passing, one should note an emphasis on a mammal-like occlusion in the teeth of a much more primitive group of chordates, the extinct conodonts; see P. C. J. Donoghue and M. A. Purnell in *Paleobiology*, vol. 25, pp. 58–74, 1999. This example is interesting because the marine conodonts, whose precise taxonomic position is controversial, lack anything like a jaw. See also P. C. J. Donoghue in *Proceedings of the Royal Society of London B*, vol. 268, pp. 1691–1698, 2001.

183. The tribosphenic molar is typically of the primitive marsupial and placental mammals, and their immediate forebears. Z-x. Luo *et al.* (in *Nature*, vol. 409, pp. 53–57) argue that the tribosphenic condition evolved twice, both in the aforementioned groups and independently in the ancestors of the monotremes (whose extant representatives are toothless); see also T. H. Rich *et al.* (in *Journal of Vertebrate Paleontology*, vol. 22, pp. 466–469, 2002) for possible qualification of this conclusion. The arrangement of interlocking cusps permits precise occlusion and shearing, and not surprisingly there are further examples of convergence, J. P. Hunter and J. Jernvall, in *Proceedings of the National Academy of Sciences, USA*, vol. 92, pp. 10718–10722, 1995, estimate that the evolution of an additional cusp, the hypocone, happened more than 20 times, and regard it as a key pre-adaptation to evolutionary success. See also J. Jernvall's follow-up paper in the same journal (vol. 97, pp. 2641–2645, 2000).

184. See the paper by R. L. Nydam *et al.* in *Journal of Vertebrate Paleontology*, vol. 20, pp. 628–631, 2000, although the authors are careful not to over-emphasize the rather striking similarity. See also R. L. Nydam and R. L. Cifelli in a subsequent issue, vol. 22, pp. 276–285, 2002. A somewhat similar occurrence is also known in the living lizard *Uromastyx*, which shows a dramatic change in dentition in its life as the animal moves from a diet of small animals to plants when an adult. More unusually, the enamel of these teeth has convergently evolved a prismatic structure, also with tubules, that is reminiscent of the mammals, but is otherwise unknown in the reptiles; see J. S. Cooper and D. F. G. Poole in *Journal of Zoology, London*, vol. 169, pp. 85–100, 1973.

185. See J. M. Clark *et al.* in *Science*, vol. 244, pp. 1064–1066, 1989, and X-C. Wu *et al.* in *Nature*, vol. 376, pp. 678–680, 1995.

186. See O. Kuhn's description of remarkably preserved material from the famous site at Geiseltal in Germany, in *Nova Acta Leopoldina*, vol. 6 (N.F.), pp. 311–329, 1938, p. 323.

187. See P. C. Sereno *et al.* in *Science*, vol. 282, pp. 1298–1302, 1998.

188. 'Spinosaurs as crocodile mimics' is the title of the commentary by T. R. Holtz that accompanies Sereno *et al.* (1998); see pp. 1276–1277. This 'intriguing case of convergence' (p. 1277) is evidently a result of a shift by this dinosaur to a piscivorous diet.

9. THE NON-PREVALENCE OF HUMANOIDS?

1. See L. F. Laporte's *George Gaylord Simpson: Paleontologist and evolutionist* (Columbia University Press, New York; 2000), as well as H. B. Whittington's encomium in *Biographical Memoirs of Fellows of the Royal Society*, vol. 32, pp. 527–539, 1986.

2. The article was published in *Science*, vol. 143, pp. 769–775, 1964a, and promptly provoked a vigorous discussion, see vol. 144, pp. 613–615, 1964. Effectively the same paper, but shorter, was published by Simpson as Chapter 13 in *This view of life: the world of an evolutionist* (Harcourt, Brace and World, New York, 1964b).

3. Together with the other giants of evolutionary study, notably R. A. Fisher, G. Ledyard Stebbins, Th. Dobzhansky and E. Mayr, Simpson is regarded as one of the main architects of the neo-Darwinian synthesis. An important kerygma in this programme was the book edited by G. L. Jepsen, G. G. Simpson, and E. Mayr entitled *Genetics, paleontology and evolution* (Princeton University Press, Princeton, 1949). Simpson himself published the very influential *Tempo and mode in evolution* (Columbia University Press, New York, 1944) and its effective revision, *The major features of evolution* (Columbia University Press, New York, 1953).

4. In *The meaning of evolution: a study of the history of life and of its significance for man* (New Haven, Yale University Press, revised edition, 1967) Simpson devoted a good part of Chapter 12 to a discussion of convergences.

5. Very similar views are expressed by E. Mayr in his essay 'The probability of extraterrestrial intelligent life', reprinted on pp. 67–74 (Essay four) of *Toward a new philosophy of biology: Observations of an evolutionist* (Belknap, Cambridge, MA, 1988). Mayr's view is that for all intents and purposes intelligence is a fluke; see his article 'Does it pay to acquire high intelligence?', in *Perspectives in Biology and Medicine*, vol. 37, pp. 337–338, 1994. Effectively identical sentiments were expressed in his chapter (pp. 152–156) in *Extraterrestrials: Where are they?*, edited by B. Zuckerman and M. H. Hart (Cambridge University Press, Cambridge, 1995 [Second edition]).

6. This being the title of the paper in *Annals of the New York Academy of Sciences*, vol. 950, pp. 276–288, 2001. This article, however, seems over-concerned with the specifics of history as against the emergence of general biological properties. This volume, entitled *Cosmic questions*, and edited by J. B. Miller, has a number of other interesting chapters, notably those by A. Case-Winters (pp. 154–168) and S. Via (pp. 225–240).

7. Simpson (1964a; see note 2 for citation), p. 774. Very similar sentiments are expressed by Carl Sagan in his *Cosmos* (Macdonald, London, 1981), where he writes of those who try to imagine extraterrestrial life that 'They seem to me to rely too much on forms of life we already know. Any given organism is the way it is because of a long series of individually unlikely steps. I do not think life anywhere else would look very much like a reptile, or an insect or a human' (p. 40). Subsequently, Sagan remarked in *Scientific American*, vol. 271 (4), pp. 70–77, 1994, that given life on Earth had a single origin then 'we have no way of knowing which aspects of terrestrial life are necessary (required by all living things anywhere) and which are merely contingent (the results of a particular sequence of happenstances that, had they gone otherwise, might have led to organisms having very different properties),' p. 71.

8. See George Beadle's 11th Annual Arthur Dehon Little Memorial Lecture, entitled 'The place of genetics in modern biology' (MIT Press, Boston, 1959).

9. G. Beadle (1959; citation is in note 8), p. 20.

10. The origins and genomic organization of the animal sodium channels are discussed by J. D. Spafford *et al.* in *Receptors and Channels*, vol. 6, pp. 493–506, 1999.

11. Concerning the sodium pump see the paper by W. D. Stein in *Philosophical Transactions of the Royal Society of London B*, vol. 349, pp. 263–269, 1995. See also R. D. Keynes and F. Elinder in *Proceedings of the Royal Society of London B*, vol. 266, pp. 843–852, 1999.

12. See *Sponges* by P. R. Bergquist (London, Hutchinson, 1978), pp. 63, 67–69. Even so, sponges possess contractile cells (the myocytes) and they evidently contain neurotransmitters; see for example the paper by T. L. Lentz in *Journal of Experimental Zoology*, vol. 162, pp. 171–179, 1966. In the hexactinellids there is also indirect evidence for electrical conduction, albeit non-nervous and probably controlled by calcium; see G. O. Mackie *et al.* in *Philosophical Transactions of the Royal Society of London B*, vol. 301, pp. 401–418, 1983. For a more general discussion of conductance in sponges see G. O. Mackie in *American Zoologist*, vol. 30, pp. 907–920, 1990.

13. See the paper by C. Febvre-Chevalier *et al.* in *Journal of Experimental Biology*, vol. 122, pp. 177–192, 1986. Although there is also a small calcium component, this heliozoan 'is thus the first exception to [the] generalization that sodium-requiring action potentials are unknown outside the metazoan animals' (p.189). See also the chapter by C. Febvre-Chevalier *et al.* (pp. 237–253) in *Evolution of the first nervous system*, edited by P. A. V. Anderson (Plenum, New York, 1990). Here the principal dependence of *Actinocoryne* on sodium is regarded as 'puzzling' (p. 249), yet the authors also emphasize the 'remarkable sensitivity and contractility' (p. 239) of this organism.

14. For an exhaustive review of ion transport mechanisms, including sodium, see M. H. Saier in *Microbiology and Molecular Biology Reviews*, vol. 64, pp. 354–411, 2000.

15. See the series of seven papers, forming a mini-review, on bacterial sodium energetics in *Journal of Bioenergetics and Biomembranes*, vol. 21, pp. 633–740, 1989.

16. See the paper by D. Ren *et al.* in *Science*, vol. 294, pp. 2372–2375, 2001, and accompanying commentary by W. A. Catterall on pp. 2306–2308.

17. See the helpful overview of ion channels and their possible phylogenies by P. A. V. Anderson and R. M. Greenberg in *Comparative Biochemistry and Physiology B*, vol. 129, pp. 17–28, 2001. They support the idea that the sodium channel derived from a low voltage-activated T-type calcium channel.

18. Concerning the origins of electrical excitability and the evolution of the sodium pump see the article by M. Strong *et al.* in *Molecular Biology and Evolution*, vol. 10, pp. 221–242, 1993. Interestingly, the bacterial sodium channel (note 16) also seems to have an evolutionary connection to the calcium channel, but a convergent origin seems likely.

19. See the paper by S. H. Heinemann *et al.* in *Nature*, vol. 356, pp. 441–443, 1992. For other ways of altering a sodium channel so that it allows the passage of calcium, see L. F. Santana *et al.* in *Science*, vol. 279, pp. 1027–1033, 1998 (and the commentary by D. Hanck on p. 1004).

20. See J. D. Spafford *et al.* (1999; citation is in note 10), p. 494.

21. See Milton Saier's paper in *BioEssays*, vol. 16, pp. 23–29, 1994.

22. M. Saier (1994; citation is in note 21), p. 28.

23. The Drake equation is given as $N = R_* f_p n_e f_l f_i f_t L$, where N is the number of communicative civilizations, R_* is the rate of star formation, f_p is the fraction of stars possessing a planetary system, n_e is the number of planets suitable for habitation, f_l is the number of planets on which life arises, f_i is the fraction of biospheres where intelligence emerges, f_t is the number of civilizations that develop advanced communication, and L is the lifetime of a civilization.

24. For an excellent account of this early attempt at SETI and much else concerning life in the Universe see S. J. Dick's *The biological universe: the twentieth century extraterrestrial life debate and the limits of science* (Cambridge University Press, Cambridge, 1996).

25. Leonard Ornstein's wide-ranging and stimulating paper was published in *Physics Today*, vol. 35 (March), pp. 27–31, 1982.

26. See T. F. Smith and H. J. Morowitz in *Journal of Molecular Evolution*, vol. 18, pp. 265–282, 1982.

27. L. Ornstein (1982; citation is in note 25), p. 27 (his emphasis).

28. L. Ornstein (1982; citation is in note 25), p. 29.

29. The proposal put forward by Ornstein is biologically a bit muddled, but in essence argues that both cephalopod and vertebrate eyes derive from a common ancestor, also with a camera-eye. A corollary of this suggestion is that supposedly primitive eyes, as seen in various molluscs, are really regressive. On present evidence this seems decidedly less likely.

30. Robert Bieri's article 'Huminoids on other planets?' was published in *American Scientist*, vol. 52, pp. 452–458, 1964. Bieri's work has also caught the attention of others interested in the wider issues of evolution and the probabilities of extraterrestrial life; see, for example, R. Puccetti's absorbing discussion of humanoid convergences in *Persons: A study of possible moral agents in the universe* (Macmillan, London; 1968) and N. J. Berrill's *Worlds apart: a reflection on planets, life, and time* (Sidgwick & Jackson, London; 1966), especially chapter 10 where he remarks 'Given the right encouragement, brains and minds are as likely a natural evolutionary creation as legs, hearts, or wings ... Wherever life may be ... mind will follow.' (p. 149). See also Michael Ruse's *Can a Darwinian be a Christian? The relationship between science and religion* (Cambridge, Cambridge University Press, 2001), pp. 145–147. Ruse also addresses the ideas of Simpson, noting 'Sounding like a forerunner of Gould, Simpson argues that it is chance and contingency all the way', p. 147.

31. R. Bieri (1964; citation is in note 30), p. 452.

32. R. Bieri (1964; citation is in note 30), p. 457.

33. See Philip Morrison's chapter (pp. 571–583) in *Astronomical and biochemical origins and the search for life in the universe*, edited by C. B. Cosmovici *et al.* (Proceedings of the 5th International Conference on Bioastronomy: Editrice Compositori, Bologna, 1997).

34. P. Morrison (1997; citation is in note 33), p. 572. See also T. J. Grove and B. D. Sidell (in *Canadian Journal of Zoology*, vol. 80, pp. 893–901, 2002), who document myoglobin deficiency in the hearts of a variety of benthic fish of sluggish habit.

35. Interestingly, when a modification of the photosynthetic process evolves, notably emergence of the so-called the C_4 mechanism, it too shows rampant convergence, a topic which is returned to in Chapter 10.

36. See A. Eschenmoser in *Science*, vol. 284, pp. 2118–2124, 1999, as well as discussion in Chapter 2.

37. See S. J. Freeland and L. D. Hurst in *Journal of Molecular Evolution*, vol. 47, pp. 238–248, 1998, as well as discussion in Chapter 1.

38. In the description of a primitive rhizodont fish from Australia, Z. Johanson and P. E. Ahlberg (*Nature*, vol. 394, pp. 569–573, 1998) note how limb-like are the fins, especially in the case of the pectoral pair, and further note 'similarities between rhizodont fins and tetrapod limbs are thus probably convergent', p. 569. See also their subsequent discussion in *Nature*, vol. 395, pp. 792–794, 1998, of how the tetrapods arose out of one of several similar evolutionary experiments among the osteolepiform fish.

39. See note 81 of Chapter 6 concerning the conversion of a caecilian amphibian to the hydrostatic equivalent of a worm.

40. Concerning the distribution of the haemoglobins see the papers by R. C. Hardison in *Proceedings of the National Academy of Sciences, USA*, vol. 93, pp. 5675–5679, 1996 and *Journal of Experimental Biology*, vol. 201, pp. 1099–1117, 1998, as well as an overview in *American Scientist*, vol. 87, pp. 126–137, 1999. Although emphasizing invertebrates the review by R. E. Weber and S. N. Vinogradov in *Physiological Reviews*, vol. 81, pp. 569–628, 2001 is also helpful with updates of the occurrences of haemoglobin in prokaryotes and protistans. A recent report, by S. Hou *et al.* (in *Proceedings of the National Academy of Sciences, USA*, vol. 98, pp. 9353–9358, 2001) explores the role of these globins in bacteria, in the context of oxygen sensors.

41. See *Our molecular nature: The body's motors, machines and messages* (Copernicus, New York, 1996), pp. 76–77.

42. For an overview of acetylcholine as 'a universal cell molecule' see I. Wessler *et al.* in *Clinical and Experimental Pharmacology and Physiology*, vol. 26, pp. 198–205, 1999.

43. Concerning the cholinergic system in the ciliate *Paramecium* see, for example, the papers by F. Trielli *et al.* in *Journal of Experimental Zoology*, vol. 279, pp. 633–638, 1997 and M. U. Delmonte Corrado *et al.* in *Journal of Experimental Biology*, vol. 204, pp. 1901–1907, 2001.

44. See D. LeRoith *et al.* in *Proceedings of the National Academy of Sciences, USA*, vol. 79, pp. 2086–2090, 1982, as well as B. Zipser *et al.* in *Brain Research*, vol. 463, pp. 296–304, 1988, for further information on β-endorphin in the ciliate *Tetrahymena*. Concerning dopamine see the papers in *Biochimica et Biophysica Acta* by M. E. Goldman *et al.* and R. E. Gunderson and G. A. Thompson in vol. 676, pp. 221–225, 1981 and vol. 755, p. 186–194, 1983 respectively. Another neuropeptide, a somatostatin-like molecule, has been described by M. Berelowitz *et al.*; see *Endocrinology*, vol. 110, pp. 1939–1944, 1982. The glycoprotein choriogonadotropin, a hormone associated with mammalian pregnancy, is also reported from the womb-free bacteria, albeit showing a much lower activity; see T. Maruo *et al.* in *Proceedings of the National Academy of Sciences, USA*, vol. 76, pp. 6622–6626, 1979. A similar occurrence is reported in yeast, see E. Loumaye *et al.* in *Science*, vol. 218, pp. 1323–1325, 1982; in this case it is intriguing to note the peptide in yeast is responsible for mating and the formation of the zygote.

45. Thus, signalling mechanisms involving glutamate receptors evidently evolved before the divergence of plants and animals; see J. Chiu *et al.* in *Molecular Biology and Evolution*, vol. 16, pp. 826–838, 1999. Signalling for disease resistance is also very primitive, given the shared occurrence of a zinc-binding protein in protistans, plants, and animals; see K. Shirasu *et al.* in *Cell*, vol. 99, pp. 355–366, 1999.

46. G. O. Mackie (1990; citation is in note 12), see p. 909.
47. See B. J. Crespi's paper 'The evolution of social behavior in microorganisms', in *Trends in Ecology & Evolution*, vol. 16, pp. 178–183, 2001.
48. See, for example, the review entitled 'The languages of bacteria' by S. Schauder and B. L. Bassler, in *Genes & Development*, vol. 15, pp. 1468–1480, 2001.
49. B. J. Crespi (2001, citation is in note 47), p. 182.
50. See, for example, J. A. Shapiro's article in *Annual Review of Microbiology*, vol. 52, pp. 81–104, 1998.
51. Effectively the argument in *Rare Earth: Why complex life is uncommon in the Universe* by P. D. Ward and D. Brownlee (Copernicus, New York, 2000).
52. As well as the overview by M. Syvanen in *Annual Review of Genetics*, vol. 28, pp. 237–261, 1994, other more recent papers include those by H. M. Robertson and D. J. Lampe (in *Molecular Biology and Evolution*, vol. 12, pp. 850–862, 1995), W. F. Doolittle (in *Trends in Biochemical Sciences*, vol. 24 (12), pp. M5–M8, 1999), R. Jain *et al.* (in *Proceedings of the National Academy of Sciences, USA*, vol. 96, pp. 3801–3806, 1999), and M. J. Leaver (in *Gene*, vol. 271, pp. 203–214, 2001).
53. This general problem is usually referred to as the C-value paradox, C being the amount of DNA in a cell. *The evolution of genome size*, edited by T. Cavalier-Smith (Wiley, Chichester, 1985) provides an introduction; an overview is given by T. R. Gregory in *Biological Reviews*, vol. 76, pp. 65–101, 2001.
54. These results have been widely reported: J. F. Y. Brookfield in *Current Biology*, vol. 10, pp. R514–R515, 2000 gives a helpful overview.
55. Concerning the body axis, where the *Hox* genes play an important role, see, for example, papers by M. Akam in *Philosophical Transactions of the Royal Society of London B*, vol. 349, pp. 313–319, 1995; S. B. Carroll in *Nature*, vol. 376, pp. 479–485, 1995; J. K. Grenier *et al.* in *Current Biology*, vol. 7, pp. 547–553, 1997; and B. Schierwater and K. Kuhn in *Molecular Phylogeny and Evolution*, vol. 9, pp. 375–381, 1999.
56. The topic of dorso-ventrality and the developmental similarities between the dorsal side of vertebrate and ventral side of arthropods (and vice versa), with the implication that among other things the hollow neural tube of the former might be homologous with the nerve cord of the fly has attracted much attention. See D. Arendt and K. Nübler-Jung in *Roux's Archive for Developmental Biology*, vol. 203, pp. 357–366, 1994, and *Development*, vol. 126, pp. 2309–2325, 1999; E. M. De Robertis and Y. Sasai in *Nature*, vol. 380, pp. 37–40, 1996; J. Schmidt *et al.* in *Development*, vol. 121, pp. 4319–4328, 1995; E. L. Ferguson in *Current Opinions in Genetics and Development*, vol. 6, pp. 424–431, 1996; and S. A. Holley *et al.* in *Nature*, vol. 376, pp. 249–253, 1995 and *Cell*, vol. 86, pp. 607–617, 1996.
57. The similarities between the genetic architecture of fly wings and vertebrate limbs are reviewed by S. J. Gaunt in *Nature*, vol. 368, pp. 324–325, 1997; N. Shubin *et al.* in *Nature*, vol. 388, pp. 639–648, 1997; and C. J. Tabin *et al.* in *American Zoologist*, vol. 39, pp. 650–663, 1999. For a wider review see G. Panganiban *et al.* in *Proceedings of the National Academy of Sciences, USA*, vol. 94, pp. 5162–5166, 1997.
58. Eyes are the classic story of genetic similarity, especially via the *Pax-6* gene. See, so to speak, papers by R. Quiring *et al.* in *Science*, vol. 265, pp. 785–789, 1994; G. Halder *et al.* in *Science*, vol. 267, pp. 1788–1792, 1995; N. M. Bonini *et al.* in *Development*, vol. 124, pp. 4819–4826, 1997; S. I. Tomarev *et al.* in *Proceedings of the National Academy of Sciences, USA*, vol. 94, pp. 2421–2426, 1997; R. J. Davis *et al.* in *Development, Genes and Evolution*, vol. 209, pp. 526–536, 1999; and P.-X. Yu *et al.* in *Development*, vol. 126, pp. 383–395, 1999.
59. The gene known as *tinman* plays a key role in heart formation. See, for example, Y. Fu *et al.* in *Development, Genes and Evolution*, vol. 207, pp. 352–358, 1997.
60. See P. W. H. Holland and J. García-Fernández in *Developmental Biology*, vol. 173, pp. 382–395, 1996, and A. P. Martin in *American Naturalist*, vol. 154, pp. 111–128, 1999.

61. See her paper in *Journal of Experimental Zoology (Molecular Development and Evolution)*, vol. 285, pp. 19–26, 1999.

62. S. F. Gilbert *et al.* (in *Developmental Biology*, vol. 173, pp. 357–372, 1996) point out that hitherto curious examples of linked malformations, such as those affecting both limbs and kidneys, are now more comprehensible because of shared genetic proximity early in development.

63. Relevant papers include those by M. Koyanagi *et al.* in *FEBS Letters*, vol. 436, pp. 323–328, 1998; K. Ono *et al.* and H. Suga *et al.* respectively in *Journal of Molecular Evolution*, vol. 48, pp. 654–662 and pp. 646–653, 1999; H. Suga *et al.* in *Gene*, vol. 280, pp. 195–201, 2001; and J. Zhang *et al.* in *Development*, vol. 128, pp. 1607–1615, 2001.

64. Citation is in note 316, Chapter 7.

65. See papers in *Development* by C. Walther and P. Gruss in vol. 113, pp. 1435–1449, 1991; and T. Yamasaki *et al.* in vol. 128, pp. 3133–3144, 2001; see also note 317, Chapter 7.

66. Citation is in note 318, Chapter 7.

67. See L-I. Larsson *et al.* in *Mechanisms of Development*, vol. 79, pp. 153–159, 1998.

68. Citation is in note 319, Chapter 7.

69. See A. D. Chisholm and H. R. Horvitz (pp. 52–54) and Y. Zhang and S. W. Emmons (pp. 55–59) in vol. 377 of *Nature* (1995).

70. See E. Saló *et al.* in *Gene*, vol. 287, pp. 67–74, 2002. See also note 320, Chapter 7.

71. See A. Cvekl and J. Piatigorsky in *BioEssays*, vol. 18, pp. 621–630, 1996.

72. See G. Sheng *et al.* in *Genes & Development*, vol. 11, pp. 1122–1131, 1997.

73. See, for example, G. R. Merlo *et al.* in *International Journal of Development Biology*, vol. 44, pp. 619–626, 2000, and R. I. Schwartz and E. N. Olson in *Development*, vol. 126, pp. 4187–4192, 1999.

74. See the paper by F. Loosli *et al.* in *Mechanisms of Development*, vol. 74, pp. 159–164; 1998.

75. See, for example, C. Desplan in *Cell*, vol. 91, pp. 861–864, 1997.

76. This unease is well articulated by E. H. Davidson and G. Ruvkun in *Journal of Experimental Zoology (Molecular Development and Evolution)*, vol. 285, pp. 104–115, 1999, and is taken a bit further in a review I wrote in *Cell*, vol. 100, pp. 1–11, 2000.

77. See Lisa Nagy's article in *American Zoologist*, vol. 38, pp. 818–828, 1998.

78. L. Nagy (1998; citation is in note 77), pp. 818–819.

79. See B. Mittmann and G. Schultz in *Development, Genes and Evolution*, vol. 211, pp. 232–243, 2001.

80. See, for example, A. Simeone *et al.* in *Current Opinion in Genetics & Development*, vol. 12, pp. 409–415, 2002.

81. See Nic Williams and Peter Holland's article in *Molecular Biology and Evolution*, vol. 15, pp. 600–607, 1998.

82. N. Williams and P. Holland (1998; citation is in note 81), p. 606.

83. Again and again we see that a trivial change, sometimes a single substitution of an amino acid at a key site in a protein, can have major consequences for its function. Not surprisingly, given this sensitivity examples of molecular convergence are more frequent than sometimes realized. This topic is returned to in Chapter 10.

84. See the two papers in *Science* dealing with chimp/human immunochemistry and genes respectively by V. M. Sarich and A. C. Wilson (vol. 158, pp. 1200–1203, 1967) and M. C. King and A. C. Wilson (vol. 188, pp. 107–116, 1975). An overview is given by P. Gagneaux and A. Varki in *Molecular Phylogenetics and Evolution*, vol. 18, pp. 2–13, 2001.

85. Appropriately entitled 'Jack of all trades', further details of this amazing story can be found in *Nature*, vol. 347, p. 704, 1990. I am most grateful also to Euan for a copy of an original newspaper clipping from South Africa, from which some of the information is taken.

86. For an overview see the chapter (pp. 216–244) by H. J. Jerison in *Handbook of intelligence*, edited by R. J. Sternberg (Cambridge, Cambridge University Press, 2000).

87. See H. J. Jerison's *Evolution of the brain and intelligence* (Academic Press, New York, 1973), especially his Fig. 3.6 (p. 7) and Table 10.3 (p. 219), and chapter 21 (pp. 275–283) and Appendix 6 of J. F. Eisenberg's *The mammalian radiations: An analysis of trends in evolution, adaptation, and behaviour* (Athlone Press, London, 1981).

88. His article entitled 'Fast tracks to intelligence (considerations from neurobiology and evolutionary biology)' is on pp. 237–245 of *Bioastronomy – The next steps*, edited by G. Marx (Kluwer, Dordrecht, 1988). See also note 99.

89. Concerning the primates see S. M. Reader and K. N. Laland in *Proceedings of the National Academy of Sciences, USA*, vol. 99, pp. 4436–4441, 2002. The evidence for a correlation between brain size and behavioural flexibility in birds is given by the group led by L. Lefebvre, see articles in *Oikos*, vol. 90, pp. 599–605, 2000; *Behaviour*, vol. 139, pp. 939–973, 2002; and vol. 137, pp. 1415–1429, 2000; and *Animal Behaviour*, vol. 63, pp. 495–502, 2002.

90. Although the following section emphasizes the dolphins, it is worth mentioning that there are other fascinating convergences. One such involves an extinct Peruvian whale, probably related to the monodontids (beluga, narwhal), that has become strikingly convergent on the Northern Hemisphere walrus, referred to as 'startling' by C. de Muizon in *Nature*, vol. 365, pp. 745–748, 1993; see also C. de Muizon *et al.* in *Comptes Rendus de l'Académie des Sciences, Paris; Sciences de la Terre et des Planètes*, vol. 329, pp. 449–455, 1999, and C. de Muizon and D. P. Domning in *Zoological Journal of the Linnean Society*, vol. 134, pp. 423–454, 2002.

91. See L. Marino in *Brain, Behavior and Evolution*, vol. 51, pp. 230–238, 1998.

92. See the article by L. Marino in *Evolutionary Anthropology*, vol. 5, pp. 81–85, 1996.

93. The section that follows owes much to L. Marino's paper 'Convergence of complex cognitive abilities in cetaceans and primates' in *Brain, Behavior and Evolution*, vol. 59, pp. 21–32, 2002.

94. See L. Marino *et al.* in *Journal of Mammalian Evolution*, vol. 7, pp. 81–94, 2000.

95. See, for example, the papers by J. G. M. Thewissen *et al.* in *Nature*, vol. 413, pp. 277–281, 2001, with accompanying commentaries by C. de Muizon and H. Gee on pp. 259–260 of the same issue, and J. Gatesy and M. A. O'Leary in *Trends in Ecology & Evolution*, vol. 16, pp. 562–570, 2001. Also of direct relevance is *The emergence of whales: Evolutionary patterns in the origin of Cetacea*, edited by J. G. M. Thewissen (Plenum, New York, 1998).

96. Lori Marino, personal communication (03/01/02).

97. In humans the brain accounts for about 2% of the body weight, but consumes 20% of the total metabolic energy; see J. W. Mink *et al.* in *American Journal of Physiology*, vol. 241, pp. R202–R212, 1981 and E. Armstrong in *Science*, vol. 220, pp. 1302–1304, 1983.

98. In a molecular phylogeny of the cetaceans M. Nikaido *et al.* in *Proceedings of the National Academy of Sciences, USA*, vol. 98, pp. 7384–7389, 2001, identify a rapid radiation of this group at about 30 Ma, with the dolphins diversifying at about 20 Ma.

99. Thus, in sceptical vein, Calvin (1988; see note 88 for citation) remarks of the standard Darwinian view how it is that it continues to bask in optimism, 'And thus, in this caricature of evolutionary thought, progress towards intelligence seems inevitable (and indeed a soothing counter to the inevitable disorder predicted by entropy) as long as variations keep exploring the possibilities.', p. 238.

100. See R. E. Fordyce in *Palaeogeography, Palaeoclimatology, Palaeoecology*, vol. 31, pp. 319–336, 1980, and his chapter (pp. 368–381) on cetacean evolution in *Eocene–Oligocene climatic and biotic evolution*, edited by D. R. Prothero and W. A. Berggren (Princeton University Press, Princeton, 1992).

101. The process by which deeper, nutrient-rich waters are brought to the surface. There are several mechanisms by which this can occur, although the most familiar is persistent wind shear that displaces surface waters, so allowing the ascent of the deeper waters.

102. This possibility is reviewed by several authors in *Paleoclimate and evolution, with emphasis on human origins*, edited by E. S. Vrba *et al.* (Yale University Press, New Haven, 1995). See

also J. Chaline *et al.* in *Special Publications of the Geological Society of London* vol. 181, pp. 185–198, 2000.

103. See the chapter by R. C. Connor *et al.* (pp. 415–443) in *Coalitions and alliances in humans and other animals*, edited by A. H. Harcourt and F. B. M. de Waal (Oxford University Press, Oxford, 1992), and *Cetacean societies: Field studies of dolphins and whales*, edited by J. Mann *et al.* (University of Chicago Press, Chicago, 2000).

104. D. Reiss *et al.*, in *Trends in Cognitive Science*, vol. 1, pp. 140–145, 1997, p. 141. The extent of the convergence between dolphin and chimp societies is reviewed by R. C. Connor *et al.* in their chapters (pp. 91–126 and 247–269) in Mann *et al.* (2000).

105. See p. 91 of R. C. Connor *et al.* in J. Mann *et al.* (2000; citation is in note 103).

106. A similar fission–fusion society is also found in the northern bottlenose whale; see p. 217 of R. C. Connor (in J. Mann *et al.* 2000; citation is in note 103).

107. L. Marino (2002; citation is in note 93), p. 27.

108. See, for example, the paper by R. C. Connor *et al.* in *Behaviour*, vol. 133, pp. 37–69, 1996.

109. See B. Würsig in *Biological Bulletin*, vol. 154, pp. 348–359, 1978, p. 355.

110. See the paper by R. C. Connor *et al.* in *Proceedings of the Royal Society of London B*, vol. 268, 263–267, 2001.

111. Thus R. C. Connor *et al.* (in J. Mann *et al.* 2000; citation is in note 103) remark that at one spot 'in southern Brazil fishermen line up in murky thigh-deep water, holding weighted throw nets, in pursuit of fish they cannot see. One or two dolphins, facing seaward several metres offshore of the fisherman ... come to an abrupt halt 5–7 m from the fisherman, diving with a surging roll that cues the fisherman to toss their nets' (p. 98). On pp. 211–212 R. C. Connor notes similar cooperation on the coast of Mauritania.

112. See the paper by L. Weilgart *et al.* appropriately entitled 'A colossal convergence' in *American Scientist*, vol. 84, pp. 278–287, 1996; see also R. C. Connor *et al.* in *Trends in Ecology & Evolution*, vol. 13, pp. 228–232, 1998, and H. Whitehead and L. Weilgart (in J. Mann *et al.* 2000; citation is in note 103). This example has attracted the attention of those interested in extraterrestrial intelligences; see P. Morrison on p. 573 of his chapter in *Astronomical and biochemical origins and the search for life in the Universe*, pp. 571–583, edited by C. B. Cosmovici *et al.* (Editrice Compositori, Bologna, 1997). Morrison here emphasizes the similarities in their long-distance communication.

113. L. Weilgart *et al.* (1996; citation is in note 112), p. 278.

114. Concerning the long-distance communication by elephants (and some other large mammals) see C. E. O'Connell-Rodwell *et al.* in *American Zoologist*, vol. 41, pp. 1157–1170, 2001. See also discussion by B. T. Arnason *et al.* of elephants' response to various geophysical fields in *Journal of Comparative Psychology*, vol. 116, pp. 123–132, 2002.

115. H. P. Whitehead in *Behavioral Ecology and Sociobiology*, vol. 38, pp. 237–244, 1996.

116. See B. McCowan and D. Reiss on pp. 178–207 of *Social influences on vocal development*, edited by C. Snowdon and M. Hausberger (Cambridge University Press, Cambridge, 1997). The likely importance of communication in the fission–fusion communities is emphasized by R. A. Smolker *et al.* in *Behavioral Ecology and Sociobiology*, vol. 33, pp. 393–402, 1992, and L. S. Sayigh *et al.* in *Animal Behaviour*, vol. 57, pp. 41–50, 1998, who remark 'Overall, what is known of bottlenose dolphin behaviour and social structure supports the idea that individuals do have concepts of one another as individuals and that they track the history of their individual relationships,' p. 48.

117. See D. Reiss and B. McCowan in *Journal of Comparative Psychology*, vol. 107, pp. 301–312, 1993, as well as P. L. Tyack in *Behavioral Ecology and Sociobiology*, vol. 18, pp. 251–257, 1986, and D. G. Richards *et al.* in *Journal of Comparative Psychology*, vol. 98, pp. 10–28, 1984.

118. See V. M. Janik in *Science*, vol. 289, pp. 1355–1357, 2000, and commentary by P. L. Tyack on pp. 1310–1311.

119. Their paper on 'The fallacy of the signature whistle hypothesis' (p. 1159) is published in *Animal Behaviour*, vol. 62, pp. 1151–1162, 2001.

120. This is briefly reviewed by P. L. Tyack *et al.* in their chapter (pp. 333–339) in J. Mann *et al.* (2000; citation is in note 103).

121. Thus Tayler and Saayman in *Annals of the Cape Provincial Museums (Natural History)*, vol. 9, pp. 11–49, 1972 note 'It is remarkable that our dolphins [held in the Port Elizabeth Oceanarium] have not learned to receive or to repeat – and the latter is possible by means of their sonar system, there being no vocal chords – the complex harmonic composition of human Afrikaans words,' p. 46.

122. See K. Ralls *et al.* in *Canadian Journal of Zoology*, vol. 63, pp. 1050–1056, 1985.

123. K. Ralls *et al.* (1985; citation is in note 122), p. 1051.

124. See the article 'A Beluga whale imitates human speech' by R. L. Eaton in *Carnivore*, vol. 2, pp. 22–23, 1979.

125. Among aquatic mammals at least two whales (beluga, humpback) and the Harbour seal show evidence of vocal learning; see P. L. Tyack's chapter (pp. 270–307) on cetacean communication in J. Mann *et al.* (2000; citation is in note 103). He remarks, 'Evidence for vocal learning among seals . . . is particularly interesting from an evolutionary perspective, because the pinnipeds evolved from a different terrestrial ancestor than the cetaceans. This suggests that there were at least two independent origins of vocal learning among marine mammals,' p. 306.

126. See B. McCowan and D. Reiss in *Journal of Comparative Psychology*, vol. 109, pp. 242–260, 1995, and B. McCowan *et al.* in *Animal Behaviour*, vol. 57, pp. 409–419, 1999.

127. B. McCowan and D. Reiss in *Ethology*, vol. 100, pp. 194–209, 1995, as well as comparisons to Zipf's Law by McCowan *et al.* (1999). See also the chapter by R. C. Connor *et al.* (pp. 91–126) in J. Mann et al. 2000; citation is in note 103.

128. Thus in the pygmy marmoset, a New World monkey, the young also engage in a similar 'babbling'. See A. M. Elowson *et al.* and C. T. Snowdon and A. M. Elowson in *Behaviour*, vol. 135, pp. 643–664, 1998, and vol. 138, pp. 1235–1248, 2001 respectively. In an overview in *Trends in Cognitive Sciences*, vol. 2, pp. 31–37, 1998 A. M. Elowson *et al.* remark 'social parallels make the Callitrichids [the group that includes the pygmy marmoset] a more compelling analogous group [to humans] to study than the phylogenetically closer, but socially dissimilar, apes,' p. 32.

129. See their chapter (pp. 663–680) in *Phonological development: Models, research, implications*, edited by C. A. Ferguson *et al.* (York Press, Timonium, MA, 1992); quotation is on p. 669.

130. Quotation is on p. 253 of the chapter (pp. 253–264) by D. Reiss in Marx (1988; citation is in note 88) entitled 'Can we communicate with other species on this planet? (Pragmatics of communication between humanoid and non-humanoid species)'. See also B. McCowan *et al.* 1999 (citation is in note 126).

131. These famous 'songs' are produced only by the male. P. M. Gray *et al.* (2001; citation is in note 171, Chapter 8) remark, 'The undersea songs of humpback whales are similar in structure to bird and human songs and prove that these marine mammals are inveterate composers . . . humpback whale songs are constructed according to laws that are strikingly similar to those adopted by human composers,' p. 52.

132. Thus P. M. Gray *et al.* (2001; citation is in note 171, Chapter 8) comment 'Is there a universal music [as part of Mathematical Platonism] awaiting discovery, or is all music just a construct of whatever mind is making it, human, bird, whale? The similarities among human music, bird song, and whale song tempt one to speculate that the Platonic alternative may exist – that there is a universal music awaiting discovery,' p. 54.

133. The paper on 'Comprehension of sentences by bottle-nosed dolphins' is by L. M. Herman *et al.*, and is published in *Cognition*, vol. 16, pp. 129–219, 1984, while in *Journal of Experimental Psychology: General*, vol. 122, pp. 184–194, 1993, L. M. Herman *et al.* report responses by dolphins to syntactically anomalous sequences. In the same year L. M. Herman *et al.* published a parallel paper, as Chapter 20 (pp. 403–442) in *Language and communication: Comparative perspectives*, edited by H. L. Roitblat *et al.* (Lawrence Erlbaum Associates, Hillsdale, NJ, 1993), where they remarked that it appeared that dolphins can

generate 'a mental representation of the grammatical structure of their language' (p. 408) in a way directly analogous to humans who when presented with grammatical novelties in an artificial language are then in a position to decode other novel strings to at least a reasonable degree of accuracy.

134. See L. M. Herman *et al.* in *Animal Learning and Behavior*, vol. 29, pp. 250–264, 2001.

135. L. Marino (2002, citation is in note 93), p. 28.

136. In a more general review on the evolution of intelligence E. MacPhail and J. J. Bolhuis, in *Biological Reviews*, vol. 76, pp. 341–364, 2001, remark 'that there are no qualitative differences in cognition between animal species in the processes of learning and memory' (p. 341). Interestingly, they also conclude that human language is a unique attribute, concluding, 'Unlike Darwin, we believe that there is a qualitative difference between human and non-human intelligence; but we know of no convincing evidence that language evolved to "solve" some novel "problem" posed by a niche invaded by an initially pre-linguistic ancestor, nor any analysis of that niche's demands that promises to throw light on the mechanisms of language,' p. 361.

137. See the discussion by E. Kako in *Animal Behavior and Learning*, vol. 27, pp. 1–14, 1999, where he also addresses the question, in a moderately sceptical tone, of similar comprehension by African Grey parrots and chimps. The response, on pp. 18–23, by L. M. Herman and R. K. Uyeyama on the grammatical competency of bottlenose dolphins is an apt counterpoint to Kako's paper, to which he replies (pp. 26–27). The cases for African Grey parrots and chimps are put forward respectively by I. M. Pepperberg (pp. 15–17) and S. G. Shanker *et al.* (pp. 24–25).

138. Nor should we forget other attributes in birds such as tool-making (notes 176–178), and remarkable accounts of coordinated hunting, such as in pied currawongs summoning assistance to dispatch a rat; see M. G. O'Neill and R. J. Taylor in *Corella*, vol. 8l, pp. 95–96, 1984.

139. A point stressed by both I. M. Pepperberg (1999) and L. M. Herman and R. K. Uyeyama (1999).

140. See the paper 'Evolution of a universal grammar' by M. A. Nowak *et al.*, in *Science*, vol. 291, pp. 114–118, 2001. B. McCowan *et al.* (in *Journal of Comparative Psychology*, vol. 116, pp. 166–172, 2002) emphasize basic similarities in Zipf's coefficient (see note 5, Chapter 1) and Shannon's measure of entropy between human words, bottlenose whistles, and squirrel monkey calls at different stages of development. In each case communication has arisen for specific adaptational reasons, but it is also convergent.

141. See R. K. Thompson and L. M. Herman in *Science*, vol. 195, pp. 501–503, 1977.

142. See E. Mercado *et al.* in *Animal Learning & Behavior*, vol. 26, pp. 210–218, 1998.

143. Their paper is in *Journal of Experiment Psychology: General*, vol. 124, pp. 391–408, 1995; quotation is on p. 391.

144. 'Half-awake to the risk of predation' is the title of the paper by N. C. Rattenborg *et al.* in *Nature*, vol. 397, pp. 397–398, 1999; a more general review of unihemispheric sleep in birds (as well as cetaceans and some other marine mammals) is given by the same group in *Neuroscience and Biobehavioural Reviews*, vol. 24, pp. 817–842, 2000. The origins of such sleep may lie deeper than the birds, as some evidence suggests equivalent states in the reptiles. Its evolution in two or three groups of marine mammals (and possibly the elephants, p. 833), however, is almost certainly independent and regained from primitive Mesozoic mammals that may have lost this ability when their mode of life was largely nocturnal and/or burrowing.

145. See the paper by P. D. Goley in *Marine Mammal Sciences*, vol. 15, pp. 1054–1064, 1999, as well as Rattenborg *et al.* (2000).

146. As reported by H. Elias and P. Swartz in *Science*, vol. 166, pp. 111–113, 1969; see also Jerison (2000; citation is in note 86, pp. 222–223).

147. See L. Marino *et al.* in *Brain, Behavior and Evolution*, vol. 56, pp. 204–211, 2000.

148. Thus in assessing the question of convergences in the dolphin brain, with the subtitle 'Does evolutionary history repeat itself?' P. Manger et al. (in Journal of Cognitive Neuroscience, vol. 10, pp. 153–166, 1998) emphasize how both the modularity of the brain and the underlying constraints of neural architecture may lead to similar end points.

149. Helpful summaries are given by L. Marino (1996, 2002; see notes 92 and 93 for citations). See also L. Marino et al. in Anatomical Record, vol. 264, pp. 397–414, 2001. Other key references are by P. J. Morgane et al. in Brain Research Bulletin, vol. 5 (Supplement 3), pp. 1–107, 1980, and I. I. Glezer et al. in Behavioral and Brain Sciences, vol. 11, pp. 75–89 (and succeeding discussion on pp. 89–116), 1988.

150. L. Marino (2002, citation is in note 93), p. 24.

151. See L. Marino (1996; citation is in note 92), pp. 84–85.

152. L. Marino (2002; citation is in note 93), p. 25.

153. See the paper on the scalable architecture of mammal brains by D. A. Clark et al. in Nature, vol. 411, pp. 189–193, 2001, as well as the commentary by J. H. Kaas and C. E. Collins on pp. 141–142.

154. L. Marino (2002, citation is in note 93), p. 25.

155. See, for example, the paper on mosaic evolution of brain structure in mammals by R. A. Barton and P. H. Harvey in Nature, vol. 405, pp. 1055–1058, 2000. The thrust of their argument is that the mosaic evolution of the brain, with emphasis on those sections, e.g. neocortex, that are functionally significant, will transcend developmental constraints.

156. See I. I. Glezer et al. on pp. 1–38 of Marine mammal sensory systems, edited by J. A. Thomas et al. (Plenum, New York, 1992).

157. See the paper by B. McCowan et al. in Journal of Comparative Psychology, vol. 114, pp. 98–106, 2000.

158. See the two chapters in the book Self-awareness in animals and humans: Developmental perspectives, edited by S. T. Parker et al. (Cambridge University Press, Cambridge, 1994), respectively by K. Marten and S. Psarakos (pp. 361–379) and L. Marino et al. (pp. 380–391).

159. See D. Reiss and L. Marino in Proceedings of the National Academy of Sciences, USA, vol. 98, pp. 5937–5942, 2001.

160. D. Reiss and L. Marino (2001; citation is in note 159), p. 5942.

161. See, for example, L. von Ferson et al. on pp. 753–762 in Thomas et al. (1992; see note 156 for citation), and also B. McCowan et al. (2000; citation is in note 157).

162. Particular attention has been paid to the killer whales (see, for example, J. K. B. Ford in Canadian Journal of Zoology, vol. 69, pp. 1454–1483, 1991, and more recently P. J. O. Miller and D. E. Bain in Animal Behaviour, vol. 60, pp. 617–628, 2000), as well as the sperm-whales (see L. Weilgart and H. Whitehead in Behavioral Ecology and Sociobiology, vol. 40, pp. 277–285, 1997). Information on elephant seals and Weddell seals can be found in papers in Animal Behaviour, vol. 22, pp. 656–663, 1974 by B. J. Le Boeuf and L. F. Petrinovich (elephant seals) and Canadian Journal of Zoology, vol. 61, pp. 2203–2214, 1984 by J. A. Thomas and I. Stirling (Weddell seals).

163. The extent to which birds can be said to have cultural attributes is another fascinating (and contentious) topic; the brief remarks (pp. 341–342) by S. K. Lynn and I. M. Pepperberg in L. Rendell and H. Whitehead (2001, citation is in 164) are a forceful plea to take some birds as seriously as killer whales and humans. The existence of dialects is also well known. T. F. Wright and G. S. Wilkinson, in Proceedings of the Royal Society of London B, vol. 268, pp. 609–616, 2001, present evidence in a parrot, the Yellow-naped Amazon, for emergence of dialects despite gene flow between populations. See also P. Enggist-Dueblin and U. Pfister in Animal Behaviour, vol. 64, pp. 831–841, 2002.

164. Key papers are those by H. Whitehead in Science, vol. 282, pp. 1708–1711, 1998 (with a commentary by G. Vogel on p. 1616); V.-B. Deecke et al. in Animal Behaviour, vol. 60, pp. 629–638, 2000, and L. Rendell and H. Whitehead in Behavioral and Brain Sciences, vol. 24, pp. 309–382, 2001 (which includes a series of commentaries and critiques) and

Proceedings of the Royal Society of London B, vol. 270, p. 225–231, 2003. Vertical cultural transmission in stable matrilineal groups is reviewed by H. Whitehead and J. Mann in their chapter (pp. 219–246) on female reproductive strategies in J. Mann *et al.* (2000; citation is in note 103) where they comment: 'it seems possible that cetacean societies ... such as those of killer, sperm, and pilot whales, contain cultures that are qualitatively more similar to human cultures than are those of most terrestrial mammals ... [and] may therefore explain curious attributes of these species, such as nonadaptive mass strandings, low genetic diversity, and menopause,' pp. 243–244.

165. See H. Whitehead and J. Mann in J. Mann *et al.* (2000; citation is in note 103), pp. 219–246, who remark, 'It is notable that the cetacean species in which menopause has been found, pilot and killer whales, possess matrilineally based social systems. There is some evidence that cultural processes are very important within the matrilineal groups of these animals ... If an older female's role as a source of information significantly increases her descendants' fitness, and reproduction towards the end of her life decreases survival, menopause could be adaptive,' p. 233.

166. See K. McComb *et al.* in *Science,* vol. 292, pp. 491–494, 2001, as well as the commentary by E. Pennisi on pp. 417–418.

167. Moreover, as Whitehead and Weilgart (2000; citation is in note 112) note, 'Evidence is beginning to accumulate that sperm whale units have distinctive vertically transmitted cultures: functionally important information learned maternally or within matrilineal units. Units have characteristic coda repertoires, which are unlikely to be genetically determined,' p. 169.

168. Whitehead and Weilgart (2000) remind us the sperm whale 'was the subject of two massive hunts. The first of these, peaking in about 1840, greatly reduced sperm whale populations throughout the world ... while providing the oil that lubricated the Industrial Revolution' (p. 155), while in the 1960s more than 20 000 a year were killed.

169. Their paper is published in *Animal Behaviour,* vol. 60, pp. 617–628, 2000.

170. Miller and Bain (2000), p. 626.

171. 'Cultural revolution in whale songs' is the title of the paper by M. J. Noad *et al.* in *Nature,* vol. 408, p. 537, 2000.

172. For this, the first report of tool use by dolphins, see the paper by R. Smolker *et al.* in *Ethology,* vol. 103, pp. 454–464, 1997.

173. This 'bubble-feeding' is reviewed by P. J. Clapham in his chapter (pp. 173–196) on the humpback whale in J. Mann *et al.* (2000; citation is in note 103). There is also some suggestion that unweaned calves blowing small bubble-clouds next to their feeding mothers may be learning the technique.

174. His book is entitled *Nature's destiny: How the laws of biology reveal purpose in the Universe* (The Free Press, New York, 1998); see his Chapter 11. Denton's work shows a number of valuable parallels to this book but *Nature's destiny* has, however, a distinct anti-evolutionary bias (see my review 'Awe, wonder and gene committees' in the *Times Higher Education Supplement,* p. 25 of the December 24/31, 1999 issue).

175. See, for example, J. Boswall in *Avicultural Magazine,* vol. 89, pp. 94–108 and 170–181, 1983.

176. See G. R. Hunt in *Nature,* vol. 379, pp. 249–251, 1996 (with commentary by C. Boesch on pp. 207–208), and *Emu,* vol. 100, pp. 109–114, 2000 and vol. 102, pp. 349–353, 2002.

177. See also the report of hook manufacture in laboratory-kept New Caledonian crows by A. A. S. Weir *et al.* in *Science,* vol. 297, p. 981, 2002.

178. Concerning this, see G. R. Hunt in *Proceedings of the Royal Society of London B,* vol. 267, pp. 403–413, 1999, and G. R. Hunt *et al.* in *Nature,* vol. 414, p. 707, 2001.

179. G. R. Hunt (1999; citation is in note 178), p. 412. See also G. R. Hunt and R. D. Gray in *Proceedings of the Royal Society of London B,* vol. 270, pp. 867–874, 2003.

180. See I. M. Pepperberg's *The Alex studies: cognitive and communicative abilities of Grey Parrots* (Harvard University Press, Cambridge, MA, 1999).

181. See C. Caffrey in *The Wilson Bulletin*, vol. 112, pp. 283–284, 2000 and vol. 113, pp. 114–115, 2001.

182. See *The use of tools by human and non-human primates*, edited by A. Berthelet and J. Chavaillon (Clarendon Press, Oxford, 1993), especially the chapters by C. Boesch and H. Boesch (pp. 158–174) and Y. Sugiyama (pp. 175–190), and also W. M. McGrew's *Chimpanzee material culture: Implications for human evolution* (Cambridge University Press, Cambridge, 1992).

183. See N. Toth *et al.* in *Journal of Archaeological Science*, vol. 20, pp. 81–91, 1993.

184. N. Toth *et al.* (1993; citation is in note 183), p. 89.

185. See W. C. McGrew in *Science*, vol. 288, p. 1747, 2000.

186. See T. S. Stoinski and B. B. Beck in *Primates*, vol. 42, pp. 319–326, 2001.

187. See, for example, G. C. Westergaard and S. J. Suomi in *Current Anthropology*, vol. 35, pp. 75–79 and pp. 468–470, 1994; *International Journal of Primatology*, vol. 16, pp. 1017–1024, 1995; and *Journal of Archaeological Sciences*, vol. 22, pp. 677–681, 1995; as well as E. Visalberghi on pp. 118–135 of A. Berthelet and J. Chavaillon (1993; citation is in note 182). In addition, as an extension of S. M. Reader and K. N. Laland (2002; citation is in note 89), S. M. Reader has looked at plots of behavioural plasticity among the primates, including tool use. Here too, with chimp and bonobo *Cebus* stands out. See his paper in *Towards a biology of traditions: models and evidence*, edited by D. Fragaszy and S. Perry (Cambridge University Press, Cambridge, in press).

188. See S. Chevalier-Skolnikoff in *Behavioral and Brain Sciences*, vol. 12, pp. 561–627, 1989.

189. See G. C. Westergaard and S. J. Suomi in *Journal of Human Evolution*, vol. 27, pp. 399–404, 1994.

190. G. C. Westergaard and S. J. Suomi (1994; citation is in note 189), p. 403.

191. See their paper in *Journal of Human Evolution*, vol. 6, pp. 623–641, 1977, see p. 634.

192. See Gregory Westergaard and colleagues' paper in *International Journal of Primatology*, vol. 20, pp. 153–162, 1999.

193. G. C. Westergaard *et al.* (1999; citation is in note 192), p. 161.

194. See S. Chevalier-Skolnikoff in *Primates*, vol. 31, pp. 375–383, 1990. K. A. Phillips (in *American Journal of Primatology*, vol. 46, pp. 259–261, 1998) reports the use of 'leaves as cups to retrieve water from tree cavities', p. 259.

195. E. Visalberghi (1993; citation is in note 187; see also note 182), see pp. 127–128.

196. The nature of the food resources may also be important; see M. A. Panger in *American Journal of Physical Anthropology*, vol. 106, pp. 311–321, 1998.

197. See G. C. Westergaard and S. J. Suomi in *Journal of Human Evolution*, vol. 30, pp. 291–298, 1996.

198. G. C. Westergaard and S. J. Suomi (1996; citation is in note 197), p. 296.

199. See E. Visalberghi and L. Limongelli in their chapter (pp. 57–79) of *Reaching into thought: The minds of the great apes*, edited by A. E. Russon *et al.* (Cambridge University Press, Cambridge, 1996). Capuchins in the context of primate mentalities are reviewed by R. Byrne in his chapter (pp. 110–124) in *Creativity in human evolution and prehistory*, edited by S. Mithen (Routledge, London, 1998).

200. E. Jalles-Filho *et al.*, in *Journal of Human Evolution*, vol. 40, pp. 365–377, 2001. They point out that there may be cognitive limits that militate against such key activities as carrying the tools to the food, although this may also be a consequence of a social structure whereby access to food is largely controlled by aggression. Hence it makes little sense to wander off looking for a suitable tool and leave the food unattended. However, food and tool exchanges are known; see G. C. Westergaard and S. J. Suomi in *American Journal of Primatology*, vol. 43, pp. 33–41, 1997, where they note, 'we propose that tool- and food-sharing came into existence through convergent evolution in large-brained, extractive foraging primate genera, including *Cebus*, *Pan* and *Homo*' (p. 40).

201. See S. W. Williston, in *Entomological News and Proceedings of the Entomological Section of the Academy of Natural Sciences of Philadelphia*, vol. 3, pp. 85–86, 1892.

202. S. W. Williston (1892; citation is in note 201), pp. 85–86.

203. For a review of parallel homoplasies in the mammalian neocortex, see R. G. Northcutt and J. H. Kaas in *Trends in Neurosciences*, vol. 18, pp. 373–379, 1995.

204. Willem de Winter and Charles Oxnard's paper is published in *Nature*, vol. 409, pp. 710–714, 2001.

205. The essay, by Thomas Nagel, entitled 'What it is like to be a bat', may be found in *The Philosophical Review*, vol. 83, pp. 433–450, 1974.

206. T. Nagel (1974; citation is in note 205), p. 438.

207. T. Nagel (1974; citation is in note 204), pp. 439–440.

208. A similar point is made by R. Dawkins on pp. 33–35 *The blind watchmaker* (Norton, New York, 1987) The emergence of novel functions in brain evolution in the context of convergence is well addressed by K. C. Nishikawa (in *Bioscience*, vol. 47, pp. 341–354, 1997). Thus he writes, 'Convergent evolution of neural circuits that serve similar functions may provide insights into the functional architecture of nervous systems. For example, features such as parallel, distributed processing and population coding (in which a signal is encoded in the activity of a population of neurons instead of a single cell) have evolved convergently in distantly related species throughout the animal kingdom; these features likely represent analogous solutions to similar problems in different animals,' p. 342.

209. T. Nagel (1974; citation is in note 204), p. 441.

210. See U. M. Norberg and M. B. Fenton in *Biological Journal of the Linnean Society*, vol. 33, pp. 383–394, 1988.

211. Concerning this, see the convergent evolution of trichromacy (Chapter 7, notes 122–124) and the enzyme lysozyme (Chapter 10, notes 88, 89).

212. W. de Winter and C. Oxnard (2001, citation is in note 204), pp. 713–714. Similar conclusions were reached by F-J. Lapointe *et al.* (in *Brain, Behavior and Evolution*, vol. 54, pp. 119–126, 1999) where they noted 'obvious adaptive convergences of the brain to trophic niches ... one could clearly distinguish a folivorous, an insectivorous/carnivorous, and a frugivorous/nectarivorous clade based on size-corrected brain data,' p. 122.

213. See also the example of convergence in echolocation between a Neotropical species of *Myotis* and the temperate pipistrellids by M. Siemers *et al.* in *Behavioural Ecology and Sociobiology*, vol. 50, pp. 317–328, 2001, who emphasize evolutionary flexibility over phylogenetic constraint in response to particular ecological challenges.

214. See L. Krubitzer's discussion of convergences in the mammalian neocortex in *Trends in Neurosciences*, vol. 18, pp. 408–417, 1995.

215. L. Krubitzer (1995; citation is in note 214), p. 416.

216. The convergence between locomotory gaits of primates and the South American didelphid *Caluromys* (the woolly opossum) is addressed by D. Schmitt and P. Lemelin in *American Journal of Physical Anthropology*, vol. 118, pp. 231–238, 2002, and thus reinforces the various other convergences already noted between this marsupial and the prosimian primates; see, for example, P. Lemelin in *Journal of Zoology*, vol. 247, pp. 165–175, 1999 and D. T. Rasmussen in *American Journal of Primatology*, vol. 22, pp. 263–277, 1990.

217. See the paper by J. G. Fleagle in *Folia Primatologica*, vol. 26, pp. 245–269, 1976.

218. J. C. Fleagle (1976; citation is in note 217), p. 264.

219. These are the platyrrhines, and include the capuchins (*Cebus*), the spider monkeys (*Ateles*), the howlers (*Alouatta*), marmosets (*Callithrix*), and owl monkeys (*Aotus*); see M. Moynihan's *The New World primates; Adaptive radiation and the evolution of social behavior, languages, and intelligence* (Princeton University Press, Princeton, 1976). These monkeys evidently represent invaders from the Old World, arriving about 30 Ma ago (see R. F. Kay *et al.* in *Journal of Vertebrate Paleontology*, vol. 18, pp. 189–199, 1998), most probably by rafting across the Atlantic from Africa; see A. Houle in *American Journal of Physical Anthropology*,

vol. 109, pp. 541–559, 1999. Concerning their interrelationships, see I. Horowitz and A. Meyer on pp. 189–224 of *Molecular evolution and adaptive radiation*, edited by T. J. Givnish and K. J. Sytsma (Cambridge University Press, Cambridge, 1997) and I. Horovitz in *American Museum Novitates*, no. 3269, pp. 1–40, 1999, with inclusion of more fossil taxa. The present-day fauna may be markedly impoverished in comparison to the fossil diversity. W. C. Hartwig and C. Cartelle describe (in *Nature*, vol. 381, pp. 307–311, 1996) a giant Pleistocene Brazilian monkey (*Protopithecus*) with a rather puzzling mixture of howler monkey-like (cranial) and spider monkey-like (postcranial) anatomy.

220. See the papers in *Folia Primatologica* by R. A. Mittermeier in vol. 30, pp. 161–193, 1978, and J. G. H. Cant in vol. 46, pp. 1–14, 1986. Concerning convergent evolution within the ateline monkeys see C. A. Lockwood in *American Journal of Physical Anthropology*, vol. 108, pp. 459–482, 1999. This paper is important, not only because of its implications for convergence with the Old World apes, but also because of Lockwood's emphases on the importance of adaptation (Chapter 10, note 1) and the difficulty of discerning a reliable phylogenetic signal in the face of rampant convergence. The evolutionary position of the New World monkeys (platyrrhines) in the general scheme of primate phylogenies is addressed by S. M. Ford on pp. 595–673 of *Anthropoid origins*, edited by J. G. Fleagle *et al.* (Plenum, New York; 1994). She notes that the ancestral form was very generalized, readily diversified in a number of directions, and that 'The hominoids, in particular, and *Pliopithecus* [a Miocene primate], demonstrate numerous convergences with atelines' (p. 656); see also D. R. Begun in *Journal of Human Evolution*, vol. 24, pp. 373–402, 1993.

221. See D. A. Clark *et al.* (2001; citation is in note 153), where they note that 'neocortical volume fractions have become successively larger in lemurs and lorises, New World monkeys, Old World monkeys, and hominoids, lending support to the idea that primate brain architecture has been driven by directed selective pressure,' p.189.

222. Details of the convergences (and divergences) between atelines (and capuchins and squirrel monkeys) with Old World primates is given by J. G. Robinson and C. H. Janson (pp. 69–82) in *Primate Societies*, edited by B. B. Smuts *et al.* (University of Chicago Press, Chicago; 1987), see pp. 80–82. A more general overview of convergences and divergences is in the chapter (pp. 158–170) by P. M. Kappeler in *Primate communities*, edited by J. G. Fleagle *et al.* (Cambridge University Press, Cambridge, 1999).

223. See the paper by C. A. Chapman *et al.* in *Behavioral Ecology and Sociobiology*, vol. 36, pp. 59–70, 1995. Quotation is on p. 59. These workers are careful to point out that the similarities, while strong, are not precise.

224. M. Moynihan (1976, citation is in note 219); quotations on pp. 107 and 108 respectively.

225. Thus G. C. Westergaard and S. J. Suomi (1999; citation is in note 192) draw attention to bipedalism in capuchin monkeys, interestingly in connection with food loads.

226. One of the curiosities of tool use in the primates is that given they are all pretty intelligent, then why is the employment of tools so sporadic? Thus W. C. McGrew has observed that 'All great apes are smart enough to use tools but they only do so in useful circumstances', see p. 470 of his chapter (pp. 457–472) in *Comparative socioecology: The behavioural ecology of humans and other mammals*, edited by V. Standen and R. A. Foley (Blackwell, Oxford, 1989). And these 'useful circumstances' seem to be best correlated with access to animal material, a point returned to in his chapter (pp. 143–157) in A. Berthelet and J. Chavaillon (1993; citation is in note 182).

227. See the paper by F. Spoor *et al.* in *Nature*, vol. 369, pp. 645–648, 1994, as well as further discussion in *Journal of Human Evolution*, vol. 30, pp. 183–187, 1996. The crux of their argument involves the structure of the vestibular system (semicircular canals) of the inner ear, which in the obligatory bipeds has to be very finely tuned to permit balance (not least on a bicycle). They conclude that bipedality was less well developed in early hominids, but certainly present by *Homo erectus* times.

228. See P. E. Wheeler in *Journal of Human Evolution*, vol. 13, pp. 91–98, 1984, vol. 14, pp. 23–28, 1985, vol. 21, pp. 107–116 and 117–136, 1991, and vol. 23, pp. 379–388, 1992. A critique of some of these ideas is presented by L. Q. Amaral in vol. 30, pp. 357–366, 1996.

229. For an overview, see A. Gibbons in *Science*, vol. 295, pp. 1214–1219, 2002.

230. B. Wood in *Nature*, vol. 418, pp. 145–151, 2002.

231. B. Wood (2002; citation is in note 230), p. 134.

232. Dating of the beds that yield *Oreopithecus* is reported by L. Rook *et al.* in *Journal of Human Evolution*, vol. 39, pp. 577–582, 2000.

233. The phylogenetic position of this animal is difficult to resolve on account of its peculiar anatomy. S. Moyà Solà and M. Köhler, in *Comptes Rendus de l'Académie des Sciences, Paris*, vol. 324 (Ser. IIa), pp. 141–148, 1997, place *Oreopithecus* close to such apes as *Dryopithecus*. On the other hand T. Harrison and L. Rook in their chapter (pp. 327–362) in *Function, phylogeny and fossils: Miocene hominoid evolution and adaptations*, edited by D. R. Begun *et al.* (Plenum, New York, 1997), tentatively conclude that *Oreopithecus* is very close to, if not actually, a hominid, but they emphasize also specific features that make convergence an alternative possibility; see also E. E. Sarmiento in *American Museum Novitates*, no. 2881, pp. 1–44, 1987.

234. See R. L. Bernor *et al.* in *Bollettino della Società Paleontologia Italiana*, vol. 40, pp. 139–148, 2001.

235. See M. Köhler and S. Moyà-Solà in *Proceedings of the National Academy of Sciences, USA*, vol. 94, pp. 11747–11750, 1997.

236. M. Köhler and S. Moyà-Solà (1997; citation is in note 235), p. 11747.

237. See L. Rook *et al.* in *Proceedings of the National Academy of Sciences, USA*, vol. 96, pp. 8795–8799, 1999, which examines the bone structure of the hip bone known as the iliac, and specifically the cancellous bone architecture that orientates itself according to the prevailing stress regime, which is characteristic for bipedal activity.

238. See S. Moyà-Solà *et al.* in *Proceedings of the National Academy of Sciences, USA*, vol. 96, pp. 313–317, 1999, where they remark 'The functional resemblances with the australopithecine pattern suggests for *Oreopithecus* similar manipulative skills, with improved finger control and the capability to hold objects securely and steadily between the pads of thumb and index finger ... This strongly suggests that the hand morphology of *Oreopithecus* is derived for apes and convergent with that of early hominids,' p. 316.

239. See M. Köhler and S. Moyà-Solà (1997, citation is in note 235). They remark 'insular ecosystems ... are characterized by lack of predators and limitation of space and thus of trophic resources ... Whereas the absence of predation removes the need for adaptations related to predator avoidance, intraspecific and interspecific competition for food resources increases ... Both factors impose specific selective pressures that favor ... adaptations ... related to energetically less expensive locomotor activities ... and to reduction of bone mass in the locomotor apparatus at the expense of mobility and speed. On the other hand they select for feeding strategies that increase the efficiency of resource utilization ... These adaptations are universally found in all mammal faunas of small islands. These selective pressures probably played a crucial role in the evolution of *Oreopithecus* ... the lack of predators may have led to a decrease of energetically expensive ... and risky ... climbing activities, while favoring significant terrestriality' (pp. 11749–11750). Some support for this also comes from the reduction in canine size. As D. M. Alba *et al.* (in *Journal of Human Evolution*, vol. 40, pp. 1–16, 2001) discuss whether this decrease could represent diminished levels of agonistic behaviour, especially in the males, itself a reflection of the relaxation of selective pressures that result from insularity and the absence of predators.

240. Being marooned on islands has other evolutionary consequences. Such environments, most famously in the case of the Galápagos and Hawaiian island chain, have long been famous as natural evolutionary 'laboratories'. Indeed, while the island habitat is a rich source of evolutionary novelty, so, too, are they of convergence. The mammal-like kiwis and mice-like

wetas of New Zealand (Chapter 8) are splendid examples. Island faunas show other types of convergence, of which one of the most fascinating examples is how big animals get small, and small animals big. Thus, in the various Mediterranean islands, for example, the fossil record shows dwarf elephants (a metre high and weighing little more than a large human) and hippopotamuses, along with mice the size of rabbits. For a further discussion of this topic, see for example, the papers by P. Y. Sondaar on pp. 671–707 of *Major patterns of vertebrate evolution*, edited by M. K. Hecht *et al.* (New York, Plenum, 1977); L. R. Heaney in *Evolution*, vol. 32, pp. 29–44, 1978; M.V. Lomolino in *American Naturalist*, vol. 125, pp. 310–316, 1985; V. L. Roth in *Oxford Surveys in Evolutionary Biology*, vol. 8, pp. 259–288, 1992; and J. Damuth in *Nature*, vol. 365, pp. 748–750, 1993.

241. Part of the problem of assigning an appropriate phylogenetic position (citation is in note 233) is the substantial modification of the skull that shows *Oreopithecus* had a jaw structure and teeth consistent with it being a dynamic vegetarian.

242. The origin of skilled forelimb movement is reviewed by A. N. Iwaniuk and I. Q. Whishaw in *Trends in Neurosciences*, vol. 23, pp. 372–376, 2000. They produce an impressive catalogue of examples, and argue that these manipulative abilities are primitive and drawn upon by various tetrapod vertebrates as and when appropriate.

243. Enthusiasts for the intriguing 'aquatic ape' hypothesis, originally put forward by Alister Hardy and revived by Elaine Morgan, may find an echo in this idea of an insular origination.

244. See M. G. Leakey *et al.* in *Nature*, vol. 410, pp. 433–440, 2001, with a thoughtful commentary on pp. 419–420 by D. E. Lieberman. Interestingly, in discussing where best to place this fossil, Lieberman remarks 'At present, it is hard to believe any reconstruction of hominin relationships because of the abundance of independently evolved similarities in the hominin fossil record. The complex mosaic of features seen in the new fossil will only exacerbate the problem' (p. 419).

245. See M. Brunet *et al.* in *Nature*, vol. 418, pp. 145–151, 2002; see also note 229.

246. See B. Senut *et al.* in *Comptes Rendus de l'Académie des Sciences, Paris, Sciences de la Terre et des Planètes*, vol. 332, pp. 137–144, 2001. See also the commentary on both *Kenyanthropus* and *Orrorin* by L. E. Aiello and M. Collard in *Nature*, vol. 410, pp. 526–527, 2001.

247. The importance of homoplasy in hominid evolution is addressed by H. M. McHenry and L. R. Berger in *Journal of Human Evolution*, vol. 35, pp. 1–22, 1998.

248. Robust convergence is documented by R. R. Skelton and H. M. McHenry in *Journal of Human Evolution*, vol. 23, pp. 309–349, 1992. They remark that a robust chewing habit also emerged in a number of extinct primates, including the Madagascan *Hadropithecus*, and *Gigantopithecus*, apparently as adaptations to xeric (i.e. desert) vegetation.

249. See Rob Foley's commentary on Skelton and McHenry (1992), published in *Trends in Evolution & Ecology*, vol. 8, pp. 196–197, 1993.

250. R. Foley (1993; citation is in note 249), p. 196.

251. See pp. 31–42 of *Structure and contingency: Evolutionary processes in life and human society*, edited by J. Bintliff (Leicester University Press, London, 1999).

252. R. Foley (1999; citation is in note 249), p. 40.

253. R. Foley (1999; citation is in note 249), p. 36.

254. R. Foley (1999; citation is in note 249), pp. 40–41.

255. See the paper by S. Semaw *et al.* in *Nature*, vol. 385, pp. 333–336, 1997 with accompanying commentary by B. A. Wood on pp. 292–293, as well as S. Semaw in *Journal of Archaeological Science*, vol. 27, pp. 1197–1214, 2000, where he suggests that the oldest of the artefacts may have been made by an australopithecine species (*A. garhi*).

256. See H. Roche *et al.* in *Nature*, vol. 399, pp. 57–60, 1999 (with an accompanying review by J. Steele on pp. 24–25). Although written before these important discoveries were made an interesting overview of the earliest hominid stone cultures is given by J. W. K. Harris and S. D. Capaldo on pp. 196–220 of Berthelet and Chavaillon (1993; citation is in note 182). While the vast bulk of our understanding of early hominid technology is based on lithic

material, evidence dating back to c. 1.8 Ma suggests that bone tools were being used for termite foraging; see L. R. Backwell and F. d'Errico in *Proceedings of the National Academy of Sciences, USA*, vol. 98, pp. 1358–1363, 2001, with commentary by P. Shipman on pp. 1335–1337.

257. Fragmentary remains of early hominids and frequently complex cave stratigraphies with recurrent roof falls make definitive conclusions difficult, not to mention that rather too seldom is a skeleton found grasping a stone tool. Even so, K. Kuman and R. J. Clarke, in *Journal of Human Evolution*, vol. 38, pp. 827–847, 2000, present evidence for tool-making by late (c. 2 Ma) *Paranthropus*. Somewhat older evidence for use of tools is given by T. R. Pickering *et al.* in *American Journal of Physical Anthropology*, vol. 111, pp. 579–584, 2000. Substantially older finds, from about 2.5 Ma ago, include bones with cut marks and percussive damage that are almost certainly the product of intentional fleshing and marrow extraction. Whether, however, these activities were at the hands of early *Homo* or australopithecines is not known; see J. de Heinzelin *et al.* in *Science*, vol. 284, pp. 625–629, 1999.

258. See S. Elton *et al.* in *Journal of Human Evolution*, vol. 41, pp. 1–27, 2001.

259. S. Elton *et al.* (2001; citation is in note 258), pp. 23–24.

260. D. E. Lieberman *et al.* in *Journal of Human Evolution*, vol. 30, pp. 97–120, 1996.

261. D. E. Lieberman *et al.* (1996; citation is in note 260), p. 115.

262. See Gerrit van Vark's paper in *Perspectives in Human Biology*, vol. 4, pp. 237–243, 1999.

263. G. van Vark (1999; citation is in note 262), p. 241.

264. See, however, H. Roche *et al.* (1999; citation is in note 256), who question the notion of technological stasis in very early tool material; see also Semaw (2000; citation is in note 255) for a somewhat different view (p. 1209).

265. See H. Thieme in *Nature*, vol. 385, pp. 807–810, 1997; and commentary by R. Dennell on pp. 767–768.

266. A useful overview concerning the difficulties in establishing the use of fire by ancient hominids is given by S. J. James in *Current Anthropology*, vol. 30, pp. 1–26, 1989. It is important also to distinguish between levels of utility in fire use, and not to assume that fire indicates spit-roasts, gravy pans, and a neat stack of logs.

267. Evidence from the famous Swartkrans Cave in South Africa is used by C. K. Brain and A. Sillen in *Nature*, vol. 336, pp. 464–466, 1988 for use of fire as far back as 1–1.5 Ma. Similarly R. V. Bellomo, in *Journal of Human Evolution*, vol. 27, pp. 173–195, 1994, identifies the controlled use of fire by hominids from 1.6 Ma sediments in Kenya, but notes that they 'did not use fire for the purpose of hunting, cooking, preserving food, intentional plant selection, vegetation clearing, or improving the flaking characteristics of lithic materials ... [but] primarily as a source of protection against predators, as a source of light, and/or a source of heat' (p. 173). However, S. Weiner *et al.* argue against the controlled use of fire by the half-million-year-old *Homo erectus* inhabitants of the Zhoukoudian caves near Beijing in *Science*, vol. 281, pp. 251–253, 1998; see also the accompanying commentary by B. Wuethrich on pp. 165–166, as well as the interesting article by N. T. Boaz and R. L. Ciochon in *Natural History*, vol. 110(2), pp. 46–51, 2001.

268. See A. Walker *et al.* in *Nature* vol. 296, pp. 248–250, 1982. An accessible account is also given in *The wisdom of the bones: In search of human origins* by A. Walker and P. Shipman (Knopf, New York, 1996); see pp. 158–167.

269. O. Bar-Yosef, in *Cambridge Archaeological Journal*, vol. 8, pp. 141–163; 1998, provides a useful overview. The best evidence for this dramatic event is from Europe, but its origins were possibly in either East Africa or the Levant, unless these are independent?

270. These, of course, are almost entirely lithic, but rare, and to some extent enigmatic, use of bone tools is also known; see S. Gaudzinski in *Journal of Archaeological Science*, vol. 26, pp. 125–141, 1999.

271. See the report by I. Turk *et al.* report (in *L'Anthropologie*, vol. 101, pp. 531–540, 1997) of a possible flute from a Mousterian culture in Slovenia dated at c. 45 000 years BP. A common

explanation, that the perforations, in this case in the femur of a cave bear, are the product of a chomping carnivore, are dismissed by these researchers; see also M. Otte in *Current Anthropology*, vol. 41, pp. 271–272, 2000, who supports the musical interpretation.

272. P. M. Gray *et al.* (2001; citation is in note 171, Chapter 8), p. 53. Recall also that the parallels in the songs of birds and humpback whales suggest convergence, and possibly access to a universal music.

273. Evidence for rather advanced cultures, including the manufacture of bone points, and possibly dating back to *c.* 90 000 years BP, from Katanda in Zaire is reported by A. S. Brooks *et al.* and J. F. Yellen *et al.* in *Science*, vol. 268, pp. 548–553 and pp. 553–556 respectively, 1995 (see also accompanying commentary by A. Gibbons on pp. 495–496, which casts some doubts on the reliability of these dates). Certainly the appearance of some of the artefacts, such as the well-crafted harpoon points, is astonishingly modern, and J. E. Yellen (in *African Archaeological Review*, vol. 15, pp. 173–198, 1998) argues for an effective continuity in an African context, albeit not in terms of a specific culture.

274. See the lengthy arguments by S. McBrearty and A. S. Brooks in *Journal of Human Evolution*, vol. 39, pp. 453–563, 2000. See also the overview by R. Foley and M. M. Lahr in *Cambridge Archaeological Journal*, vol. 7, pp. 3–36, 1997, with an emphasis on the so-called Mode 3 (*c.* 250 000 years ago) technology as a key step in hominid behaviour.

275. Sites in southern Africa, such as Blombos Cave, located east of Cape Town, are of particular importance in providing evidence for the emergence of modern human behaviour, and by implication language, *c.* 70 000 years ago. See C. S. Henshilwood *et al.* in *Journal of Human Evolution*, vol. 41, pp. 631–678, 2001.

276. A. Marshack in *Current Anthropology*, vol. 37, pp. 357–364, 1996, describes an intriguing find from the Quneitra Mousterian site on the Golan Heights, consisting of a flat flint cortex with regular incised lines, including nested semicircles. And its significance? Removing our preconceptions is practically impossible, but Marshack tentatively suggests the pattern represents a rainbow, but of 'spiritual' rather than pictorial significance.

277. See F. d'Errico *et al.* in *Antiquity*, vol. 75, pp. 309–315, 2001). These authors argue that the striations were not the result of butchery, but rather were constructed as a series of intentional strokes.

278. See C. S. Henshilwood *et al.* in *Science*, vol. 295, pp. 1278–1280, 2002.

279. Excellent introductions are: *The Neandertals: Changing the image of mankind* by E. Trinkaus and P. Shipman (Pimlico, London, 1994); *The Neanderthal legacy: an archaeological perspective from western Europe* by P. Mellars (Princeton University Press, Princeton, 1996); and *In search of the Neanderthals: solving the puzzle of human origins* by C. Stringer and C. Gamble (Thames and Hudson, London, 1993).

280. See, for example, J. H. Schwartz and I. Tattersall, in *Proceedings of the National Academy of Sciences, USA*, vol. 93, pp. 10852–10854, 1996, on a peculiarity of the nasal structure in the form of a prominent bony projection extending into the cavity; a commentary by J. T. Laitman *et al.* on pp. 10543–10545, focuses on implications of air-flow through the nose, and what this might tell us about respiratory adaptations and vocalizations.

281. See M. S. Ponce de Leon and C. P. E. Zollikofer in *Nature*, vol. 412, pp. 534–538, 2001; as well as remarks on cranial form by D. E. Liebermann *et al.* in *Proceedings of the National Academy of Sciences, USA*, vol. 99, pp. 1134–1139, 2002.

282. See M. Krings *et al.* in *Proceedings of the National Academy of Sciences, USA*, vol. 96, pp. 5581–5585, 1999, and *Nature Genetics*, vol. 26, pp. 144–146, 2000, as well as I. V. Ovchinnikov *et al.* in *Nature*, vol. 404, pp. 490–493, 2000, and R. W. Schmitz *et al.* in *Proceedings of the National Academy of Sciences, USA*, vol. 99, pp. 13342–13347, 2002. A somewhat critical review of the present state of play, however, is offered by N. Caldararo and S. Gabow in *Ancient Biomolecules*, vol. 3, pp. 135–158, 2000, and G. Gutiérrez *et al.* in *Molecular Biology and Evolution*, vol. 19, pp. 1359–1366, 2002. Rob Foley, who agrees that Neanderthals are probably a separate species, reminds me that the DNA evidence is equivocal

inasmuch as the comparison refers to that of modern human genetic variation, which is remarkably limited and probably reflects an evolution bottleneck ('mitochondrial Eve'), c. 150 000 years ago.

283. C. W. Marean in *Journal of Human Evolution*, vol. 35, pp. 111–136, 1998 questions the widely held view that these hominids were principally scavengers, a view broadly shared by J. J. Shea, who argues in *Current Anthropology*, vol. 39 (Supplement), S45–S78, 1998, that the Neanderthals had a variety of hunting strategies but were possibly greater carnivores than modern humans, albeit replying on close-quarter intercepts (see also note 285). Their role as top-level carnivores is reinforced by isotopic study of their bones, especially with respect to nitrogen (δ^{15}N), see M. P. Richards *et al.* in *Proceedings of the National Academy of Sciences, USA*, vol. 97, pp. 7663–7666, 2000; see also H. Bocheren *et al.* in *Journal of Archaeological Science*, vol. 26, pp. 599–607, 1999, and V. Balter in *Comptes Rendus de l'Académie des Sciences, Paris: Sciences de la Terre et des Planètes*, vol. 332, pp. 59–65, 2001.

284. A. Defleur *et al.* present convincing evidence from a site in southern France, reported in *Science*, vol. 286, pp. 128–131, 1999; see also the accompanying commentary on pp. 18–19 by E. Culotta.

285. For an engaging account of the excavations at Shanidar, see R. S. Solecki's *Shanidar: The humanity of Neanderthal man* (Penguin [Allen Lane], London, 1972), as well as the more technical account in E. Trinkaus's *The Shanidar Neandertals* (Academic Press, New York, 1983).

286. See E. Trinkaus and M. R. Zimmerman in *American Journal of Physical Anthropology*, vol. 57, pp. 61–76, 1982. Evidence for care in more primitive Neanderthals, dating back to about 170 000 years, is given by S. Lebel *et al.* in *Proceedings of the National Academy of Sciences, USA*, vol. 98, pp. 11097–11102, 2001, where the discovery of a mandible with evidence for an abscess may indicate preparation of softer food for the sufferer, a reasonable assumption given that Neanderthal teeth typically show extensive wear suggesting heavy-duty use in daily activities.

287. See T. D. Berger and E. Trinkaus in *Journal of Archaeological Science*, vol. 22, pp. 841–852, 1995.

288. T. D. Berger and E. Trinkaus (1995; citation is in note 287); both quotations are on p. 849.

289. For evidence of interpersonal violence and the likelihood of subsequent assistance, see C. P. E. Zollikofer *et al.* in *Proceedings of the National Academy of Sciences, USA*, vol. 99, pp. 6444–6448, 2002.

290. The paper, whose full title is 'Grave shortcomings: the evidence for Neandertal burial', by R. H. Gargett in *Current Anthropology*, vol. 30, pp. 157–190, 1989. Gargett remounts his sceptical attack in *Journal of Human Evolution*, vol. 37, pp. 27–90, 1999. See also note 291.

291. The claim for flowers in a Neanderthal burial in the Shanidar caves in Iraq is made by A. Leroi-Gourhan in *Science*, vol. 190, pp. 562–565, 1975, but the evidence is based on the abundance of flower pollen in the associated sediments, for which there could be alternative explanations. The description of the excavation of the skeleton (Shanidar IV) is well described in R. S. Solecki (1972; citation is in note 285), see pp. 173–178. Scepticism of this claim can be found in Gargett (1989), p. 176, although in the discussion Leroi-Gourhan provides a short response (p. 182), to which Gargett replies on p. 185.

292. B. Vandermeersch in *Comptes Rendus de l'Académie des Sciences, Paris D*, vol. 270, pp. 298–301, 1970, however, interprets deer bones in association with a Neanderthal child at Qafzeh as evidence of grave goods.

293. This includes the famous Kebara burial in Israel, see O. Bar-Yosef *et al.* in *Current Anthropology*, vol. 27, pp. 63–64, 1986, and H. Valladas *et al.* in *Nature*, vol. 330, pp. 159–160, 1987. The latter paper provides an estimated age of c. 60 000 years. Other reports include those by L. V. Golovanova *et al.* from the Mezmaiskaya cave in the northern Caucasus, reported in *Current Anthropology*, vol. 40, pp. 77–86, 1999.

294. The hyoid is located in front of the larynx, and muscles attached to it help to control the shape of this part of the vocal tract and thereby sound production. The Neanderthal hyoid is

very similar to our own, leading some, e.g. A. B. Arensburg *et al.* in *American Journal of Physical Anthropology*, vol. 83, pp. 137–146, 1990, to argue that their speech was effectively the same as ours. In addition to the hyoid bone, much has been made of the so-called 'descended larynx' and its importance in the production of articulate speech. Interestingly, this feature, thought to be unique to humans, is now known to occur in such mammals as the red deer, famous for the stag's roaring, and as such is another example of independent evolution; see W. T. Fitch and D. Reby in *Proceedings of the Royal Society of London B*, vol. 268, pp. 1669–1675, 2001. This is not to say that one day deer will speak, but as Fitch and Reby point out it is possible that originally the human descended larynx was associated with sexual display, specifically so-called size exaggeration (the bull-roarer effect), even though we now associated this feature with the pleasures of hearing the male baritone.

295. See for example D. Lieberman *et al.* in *Journal of Human Evolution*, vol. 23, pp. 447–467, 1992. For an accessible and more popular introduction to the topic, *inter alia*, of Neanderthal fluency, see P. Lieberman's *Eve spoke: Human language and human evolution* (Macmillan [Picador], London, 1998).

296. See R. F. Kay *et al.* in *Proceedings of the National Academy of Sciences, USA*, vol. 95, pp. 5417–5419, 1998.

297. See A. M. MacLarnon and G. P. Hewitt in *American Journal of Physical Anthropology*, vol. 109, pp. 341–363, 1999.

298. That may be true to the first approximation, but still requires qualification. Marshack (1996; citation is in note 276), for example, draws attention to the example of a Neanderthal carving of 'an exquisite nonutilitarian oval plaque from a lamella of a compound mammoth molar at Tata, Hungary, dated ca. 100,000 B.P.' (p. 361); see also the article in *Yearbook of Physical Anthropology*, vol. 32, pp. 1–34, 1989.

299. While taking a somewhat different slant on various matters the review on the origins of the Aurignacian by S. E. Churchill and F. H. Smith in *Yearbook of Physical Anthropology*, vol. 43, pp. 61–115, 2000 is of great value. In addition, what appear to be shell beads dating back to about 40 000 years have been described from caves near the Mediterranean coast of Turkey and Lebanon; see S. L. Kuhn *et al.* in *Proceedings of the National Academy of Sciences, USA*, vol. 98, pp. 7641–7646, 2001.

300. The youngest Neanderthals are dated at about 28 000 years BP, see F. H. Smith *et al.* in *Proceedings of the National Academy of Sciences, USA*, vol. 96, pp. 12281–12286, 1999.

301. For a helpful introduction see S. E. Churchill and F. H. Smith (2000; citation is in note 299), pp. 76–77.

302. These spectacular examples of Châtelperronian expertise are discussed by A. Leroi-Gourhan and A. Leroi-Gourhan in *Gallia Préhistoire*, vol. 7, pp. 1–64, 1965, and Marshack (1989; citation is in note 298); other occurrences from Lot and Dordogne are reviewed by J. Pelegrin in *Cahiers du Quaternaire* 20 (Technologie lithique le Châtelperronien de Roc-de-Combe (Lot) et de la Côte (Dordogne) (CNRS Editions, Paris, 1995)). Unequivocal evidence of the association of the Châtelperronian culture with Neandertal skeletal material is given by J-J. Hublin *et al.* in *Nature*, vol. 381, pp. 224–226, 1996.

303. See J. J. Hublin *et al.* in *Comptes Rendus d'Académie des Sciences, Paris*, vol. 321 (IIa), pp. 931–937, 1995. R. N. E. Barton *et al.* in *Antiquity*, vol. 73, pp. 13–23, 1999 provide a preliminary report on the Neanderthals from Gibraltar, with dates as young as *c.* 32 000 years BP, but still showing Mousterian technology. An overall review of the Iberian occurrences is also given by F. d'Errico *et al.* in *Current Anthropology*, vol. 39 (Supplement), 1998, pp. S19–S20 in the context of the identification of the Châtelperronian cultures. It is evident that in Iberia the newly arriving Upper Palaeolithic industries of *Homo sapiens* made little, if any impact, and d'Errico *et al.* use this finding to support their thesis (citation is in note 308) that elsewhere in Europe the shift by Neanderthals to the Châtelperronian cultures was independent of our species' technological progression. In passing one should also remember that occasional Châtelperronian artefacts have been found in Iberia, and as Hublin *et al.*

(1995) remark 'The occurrence of these artefacts is still puzzling' (p. 934); perhaps Neanderthal trade?

304. As suggested, for example, by R. White (in *Annual Review of Anthropology*, vol. 21, pp. 537–564, 1992).

305. In his response as part of the discussion concerning Neanderthal acculturation in *Current Anthropology*, vol. 39 (Supplement), pp. S1–S44, 1998, where Mellars remarks, 'no one has ever suggested that the copying of airplane forms in the New Guinea cargo cults implied a knowledge of aeronautics or international travel' (p. S26), a point he reiterates in a continuation of the debate in a subsequent issue, vol. 40, pp. 341–364, 1999; see p. 350.

306. For an engaging and intriguing account of the John Frum cult in Tanna see Chapter 13 of David Attenborough's *Quest in paradise* (Lutterworth, London, 1960), with an illustration of a cultic turboprop opposite p. 154. A useful overview of cargo cults is given by I. C. Jarvie on pp. 133–137 of *Encyclopedia of Papua and New Guinea* (Melbourne University Press, Melbourne, 1972).

307. From my chapter (pp. 329–347) in *God and design: The teleological argument and modern science*, edited by N. A. Manson (Routledge, London, 2003), p. 340.

308. The key paper is by J. Zilhâo and F. d'Errico in *Journal of World Prehistory*, vol. 13, pp. 1–60, 1999; see also the short article by J. Zilhâo in the July/August 2000 issue of *Archaeology* on pp. 24–31.

309. See, for example, the extended debates in two issues of *Current Anthropology*, vol. 39 (Supplement), pp. S1–S44, 1998 and vol. 40, pp. 341–364, 1999. See also D. Richter *et al.* in *Journal of Archaeological Science*, vol. 27, pp. 71–89, 2000 (see pp. 84–86), and F. B. Harrold in *Journal of Anthropological Research*, vol. 56, pp. 59–75, 2000, who provides a historical overview (see especially pp. 68–69).

310. See E. Trinkhaus *et al.* in *Journal of Archaeological Science*, vol. 26, pp. 753–773, 1999, who suggest that the massive proportions of the body are a reflection of a continuing existence in a near-Arctic environment rather than a phylogenetic 'baggage'.

311. This point is emphasized by d'Errico *et al.* in *Current Anthropology*, vol. 39 (Supplement), 1998, see pp. S11–S13; see also note 314.

312. See I. Karavanić and F. H. Smith in *Journal of Human Evolution*, vol. 34, pp. 223–248, 1998. Subsequently these authors, with colleagues (citation is in note 264) obtained new, and younger, dates. These indicated that in this region of Europe (Croatia) Neanderthals and modern humans had not overlapped. Moreover, re-dating of an associated bone of a modern human (a mere 5000 years BP) together with consideration of other dates, suggest that modern humans arrived even later (*c.* 32 000 years BP) in Europe than thought. While this does not rule out trade, it certainly strengthens the case for cultural independence.

313. A view in support of Neanderthal genetic assimilation into European populations is given by S. E. Churchill and F. H. Smith (2000; citation is in note 299).

314. D. Kaufman weighs up the possibilities of contacts between *H. sapiens* and Neanderthals, especially in the Levant, in *Oxford Journal of Archaeology*, vol. 20, pp. 219–240, 2001. He notes that the Châtelperronian culture has certain hallmarks such as abundant use of colouring and particular types of incised stones and pendants that were possibly adopted by our ancestors.

315. S. E. Churchill and F. H. Smith (2000; citation is in note 299), pp. 105–106.

316. What L. G. Straus tartly refers to as 'A review of reality' concerning the likelihood that Solutrean people crossed to North America is reviewed in *American Antiquity*, vol. 65, pp. 219–226, 2000.

317. This strange occurrence from Tecaxic-Calixtlahuaca has been known for many years (see the chapter (pp. 5–53) by S. C. Jeff in *Man across the sea: Problems of Pre-Columbian contacts*, edited by C. L. Riley *et al.* (University of Texas Press, Austin, 1971)), and was brought back to prominence by R. Hristov and S. Genovés, in *Ancient Mesoamerica*, vol. 10, pp. 207–213, 1999. Questions of the dating, using the techniques of thermoluminescence (TL), were

subsequently raised by P. Schaaf and G. A. Wagner in the same journal (vol. 12, pp. 79–81, 2001), to which the original authors robustly reply (pp. 83–86). The principal point is that the TL date is of relatively little use in establishing the authenticity of this find, but it still does not support a colonial, i.e. a post 1492 AD date. Hristov and Genovés emphatically reject the suggestion that the head was 'planted' and remind us that, given that the Canary Islands were colonized by at least the fifth century BC, a chance crossing of the Atlantic more than a thousand years in advance of Columbus is not inconceivable.

318. Interestingly, this is not the only possible evidence for transatlantic pre-Columbian contact (excluding of course the well-known Viking forays). Jeff (1971) reviews a number of other possible examples. The whole question of cultural contacts, 'Diffusion versus independent development' to use Jeff's chapter title, is, of course, relevant to cultural convergences. Do the many striking similarities found between far-flung societies indicate independent innovation, which in turn might reflect a 'limitation of possibilities' (p. 32), or the sharing of ideas?

319. In his justly famous introduction to *English literature in the sixteenth century excluding drama* (Clarendon Press, Oxford, 1954), C. S. Lewis recalls how the great essayist Montaigne 'passionately asks why so noble a discovery [of the Americas] could not have fallen to the Ancients who might have spread civility where we have spread only corruption' (p. 17).

10. EVOLUTION BOUND: THE UBIQUITY OF CONVERGENCE

1. Now is the time to avoid that old chestnut of whether it is convergent evolution as against parallel evolution. As W. R. Scott, in his prescient discussion of mammal evolution (in *Journal of Morphology*, vol. 5, pp. 301–406, 1891) remarked 'The problems of parallelism and convergence ... open up a discussion of far-reaching extent and importance ... The distinction between the two classes of phenomena is obviously one of degree rather than of kind, and it will therefore be convenient to consider them together' (p. 363). For a level-headed comparison of convergent and parallel evolution based on taxonomy, see J. Kaster and J. Berger in *BioSystems*, vol. 9, pp. 195–200, 1977. C. A. Lockwood and J. G. Fleagle (in *Yearbook of Physical Anthropology*, vol. 42, pp. 189–232, 1999) offer a valuable review of this topic.

2. R. Bieri (1964, citation is in note 30, Chapter 9), p. 457.

3. Within the arthropods themselves arthropodization from a soft-bodied onychophore-like ancestor may have occurred twice; see G. E. Budd in *Transactions of the Royal Society of Edinburgh: Earth Sciences*, vol. 89, pp. 249–290, 1999; see p. 286; see also G. Fryer in *Biological Journal of the Linnean Society*, vol. 58, pp. 1–55, 1996. Insects may be convergent; see F. Nardi *et al.* in *Science*, vol. 299, pp. 1887–1889, 2002. Arthropods belong to a superclade known as the Ecdysozoa, of which the other principal phyla are the priapulids and nematodes. The rather obscure kinorhynchs, related to the priapulids have also developed arthropodization, presumably independently. Finally an extinct group of Cambrian fossils, the vetulicolians that are believed to be related to the chordates, show what may be a trend towards a sort of arthropodization; see D. G. Shu *et al.* in *Nature*, vol. 414, pp. 419–424, 2001 (and commentary by H. Gee on pp. 407, 409).

4. See, for example, W. D. I. Rolfe's paper on pp. 117–152 of *The terrestrial environment and the origin of land vertebrates*, edited by A. L. Panchen (Academic Press, London, 1980), where he writes '*Hibbertopterus* ... has short, stubby limb segments and is noticeably hexapodous: it would seem well adapted for movement on land despite its bulky opithosoma,' p. 145.

5. Tracheal systems have evolved independently in winged insects, isopod crustaceans (wood-lice and their relatives), arachnids (spiders), and onychophorans; see p.14 of J. Moore and P. Willmer's outstanding paper on convergence in *Biological Reviews*, vol. 72, pp. 1–60, 1997. See also W. Ripper in *Zeitschrift für Wissenschaftliche Zoologie*, vol. 138, pp. 303–369, 1931, and G. Pritchard *et al.* in *Biological Journal of the Linnean Society*, vol. 49, pp. 31–44, 1993.

6. See C. J. Klok *et al.* in *Journal of Experimental Biology*, vol. 205, pp. 1019–1029, 2002.

7. See L. M. Meffert *et al.* in *Journal of Evolutionary Biology*, vol. 12, pp. 859–868, 1999. Their principal conclusion is also of wider interest, because it appears that the convergence overrides the founder-flush effect, whereby tiny populations pass through evolutionary 'bottlenecks' before rediversifying on the basis of a restricted genetic diversity.

8. See C. S. Henry *et al.* in *Evolution*, vol. 53, pp. 1165–1179, 1999, where they demonstrate 'songs that are strikingly similar' (p. 1165) in species of the green lacewing *Chrysoperla* in western North America and Kyrgyzstan in central Asia. The degree of similarity is such that North American representatives can be fooled by an Asian song.

9. For a possible example in the Hawaiian cricket *Laupala*, which like many groups in this mid-ocean archipelago demonstrates a major adaptive radiation, see K. L. Shaw in *Evolution*, vol. 50, pp. 237–255, 1996. Convergence of song in the more familiar field-cricket *Gryllus* is discussed by R. G. Harrison in *Evolution*, vol. 33, pp. 1009–1023, 1979; see also R. G. Harrison and S. M. Bogdanowicz (in *Journal of Evolutionary Biology*, vol. 8, pp. 209–232, 1995).

10. See R. Kusmierski *et al.* in *Proceedings of the Royal Society of London B*, vol. 264, pp. 307–313, 1997, with an emphasis on the lifting of phylogenetic constraints.

11. A helpful overview of the bowers is given by M. Hansell in *Bird nests and construction behaviour* (Cambridge University Press, Cambridge, 2000). Hansell draws attention (pp. 212–213) to the independent development of elaborate display and so-called court objects in other birds, as well as cichlid fish. In this context it is also interesting to note the positive correlation between bower complexity and brain size; see J. Madden in *Proceedings of the Royal Society of London B*, vol. 268, pp. 833–838, 2001.

12. See Chapter 9 of *The rise and fall of the third chimpanzee* (Radius, London, 1991).

13. See C. Sturmbauer *et al.* in *Proceedings of the National Academy of Sciences, USA*, vol. 93, pp. 10855–10857, 1996. They remark that 'adaptations to higher intertidal life, such as excellent vision, deep burrowing, rapid locomotion, and water retention preceded and allowed the rise of extensive periods of subaerial reproductive displays. Sexual selection on morphology and behavior may also have been directional, resulting in strikingly similar solutions irrespective of the phylogenetic position' (p. 10857).

14. See G. F. Striedter and R. G. Northcutt in *Brain, Behavior and Evolution*, vol. 38, pp. 177–189, 1991.

15. A. M. Paterson *et al.* in *Evolution*, vol. 49, pp. 974–989, 1995, take a positive, if not upbeat, view of the use of behavioural data in helping to establish bird phylogenies. So, too, do M. Kennedy *et al.* in *Animal Behaviour*, vol. 51, pp. 273–291, 1996, although some of their results are qualified by subsequent work; see Kennedy *et al.* in *Molecular Phylogenetics and Evolution*, vol. 17, pp. 345–359, 2000.

16. 'Why are some protein structures so common?' ask S. Govindarajan and R. A. Goldstein in *Proceedings of the National Academy of Sciences, USA*, vol. 93, pp. 3341–3345, 1996, and provide an answer in the context of highly optimizable structures.

17. See, for example, C-I. Bränden in *Current Opinion in Structural Biology*, vol. 1, pp. 978–983, 1991, as well as G. K. Farber in the same journal, vol. 3, pp. 409–412, 1993.

18. See, for example, the essay by J-F. Gibrat *et al.* in *Current Opinion in Structural Biology*, vol. 6, pp. 377–385, 1996.

19. See, for example, the review by T. Takagi in *Current Opinion in Structural Biology*, vol. 3, pp. 413–418, 1993. See also R. A. Watts *et al.* in *Proceedings of the National Academy of Sciences, USA*, vol. 98, pp. 10119–10124, 2001, who document a haemoglobin common to plants and various microorganisms. This protein possesses unique biochemical properties, and is absent from animals (and fungi).

20. See the paper by L. Moens *et al.* in *Molecular Biology and Evolution*, vol. 13, pp. 324–333, 1996.

21. See A. Pesce *et al.* in *EMBO Journal*, vol. 19, pp. 2424–2434, 2000, p. 2432.

22. See T. Burmester in *Journal of Comparative Physiology*, vol. 172B, pp. 95–107, 2002. Although the arthropod haemocyanin appears to have a single origin, there is a diversity of

sub-unit forms and evidently these have emerged by convergence; see J. Markl and H. Decker in *Advances in Comparative and Environmental Physiology*, vol. 13, pp. 325–376, 1992.

23. See C. P. Mangum in *Proceedings of the Biological Society of Washington*, vol. 103, pp. 235–247, 1990; K. E. van Holde and K. I. Miller in *Advances in Protein Chemistry*, vol. 47, pp. 1–81, 1995; and Burmester (2002). In their review K. E. van Holde *et al.* (in *Journal of Biological Chemistry*, vol. 276, pp. 15563–15566, 2001) acknowledge the differences between arthropod and molluscan haemocyanins, but suggest that despite their independent evolution they may both derive from a primitive copper protein (see also A. Volbeda and W. G. M. Hol in *Journal of Molecular Biology*, vol. 206, pp. 531–546, 1989).

24. For a review of the evolution of myoglobin see the paper by T. Suzuki and K. Imai in *Cellular and Molecular Life Sciences*, vol. 54, pp. 979–1004, 1998.

25. See, for example, the paper by S-G. Hou *et al.* in *Nature*, vol. 403, pp. 540–544, 2000, where they report the presence of myoglobin-like proteins in both the Archaea and Eubacteria that serve to monitor oxygen levels.

26. The report of myoglobin in cyanobacteria is by M. Potts *et al.* in *Science*, vol. 256, pp. 1690–1692, 1992. For its occurrence in the protistan *Tetrahymena*, and an overview of the earlier literature, see S. Korenaga *et al.* in *Biochimica et Cosmochimica Acta*, vol. 1543, pp. 131–145, 2000. Korenaga *et al.* conclude 'that the contracted or truncated globins from various types of unicellular organisms have a separate, distinct origin from conventional globins', p. 143.

27. See, for example, C. Busch in *Comparative Biochemistry and Physiology*, vol. 86A, pp. 461–463, 1987. A. J. Lechner (in *Journal of Applied Physiology*, vol. 41, pp. 168–173, 1976) also documents high levels of muscle myoglobin in the burrowing pocket gophers, and remarks on the parallels with the deep-diving mammals.

28. See K. A. Joysey *et al.* on pp. 167–178 of *Myoglobin*, edited by A. G. Schnek and C. Vandecasserie (Editions de l'Université de Bruxelles, Brussels, 1977); A. M. Gurnett *et al.* in *Journal of Protein Chemistry*, vol. 3, pp. 445–454, 1984; and K. A. Joysey on pp. 34–48 of *Molecular evolution and the fossil record; Short Courses in Paleontology 1*, edited by B. Runnegar and J. W. Schopf (Paleontological Society, Knoxville, TN, 1988). These data are preliminary, but we should note Joysey's (1988) prescient remark, 'it is my view that many other examples of adaptation in molecular evolution will emerge in due course', p. 47.

29. See T. Suzuki and T. Takagi in *Journal of Molecular Biology*, vol. 228, pp. 698–700, 1992; T. Suzuki and K. Imai in *Comparative Biochemistry and Physiology*, vol. 117B, pp. 599–604, 1997; and the review by T. Suzuki *et al.* in *Comparative Biochemistry and Physiology B*, vol. 121, pp. 117–128, 1998.

30. See T. Shimizu *et al.* in *Journal of Biological Chemistry*, vol. 253, pp. 4700–4706, 1978.

31. See F. Hirata *et al.* in *Journal of Biological Chemistry*, vol. 252, pp. 4637–4642, 1977.

32. The details of the gene structure of the IDO are reviewed by T. Suzuki *et al.* in *Biochimica et Biophysica Acta*, vol. 1308, pp. 41–48, 1996. They posit a gene duplication in the IDO gene that allowed the protein to be recruited for oxygen carrying.

33. See several papers by T. Suzuki *et al.* in *Journal of Protein Chemistry*, vol. 14, pp. 9–13, 1994, and vol. 17, pp. 651–656 and 817–826, 1998 respectively.

34. See N. Maeda and W. M. Fitch in *Journal of Biological Chemistry*, vol. 257, pp. 2806–2815, 1982. See also T. J. Grove and B. D. Sidell in *Canadian Journal of Zoology*, vol. 80, pp. 893–901, 2002.

35. The knock-out experiments in mice are described by D. J. Garry *et al.* in *Nature*, vol. 395, pp. 905–908, 1998.

36. See J. Gatesy *et al.* in *Science*, vol. 291, pp. 2603–2605, 2001.

37. See, for example, K. B. Storey and J. M. Storey in *Annual Review of Physiology*, vol. 54, pp. 619–637, 1992.

38. Notothenioid fish show spectacular evidence for an adaptive radiation, as reviewed by J. Montgomery and K. Clements in *Trends in Ecology & Evolution*, vol. 15, pp. 267–271, 2000.

These authors also touch on such topics as loss of haemoglobin and myoglobin (see above p. 289).

39. See L. Chen *et al.* in *Proceedings of the National Academy of Sciences, USA*, vol. 94, pp. 3817–3822, 1997. This example of molecular convergence has received justifiably wide attention, and J. M. Logsdon and W. F. Doolittle provide a thoughtful commentary of what they describe as 'a cool tale', on pp. 3485–3487.

40. Details of this gene are given by L. Chen *et al.* in the paper preceding L. Chen *et al.* (1997), on pp. 3811–3816. Further information on the glycoprotein itself is provided by A. N. Lane *et al.* in *Biophysics Journal*, vol. 78, pp. 3195–3207, 2000.

41. Further details of the evolutionary steps from a trypsinogen-like protein to one with an antifreeze function are given by C. H. C. Cheng and L. Chen in *Nature*, vol. 401, pp. 443–444, 1999.

42. G. L. Fletcher *et al.* in *Annual Review of Physiology*, vol. 63, pp. 359–390, 2001. Fletcher *et al.* also suggest another example of molecular convergence among the antifreeze proteins, this time involving C-type lectins in sea raven, herring, and smelt.

43. See P. L. Davies *et al.* in *Philosophical Transactions of the Royal Society of London B*, vol. 357, pp. 927–935, 2002. As part of a general review Davies *et al.* draw attention to a striking convergence in spruce budworm and mealworm beetle antifreeze proteins, specifically at the ice-binding sites. Concerning antifreeze proteins in insects, with comments on the equivalents in the fish, see also N. Li *et al.* in *Journal of Experimental Biology*, vol. 201, pp. 2243–2251, 1998.

44. See W. J. Swanson and C. F. Aquadro in *Journal of Molecular Evolution*, vol. 54, pp. 403–410, 2002.

45. See M. Pitts and M. Roberts's *Fairweather Eden: Life in Britain half a million years ago as revealed by the excavations at Boxgrove* (Century, London, 1997).

46. See, for example, the papers by J. R. Petit *et al.* and K. M. Cuffey and F. Vimeux in *Nature*, vol. 399, pp. 429–436, 1999 and vol. 412, pp. 523–527, 2001 respectively.

47. See, for example, the papers by S. Bains *et al.* in *Science*, vol. 285, pp. 724–727, 1999; G. R. Dickens *et al.* in *Geology*, vol. 25, pp. 259–262, 1997; and M. E. Katz *et al.* in *Science*, vol. 286, pp. 1531–1533, 1999. It is sobering to realize that the recovery time of the planet after the LPTM was about 100 000 years. Massive methane release has also been implicated in Quaternary events (see J. P. Kennett *et al.* in *Science*, vol. 288, pp. 128–133, 2000 (with commentary by T. Blunier on pp. 68–69)) and the end-Cretaceous firestorm (see M. D. Max *et al.* in *Geo-Marine Letters*, vol. 18, pp. 285–291, 1999).

48. See C_4 *plant biology*, edited by R. F. Sage and R. K. Monson (Academic Press, San Diego, 1999), especially the introductory chapter (pp. 3–16) by the senior editor, and the overview by R. F. Sage in *Plant Biology*, vol. 3, pp. 203–213, 2001.

49. The expansion of C_4 ecosystems is documented by T. E. Cerling *et al.* in *Nature*, vol. 361, pp. 344–345, 1993 and vol. 389, pp. 153–158, 1997, and *Journal of Vertebrate Paleontology*, vol. 16, pp. 103–115, 1996.

50. See R. F. Sage (2001; citation is in note 48), and Y. Huang *et al.* in *Science*, vol. 293, pp. 1647–1651, 2001, with commentary by R. A. Kerr on pp. 1572–1573.

51. See N. R. Sinha and E. A. Kellogg, in *American Journal of Botany*, vol. 83, pp. 1458–1470, 1996, where they remark, 'Such complexity should be difficult to evolve, yet the [C_4] pathway has evolved multiple times in the history of the flowering plants' (p. 1458). See also Kellogg's chapter on pp. 411–444 in R. F. Sage and R. K. Monson (1999; citation is in note 48).

52. See E. V. Voznesenskaya *et al.*, in *Nature*, vol. 414, pp. 543–546, 2001; and H. Freitag and W. Stichler in *Plant Biology*, vol. 2, pp. 154–160, 2000, as well as an overview by R. F. Sage in *Trends in Plant Science*, vol. 7, pp. 283–285, 2002.

53. See the chapter by R. K. Monson, on pp. 377–410 of R. F. Sage and R. K. Monson (1999; citation is in note 48). Some of the necessary antecedents to C_4 photosynthesis are discussed by J. Hibberd and W. P. Quick in *Nature*, vol. 415, pp. 451–453, 2002 (with commentary by J. A. Raven on pp. 375 and 377).

54. See R. F. Sage (2001; citation is in note 48).

55. See J. R. Reinfelder *et al.* in *Nature*, vol. 407, pp. 996–999, 2000, as well as the preceding paper by U. Riebesell on pp. 959–960. Scepticism was expressed by A. M. Johnston *et al.* in a subsequent issue of *Nature* (vol. 412, pp. 40–41, 2001), to which Reinfelder offered a reply (p. 41).

56. See M. M. M. Kuypers *et al.* in *Nature*, vol. 399, pp. 342–345, 1999. They also note that increasing aridity may have played a part in this shift to C_4 photosynthesis (see also note 50).

57. See T. E. Cerling *et al.* (1997; citation is in note 49) where in passing the authors note that although the history of terrestrial plants has been dominated by C_3 photosynthesizers 'A possible exception to this could have been during the late Carboniferous to Permian glaciation if P_{CO_2} levels were low enough and some plants independently evolved the C_4 pathway, which was subsequently lost in the Mesozoic when CO_2 levels were again high', p. 157.

58. See N. R. Sinha and E. A. Kellogg (1996; citation is in note 51); see also L. M. Giussani *et al.* in *American Journal of Botany*, vol. 88, pp. 1993–2012, 2001.

59. See J. Kraut in *Annual Review of Biochemistry*, vol. 46, pp. 331–358, 1977, and I-M. Frick *et al.* in *Proceedings of the National Academy of Sciences USA*, vol. 89, pp. 8532–8536, 1992. Kraut notes the independent evolution of serine proteases in eukaryotes (e.g. trypsin) and bacteria (subtilisin) (see also D. L. Ollis *et al.* in *Protein Engineering*, vol. 5, pp. 197–211, 1992). Their function is to break the peptide and ester bonds of other compounds, so making them available for digestion. Concerning these two families Kraut remarks that 'the two enzyme families have entirely different overall three-dimensional structures and are therefore probably descended from unrelated ancestral enzymes. Thus nature appears to have invented the same biochemical mechanism on at least two separate occasions' (p. 332). In a similar vein Ollis *et al.* (1992) comment, 'It is remarkable that there are now four different examples of the catalytic triad which are related by convergent, not divergent evolution ... We believe that this reflects the primordial nature of hydrolysis ... It also reflects how central hydrolysis is to biochemical pathways, and how few solutions are possible ... to the problem of hydrolyzing esters and amides' (p. 210). See also R. M. Garavito *et al.* in *Biochemistry*, vol. 16, pp. 5065–5071, 1977, who also draw attention to convergences in terms of the active sites of various enzymes, including lactate dehydrogenase, glyceraldehyde-3-phosphate dehydrogenase, and papain.

60. N. D. Rawlings and A. J. Barrett in *Biochemical Journal*, vol. 290, pp. 205–218, 1993 remark that the peptidases that catalyse the hydrolysis of peptide bonds, could be represented by up to 60 evolutionary lines and that there may have been 'many separate evolutionary origins of peptidases,' p. 216.

61. See T. Terada *et al.* in *Nature Structural Biology*, vol. 9, pp. 257–262, 2002. The details concern two lysyl-tRNA synthetases which in an archaeal bacteria has a completely different protein architecture from the 'normal' synthetase, but converges in terms of strategies for substrate recognition.

62. See A. Beschin *et al.* in *Nature*, vol. 400, pp. 627–628, 1999.

63. See K. J. H. Robson *et al.* in *Nature*, vol. 335, pp. 79–82, 1988. R. F. Doolittle, in *Trends in Biochemical Sciences*, vol. 19, pp. 15–18, 1994, expressed some caution about this example, but conceded that if the sequences are 'truly unrelated, this case would have to rate as the nearest thing to sequence convergence yet reported', p. 18.

64. See Y. Cao *et al.* in *Journal of Molecular Evolution*, vol. 47, pp. 307–322, 1998.

65. See G. Wu *et al.* in *Proceedings of the National Academy of Sciences, USA*, vol. 96, pp. 6285–6290, 1999.

66. See S. Schenk and K. Decker in *Journal of Molecular Evolution*, vol. 48, pp. 178–186, 1999.

67. See S. J. Charnock *et al.* in *Proceedings of the National Academy of Sciences, USA*, vol. 99, pp. 12067–12072, 2002, a convergence that the authors regard as 'stunning'.

68. See J. La Roche *et al.* in *Proceedings of the National Academy of Sciences, USA*, vol. 93, pp. 15244–15248, 1996.
69. See P. Robson *et al.* in *Molecular Biology and Evolution*, vol. 17, pp. 1739–1752, 2000.
70. See Z. C. Shen and M. Jacobs-Lorena in *Journal of Molecular Evolution*, vol. 48, pp. 341–347, 1999.
71. See K. A. Crandall *et al.* in *Molecular Biology and Evolution*, vol. 16, pp. 372–382, 1999.
72. See K. H. Roux *et al.* in *Proceedings of the National Academy of Sciences, USA*, vol. 95, pp. 11804–11809, 1998.
73. See G. E. Schulz in *Current Opinion in Structural Biology*, vol. 2, pp. 61–67, 1992; see also the discussion of convergence among nucleic acid binding molecules by P. Graumann and M. A. Maraherl in *BioEssays*, vol. 18, pp. 309–315, 1996.
74. See J. Hodgkin in *Genes & Development*, vol. 16, pp. 2322–2326, 2002, forming a commentary on the paper by R. Lints and S. W. Emmons on pp. 2390–2407.
75. See C. S. Thummel and J. Chory in *Genes & Development*, vol. 16, pp. 3113–3129, 2002.
76. See E. Stebbins and J. E. Galán in *Nature*, vol. 412, pp. 701–705, 2001.
77. This is just one example, referring to the pathogen *Yersinia pseudotuberculosis*. In the case of another bacterium, *Salmonella*, not only does the pathogen use a molecular convergence to allow it to enter through the cell wall, but once inside the host cell the bacterium helps to repair the damage to protect its new niche, again using methods of molecular mimicry.
78. E. Stebbins and J. E. Galán (2001; citation is in note 76), p. 703.
79. E. Stebbins and J. E. Galán (2001; citation is in note 76), p. 705.
80. See Y. Nakamura in *Journal of Molecular Evolution*, vol. 53, pp. 282–289, 2001. He comments, 'What's so remarkable in molecular mimicry is the fact that the three proteins structurally known as a tRNA mimic possesses completely different protein folds with unrelated primary and secondary structures of protein' (p. 287). Protein-DNA mimics have also been recognized, see for example, the papers in *Cell* by D. C. Mol *et al.*, vol. 82, pp. 701–708, 1995 and D. Liu *et al.*, vol. 94, pp. 573–583, 1998.
81. See K. Salehi-Ashtiani and J. W. Szostak in *Nature*, vol. 414, pp. 82–84, 2001. (See also note 46, Chapter 4).
82. K. Salehi-Ashtiani and J. W. Szostak (2001; citation is in note 81), p. 84.
83. See, in particular, the article by Doolittle (1994; citation is in note 63), with the title 'Convergent evolution – the need to be explicit'.
84. See T. Bauchop and R. W. Martucci in *Science*, vol. 161, pp. 698–700, 1968. Ruminant-like digestion is also known in the macropod marsupials, e.g. kangaroos (see D. W. Dellow *et al.* in *Australian Journal of Zoology*, vol. 31, pp. 433–443, 1983), and possibly the sloths (see the chapter (pp. 329–359) by G. G. Montgomery and M. E. Sunquist in *The ecology of arboreal folivores*, edited by G. G. Montgomery (Smithsonian Institution Press, Washington, DC, 1978)). So far as I am aware there is no evidence whether the digestive system of at least the macropods shows the molecular convergence of lysozymes.
85. The hoatzin is evidently fairly closely related to the cuckoos; see S. B. Hedges *et al.* in *Proceedings of the National Academy of Sciences, USA*, vol. 92, pp. 11662–11665, 1995. E. S. Morton, in his chapter (pp. 123–130) in Montgomery (1978) also draws attention to various similarities between the hoatzin and New Zealand Owl Parrot, remarking 'Both species are highly convergent in ways directly attributable to their leaf-eating habits,' p. 125.
86. Concerning its fore-gut fermentation systems, see A. Grajal *et al.* in *Science*, vol. 245, pp. 1236–1238, 1989. The parallels with the system in ruminant mammals is emphasized by A. Grajal in *The Auk*, vol. 112, pp. 20–28, 1995, and his remark that 'Fore-gut fermentation in a 680-g flying endotherm is theoretically unexpected' (p. 26) is an important reminder of how principles of convergence leap both phylogenetic barriers and sometimes our expectations.
87. These similarities include functional wing claws in the juveniles that enable them to climb trees. It now seems that the limited flight abilities of the hoatzin are more to do with the need to accommodate the massive fermentation chambers at the expense of the flight muscles.

88. See C. B. Stewart and A. C. Wilson in *Cold Spring Harbor Symposia on Quantitative Biology*, vol. 52, pp. 891–899, 1987.

89. See J. R. Kornegay *et al.* in *Molecular Biology and Evolution*, vol. 11, pp. 921–928, 1994.

90. See R. F. Doolittle (1994; citation is in note 63).

91. See C. B. Stewart and A. C. Wilson (1987; citation is in note 88), and also further evidence for adaptive evolution of these langur monkey lysozymes in the paper by W. Messier and C. B. Stewart in *Nature*, vol. 385, pp. 151–154, 1997 (with a commentary by P. M. Sharp on pp. 111–112) and (with certain provisos) by Z. Yang in *Molecular Biology and Evolution*, vol. 15, pp. 568–573, 1998 (see also the subsequent paper by Z. Yang and R. Nielsen in the same journal, vol. 19, pp. 908–917, 2002). In addition, evidence exists in the langur monkey for adaptive evolution in the ribonuclease enzyme, produced in the pancreas and involved with digestion of the fore-gut bacteria. See J. Zhang *et al.* in *Nature Genetics*, vol. 30, pp. 411–415, 2002, and commentary by S. Yokoyama on pp. 350–351. The evolution of the enzyme is due to a few critical amino acid substitutions that allow the ribonuclease to process large quantities of bacterial RNA in a lower pH environment. It will be interesting to see if parallel substitutions occur in other ruminant species, in a way analogous to the 'five site rule' of colour vision (note 168, Chapter 7).

92. See, for example, N. G. C. Smith and A. Eyre-Walker and J. C. Fay *et al.* in *Nature*, vol. 415, pp. 1022–1024 and pp. 1024–1026 respectively, 2002, as well as overviews by Z. Yang and J. P. Bielawksi in *Trends in Ecology & Evolution*, vol. 15, pp. 496–507, 2000 and M. Kreitman and H. Akashi in *Annual Review of Ecology and Systematics*, vol. 26, pp. 402–422, 1995.

93. See M. Harry *et al.* in *Molecular Phylogenetics and Evolution*, vol. 9, pp. 542–551, 1998.

94. See T. B. Patterson and T. J. Givnish in *Evolution*, vol. 56, pp. 233–252, 2002.

95. Their paper is published in *American Naturalist*, vol. 146, pp. 349–364, 1995, where they specifically address the topic of seed size.

96. See S. J. Gould and R. C. Lewontin in *Proceedings of the Royal Society of London B*, vol. 205, pp. 581–598, 1979.

97. Not that every example of convergence is necessarily adaptive; see for example the important paper by D. B. Wake in *American Naturalist*, vol. 138, pp. 543–567, 1991.

98. See Kirk Winemiller's paper in *Ecological Monographs*, vol. 61, pp. 343–365, 1991; See also note 134; Chapter 6.

99. K. Winemiller (1991; citation is in note 98), p. 361.

100. See Rob Foley's chapter *Pattern and process in hominid evolution*, on pp. 31–42 of *Structure and contingency: Evolutionary processes in life and human society*, edited by J. Bintliff (Leicester University Press, London, 1999).

101. R. Foley (1999; citation is in note 100), p. 40. Oddly, in the same volume S. J. Gould writes of Foley's work that it 'suggests a much chancier, much less guaranteed, much less repeatable story replete with dominating contingency,' p. xx. Did Gould actually read what Foley wrote, I wonder?

102. See Eviator Nevo's *Mosaic evolution of subterranean mammals: Regression, progression and global convergence* (Oxford University Press, Oxford, 1999).

103. E. Nevo (1999), p. 208.

104. Notably his *The blind watchmaker: why the evidence of evolution reveals a universe without design* (Norton, New York, 1987). Given Dawkins's emphasis on adaptation and a specific section on convergence, it is not surprising that a number of the examples given here are also discussed in his book.

105. See for example, the papers by B. D. Patterson (in *Journal of Mammalogy*, vol. 80, pp. 345–360, 1999) and S. B. Emerson (in *Biological Journal of the Linnean Society*, vol. 73, pp. 139–151, 2001).

106. See, for example, G. Balavoine in *Compte Rendu d'Académie des Sciences, Paris, Science de la Vie*, vol. 320, pp. 83–94, 1997. It should be noted, however, that the acoels are probably

genuinely primitive within the triploblastic metazoans; see I. Ruiz-Trillo *et al.* in *Proceedings of the National Academy of Sciences, USA*, vol. 99, pp. 11246–11251, 2002.

107. See the paper in *Nature*, vol. 401, p. 762, 1999, by M. Kobayashi *et al.*

108. See note 113, chapter 7.

109. See, for example, M. H. Tai *et al.* in *Proceedings of the National Academy of Sciences, USA*, vol. 90, pp. 1852–1856, 1993. Similarly, in reviewing the Wilms' tumour suppressor gene (WT1), N. D. Hastie (in *Cell*, vol. 106, pp. 391–394, 2001) noted that the two principal isoforms (−KTS, +KTS) have very distinct functions, but differ by only three amino acids.

110. See M. Rosenquist *et al.* in *Journal of Molecular Evolution*, vol. 51, pp. 446–458, 2000.

111. See the overview by G. B. Golding and A. M. Dean in *Molecular Biology and Evolution*, vol. 15, pp. 355–369, 1998, and more specifically papers by S. Brogna *et al.* in the same journal, vol. 18, pp. 322–329, 2001; Y-H. Lee and V. D. Vacquier in *Biological Bulletin*, vol. 182, pp. 97–104, 1992; W. J. Swanson and V. D. Vacquier in *Proceedings of the National Academy of Sciences, USA*, vol. 92, pp. 4957–4961, 1995 and *Science*, vol. 281, pp. 710–712, 1998; J. Vieira and B. Charlesworth in *Genetics*, vol. 155, pp. 1701–1709, 2000; J. Vieira *et al.* in *Genetics*, vol. 158, pp. 279–290, 2001; X. Gu *et al.* in *Genetica*, vol. 102/103, pp. 383–391, 1998; L. S. Jermiin *et al.* in *Molecular Biology and Evolution*, vol. 12, pp. 558–563, 1995; N. A. Singhania *et al.* in *Journal of Molecular Evolution*, vol. 49, pp. 721–728, 1999; C. S. Willett in *Molecular Biology and Evolution*, vol. 17, pp. 552–562, 2000. See also note 92.

112. See, for example, *Evolutionary trends*, edited by K. J. McNamara (Belhaven, London, 1990), where a variety of views are expressed, as well as specific examples such as those concerning the hinge mechanism of articulate brachiopods (S. J. Carlson in *Paleobiology*, vol. 18, pp. 344–366, 1992) and suture complexity in Palaeozoic ammonoids (W. B. Saunders *et al.* in *Science*, vol. 286, pp. 760–763, 1999). Also germane to this area is G. J. Vermeij's interesting essay in *American Naturalist*, vol. 153, pp. 243–253, 1999.

113. Janis and Damuth in K. J. McNamara (1990; citation is in note 112), p. 313.

114. Janis and Damuth in K. J. McNamara (1990; citation is in note 112), p. 337.

115. See his paper in *Journal of Paleontology*, vol. 62, pp. 319–329, 1988; see also S. C. Wang in *Evolution*, vol. 55, pp. 849–858, 2001 for an analysis of passive and driven evolutionary trends. A. H. Knoll and R. K. Bambach, in *Paleobiology*, vol. 26 (Supplement), pp. 1–14, 2000, offer an outstanding critique on this issue.

116. See his paper in *Nature*, vol. 385, pp. 250–252, 1997, as well as the chapter (pp. 256–289) in *Evolutionary biology: in honor of James W. Valentine*, edited by D. Jablonksi *et al.* (University of Chicago Press, Chicago, 1996).

117. See J. Trammer and A. Kaim in *Historical Biology*, vol. 13, pp. 113–125, 1999.

118. See J. Alroy in *Science*, vol. 280, pp. 731–734, 1998.

119. J. Alroy (1998; citation is in note 118), p. 732.

120. See B. A. Maurer in *Evolutionary Ecology*, vol. 12, pp. 925–934, 1998.

121. B. A. Maurer (1998; citation is in note 120), p. 925.

122. Their paper is published in *Evolution*, vol. 46, pp. 939–953, 1992.

123. Maurer *et al.*, p. 951; on p. 949 they argue that 'Contrary to Gould's (1988; citation is in note 115) assertion, more than random cladogenetic events are required to account for evolutionary trends such as Cope's rule.'

124. See Peter Wagner's paper in *Evolution*, vol. 54, pp. 365–386, 2000. Also directly relevant is the exploration by C. K. Boyce and A. H. Knoll (in *Paleobiology*, vol. 28, pp. 70–100, 2002) of plant leaf form and the exhaustion of potentiality, leading to widespread convergence.

125. P. Wagner (2000; citation is in note 124), p. 382.

126. See S. J. Gould's chapter (pp. 319–338) in *Evolutionary progress*, edited by M. H. Nitecki (University of Chicago Press, Chicago, 1988).

127. Or more specifically 'Progress is a noxious, culturally embedded, untestable, nonoperational, intractable idea that must be replaced if we wish to understand the patterns of history,'

p. 319. For a polite, short, and devastating demolition of this canard of Gould, see the letter by F. K. McKinney in *Science*, vol. 237, p. 575, 1987.

128. In this context see the discussion by D. H. Geary *et al.* (in *Paleobiology*, vol. 28, pp. 208–221, 2002) discussion of iterative evolution in gastropods, with a wide-ranging and intelligent analysis of the alternative explanations.

129. See, for example, G. B. West *et al.* in *Science*, vol. 284, pp. 1677–1679, 1999, and papers by K. J. Niklas and B. J. Enquist and J. H. Marden and L. R. Allen, in *Proceedings of the National Academy of Sciences, USA*, vol. 98, pp. 2922–2927; 2001 and vol. 99, pp. 4161–4166, 2002 respectively.

130. See also my paper in *Astronomical Society of the Pacific Conference Series*, (IAU Symposium 213) (in press).

131. See Lee Cronk *That complex whole: Culture and the evolution of human behaviour* (Westview, Boulder, CO, 1999).

132. L. Cronk (1999; citation is in note 131), p. 26.

133. The question of such 'navigation' is not, of course, new; see, for example, K. J. Niklas in *Proceedings of the National Academy of Sciences, USA*, vol. 91, pp. 6772–6779, 1994, and a series of papers by S. Gavrilets, e.g. in *Trends in Ecology & Evolution*, vol. 12, pp. 307–312, 1997; *American Naturalist*, vol. 154, pp. 1–22, 1999; and *Proceedings of the Royal Society of London B*, vol. 266, pp. 817–824, 1999.

134. L. Cronk (1999; citation is in note 131) invokes such an analogy when he writes, 'we are in search of the Great Attractor of human culture, the unseen mass that pulls human cultures toward it and so limits their diversity,' p. 26.

135. So L. Cronk (1999; citation is in note 131) writes 'Maybe the problem is that only certain pathways through ethnographic hyperspace are actually possible,' p. 26.

II. TOWARDS A THEOLOGY OF EVOLUTION?

1. The predicament is clear for all to see. As R. Helms in *Tolkien's world* (Thames and Hudson, London, 1974) remarks, 'From the end of the Middle Ages to the first nuclear explosion (to be overly precise) our deepest spiritual urges have been Faustian, directing our emotional and intellectual energies in an endless quest for knowledge of and power over nature, over our world. Now we have come like Sauron; we *can* control nature, but we find in the process that every controlling touch spoils and corrupts. Like Sauron, we can darken the sky, blast the vegetation, pervert and control even the minds of men; and again like Sauron, we remain the prisoners of our own assumptions, seeing no alternative to the constant expansion of our corrupting control,' p. 68.

2. The curious thing is that while animals, be they termites or cows, rely on symbionts to break down the refractory cellulose, a few animals (arthropods) actually possess enzymes (cellulases) that can break down cellulose. Most probably this ability was acquired by gene transfer from bacteria, and it would be interesting to know why this apparently has not happened more often; see H. Watanabe and G. Tokuda in *Cellular and Molecular Life Sciences*, vol. 58, pp. 1167–1178, 2001.

3. S. Garcia Vallvé *et al.* (in *Molecular Biology and Evolution*, vol. 17, pp. 352–361, 2000), however, report evidence for the lateral transfer of a gene involved with degradation of cellulose, from a bacterium to a fungus. Both these organisms occur in mammalian rumens, such as those of the cow.

4. See M. Carrier's chapter (pp. 83–97) in *Concepts, theories, and rationality in the biological sciences*, edited by G. Wolters and J. G. Lennox (Universitätsverlag and University of Pittsburgh Press, Konstanz and Pittsburgh, 1995).

5. M. Carrier (1995; citation is in note 4), p. 95.

6. See D. Duboule and A. Wilkins in *Trends in Genetics*, vol. 14, pp. 54–59, 1998.

7. See R. N. Brandon's essay on biological teleology in *Studies in the History and Philosophy of Science*, vol. 12, pp. 91–105, 1981.

8. *The blue cross*, the first story in *The innocence of Father Brown* (Cassell, London, 1940), p. 22.

9. See A. Peacocke's chapter (pp. 101–130) of *Darwinism and divinity: Essays on evolution and religious belief*, edited by J. Durant (Blackwell, Oxford, 1985).

10. A. Peacocke (1985; citation is in note 9), p. 123

11. An example comes from a review (published in the *Times Literary Supplement*, 24 December, 1999) by Philip Kitcher of Matt Ridley's book *Genome* where he remarks, 'Ridley obviously has a fine time sharing his delight [of the genome]. Indeed, perhaps he has too good a time. For the booming voice of conviction that sounds through the chapters, from the initial discussion of the origins of life to the philosophically limp conclusions about free will, is utterly certain about everything. Like the village squire to the Victorian parson with "doubts", Ridley prescribes fresh air and exercise. He seems quite unworried by the thought that some of the scientific claims he reports might be controversial or even unfounded, and even less disconcerted by the possibility that ... scientific truths might lead to social harm,' p. 24.

12. J. C. Greene's *Debating Darwin: Adventures of a scholar* (Regina Books, Claremont, 1999).

13. J. C. Greene (1999; citation is in note 12), p. 42.

14. J. C. Greene (1999; citation is in note 12), p. 43.

15. See, in particular, J. C. Greene (1999; citation is in note 12) and his essay 'Huxley to Huxley' in his *Science, ideology, and world view. Essays in the history of evolutionary ideas* (University of California Press, Berkeley, 1981).

16. Published by Penguin (London, 1998).

17. See my review in *Geological Magazine*, vol. 136, pp. 601–603, 1999.

18. Julian Huxley *Evolution in action: Based on the Patten Foundation Lectures delivered at Indiana University in 1951* (Chatto & Windus, London, 1953).

19. J. Huxley (1953; citation is in note 15), p.12; for its discussion in the context of process theology, see J. C. Greene (1981; citation is in note 15), pp. 164–165.

20. J. Huxley (1953; citation is in note 19), pp. 152–153.

21. J. C. Greene (1981; citation is in note 15); both quotations are on p. 165.

22. J. C. Greene (1981; citation is in note 15), p. 168; see also in note 1.

23. Daniel Gasman *The scientific origins of national socialism: Social Darwinism in Ernst Haeckel and the German Monist League* (Macdonald and American Elsevier, London and New York, 1971).

24. D. Gasman (1971; citation is in note 23), explanation to Plate I, facing p. 8.

25. The close connection between Haeckel's monistical phantasies and the rise of European fascism, including not only Germany but Italy and France, is returned to by D. Gasman in *Haeckel's Monism and the birth of fascist ideology*, Studies in modern European history, vol. 33, edited by F. J. Coppa (Peter Lang, New York, 1998).

26. Chapter 1 (pp. 9–32) of *Structure: In science and art*, edited by W. Pullan and H. Bhadeshia (Cambridge University Press, Cambridge, 2001).

27. D. Gasman (1971; citation is in note 23), p. 61.

28. S. Conway Morris (2001; citation is in note 26), pp. 30–31.

29. Concerning this trial see K. Tierney's *Darrow: A biography* (Crowell, New York, 1979) and E. J. Larson's *Summer for the gods: The Scopes trial and America's continuing debate over science and religion* (Harvard University Press, Cambridge, 1997).

30. K. Tierney (1979; citation is in note 29), p. 341.

31. K. Tierney (1979; citation is in note 29), p. 357.

32. K. Tierney (1979; citation is in note 29), p. 359.

33. K. Tierney (1979; citation is in note 29), p. 393.

34. K. Tierney (1979; citation is in note 29), p. 358, taking a quote from W. Herberg.

35. K. Tierney (1979; citation is in note 29), p. 365

36. K. Tierney (1979; citation is in note 29), p. 393.

37. See Susan Oyama's article 'The accidental chordate: contingency in developmental systems' in *South Atlantic Quarterly*, vol. 94, pp. 509–526, 1995.
38. S. Oyama (1995; citation is in note 37), p. 512.
39. See Peter Koslowski's chapter on pp. 301–328 in *Sociobiology and bioeconomics: The theory of evolution in biological and economic theory*, edited by P. Koslowski (Springer, Berlin, 1999).
40. P. Koslowski (1999; citation is in note 39), p. 308.
41. But see W. J. Alonso and C. Shuck-Paim in *Proceedings of the National Academy of Sciences, USA*, vol. 99, pp. 6843–6847, 2002.
42. See Eva Neumann-Held's article on pp. 105–137 of Kowslowski (1999; citation is in note 39).
43. *Consilience: The unity of knowledge* (Knopf, New York, 1998).
44. As one reviewer of *Consilience*, the philosopher John Dupré, wrote, 'the central thesis of the book is vague, and many of the opinions expressed are quite eccentric'; see *Science*, vol. 280, pp. 1395–1396, 1998.
45. See his *Beyond evolution: Human nature and the limits of evolutionary explanation* (Clarendon Press, Oxford, 1997); in particular pp. 156–158.
46. The title of one of C. S. Lewis's most influential books, first published in 1943.
47. See J. C. Greene's article on Darwin and religion in *Proceedings of the American Philosophical Society*, vol. 103, pp. 716–725, 1959.
48. J. C. Greene (1959; citation is in note 47), p. 725.
49. J. C. Greene (1959; citation is in note 47), p. 725.
50. See T. F. Smith and H. J. Morowitz in *Journal of Molecular Evolution*, vol. 18, pp. 265–282, 1982.
51. T. F. Smith and H. J. Morowitz (1982; citation is in note 50), p. 281.
52. C. S. Lewis's *That hideous strength: a modern fairy tale for grown-ups* (Bodley Head, London, 1946).
53. C. S. Lewis (1946; citation is in note 52), pp. 46–48.
54. The title of his extraordinary book, the full title of which is *Forbidden knowledge: From Prometheus to pornography* (Harcourt, Brace [Harvest], San Diego, 1997).
55. See R. S. Noel's *The mythology of Middle-Earth* (Thames and Hudson, London, 1977), and recall the power of the One Ring, whose property is to enthrall its owner and to lead him or her to a damnation that ultimately may not be of their own choosing.
56. See Howard Van Till's contribution on pp. 188–194 in *Science & Christianity: Four views*, edited by R. F. Carlson (InterVarsity Press, Downers Grove, 2000).
57. H. Van Till (2000; citation is in note 56), p. 192 (his emphases).
58. Michael Polanyi *Personal knowledge: Towards a post-critical philosophy* (Routledge and Kegan Paul, London, 1962).
59. M. Polanyi (1962; citation is in note 58), pp. 284–285.
60. See my book *The crucible of creation: The Burgess Shale and the rise of animals* (Oxford University Press, Oxford, 1998), p. 223, and S. R. Taylor's *Destiny or chance: our solar system and its place in the cosmos* (Cambridge University Press, Cambridge, 1998), p. 204.
61. See notably D. A. Griffin's *Animal minds: Beyond cognition to consciousness* (Chicago, University of Chicago Press, 2001), as well as the discussion in Chapter 9.
62. See G. K. Chesterton's 'A defence of humility' in *The defendant* (Dent, London, 1922).
63. G. K. Chesterton (1922; citation is in note 62), pp. 134–135.
64. Asked what most struck him of God's creation, J. B. S. Haldane is said to have replied: an inordinate fondness for beetles.

General index

Page numbers in italics indicate Figures.

Convergences index

Page numbers in italics indicate Figures.

Index of authors and other individuals

Page numbers in italics indicate Figures. [Notes] are listed by [page number].